MW00388282

Rethinking Technology is an essential reference for all students of architecture, design and the built environment; providing a convenient single source for all the key texts in the recent literature on architecture and technology.

The essays included are chronicles, manifestos, reflections, and theories produced by architects and architectural writers. Arranged in chronological order of original publication, these essays allow comparisons to be made between writings produced in a similar historical context and reveal the discipline's long and close attention to the experience and effects of new technologies, from the early twentieth century to the present day.

With the ever increasing pace of technological change, the fact and condition of change itself has become the subject of architectural discussion, made manifest in organic and dynamic analogies and the use of terms like process, flow, and emergence. Most architects still use the word technology to refer to the different means and methods of building, however in recent years the term has become synonymous with the digital realm and the whole apparatus of computerized information flow. With that change, the tools of design and construction have become a matter of processes, networks, and systems.

The editors preface each text with a short introduction explaining the significance of the essay in relation to the broader developments charted by the book. Cross-references are also made between individual texts in order to highlight important thematic connections across time.

William W. Braham is associate professor of architecture at the University of Pennsylvania. He has written widely on environmental technologies, combining technical analysis with historical and theoretical accounts. He is the author of *Modern Color/Modern Architecture: Amedee Ozenfant and the genealogy of color in modern architecture* (2002). **Jonathan A. Hale** is associate professor and director of research in architecture at the School of the Built Environment, University of Nottingham. He is the author of *Building Ideas: An Introduction to Architectural Theory* (2000). **John Stanislav Sadar** is an architect and partner in the multidisciplinary design firm *little wonder*.

RE-THINK-ING TECH-NOL-OGY

RE-THINK-ING TECH-NOL-OGY

A READER IN
ARCHITECTURAL
THEORY

EDITED BY
WILLIAM W. BRAHAM
AND
JONATHAN A. HALE
WITH
JOHN STANISLAV SADAR

Routledge
Taylor & Francis Group

LONDON AND NEW YORK

First published 2007
by Routledge
2 Park Square, Milton Park, Abingdon, Oxon OX14 4RN

Simultaneously published in the USA and Canada
by Routledge
270 Madison Ave, New York, NY 10016

Routledge is an imprint of the Taylor & Francis Group, an informa business

© 2007 Selection and editorial matter, William W. Braham and Jonathan A. Hale; individual chapters, the contributors

Typeset in Frutiger by Wearset Ltd, Boldon, Tyne and Wear
Printed and bound in Great Britain by TJ International Ltd, Padstow, Cornwall

All rights reserved. No part of this book may be reprinted or reproduced or utilized in any form or by any electronic, mechanical, or other means, now known or hereafter invented, including photocopying and recording, or in any information storage or retrieval system, without permission in writing from the publishers.

British Library Cataloguing in Publication Data
A catalogue record for this book is available from the British Library

Library of Congress Cataloging in Publication Data
Rethinking technology: a reader in architectural theory/edited by William W. Braham and Jonathan A. Hale.
p. cm.
Includes bibliographical references and index.
1. Architecture and technology. 2. Architecture–Philosophy. I. Braham, William W., 1957– II. Hale, Jonathan A.
NA2543.T43R47 2007
720.1–dc22

2006014792

ISBN10: 0-415-34653-3 (hbk)
ISBN10: 0-415-34654-1 (pbk)
ISBN10: 0-203-62433-5 (ebk)

ISBN13: 978-0-415-34653-5 (hbk)
ISBN13: 978-0-415-34654-2 (pbk)
ISBN13: 978-0-203-62433-3 (ebk)

CONTENTS

PREFACE

The possibility of this volume grew out of conversations at the University of Pennsylvania over a decade ago. For a brief and remarkably intense period Ivan Illich taught a weekly seminar in the PhD program in Architecture headed by Joseph Rykwert. Like so many moments of intensity, it was surprisingly short lived, though its topics and debates continue to reverberate among those fortunate enough to have participated. Illich brought his broad experience to questions fostered by Marco Frascari's studies of representation, David Leatherbarrow's writing about materials and assemblies, Peter McCleary's seminar on the philosophy of technology, and Rykwert's depth of knowledge and curiosity about everything.

Our initial proposal was to prepare a reader of essays explicitly on the philosophy of technology, but the recent and rapid appearance of several excellent anthologies on this subject inspired us to focus more directly on the architectural literature, which offered its own variations on the question of technology. The first lists of essays numbered in the hundreds, so for each text included in this reader there were at least five equally compelling pieces that had to be left out.

The project would not have been possible without the patient and fastidious work of John Sadar, a PhD candidate at the University of Pennsylvania. His own doctoral research on the effects of technological innovations in early twentieth-century architecture has added another important dimension to this collection. Some credit also must go to all of our students at the University of Pennsylvania and the University of Nottingham, whose questions, enquiries, and interests have often led us to places we might not otherwise have considered.

For their enthusiasm, patience, and above all confidence in this project our thanks go to Caroline Mallinder, Georgina Johnson, and the publishing team at Routledge/Taylor and Francis. On a more personal note, appreciation for support, ideas, and inspiration should also be expressed to: Andrew Ballantyne, Iain Borden, Ted Cullinan, Jocelyn Dodd, Thomas Hughes, Don Ihde, David Leatherbarrow, Detlef Mertins, Peter McCleary, Jane Rendell, Joseph Rykwert, Adam Sharr, and Jeremy Till.

Thanks always to the staff at the Fisher Fine Arts Library at the University of Pennsylvania, the finest circulating library of architecture.

William W. Braham
Jonathan A. Hale

ACKNOWLEDGMENTS

The editors and publishers gratefully acknowledge the following for permissions to reproduce material in this book.

Alexander, Christopher, *Notes on the Synthesis of Form*, copyright © 1964 by the President and Fellows of Harvard College. Copyright © renewed 1992 by Christopher Alexander.

Banham, Reyner "A Home is Not a House," first published in *Art in America*, April 1965. Copyright © *Art in America*, Brandt Publications.

Banham, Reyner, *Theory and Design in the First Machine Age*, copyright © 1980 MIT Press.

Berkel, Ben van and Caroline Bos, *Move: Techniques*, copyright © 1999. UN Studio and Goose Press.

Boyce, James R., "What is the Systems Approach?" copyright © 1969 Architecture/VNU Publications.

Brand, Stewart, *How Buildings Learn*, copyright © 1994 Viking Penguin, a division of Penguin Group (USA) Inc.

Buckminster Fuller, R., *4D Time Lock*, copyright © 1929 the Estate of R. Buckminster Fuller.

Buckminster Fuller, R., *Operating Manual for Spaceship Earth*, copyright © 1969 the Estate of R. Buckminster Fuller.

Cache, Bernard, "Digital Semper," copyright © 2000 MIT Press, Anyone Corporation.

Castells, Manuel, "Space of Flows, Space of Places: Materials for a Theory of Urbanism in the Information Age," copyright © 2004 Routledge Publishing, Inc.

Collins, Peter, "The Biological Analogy," first published in the *Architectural Review*, December 1959. Copyright © *Architectural Review*/EMAP Communications.

Colquhoun, Alan, *Essays in Architectural Criticism: Modern Architecture and Historical Change*, copyright © 1981 MIT Press.

Cook, Peter, *Experimental Architecture*, copyright © 1970 Rizzoli.

Cowan, Ruth Schwartz, "The 'Industrial Revolution' in the Home: Household Technology and Social Change in the Twentieth Century," copyright © 1976 Johns Hopkins University Press – Journals.

De Landa, Manuel, "Deleuze and the Use of the Genetic Algorithm in Architecture," copyright © 2002 Manuel De Landa.

Fernández-Galiano, Luis, *Fire and Memory: On Architecture and Energy*, copyright © 2000 MIT Press. This work originally appeared in Spanish under the title *El fuego y la memoria. Sobre arquitectura y energiá*, copyright © 1991 Alianza Editorial, S.A., Madrid.

Giedion, Siegfried, *Building in France, Building in Iron, Building in Ferro-Concrete*, copyright © 1995 J. Paul Getty Trust.

Giedion, Siegfried, *Mechanization Takes Command: A Contribution to Anonymous History*, copyright © 1948 Oxford University Press, Inc.

Giedion, Siegfried, *Time and Architecture: The Growth of a New Tradition*, copyright © 1941, 1949, 1954, 1962, 1967, 1969, by the President and Fellows of Harvard College, copyright © 1997 by William J. Callaghan, copyright © 1982 by Andreas Giedion and Verena Clay-Giedion.

Guattari, Félix, *Chaosmosis: An Ethico-aesthetic Paradigm*, English translation © 1995 Power Institute, Paul Bains, and Julian Pefanis. Reprinted by permission of Indiana University Press.

Häring, Hugo, "The House as an Organic Structure," from Ulrich Conrads, *Programs and Manifestoes on 20th-Century Architecture*, copyright © 1964 by Verlag Ullstein GmbH, Frankfurt/M-Berlin. English translation copyright © 1970 Lund Humphries, London, and MIT Press, Cambridge, MA.

Honzík, Karel, "Biotechnics: Functional Design and the Vegetable World," first published in the *Architectural Review*, January 1937. Copyright © *Architectural Review*/EMAP Communications.

Katavolos, William, "Organics." from Ulrich Conrads, *Programs and Manifestoes on 20th-Century Architecture*, copyright © 1964 by Verlag Ullstein GmbH, Frankfurt/M-Berlin. English translation copyright © 1970 Lund Humphries, London, and MIT Press, Cambridge, MA.

Kiesler, Frederick, "On Correlation and Biotechnique: A Definition and Test of a New Approach to Building Design," *The Architectural Record*, September 1939, 60–75.

Kohr, Leopold, *The Inner City*, copyright © 1989 Y Lolfa Cyf, Talybont, Dyfed, Wales. www.ylolfa.com

Koolhaas, Rem, *Small, Medium, Large, Extra-large*, copyright © 1995 Monacelli Press and Rem Koolhaas.

Kurokawa, Kisho, *Metabolism in Architecture*, copyright © 1977.

Latour, Bruno, "Mixing Humans and Nonhumans Together: The Sociology of a Door-Closer," copyright © 1988 University of California Press – Journals.

Le Corbusier, "Architecture: The Expression of the Materials and Methods of our Times," copyright © 2006 Fondation Le Corbusier, Paris.

Le Corbusier, *Towards a New Architecture*, copyright © 1960. Greenwood Publishing Group, Inc., Westport, CT.

Leatherbarrow, David and Mohsen Mostafavi, *Surface Architecture*, copyright © 2002 MIT Press.

McCleary, Peter, "Some Characteristics of a New Concept of Technology," copyright © 1988.

McDonough, William and Michael Braungart, *Cradle to Cradle: Remaking the Way We Make Things*, copyright © 2002 Melanie Jackson Agency, LLC.

McLuhan, Marshall, *Understanding Media: The Extensions of Man*, copyright © 1964 MIT Press.

Mies van der Rohe, Ludwig, "Technology and Architecture," from Ulrich Conrads, *Programs and Manifestoes on 20th-Century Architecture*, copyright © 1964 by Verlag Ullstein GmbH, Frankfurt/M-Berlin. English translation copyright © 1970 Lund Humphries, London, and MIT Press, Cambridge, MA.

Mitchell, William, "E-Bodies, E-Buildings, E-Cities," copyright © 2002 John Wiley & Sons.

Mumford, Lewis, *Technics and Civilization*, copyright © 1934 Harcourt. Copyright renewed © 1962 by Lewis Mumford.

Neutra, Richard, *Survival Through Design*, copyright © 1954 Dion Neutra, Architect and © Richard and Dion Neutra Papers, Department of Special Collections, Charles E. Young Research Library, UCLA.

Pawley, Martin, "Technology Transfer," first published in the *Architectural Review*, September 1987. Copyright © *Architectural Review*/EMAP Communications.

Rykwert, Joseph, "Organic and Mechanical," copyright © 1992.

Sant' Elia, Antonio and Filippo Tommaso Marinetti, "Manifesto of Futurist Architecture," from Ulrich Conrads, *Programs and Manifestoes on 20th-Century Architecture*, copyright © 1964 by Verlag Ullstein GmbH, Frankfurt/M-Berlin. English translation copyright © 1970 Lund Humphries, London, and MIT Press, Cambridge, MA.

Soleri, Paolo, *Bridge Between Matter and Spirit is Matter Becoming Spirit*, copyright © 1973 by Cosanti Foundation. Reprinted by permission of Doubleday, a division of Random House, Inc.

Steadman, Philip, *The Evolution of Designs: Biological Analogy in Architecture and the Applied Arts*, copyright © 1991.

Superstudio, "Description of Microevent/Microenvironment," reprinted by permission from *Italy: The New Domestic Landscape*, copyright © 1972 Museum of Modern Art and Gian Piero Frasinelli.

Virilio, Paul, "The Third Interval," copyright © 1997 Editions Galilée.

Wachsmann, Konrad, "Seven Theses," from Ulrich Conrads, *Programs and Manifestoes on 20th-Century Architecture*, copyright © 1964 by Verlag Ullstein GmbH, Frankfurt/M-Berlin. English translation copyright © 1970 Lund Humphries, London, and MIT Press, Cambridge, MA.

Wright, Frank Lloyd, "The Art and Craft of the Machine," copyright © 1992 Frank Lloyd Wright Foundation, Taliesin West, Scottsdale, AZ.

Yeang, Ken, *The Green Skyscraper: The Basis for Designing Sustainable Intensive Buildings*, copyright © 1999.

Considerable effort has been made to trace and contact copyright holders and to secure replies prior to publication. However, this has not always been possible. The editor and publishers apologize for any errors and omissions. If notified, the publishers will endeavor to correct these at the earliest opportunity

INTRODUCTION

William W. Braham
Jonathan A. Hale

> "sometime during the 1980s the technological society which began in the fourteenth century came to an end. Now I recognize that dating epochs involves interpretation and perhaps some fuzziness in assigning beginnings and endings; but, nevertheless, it appears to me that the age of tools has now given way to the age of systems, exemplified in the conception of the earth as an ecosystem, and the human being as an immune system."[1]
>
> Ivan Illich

› A reader in architectural theory

This collection of essays provides an introduction to the literature on architecture and technology. It is offered to architects and architecture students for whom technology and design have largely been separated in school curricula, in trades and professional associations, and in design practice itself. It is intended to support courses in architectural technology, architectural theory, and also the philosophy of technology. It should also be of interest to professionals involved in teaching, or reconsidering their work in the light of current theoretical debates. These essays reveal the discipline's long and close attention to the experience and effects of contemporary technology, from the early twentieth century to the current moment.

The collection is also intended for those interested in technology as a discrete historical, philosophical, or sociological subject, and for whom the architectural literature will be less well known. It is notable that in the time this architectural reader was being prepared two large new readers in the philosophy of technology were published, suggesting the growing maturity of the field, and also making the purpose of this collection even clearer.[2] With all the historical and philosophical writing on technology contained in those publications, architecture is only occasionally considered, and usually as an example among other examples.

The essays included in the collection are chronicles, manifestos, reflections, and theories offered by architects and architectural writers in their encounter with technology. However, a comprehensive collection of such writings would fill many such volumes, and so much of our task has been to rethink the topic itself.

› Why rethink technology?

Through the course of the nineteenth and twentieth centuries architects have largely become technocrats. Even as the elite of the profession have secured their status as visionaries and artists, the majority of architects spend their time processing the flow of information that guides the assembly of complex technical constructions. In that respect, an architect of the early nineteenth century would have more in common with the Roman architect Vitruvius than he would with a practitioner of the twentieth or twenty-first century.

Properly speaking, architects have always been concerned with technology, but since the effects of the first industrial revolution became widespread in the early nineteenth century, the technology encountered by architects has changed in scope and kind, becoming a restless and accelerating process of transformation. To rethink technology at the beginning of the twenty-first century means reconsidering the strong claims made about technology – utopian and dystopian – by the modernist and postmodernist architects and historians of the twentieth century, as the actual impacts of that technology were encountered.

With the ever increasing pace of technological change, the fact and condition of change itself has become the subject of architectural discussion, made manifest in organic analogies and the use of terms like process, flow, and emergence. For most architects the word technology still means the different means and methods of building, however, in recent decades, the term has become synonymous with computers and the whole apparatus of networked information flow. With that change, the shift described by Ivan Illich in the opening quotation has become wholly palpable: architecture and technology or, more precisely, the tools of design and construction, have become a matter of systems.

The goal of this collection is to chart the emergence of that "age of systems" within the architectural discourse. But even that task would extend well into the pre-modern period and would require many more essays than seemed manageable, so we settled on a selective survey beginning at the turn of the twentieth century.

All of this begs a central question: in what ways is architecture technological? Certainly the process of building is now wholly technological, as is the society in which buildings are conceived, financed, and evaluated. Ultimately, rethinking technology and architecture in the age of systems means rethinking the practical and ethical dimensions of change, development, and evolution in architecture.

› Philosophy of technology

As technology changes so do society, the environment, and the practice of architecture. The globalizing "network society" has certainly forced architects to rethink the relationship of their work to new modes of production and construction, new patterns of movement and settlement, and new cultural priorities. Since at least the mid-nineteenth century – Karl Marx provides the classic example – it has been commonly assumed that technologies change society in more or less predictable ways; that technology is both autonomous (evolving) and deterministic in its effects. Through the twentieth century philosophers and historians have debated the nature of that relationship, leading in recent decades to a more nuanced view about their interaction and the degree to which technology itself is "socially constructed," or at least culturally embedded and co-evolving.

It is usual to describe these twentieth-century developments in terms of a first and second generation of philosophers of technology. The first generation could be traced back to Comte, Marx, Ruskin, and other nineteenth-century figures who were all in some way reacting – both positively and negatively – to the impact of the industrial revolution on the social and cultural conditions of the time. By the early twentieth century these ideas were beginning to harden into a sustained critique of technological utopianism, most famously in the writings of the German philosopher Martin Heidegger. Heidegger was the author of perhaps the single most important – though not always the most popular – statement on the "essence" of modern technology, in his 1953 essay *The Question Concerning Technology*. Subsequent writing in the 1950s and 1960s led to the emergence of a distinct and identifiable field, including the urban and architectural writers Lewis Mumford and Siegfried Giedion, alongside philosophers and sociologists like Jacques Ellul, Herbert Marcuse, Ivan Illich, and Jurgen Habermas. What unites this disparate group of thinkers is the belief in the autonomy of technological development – a sense that society's tools had turned against their creators in a kind of Frankenstein scenario – locking us irrevocably within a technological "system" (to use Ellul's term) or a "megamachine," as Mumford described it.

Another common factor within these broadly dystopian critiques was their neglect of empirical evidence in favor of a "high-altitude" theoretical analysis. The second generation of philosophers of technology sought to correct this imbalance by delving inside the "black box" in an attempt to uncover the complex interactions between technical and cultural factors. This reaction took many different forms. Authors such as Albert Borgmann have extended Heidegger's insights into the details of everyday life, while others have extended the insights of Dewey or Pierce. What has since become known as the "social constructivist" approach to technology grew out of a series of specific social and anthropological case-studies: writings based on the historical and ethnographic analysis of particular technological developments which attempt to show the extent to which they are driven by social and cultural forces. Significant figures in this field – which first came to prominence in the early 1980s – include Ruth Schwartz Cowan, Thomas Hughes, Michel Callon, and Bruno Latour. What they sought to chart was the often unpredictable and occasionally counterintuitive rise of new technical innovations and their subsequent success or failure. The adoption and popularity of novel technologies may often be based more on social, cultural, and psychological factors than on "pure" scientifically testable qualities such as efficiency, economy, and reliability.

Current thinking in this expanding field is still influenced by both of these approaches, whose differences might simply be characterized as philosophical versus historical or sociological, though they are equally distinguished by the object of their study, whether it is technology, modern science, engineering, industry, or society itself. Such differences are legible in the names of the different academic departments, societies, and journals that embrace these subjects. For example: the Society for the History of Technology, founded in 1958, publishes the journal *Technology and Culture*; the American Sociological Association organized a research committee on the Sociology of Science and Technology in 1966; the Society for Philosophy and Technology, formed in 1976, publishes the journal *Téchne*; while the European Association for the Study of Science and Technology, formed in 1981, publishes the EASST Review.

The Dutch philosopher of technology, Hans Achterhuis, and his colleagues have argued the field has taken an "empirical turn," resisting large philosophical statements in favor of investigations of the details and complex interactions surrounding even the smallest technological artifact or condition.[3] It is in that spirit that this collection from the architectural literature might fit into the broader history and philosophy of technology.

› What is a system?

The opening quotation by Ivan Illich was drawn from his discussion about the changing notion of contingency, particularly the idea of instrumental causality introduced in the twelfth century, which he argued had inaugurated and characterized the age of tools or of technology. However, he did not welcome the age of systems. He viewed it as an even more difficult condition within which to live a good life, though he saw its emerging attributes clearly and also recognized the degree to which ideas often precede their realization. For him, a system was different from a tool because "when you became the user of a system, you became part of the system."[4] The groundwork for the description of interconnected, bottom-up, self-organizing entities has been emerging for generations – strongly visible in concepts such as the "eco-system" which appear in the 1930s, or more recently in the understanding of the human immune system – but this idea is actually discernible as far back as Adam Smith's eighteenth-century notion of the "invisible hand" of the free market economy (conceived during the Scottish Enlightenment).[5]

The immediate importance of a collection of essays dedicated to tracing the changing nature of technology in architecture is to penetrate beyond broad generalizations about technology, society, and architecture. It is necessary to understand the broad historical conditions and actual processes of their realization. Radical changes were encountered by architects in every aspect of their work and they tried many different formulations to manage and understand them. Principle among them were various kinds of organic and biological analogies, which gained increasing precision as cybernetics, general systems theory, and complexity analysis matured. It is important at the outset to recognize the degree to which those same developments were changing the understanding of organic life itself. In other words, as new paradigms of explanation develop they are applied equally to buildings, bodies, and machines.

The collection is also meant to help reframe the architect's question of how best to work in such conditions. That becomes both an ethical and a practical question. As Reyner Banham requested in the second edition of *The Architecture of the*

Well-Tempered Environment (1984), "this book must no longer be filed under Technology."

1   Ivan Illich, *The River North of the Future: the Testament of Ivan Illich*, as told to David Cayley (Toronto: House of Anasazi Press, 2005), p. 77.
2   David M. Kaplan (ed.), *Readings in the Philosophy of Technology* (Lanham, MD: Rowman & Littlefield, 2004); Robert C. Scharff and Val Dusek (eds), *Philosophy of Technology: The Technological Condition: An Anthology* (Malden, MA: Blackwell, 2003).
3   Hans Achterhuis (ed.), *American Philosophy of Technology: The Empirical Turn* (Bloomington, IN: Indiana University Press, 2001).
4   Illich, *River North*, p. 78.
5   Ronald Hamowy, "The Scottish Enlightenment and the Theory of Spontaneous Order." *Journal of the History of Philosophy Monographs* (Carbondale, IL: Southern Illinois University Press, 1987).

We chose to begin this collection with an essay by the American architect Frank Lloyd Wright (1867–1959), because it was conveniently given as a lecture at the very beginning of the twentieth century. It also provides a useful introduction to the themes encountered by designers in the late nineteenth century, for whom the term "Machine" served as shorthand for the social and aesthetic effects of the first technological revolution.

The lecture was Wright's original manifesto, and he returned to its themes and phrases throughout his long career. It was delivered as a lecture on March 1, 1901 at the Hull House in Chicago at the height of his first period of productivity and fame, and offers his critique of the Arts and Craft movement. He summarizes their protest against the machine and even paraphrased Victor Hugo on the effect of the printing press on architecture – "The One Will Kill the Other" – from *The Hunchback of Notre-Dame*.

In the context of this collection, the biological themes that he uses toward the end of the essay are critical, extending the ancient analogy between bodies and buildings into the dynamic processes and flows of the modern, industrial city. For the rest of his career Wright invoked the notion of an organic architecture to explain his work, using the term in many different senses. However, in the compelling phrase, "blind obedience to organic law," we see the first glimmer of the age of systems.

1901

Frank
Lloyd Wright

The Art and Craft
of the Machine

As we work along our various ways, there takes shape within us, in some sort, an ideal – something we are to become – some work to be done. This, I think, is denied to very few, and we begin really to live only when the thrill of this ideality moves us in what we will to accomplish. In the years which have been devoted in my own life to working out in stubborn materials a feeling for the beautiful, in the vortex of distorted complex conditions, a hope has grown stronger with the experience of each year, amounting now to a gradually deepening conviction that in the Machine lies the only future of art and craft – as I believe, a glorious future; that the Machine is, in fact, the metamorphosis of ancient art and craft; that we are at last face to face with the machine – the modern Sphinx – whose riddle the artist must solve if he would that art live – for his nature holds the key. For one, I promise "whatever god may be"[1] to lend such energy and purpose as I may possess to help make that meaning plain; to return again and again to the task whenever and where need be; for this plain duty is thus relentlessly marked out for the artist in this, the Machine age, although there is involved an adjustment to cherished gods, perplexing and painful in the extreme; the fire of many long-honored ideals shall go down to ashes to reappear, phoenix like, with new purposes.

The great ethics of the Machine are yet, in the main, beyond the ken of the artist or student of sociology; but the artist mind may now approach the nature of this thing from experience, which has become the commonplace of his field, to suggest, in time, I hope, to prove, that the machine is capable of carrying to fruition high ideals in art – higher than the world has yet seen!

Disciples of William Morris cling to an opposite view. Yet William Morris himself deeply sensed the danger to art of the transforming force whose sign and symbol is the machine, and though of the new art we eagerly seek he sometimes despaired, he quickly renewed his hope.

He plainly foresaw that a blank in the fine arts would follow the inevitable abuse of new-found power, and threw himself body and soul into the work of bridging it over by bringing into our lives afresh the beauty of art as she had been, that the new art to come might not have dropped too many stitches nor have unraveled what would still be useful to her.

That he had abundant faith in the new art his every essay will testify.

That he miscalculated the machine does not matter. He did sublime work for it when he pleaded so well for the process of elimination its abuse had made necessary; when he fought the innate vulgarity of theocratic impulse in art as opposed to democratic; and when he preached the gospel of simplicity.

All artists love and honor William Morris.

He did the best in his time for art and will live in history as the great socialist, together with Ruskin, the great moralist: a significant fact worth thinking about, that the two great reformers of modern times professed the artist.

The machine these reformers protested, because the sort of luxury which is born of greed had usurped it and made of it a terrible engine of enslavement, deluging the civilized world with a murderous ubiquity, which plainly enough was the damnation of their art and craft.

It had not then advanced to the point which now so plainly indicates that it will surely and swiftly, by its own momentum, undo the mischief it has made, and the usurping vulgarians as well.

Nor was it so grown as to become apparent to William Morris, the grand democrat, that the machine was the great forerunner of democracy.

The ground plan of this thing is now grown to the point where the artist must take it up no longer as a protest: genius must progressively dominate the work of the contrivance it has created; to lend a useful hand in building afresh the "Fairness of the Earth."

That the medicine has dealt Art in the grand old sense a death-blow, none will deny.

The evidence is too substantial.

Art in the grand old sense – meaning Art in the sense of structural tradition, whose craft is fashioned upon the handicraft ideal, ancient or modern; an art wherein this form and that form as structural parts were laboriously joined in such a way as to beautifully emphasize the manner of the joining: the million and one ways of beautifully satisfying bare structural necessities, which have come down to us chiefly through the books as "Art." For the purpose of suggesting hastily and therefore crudely wherein the machine has sapped the vitality of this art, let us assume Architecture in the old sense as a fitting representative of Traditional-art and Printing as a fitting representation of the Machine.

What printing – the machine – has done for architecture – the fine art – will have been done in measure of time for all art immediately fashioned upon the early handicraft ideal.

With a masterful hand, Victor Hugo, a noble lover and a great student of architecture, traces her fall in *Notre-Dame*.

The prophecy of Frollo, that "the book will kill the edifice," I remember was to me as a boy one of the grandest sad things of the world.

After seeking the origin and tracing the growth of architecture in superb fashion, showing how in the Middle Ages all the intellectual forces of the people converged to one point – architecture – he shows how, in the life of that time, whoever was born poet became an architect. All other arts simply obeyed and placed themselves under the discipline of architecture. They were the workmen of the great work. The architect, the poet, the master summed up in his person the sculpture that carved his façades, painting which illuminated his walls and windows, music which set his bells to pealing and breathed into his organs – there was nothing which was not forced in order to make something of itself in that time, to come and frame itself in the edifice.

Thus, down to the time of Gutenberg, architecture is the principal writing – the universal writing of humanity.[2]

In the great granite books begun by the Orient, continued by Greek and Roman antiquity, the Middle Ages wrote the last page.

So to enunciate here only summarily a process, it would require volumes to develop; down to the fifteenth century the chief register of humanity is architecture.

In the fifteenth century everything changes.

Human thought discovers a mode of perpetuating itself, not only more resisting than architecture, but still more simple and easy.

Architecture is dethroned.

Gutenberg's letters of lead are about to supersede Orpheus' letters of stone.

The book is about to kill the edifice.

The invention of printing was the greatest event in history.

It was the first great machine, after the great city.

It is human thought stripping off one form and donning another.

Printed, thought is more imperishable than ever – it is volatile, indestructible.

As architecture it was solid; it is now alive; it passes from duration in point of time to immortality.

Cut the primitive bed of a river abruptly, with a canal hollowed out beneath its level, and the river will desert its bed.

See how architecture now withers away, how little by little it becomes lifeless and bare. How one feels the water sinking, the sap departing, the thought of the times and people withdrawing from it. The chill is almost imperceptible in the fifteenth century, the press is yet weak, and at most draws from architecture a superabundance of life, but with the beginning of the sixteenth century, the malady of architecture is visible. It becomes classic art in a miserable manner; from being indigenous, it becomes Greek and Roman; from being true and modern, it becomes pseudo-classic.

It is this decadence which we call the Renaissance.

It is the setting sun which we mistake for dawn.

It has now no power to hold the other arts; so they emancipate themselves, break the yoke of the architect, and take themselves off, each in its own direction.

One would liken it to an empire dismembered at the death of its Alexander, and whose provinces become kingdoms.

Sculpture becomes statuary, the image trade becomes painting, the canon becomes music. Hence Raphael, Angelo, and those splendors of the dazzling sixteenth century.

Nevertheless, when the sun of the Middle Ages is completely set, architecture grows dim, becomes more and more effaced. The printed book, the gnawing worm of the edifice, sucks and devours it. It is petty, it is poor, it is nothing.

Reduced to itself, abandoned by other arts because human thought is abandoning it, it summons bunglers in place of artists. It is miserably perishing.

Meanwhile, what becomes of printing?

All the life, leaving architecture, comes to it. In proportion as architecture ebbs and flows, printing swells and grows. That capital of forces which human thought had been expending in building is hereafter to be expended in books; and architecture, as it was, is dead, irretrievably slain by the printed book; slain because it endures for a shorter time; slain because human thought has found a more simple medium of expression, which costs less in human effort; because human thought has been rendered volatile and indestructible, reaching uniformly and irresistibly the four corners of the earth and for all.

Thenceforth, if architecture rise again, reconstruct, as Hugo prophesies she may begin to do in the latter days of the nineteenth century, she will no longer be mistress, she will be one of the arts, never again the art; and printing – the Machine – remains "the second Tower of Babel of the human race."

So the organic process, of which the majestic decline of Architecture is only one case in point, has steadily gone on down to the present time, and still goes on, weakening the hold of the artist upon the people, drawing off from his rank poets and scientists until architecture is but a little, poor knowledge of archeology, and the average of art is reduced to the gasping poverty of imitative realism; until the whole letter of Tradition, the vast fabric of precedent, in the flesh, which has increasingly confused the art ideal while the machine has been growing to power, is a beautiful corpse from which the spirit has flown. The spirit that has flown is the spirit of the new art, but has failed the modern artist, for he has lost it for hundreds of years in his lust for the letter, the beautiful body of art made too available by the machine.

So the Artist craft wanes.

Craft that will not see that "human thought is stripping off one form and donning another," and artists are everywhere, whether catering to the leisure class of old England or ground beneath the heel of commercial abuse here in the great West, the unwilling symptoms of the inevitable, organic nature of the machine, they combat, the hell-smoke of the factories they scorn to understand.

And, invincible, triumphant, the machine goes on, gathering force and knitting the material necessities of mankind ever closer into a universal automatic fabric; the engine, the motor, and the battleship, the works of art of the century!

The Machine is Intellect mastering the drudgery of earth that the plastic art may live; that the margin of leisure and strength by which man's life upon the earth can be made beautiful, may immeasurably widen; its function ultimately to emancipate human expression!

It is a universal educator, surely raising the level of human intelligence, so carrying within itself the power to destroy, by its own momentum, the greed which in Morris' time and still in our own time turns it to a deadly engine of enslavement. The only comfort left the poor artist, sidetracked as he is, seemingly is a mean one; the thought that the very selfishness which man's early art idealized, now reduced to its lowest terms, is swiftly and surely destroying itself through the medium of the Machine.

The artist's present plight is a sad one, but may he truthfully say that society is less well off because Architecture, or even Art, as it was, is dead, and printing, or the Machine, lives? Every age has done its work, produced its art with the best tools or contrivances it knew, the tools most successful in saving the most precious thing in the world – human effort. Greece used the chattel slave as the essential tool of its art and civilization. This tool we have discarded, and we would refuse the return of Greek art upon the terms of its restoration, because we insist now upon a basis of Democracy.

Is it not more likely that the medium of artistic expression itself has broadened and changed until a new definition and new direction must be given the art activity of the future, and that the Machine has finally made for the artist, whether he will yet own it or not, a splendid distinction between the Art of old and the Art to come? A distinction made by the tool which frees human labor, lengthens and broadens the life of the simplest man, thereby the basis of the Democracy upon which we insist.

To shed some light upon this distinction, let us take an instance in the field naturally ripened first by the machine – the commercial field.

The tall modern office building is the machine pure and simple.

We may here sense an advanced stage of a condition surely entering all art for all time; its already triumphant glare in the deadly struggle taking place here between the machine and the art of structural tradition reveals "art" torn and hung upon the steel frame of commerce, a forlorn head upon a pike, a solemn warning to architects and artists the world over.

We must walk blindfolded not to see that all that this magnificent resource of machine and material has brought us so far is a complete, broadcast degradation of every type and form sacred to the art of old; a pandemonium of tin masks, huddled deformities, and decayed methods; quarreling, lying, and cheating, with hands at each other's throat – or in each other's pockets; and none of the people who do these things, who pay for them or use them, know what they mean, feeling only – when they feel at all – that what is most truly like the past is the safest and therefore the best; as typical Marshall Field, speaking of his new building, has frankly said: "A good copy is the best we can do."[3]

A pitiful insult, art and craft!

With this mine of industrial wealth at our feet we have no power to use it except to the perversion of our natural resources? A confession of shame which the merciful ignorance of the yet material frame of things mistakes for glorious achievement.

We half believe in our artistic greatness ourselves when we toss up a pantheon to the god of money in a night or two, or pile up a mammoth aggregation of Roman monuments, sarcophagi, and Greek temples for a post office in a year or two – the patient retinue of the machine pitching in with terrible effectiveness to consummate this unhallowed ambition – this insult to ancient gods. The delicate, impressionable facilities of terra-cotta becoming imitative blocks and voussoirs of toolmarked stone, badgered into all manner of structural gymnastics, or else ignored in vain endeavor to be honest; and granite blocks, cut in the fashion of the followers of Phidias, cunningly arranged about the steel beams and shafts, to look "real" – leaning heavily upon an inner skeleton of steel for support from floor to floor, which strains beneath the "reality" and would fain, I think, lie down to die of shame.

The "masters" – ergo, the fashionable followers of Phidias – have been trying to make this wily skeleton of steel seem seventeen sorts of "architecture" at once, when all the world knows – except the "masters" – that it is not one of them.

See now, how an element – the vanguard of the new art – has entered here, which the structural–art equation cannot satisfy without downright lying and ignoble cheating.

This element is the structural necessity reduced to a skeleton, complete in itself without the craftsman's touch. At once the million and one little ways of satisfying this necessity beautifully, coming to us chiefly through the books as the traditional art of building, vanish away – become history.

The artist is emancipated to work his will with a rational freedom unknown to the laborious art of structural tradition – no longer tied to the meagre unit of brick arch and stone lintel, nor hampered by the grammatical phrase of their making – but he cannot use his freedom.

His tradition cannot think.

He will not think.

His scientific brother has put it to him before he is ready.

The modern tall office-building problem is one representative problem of the machine. The only rational solutions it has received in the world may be counted upon the fingers of one hand. The fact that a great portion of our "architects" and "artists" are shocked by them to the point of offense is as valid an objection as that of a child refusing wholesome food because his stomach becomes dyspeptic from over-much unwholesome pastry – albeit he be the cook himself.

We may object to the mannerism of these buildings, but we take no exception to their manner nor hide from their evident truth.

The steel frame has been recognized as a legitimate basis for simple, sincere clothing of plastic material that idealizes its purpose without structural pretense.

This principle has at last been recognized in architecture, and though the masters refuse to accept it as architecture at all it is a glimmer in a darkened field – the first sane word that's been said in Art for the Machine.

The Art of old idealized a Structural Necessity – now rendered obsolete and unnatural by the Machine – and accomplished it through man's joy in the labor of his hands.

The new will weave for the necessities of mankind, which his Machine will have mastered, a robe of ideality no less truthful but more poetical, with a rational freedom made possible by the machine, beside which the art of old will be as the sweet plaintive wail of the pipe to the outpouring of full orchestra.

It will clothe Necessity with the living flesh of virile imagination, as the living flesh lends living grace to the hard and bony human skeleton.

The new will pass from the possession of kings and classes to the everyday lives of all – from duration in point of time to immortality.

This distinction is one to be felt now rather than clearly defined.

The definition is the poetry of this Machine Age, and will be written large in time; but the more we, as artists, examine into this premonition, the more we will find the utter helplessness of old forms to satisfy new conditions, and the crying need of the machine for plastic treatment – a pliant, sympathetic treatment of its needs that the body of structural precedent cannot yield.

To gain further suggestive evidence of this, let us turn to the Decorative Arts – the immense middle ground of all art now mortally sickened by the Machine – sickened that it may slough the art ideal of the constructural art for the plasticity of the new art – the Art of Democracy.

Here we find the most deadly perversion of all – the magnificent prowess of the machine bombarding the civilized world with the mangled corpses of strenuous horrors that once stood for cultivated luxury – standing now for a species of fatty degeneration simply vulgar.

Without regard to first principles or common decency, the whole letter of tradition – that is, ways of doing things rendered wholly obsolete and unnatural by the machine – recklessly fed into its rapacious maw until you may buy reproductions for ninety-nine cents at "The Fair" that originally cost ages of toil and cultivation, worth now intrinsically nothing – that are harmful parasites befogging the sensibilities of our natures, belittling and falsifying any true perception of normal beauty the Creator may have seen fit to implant in us.

The idea of fitness to purpose, harmony between form, and use with regard to any of these things, is possessed by very few, and utilized by them as a protest chiefly – protest against the machine! As well blame Richard Croker for the political iniquity of America.[4]

As "Croker is the creature and not the creator" of political evil, so the machine is the creature and not the creator of this iniquity; and with this difference – that the machine has noble possibilities unwillingly forced to degradation in the name of the artistic; the machine, as far as its artistic capacity is concerned, is itself the crazed victim of the artist who works while he waits, and the artist who waits while he works.

There is a nice distinction between the two.

Neither class will unlock the secrets of the beauty of this time.

They are clinging sadly to the old order and would wheedle the giant frame of things back to its childhood or forward to its second childhood, while this Machine Age is suffering for the artist who accepts, works, and sings as he works, with the joy of the here and now!

We want the man who eagerly seeks and finds, or blames himself if he fails to find, the beauty of this time; who distinctly accepts as a singer and a prophet; for no man may work while he waits or wait as he works in the sense that William Morris' great work was legitimately done – in the sense that most art and craft of today is an echo; the time when such work was useful has gone.

Echoes are by nature decadent.

Artists who feel toward Modernity and the Machine now as William Morris and Ruskin were justified in feeling then, had best distinctly wait and work sociologically where great work may still be done by them. In the field of art activity they will do distinct harm. Already they have wrought much miserable mischief.

If the artist will only open his eyes he will see that the machine he dreads has made it possible to wipe out the mass of meaningless torture to which mankind, in the name of the artistic, has been more or less subjected since time began; for that matter, has made possible a cleanly strength, an ideality and a poetic fire that the art of the world has not yet seen; for the machine, the process now smooths away the necessity of petty structural deceits, soothes this wearisome struggle to make things seem what they are not, and can never be; satisfies the simple term of the modern art equation as the ball of clay in the sculptor's hand yields to his desire – comforting forever this realistic, brain-sick masquerade we are wont to suppose art.

William Morris pleaded well for simplicity as the basis of all the art. Let us understand the significance to art of that word – SIMPLICITY – for it is vital to the Art of the Machine.

We may find, in place of the genuine thing we have striven for, an affectation of the naive, which we should detest as we detest a full-grown woman with baby mannerisms.

English art is saturated with it, from the brand-new imitation of the old house that grew and rambled from period to period to the rain-tub standing beneath the eaves.

In fact, most simplicity following the doctrines of William Morris is a protest; as a protest, well enough, but the highest form of simplicity is not simple in the sense that the infant intelligence is simple – nor, for that matter, the side of a barn.

A natural revulsion of feeling leads us from the meaningless elaboration of today to lay too great stress on mere platitudes, quite as a clean sheet of paper is a relief after looking at a series of bad drawings – but simplicity is not merely a neutral or a negative quality.

Simplicity in art, rightly understood, is a synthetic, positive quality, in which we may see evidence of mind, breadth of scheme, wealth of detail, and withal a sense of completeness found in a tree or a flower. A work may have the delicacies of a rare orchid or the stanch fortitude of the oak, and still be simple. A thing to be simple needs only to be true to itself in organic sense.

With this ideal of simplicity, let us glance hastily at a few instances of the machine and see how it has been forced by false ideals to do violence to this simplicity; how it has made possible the highest simplicity, rightly understood and so used. As perhaps wood is the most available of all homely materials and therefore, naturally, the most abused let us glance at wood.

Machinery has been invented for no other purpose than to imitate, as closely as possible, the wood carving of the early ideal – with the immediate result that no ninety-nine-cent piece of furniture is salable without some horrible botchwork meaning nothing unless it means that art and craft have combined to fix in the mind of the masses the old handcarved chair as the *ne plus ultra* of the ideal.

The miserable, lumpy tribute to this perversion which Grand Rapids alone yields would mar the face of art beyond repair; to say nothing of the elaborate and fussy joinery of posts, spindles, jigsawed beams and braces, butted and strutted, to outdo the sentimentality of the already overwrought antique product.

Thus is the woodworking industry glutted, except in rarest instances. The whole sentiment of early craft degenerated to a sentimentality having no longer decent significance nor commercial integrity; in fact all that is fussy, maudlin, and animal, basing its existence chiefly on vanity and ignorance.

Now let us learn from the Machine.

It teaches us that the beauty of wood lies first in its qualities as wood; no treatment that did not bring out these qualities all the time could be plastic, and therefore not appropriate – so not beautiful, the Machine teaches us, if we have left it to the machine that certain simple forms and handling are suitable to bring out the beauty of wood and certain forms are not; that all wood-carving is apt to be a forcing of the material, an insult to its finer possibilities as a material having in itself intrinsically artistic properties, of which its beautiful markings is one, its texture another, its color a third.

The machine, by its wonderful cutting, shaping, smoothing, and repetitive capacity, has made it possible to so use it without waste that the poor as well as the rich may enjoy today beautiful surface treatments of clean, strong forms that the branch veneers of Sheraton and Chippendale only hinted at, with dire extravagance, and which the Middle Ages utterly ignored.

The machine has emancipated these beauties of nature in wood; made it possible to wipe out the mass of meaningless torture to which wood has been subjected since the world began, for it has been universally abused and maltreated by all peoples but the Japanese.

Rightly appreciated, is not this the very process of elimination for which Morris pleaded?

Not alone a protest, moreover, for the machine, considered only technically, if you please, has placed in artist hands the means of idealizing the true nature of wood harmoniously with man's spiritual and material needs, without waste, within reach of all.

And how fares the troop of old materials galvanized into new life by the Machine?

Our modern materials are these old materials in more plastic guise, rendered so by the Machine, itself creating the very quality needed in material to satisfy its own art equation.

We have seen in glancing at modern architecture how they fare at the hands of Art and Craft; divided and subdivided in orderly sequence with rank and file of obedient retainers awaiting the master's behest.

Steel and iron, plastic cement, and terra-cotta.

Who can sound the possibilities of this old material, burned clay, which the modern machine has rendered as sensitive to the creative brain as a dry plate to the lens – a marvelous simplifier? And this plastic covering material, cement, another simplifier, enabling the artist to clothe the structural frame with a simple, modestly beautiful robe where before he dragged in, as he does still drag, five different kinds of material to compose one little cottage, pettily arranging it in an aggregation supposed to be picturesque – as a matter of fact, millinery, to be warped and beaten by sun, wind, and rain into a variegated heap of trash.

There is the process of modern casting in metal – one of the perfected modern machines, capable of any form to which fluid will flow, to perpetuate the imagery of the most delicately poetic mind without let or hindrance – within reach of everyone, therefore insulted and outraged by the bungler forcing it to a degraded seat at his degenerate festival.

Multitudes of processes are expectantly awaiting the sympathetic interpretation of the mastermind; the galvano-plastic and its electrical brethren, a prolific horde, now cheap fakirs imitating real bronzes and all manner of the antique, secretly damning it in their vitals.

Electro-glazing, a machine shunned because too cleanly and delicate for the clumsy hand of the traditional designer, who depends upon the mass and blur of leading to conceal his lack of touch.

That delicate thing, the lithograph – the prince of a whole reproductive province of processes – see what this process becomes in the hands of a master like Whistler. He has sounded but one note in the gamut of its possibilities, but that product is intrinsically true to the process, and as delicate as the butterfly's wing. Yet the most this particular machine did for us, until then in the hands of Art and Craft, was to give us a cheap, imitative effect of painting.

So spins beyond our ability to follow tonight, a rough, feeble thread of the evidence at large to the effect that the machine has weakened the artist; all but destroyed his handmade art, if not its ideals, although he has made enough miserable mischief meanwhile.

These evident instances should serve to hint, at least to the thinking mind, that the Machine is a marvelous simplifier; the emancipator of the creative mind, and in time the regenerator of the creative conscience. We may see that this destructive process has begun and is taking place that art might awaken to that power of fully developed senses promised by dreams of its childhood, even though that power may not come the way it was pictured in those dreams.

Now, let us ask ourselves whether the fear of the higher artistic expression demanded by the Machine, so thoroughly grounded in the arts and crafts, is founded upon a finely guarded reticence, a recognition of inherent weakness or plain ignorance!

Let us, to be just, assume that it is equal parts of all three, and try to imagine an Arts and Crafts Society that may educate itself to prepare to make some good impression upon the Machine, the destroyer of their present ideals and tendencies, their salvation in disguise.

Such a society will, of course, be a society for mutual education.

Exhibitions will not be a feature of its programme for years, for there will be nothing to exhibit except the shortcomings of the society, and they will hardly prove either instructive or amusing at this stage of proceedings. This society must, from the very nature of the proposition, be made up of the people who are in the work – that is, the manufacturers – coming into touch with such of those who assume the practice of the fine arts as profess a fair sense of the obligation to the public such assumption carries with it, and sociological workers whose interests are ever closely allied with art, as their prophets Morris, Ruskin, and Tolstoy evince, and all those who have as personal graces and accomplishment perfected handicraft, whether fashion old or fashion new.

Without the interest and cooperation of the manufacturers, the society cannot begin to do its work, for this is the cornerstone of its organization.

All these elements should be brought together on a common ground of confessed ignorance, with a desire to be instructed, freely encouraging talk and opinions, and reaching out desperately for anyone who has special experience in any way connected to address them.

I suppose, first of all, the thing would resemble a debating society, or something even less dignified, until someone should suggest that it was time to quit talking and proceed to do something, which in this case would not mean giving an exhibition, but rather excursions to factories and a study of processes in place – that is, the machine in processes too numerous to mention, at the factories with the men who organize and direct them, but not in the spirit of the idea that these things are all gone wrong, looking for that in them which would most nearly approximate the handicraft ideal; not looking into them with even the thought of handicraft, and not particularly looking for craftsmen, but getting a scientific ground plan of the process in mind, if possible, with a view to its natural bent and possibilities. Some processes and machines would naturally appeal to some, and some to others; there would undoubtedly be among us those who would find little joy in any of them.

This is, naturally, not child's play, but neither is the work expected of the modern artist.

I will venture to say, from personal observation and some experience, that not one artist in one hundred has taken pains to thus educate himself. I will go further and say what I believe to be true, that not one educational institution in America has as yet attempted to forge the connecting link between Science and Art by training the artist to his actual tools, or, by a process of nature-study that develops in him the power of independent thought, fitting him to use them properly.

Let us call these preliminaries then a process by which artists receive information nine-tenths of them lack concerning the tools they have to work with today – for tools today are processes and machines where they were once a hammer and a gouge.

The artist today is the leader of an orchestra, where he once was a star performer.

Once the manufacturers are convinced of due respect and appreciation on the part of the artist, they will welcome him and his counsel gladly and make any experiments having a grain of apparent sense in them.

They have little patience with a bothering about in endeavor to see what might be done to make their particular machine medieval and restore man's joy in the mere work of his hands – for this once lovely attribute is far behind.

This proceeding doubtless would be of far more educational value to the artist than to the manufacturer, at least for some time to come, for there would be a difficult adjustment to make on the part of the artist and an attitude to change. So many artists are chiefly "attitude" that some would undoubtedly disappear with the attitude.

But if out of twenty determined students a ray of light should come to one, to light up a single operation, it would have been worthwhile, for that would be fairly something; while joy in mere handicraft is like that of the man who played the piano for his own amusement – a pleasurable personal accomplishment without real relation to the grim condition confronting us.

Granting that a determined, dauntless body of artist material could be brought together with sufficient persistent enthusiasm to grapple with the Machine, would not someone be found who would provide the suitable experimental station (which is what the modern Arts and Crafts shop should be) – an experimental station that would represent in miniature the elements of this great pulsating web of the machine, where each pregnant process or significant tool in printing, lithography, galvano-electro processes, wood and steel working machinery, muffles and kilns would have its place and where the best young scientific blood could mingle with the best and truest artistic inspiration, to sound the depths of these things, to accord them the patient, sympathetic treatment that is their due?

Surely a thing like this would be worthwhile – to alleviate the insensate numbness of the poor fellows out in the cold, hard shops, who know not why nor understand, whose dutiful obedience is chained to botch work and bungler's ambition; surely this would be a practical means to make their dutiful obedience give us something we can all understand, and that will be as normal to the best of this machine age as a ray of light to the healthy eye; a real help in adjusting the Man to a true sense of his importance as a factor in society, though he does tend a machine.

Teach him that that machine is his best friend – will have widened the margin of his leisure until enlightenment shall bring him a further sense of the magnificent ground plan of progress in which he too justly plays his significant part.

If the art of the Greek, produced at such cost of human life, was so noble and enduring, what limit dare we now imagine to an Art based upon an adequate life for the individual?

The machine is his!

In due time it will come to him! Meanwhile, who shall count the slain?

From where are the trained nurses in this industrial hospital to come if not from the modern arts and crafts?

Shelley says a man cannot say – "I will compose poetry." "The greatest poet even cannot say it, for the mind in creation is as a fading coal which some invisible influence, like an inconstant wind awakens to transitory brightness; this power arises from within like the color of a flower which fades and changes as it is developed, and the conscious portions of our nature are unprophetic either of its approach or its departure"; and yet in the arts and crafts the problem is presented as a more or less fixed quantity, highly involved, requiring a surer touch, a more highly disciplined artistic nature to organize it as a work of art.

The original impulses may reach as far inward as those of Shelley's poet, be quite as wayward a matter of pure sentiment, and yet after the thing is done, showing its rational qualities, are limited in completeness only by the capacity of whoever would show them or by the imperfection of the thing itself.

This does not mean that Art may be shown to be an exact Science.

"It is not pure reason, but it is always reasonable."

It is a matter of perceiving and portraying the harmony of organic tendencies: is originally intuitive because the artist nature is a prophetic gift that may sense these qualities afar.

To me, the artist is he who can truthfully idealize the common sense of these tendencies in his chosen way.

So I feel conception and composition to be simply the essence of refinement in organization, the original impulse of which may be registered by the artistic nature as unconsciously as the magnetic needle vibrates to the magnetic law, but which is, in synthesis or analysis, organically consistent, given the power to see it or not.

And I have come to believe that the world of Art, which we are so fond of calling the world outside of Science, is not so much outside as it is the very heart quality of this great material growth – as religion is its conscience.

A foolish heart and a small conscience.

A foolish heart, palpitating in alarm, mistaking the growing pains of its giant frame for approaching dissolution, whose sentimentality the lusty body of modern things has outgrown.

Upon this faith in Art as the organic heart quality of the scientific frame of things, I base a belief that we must look to the artist brain, of all brains, to grasp the significance to society of this thing we call the Machine, if that brain be not blinded, gagged, and bound by false tradition, the letter of precedent. For this thing we call Art is it not as prophetic as a primrose or an oak? Therefore, of the essence of this thing we call the Machine, which is no more or less than the principle of organic growth working irresistibly the Will of Life through the medium of Man.

Be gently lifted at nightfall to the top of a great downtown office building, and you may see how in the image of material man, at once his glory and menace, is this thing we call a city.

There beneath, grown up in a night, is the monster leviathan, stretching acre upon acre into the far distance. High overhead hangs the stagnant pall of its fetid breath, reddened with the light from its myriad eyes endlessly everywhere blinking. Ten thousand acres of cellular tissue, layer upon layer, the city's flesh, outspreads enmeshed by intricate network of veins and arteries, radiating into the gloom, and there with muffled, persistent roar, pulses and circulates as the blood in your veins, the ceaseless beat of the activity to whose necessities it all conforms.

Like to the sanitation of the human body is the drawing off of poisonous waste from the system of this enormous creature; absorbed first by the infinitely ramifying, threadlike ducts gathering at their sensitive terminals matter destructive to its life, hurrying it to millions of small intestines, to be collected in turn by larger, flowing to the great sewer, on to the drainage canal, and finally to the ocean.

This ten thousand acres of fleshlike tissue is again knit and interknit with a nervous system marvelously complete, delicate filaments for hearing, knowing, almost feeling the pulse of its organism, acting upon the ligaments and tendons for motive impulse, in all flowing the impelling fluid of man's own life.

Its nerve ganglia! – the peerless Corliss tandems whirling their hundred ton fly-wheels, fed by gigantic rows of watertube boilers burning oil, a solitary man slowly pacing backward and forward, regulating here and there the little feed valves controlling the deafening roar of the flaming gas, while beyond, the incessant clicking, dropping, waiting – lifting, waiting, shifting of the governor gear controlling these modern Goliaths seems a visible brain in intelligent action, registered infallibly in the enormous magnets, purring in the giant embrace of great induction coils, generating the vital current meeting with instant response in the rolling cars on elevated tracks ten miles away, where the glare of the Bessemer steel converter makes a conflagration of the clouds.

More quietly still, whispering down the long, low rooms of factory buildings buried in the gloom beyond, range on range of stanch, beautifully perfected automatons, murmur contentedly with occasional click-clack, that would have the American manufacturing industry of five years ago by the throat today manipulating steel as delicately as a mystical shuttle of the modern loom manipulates a silk thread in the shimmering pattern of a dainty gown.

And the heavy breathing, the murmuring, the clangor, and the roar! – how the voice of this monstrous thing, this greatest of machines, a great city, rises to proclaim the marvel of the units of its structure, the ghastly warning boom from the deep throats of vessels heavily seeking inlet to the waterway below, answered by the echoing clangor of the bridge bells growing nearer and more ominous as the vessel cuts momentarily the flow of the nearer artery, warning the current from the swinging bridge now closing on its stately passage, just in time to receive in a rush of steam, as a streak of light, the avalanche of blood and metal hurled across it and gone, roaring into the night on its glittering bands of steel, ever faithfully encircled by the slender magic lines tick-tapping its invincible protection.

Nearer, in the building ablaze with midnight activity, the wide white band streams into the marvel of the multiple press, receiving unerringly the indelible impression of the human hopes, joys, and fears throbbing in the pulse of this great activity, as infallibly as the gray matter of the human brain receives the impression of the senses, to come forth millions of neatly folded, perfected news sheets, teeming with vivid appeals to passions, good or evil; weaving a web of intercommunication so far-reaching that distance becomes as nothing, the thought of one man in one corner of the earth one day visible to the naked eye of all men the next; the doings of all the world reflected as in a glass, so marvelously sensitive this wide white band streaming endlessly from day to day becomes in the grasp of the multiple press.

If the pulse of activity in this great city, to which the tremor of the mammoth skeleton beneath our feet is but an awe-inspiring response, is thrilling, what of this prolific, silent obedience?

And the texture of the tissue of this great thing, this Forerunner of Democracy, the Machine, has been deposited particle by particle, in blind obedience to organic law, the law to which the great solar universe is but an obedient machine.

Thus is the thing into which the forces of Art are to breathe the thrill of ideality! A SOUL!

1   From *Invictis*, by William Ernst Henley. Notes by Bruce Brooks Pfeiffer, *Frank Lloyd Wright Collected Writings* (1992).

2   From this paragraph, through the next twenty-three paragraphs Wright has paraphrased Victor Hugo's "The One Will Kill the Other" from *The Hunchback of Notre-Dame*.

3   Marshall Field (1834–1906), a Chicago merchant, commissioned both Henry Hobson Richardson and the firm of D. H. Burnham and Co. to design his stores. Richardson's wholesale store of 1885 is a building many critics consider to be one of the greatest of the nineteenth century. Field's comment, therefore, probably relates to the later Burnham building. Richardson's untimely death at the age of 48 in 1886 forced Field to choose a new architect.

4   Richard Croker (1841–1922), a New York politician of Irish birth, who rose to Tammany leadership in the mid-1880s.

1914

Antonio Sant' Elia

Manifesto of Futurist Architecture[1]

Antonio Sant' Elia (1888–1916) was an Italian architect active in the years just before the First World War. He moved to Milan in 1912 to begin his architectural practice and quickly became active among the restless avant-garde of artists, writers, and designers. His reputation rests almost entirely on a series of visionary drawings he made for the *Citta Nuova* (new city), which combined the novel elements of the industrial city with elements of the architecture of Otto Wagner and Adolf Loos. He displayed the drawings of the new city in 1914 as a member of the *Nuove Tendenze*, but immediately after the exhibition declared himself a Futurist. It is a subject of much debate how much of the manifesto was actually written by Sant' Elia, and how much was crafted by Filippo Tommaso Marinetti.

His views about the mechanization or elimination of ornament are not original and can be traced to Loos and Wagner, but like Wright's earlier, more cautious statement, the manifesto welcomes the changes brought by industrialization. And it is in the final point of his proclamation that we read the characteristic change: "the fundamental characteristics of Futurist architecture will be its impermanence and transience."

No architecture has existed since 1700. A moronic mixture of the most various stylistic elements used to mask the skeletons of modern houses is called modern architecture. The new beauty of cement and iron is profaned by the superimposition of motley decorative incrustations that cannot be justified either by constructive necessity or by our (modern) taste, and whose origins are in Egyptian, Indian or Byzantine antiquity and in that idiotic flowering of stupidity – and impotence – that took the name of NEOCLASSICISM.

These architectonic prostitutions are welcomed in Italy, and rapacious alien ineptitude is passed off as talented invention and as extremely up-to-date architecture. Young Italian architects (those who borrow originality from clandestine and compulsive devouring of art journals) flaunt their talents in the new quarters of our towns, where a hilarious salad of little ogival columns, seventeenth-century foliation, Gothic pointed arches, Egyptian pilasters, rococo scrolls, fifteenth-century cherubs, swollen caryatids, take the place of style in all seriousness, and presumptuously put on monumental airs. The kaleidoscopic appearance and reappearance of forms, the multiplying of machinery, the daily increasing needs imposed by the speed of communications, by the concentration of population, by hygiene, and by a hundred other phenomena of modern life, never cause these self-styled renovators of architecture a moment's perplexity or hesitation. They persevere obstinately with the rules of Vitruvius, Vignola, and Sansovino plus gleanings from any published scrap of information on German architecture that happens to be at hand. Using these, they continue to stamp the image of imbecility on our cities, our cities which should be the immediate and faithful projection of ourselves.

And so this expressive and synthetic art has become in their hands a vacuous stylistic jumble of ill-mixed formulae to disguise a run-of the-mill traditionalist box of bricks and stones a modern building. As if we who are accumulators and generators of movement, with all our added mechanical limbs, with all the noise and speed of our life, could live in streets built for the needs of men four, five or six centuries ago.

This is the supreme imbecility of modern architecture, perpetuated by the venal complicity of the academies, the internment camps of the intelligentsia, where the young are forced into the onanistic recopying of classical models instead of throwing their minds open in the search for new frontiers and in the solution of the new and pressing problem: THE FUTURIST HOUSE AND CITY. The house and the city that are ours both spiritually and materiality, in which our tumult can rage without seeming a grotesque anachronism.

The problem posed in Futurist architecture is not one of linear rearrangement. It is not a question of finding new mouldings and frames for windows and doors, of replacing columns, pilasters and corbels with caryatids, flies and frogs. Neither has it anything to do with leaving a facade in bare brick, or plastering it, or facing it with stone or in determining formal differences between the new building and the old one. It is a question of tending the healthy growth of the Futurist house, of constructing it with all the resources of technology and science, satisfying magisterially all the demands of our habits and our spirit trampling down all that is grotesque and antithetical (tradition, style, aesthetics, proportion), determining new forms, new lines, a new harmony of profiles and volumes, an architecture whose reason for existence can be found solely in the unique conditions of modern life, and in its correspondence with the aesthetic values of our sensibilities. This architecture cannot be subjected to any law of historical continuity. It must be new, just as our state of mind is new.

The art of construction has been able to evolve with time, and to pass from one style to another, while maintaining unaltered the general characteristics of architecture, because in the course of history changes of fashion are frequent and are determined by the alternations of religious conviction and political disposition. But profound changes in the state of the environment are extremely rare, changes that unhinge and renew, such as the discovery of natural laws, the perfecting of mechanical means, the rational and scientific use of material. In modern life the process of stylistic development in architecture has been brought to a halt. ARCHITECTURE NOW MAKES A BREAK WITH TRADITION. IT MUST PERFORCE MAKE A FRESH START.

Calculations based on the resistance of materials, on the use of reinforced concrete and steel, exclude 'architecture' in the classical and traditional sense. Modern constructional materials and scientific concepts are absolutely incompatible with the disciplines of historical styles, and are the principal cause of the grotesque appearance of 'fashionable' buildings in which attempts are made to employ the lightness, the superb grace of the steel beam, the delicacy of reinforced concrete, in order to obtain the heavy curve of the arch and the bulkiness of marble.

The utter antithesis between the modern world and the old is determined by all those things that formerly did not exist. Our lives have been enriched by elements the possibility of whose existence the ancients did not even suspect. Men have identified material contingencies, and revealed spiritual attitudes, whose repercussions are felt in a thousand ways. Principal among these is the formation of a new ideal of beauty that is still obscure and embryonic, but whose fascination is already felt even by the masses. We have lost our predilection for the monumental, the heavy, the static, and we have enriched our sensibility with a taste for the light, the practical, the ephemeral, and the swift. We no longer feel ourselves to be the men of the cathedrals, the palaces, and the podiums. We are the men of the great hotels, the railway stations, the immense streets, colossal ports, covered markets, luminous arcades, straight roads, and beneficial demolitions.

We must invent and rebuild the Futurist city like an immense and tumultuous shipyard, agile, mobile, and dynamic in every detail; and the Futurist house must be like a gigantic machine. The lifts must no longer be hidden away like tapeworms in the niches of stairwells; the stairwells themselves, rendered useless, must be abolished, and the lifts must scale the lengths of the façades like serpents of steel and glass. The house of concrete, glass, and steel, stripped of paintings and sculpture, rich only in the innate beauty of its lines and relief, extraordinarily 'ugly' in its mechanical simplicity, higher and wider according to need rather than the specifications of municipal laws. It must soar up on the brink of a tumultuous abyss: the street will no longer lie like a doormat at ground level, but will plunge many storeys down into the earth, embracing the metropolitan traffic, and will be linked up for necessary interconnections by metal gangways and swift-moving pavements.

THE DECORATIVE MUST BE ABOLISHED. The problem of Futurist architecture must be resolved, not by continuing to pilfer from Chinese, Persian, or Japanese photographs or fooling around with the rules of Vitruvius, but through flashes of genius and through scientific and technical expertise. Everything must be revolutionized. Roofs and underground spaces must be used; the importance of the façade must be diminished; issues of taste must be transplanted from the field of fussy mouldings, finicky capitals and flimsy doorways to the broader concerns of BOLD GROUPINGS AND MASSES, and LARGE-SCALE DISPOSITION OF PLANES. Let us make an end of monumental, funereal and commemorative architecture. Let us overturn monuments, pavements, arcades and flights of steps; let us sink the streets and squares; let us raise the level of the city.

I COMBAT AND DESPISE:

1   All the pseudo-architecture of the avant-garde, Austrian, Hungarian, German, and American;
2   All classical architecture, solemn, hieratic, scenographic, decorative, monumental, pretty, and pleasing;
3   The embalming, reconstruction, and reproduction of ancient monuments and palaces;
4   Perpendicular and horizontal lines, cubical and pyramidical forms that are static, solemn, aggressive, and absolutely excluded from our utterly new sensibility;
5   The use of massive, voluminous, durable, antiquated, and costly materials.

AND PROCLAIM:

1   That Futurist architecture is the architecture of calculation, of audacious temerity and of simplicity; the architecture of reinforced concrete, of steel, glass, cardboard, textile fibre, and of all those substitutes for wood, stone, and brick that enable us to obtain maximum elasticity and lightness;
2   That Futurist architecture is not because of this an arid combination of practicality and usefulness, but remains art, i.e. synthesis and expression;
3   That oblique and elliptic lines are dynamic, and by their very nature possess an emotive power a thousand times stronger than perpendiculars and horizontals, and that no integral, dynamic architecture can exist that does not include these;
4   That decoration as an element superimposed on architecture is absurd, and that THE DECORATIVE VALUE OF FUTURIST ARCHITECTURE DEPENDS SOLELY ON TILE USE AND ORIGINAL ARRANGEMENT OF RAW OR BARE OR VIOLENTLY COLOURED MATERIALS;
5   That, just as the ancients drew inspiration for their art from the elements of nature, we – who are materially and spiritually artificial – must find that inspiration in the elements of the utterly new mechanical world we have created, and of which architecture must be the most beautiful expression, the most complete synthesis, the most efficacious integration;
6   That architecture as the art of arranging forms according to pre-established criteria is finished;
7   That by the term architecture is meant the endeavour to harmonize the

environment with Man with freedom and great audacity, that is to transform the world of things into a direct projection of the world of the spirit;

8 From an architecture conceived in this way no formal or linear habit can grow, since the fundamental characteristics of Futurist architecture will be its impermanence and transience. THINGS WILL ENDURE LESS THAN US. EVERY GENERATION MUST BUILD ITS OWN CITY. This constant renewal of the architectonic environment will contribute to the victory of Futurism which has already been affirmed by WORDS-IN-FREEDOM, PLASTIC DYNAMISM, MUSIC WITHOUT QUADRATURE, AND THE ART OF NOISES, and for which we fight without respite against traditionalist cowardice.

1 Amplified from catalogue introduction, 'Nuove Tendenze'. Milan, 1914. Published in *Lacerba* (Florence), 1 August 1914.

1915

Patrick Geddes

Paleotechnic and Neotechnic

Sir Patrick Geddes (1854–1932) was a Scottish biologist, sociologist, and town planner with a strong interest in theories of education, the arts, and history. He began his professional career as a biologist in London and France, but in the late 1880s he settled in Edinburgh, where he became involved in the renewal of Edinburgh's Old Town. In 1889 he became Professor of Botany at Dundee University College, where he developed a deep fascination with the organization of human societies and their spatial manifestation in the city and the country. In 1904 Geddes published *City Development: a Study of Parks, Gardens and Culture Institutes. A Report to the Carnegie Trust Dunfermline.* After 1900 he focused his activities on London, where he co-founded the Sociological Society and displayed his Cities and Town Planning Exhibition. From 1914 to 1924 Geddes lived mainly in India, where he was involved in town planning, accepting the Chair of Sociology and Civics at the University of Bombay in 1919. During that period Geddes was also commissioned to plan the Hebrew University in Jerusalem, the garden suburbs of Jerusalem and Haifa, and a number of settlements elsewhere in Palestine, returning later to produce a plan for Tel Aviv.

Geddes approached cities like a biologist, surveying them at many scales and seeking to understand the mechanisms of their growth and form. The distinction between phases of the industrial revolution was drawn from anthropology, and helped make evident the degree to which mechanization applied to everything from explicit machines to the organization of labor to the flow of money. The following excerpt is from *Cities in Evolution* of 1915, which was broadly influential on town planning, and particularly on Lewis Mumford, who adopted Geddes's historical scope and his social-biological categories. Of course his distinction between paleotechnic and neotechnic is more than analytical, and encapsulates the most optimistic – Geddes himself called it utopian – view of the progression of technological society from wasteful, cruel, and dissipative paleotects to the life-conserving, vital budget of the neotechnic.

A new industrial age is opening. As the "Stone Age" is now distinguished into two periods, "Paleolithic" and "Neolithic," so the "Industrial Age" requires distinction into its two phases "Paleotechnic" and "Neotechnic."

Here, again, this same process is beginning – that of a new industrial age. Following James Watt, the Prometheus of steam, Glasgow gave us the very foremost of all the Prometheans of electricity in Lord Kelvin. Following upon the locomotive of Stephenson, we have motors and electric cars, and upon the marine applications of Watt's engine, we have had the gas turbine from Birmingham, from Newcastle the turbine of Parsons, already improved upon; next the application of oil fuel, the Diesel engine, and so on (1910).

Now of all the limitations of our predominant middle-class and upper-class points of view, one of the worst is not seeing how widely different are the forms of labour. Not merely in their various products, and in various rates of money wages, as economists have been wont to describe. Far beyond all these, significant in ways far too much ignored, are their effects. First on the individuals who perform these various tasks, as physicians and psychologists now observe them; second on the resultant types of family, of institutions and general civilisation, as social geographers have long been pointing out for simple societies, and as sociologists have now to work out for our complex ones.

Take a simple illustration of the first. No one surely but can see, for instance, that the practical disappearance of the legion of stokers, which oil fuel involves, is something, physiologically if not politically, comparable to the emancipation of the galley-slaves, which similarly was brought about through an improvement in modern locomotion. It is, on the whole, well to throw people out of such employment. But finer issues are less obvious, and need tracing. A great idealist, an undeniable moral force like the late John Bright felt himself logically compelled in terms of his economic creed – that of the then believed final machine and market order – to argue in Parliament against the Adulteration Acts as an interference with competition, and therefore with the life of trade! Whereas, the simplest, the least moralised or idealistic of electricians needs no public enthusiasms, no moral or social convictions, to convince him that adulteration is undesirable; since every day's work in his calling has experimentally made him feel how a trace of impurity in his copper wire deteriorates its conductivity, and how even a trifle of dirt between contact surfaces is no trifle, but may spoil contact altogether.

Such illustrations might be multiplied and developed indefinitely. But enough here if we can broadly indicate, as essential to any real understanding of the present state of the evolution of cities, that we clearly distinguish between what is characteristic of the passing industrial order, and that which is characteristic of the incipient one – the passing and the coming age. Indeed, before many years we may say the closing and the opening one.

Recall how as children we first heard of "The Stone Age"; next, how this term has practically disappeared. It was found to confuse what are really two strongly contrasted phases of civilisation, albeit here and there found mingled, in transition; in arrest or in reversion, sometimes also; frequently also in collision – hence we now call these the Old Stone Age and the New; the Paleolithic and the Neolithic.

The former phase and type is characterised by rough stone implements, the latter by skilfully chipped or polished ones; the former in common types and mostly for rougher uses, the latter in more varied types and materials, and for finer skills. The first is a rough hunting and warlike civilisation, though not without a certain vigour of artistic presentment, which later militarist or hunting types have also striven for, but seldom attained, and certainly not surpassed. The latter Neolithic folk were of gentler, agricultural type, with that higher evolution of the arts of peace and of the status of woman, which, as every anthropologist knows, is characteristic of agriculture everywhere, and is so obvious save where otherwise artificially depressed.

The record of these two different civilisations every museum now clearly shows, and they need not here be enlarged upon. Their use to us is towards making more intelligible the application of a similar analysis in our own times, and to the world around us. For although our economists have been and are in the habit of speaking of our present civilisation, since the advent of steam and its associated machinery, with all its technic strivings and masteries, as the "Industrial Age," we press for the analysis of this into two broadly and clearly distinguishable types and phases: again of older and newer, ruder and finer type, needing also a constructive nomenclature accordingly.

Simply substituting -technic for -lithic, we may distinguish the earlier and ruder elements of the Industrial Age as Paleotechnic, the newer and still often incipient elements disengaging themselves from these as Neotechnic: while the people belonging to these two dispensations we shall take the liberty of calling Paleotects and Neotects respectively.

› Illustration from synoptic view of Durham

To the former order belong the collieries, in the main as yet worked; together with the steam-engine and most of our staple manufactures; so do the railways and the markets, and above all the crowded and monotonous industrial towns to which all these have given rise. These dreary towns are, indeed, too familiar to need detailed description here; they constitute the bulk of the coalfield conurbations we were considering in the previous chapter. Their corresponding abstract developments have been the traditional political economy on the one hand, and on the other that general body of political doctrine and endeavour which was so clearly formulated, so strenuously applied by the French Revolution and its exponents, but which in this country has gone on bit by bit in association with the slower and longer Industrial Revolution.

To realise, first of all, in definite synoptic vision of a city, the change from the old regime to modern Paleotechnic conditions, there is no more vivid example perhaps in the world than the view of Durham from the railway. We see on the central ridge the great medieval castle, the magnificent cathedral, as characteristic monuments as one could wish to see of the temporal and the spiritual powers of its old County Palatine and Diocese, with its Prince and Bishop, in this case one. Next, see all around this the vast development of the modern mining town, with its innumerable mean yet decent streets, their meaner, yet decent little houses, with their main life carried on in kitchens and back courts, decent too, yet meanest of all: for here is a certain quiet and continuous prosperity, a comparative freedom from the main evils of greater cities, which makes this modern town of Durham, apart from its old cathedral and castle, altogether a veritable beauty-spot of the coal age, a paragon of the Paleotechnic order.

When we have added to this prosperous town life the Board Schools and the Carnegie Library, and to these the University Extension Lectures on Political Economy, and the Workers' Union lectures on Economic History, what is left for the heart of the collier or his "representative" to desire in the way of prosperity and education (happiness, domestic and personal, remaining his private affair), except, indeed, to make these more steady and permanent through such legislation towards relieving unemployment and sickness as may be devised? Wages, no doubt, may still perhaps be improved a little. The cathedral might be disestab-lished; and so on. But on all received principles of Paleotechnic economics or polit-ics, Durham is obviously approximately perfect.

Similarly for our larger colliery, iron, textile conurbations and towns – Amer-ican ones likewise. While the coalfield holds out, our progress seems practically assured: our chosen press shall be that which can most clearly voice this convic-tion for us, and our politicians must be those who, by this measure or that opposite one, most hopefully promise to assure its continuance.

› Interpretation of protests from "romantics" – Carlyle, Ruskin, etc.

With this growing organisation for industry in progress, and with its associated system of ideals expressed in the other industrial towns around us, who can wonder at the little success with which Carlyle, Ruskin, and Morris successively ful-minated against them? – or even of the criticisms which the politicians and econo-mists have never been able to answer? It was, of course, easier to discredit these writers as "romantic," as "aesthetic," and so on, and to assume that science and invention were all on the Paleotechnic side. But nowadays, thanks to the advance of science and of invention, we know better.

Had Carlyle or Morris but known it (Ruskin had an inkling of this, and more), their view of industry was already far more in accordance with the physicist's doc-trine of energy than is that of the conventional economists even of today. For after its prolonged darkening of counsel with economic text-books without that elementary physical knowledge which should underlie every statement of the industrial process – save perhaps at most, a reference, and that often depreciatory, to Prof. Stanley Jevons on solar crises, or on the exhaustion of our coal supplies – it is really only with President Roosevelt's "National Resources Commission" that the fundamentals of national economy became generally recognised. For this Commission began with the national forester, Gifford Pinchot, and included statesmen-agriculturists of the type of Horace Plunkett. It happily included even the economist, albeit as a brand plucked from the burning, and teaching a very different doctrine from that of his youth. These told their countrymen that to dissipate the national energies, as the

American Paleotects, of Pittsburg or where you please, have been doing, is not economics but waste; and that to go on dissipating energies for the sake of this or that individual percentage on the transaction, is no longer to be approved as "development of resources," as the mendacious euphemism for it goes, but sternly to be discouraged, as the national waste, the mischievous public housekeeping it has been all along.

As our studies of the physical realities in economic processes go on, each industrial process has to be clearly analysed – into its physical factors of material efficiency and directness on the one side, and its financial charges on the other. Thus, while we shall utilise more than ever each improvement and invention which can save energy, minimise friction, diminish waste or loss of time in transit, we shall also begin to criticise in the same spirit that commercial process which is implied in the great railway maxim of "charging what the traffic will bear," and which, in more scientific language, may be called "parasitism in transit."

The Paleotechnic mind – whether of Boards of Directors or Workers' Unions, here matters little – has been too much interested in increasing or in sharing these commercial proceeds, and too little in that of maximising physical efficiency and economy all through. And, since all this applies to more than railways, it is scarcely to be wondered at that the vast improvements of modern invention have so largely been rendered nugatory in this general Paleotechnic way, and not by any perversity peculiar to the labourer or to the capitalist alone, as they too cheaply convince themselves.

› Passage from money wages to "Vital Budget": this conception needed to build the Neotechnic town

Again, under the Paleotechnic order the working man, misdirected as he is like all the rest of us, by his traditional education towards money wages, instead of Vital Budget, has never yet had an adequate house, seldom more than half of what might make a decent one. But as the Neotechnic order comes in – its skill directed by life towards life, and for life – he, the working man, as in all true cities of the past aristo-democratised into productive citizen – he will set his mind towards house-building and town-planning, even towards city design; and all these upon a scale to rival – nay surpass – the past glories of history. He will demand and create noble streets of noble houses, gardens and parks; and before long monuments, temples of his renewed ideals, surpassing those of old.

Thus he will rapidly accumulate both civic and individual Wealth, that is, Wealth twofold, and both hereditary. It will be said – even he as yet says it, paralysed as he still is – that this is "Utopia" – that is, practically Nowhere. It is, and should be, beyond the dreams of the historic Utopists, right though they also were in their day. For their projects of real wealth were based upon the more rational use of the comparatively scanty resources and limited population of the past. But just as our Paleotechnic money-wealth and real poverty are associated with the waste and dissipation of the stupendous resources of energy and materials, and power of using them, which the growing knowledge of Nature is ever unlocking for us, so their better Neotechnic use brings with it potentialities of wealth and leisure beyond past Utopian dreams.

Utopias indispensable to social thought. The escape from Paleotechnic to Neotechnic order is thus from Kakotopia to Eutopia – the first turning on dissipating energies towards individual money gains, the other on conserving energies and organising environment towards the maintenance and evolution of life, social and individual, civic and eugenic.

This time the Neotechnic order, if it means anything at all, with its better use of resources and population towards the bettering of man and his environment together, means these as a business proposition – the creation, city by city, region by region, of its Eutopia, each a place of effective health and well-being, even of glorious, and in its way unprecedented, beauty, renewing and rivalling the best achievements of the past, and all this beginning here, there and everywhere – even where our Paleotechnic disorder seems to have done its very worst.

How can this be put yet more definitely? Simply enough. The material alternatives of real economics, which these obsessions of money economics have been too long obfuscating, are broadly two, and each is towards realising an ideal, a Utopia. These are the Paleotechnic and the Neotechnic – Kakotopia and Eutopia respectively. The first has hitherto been predominant. As Paleotechnics we make it our prime endeavour to dig up coals, to run machinery, to produce cheap cotton to clothe cheap people, to get up more coals, to run more machinery, and so on; and all this essentially towards "extending markets." The whole has been essentially organised upon a basis of "primary poverty" and of "secondary poverty" (to use Mr. Rowntree's accurate terminology, explained later), relieved by a stratum of moderate well-being, and enlivened by a few prizes, and comparatively rare fortunes – the latter chiefly estimated in gold, and after, death. But all this has been with no adequate development of real wealth, as primarily of houses and gardens, still less of towns and cities worth speaking of: our industry but maintains and multiplies our poor and dull existence. Our Paleotechnic life-work is soon physically dissipated; before long it is represented by dust and ashes, whatever our money wages may have been.

Moreover, though we thus have produced, out of all this exhaustion of the resources of Nature and of race, whole new conurbations, towns and pseudo-cities, these are predominantly, even essentially, of Slum character – Slum, Semi-slum, or Superslum – as we shall see more fully later – each, then, a Kakotopia as a whole; and in these the corresponding development of the various types of human deterioration congruent with such environment. Within this system of life there may (and do, of course) arise palliatives, and of many kinds, but these do not affect the present contrast.

The second alternative, however, also remains open, and happily has now its material beginnings everywhere – that of the nascent, Neotechnic order. Whenever – with anything like corresponding vigour and decision to that which the Paleotects have shown, once and again, as notably at the coming on of the machine age, the railway age, the financial age, and now the militarist one – we make up our minds, as some day before long we shall do, to apply our constructive skill, our vital energies, towards the public conservation instead of the private dissipation of resources, and towards the evolution instead of the deterioration of the lives of others, then we shall discern that this order of things also "pays," and this all the better for paying in kind.

That is, in having houses and gardens, and of the best, with all else that is congruent with them, towards the maintenance and the evolution of our lives, and still more of our children's. Then in a short, incredibly short, time, we, and still more they, shall have these dwellings, and with them the substantial and assured, the wholesome and delightful, contribution to the sustenance of their inhabitants which gardens, properly understood and worked, imply.

The old sociologists, in their simple societies, saw more clearly than we; but as we recover their rustic and evolutionary point of view we may see that also for ourselves – "Whatsoever a man soweth, that shall he also reap" – at any rate shall be reaped by his successors if not by himself. During the Paleotechnic period this has been usually understood and preached on as a curse. For the Neotechnic standpoint it is a blessing, manifestly rooted in the order of nature. Then why not increasingly sow what is best worth reaping?

The life and labour of each race and generation of men are but the expression and working out of their ideals. Never was this more fully done than in this Paleotechnic phase, with its wasteful industry and its predatory finance – and its consequences, (a) in dissipation of energies, (b) in deterioration of life, are now becoming manifest. Such twofold dissipation may most simply be observed upon two of its main lines; that of crude luxuries and sports, and the "dissipations" these so readily involve in the moral sense; and second through war. The crude luxury is excused, nay, psychologically demanded, by the starvation of Paleotechnic life, in well nigh every vital element of beauty or spirituality known and valued by humanity hitherto. Thus to take only one of the very foremost of our national luxuries, that of getting – more or less – alcoholised, this has been vividly defined, in a real flash of judicial wisdom, as "the quickest way of getting out of Manchester."

› Interpretation of war and of the general struggle toward survival from current point of view (1910–15)

Similarly, war and its preparations are explained, we may even say necessitated, by the accepted philosophy and the social psychology of our Paleotechnic cities, and particularly of the metropolitan ones. In the first place, war is but a generalising of the current theory of competition as the essential factor of the progress of life. For, if competition be, as we are told, the life of trade, competition must also be the trade of life. What could the simple naturalists, like Darwin and his followers, do but believe this? and thence project it upon nature and human life with a new authority!

The Paleotechnic philosophy is thus complete; and trade competition, Nature competition, and war competition, in threefold unity, have not failed to reward their worshippers. Thus the social mind, of the said cities especially, but thereafter of the whole nation they influence, is becoming characterised and dominated by an ever deepening state of diffused and habitual fear. This again is the natural accumulation, the inevitable psychological expression of certain very real evils and dangers, though not those most commonly expressed.

First, of the inefficiency and wastefulness of Paleotechnic industry, with corresponding instability and irregularity of employment, which are increasingly felt by all concerned; second, the corresponding instability of the financial system, with its pecuniary and credit illusions, which are also becoming realised; and third, the growing physical slackness or deterioration – unfitness anyhow – which we all more or less feel in our Paleotechnic town life, which therefore must more and more make us crouch behind barriers, and cry for defenders.

Hence, in fact, Tennyson's well-known eulogy of the Crimean War, and many other earlier and later ones. For as imagined military dangers become real ones, so far from increasing fear, they at once exhilarate and invigorate our ebbing courage. Of all the "Merrie England" of the past, there was but one town which habitually boasted the epithet; and that was "Merrie Carlisle," just because it guarded the marches, and stood to bear the first shock of Scottish raids or invasions; and first sent out its hardy sons, now to provoke these, now meet them with counter-initiative.

Similarly, it is not in the many coast cities lying open to bombardment, but at London – and this not simply but deeply because it is practically unattackable, besides having the assurance of immediate concentration of all the national resources of defence – that there, of all cities, the yellow journalist can always most readily exploit the popular fears (1910).

On grounds like these, which have been only too obvious In other places and times, pessimism always as naturally arises. Yet here our pessimism is but relative; for it needs no war, but only the appearance of Neotechnic art and science to evoke a corresponding courage. Hence, for instance, the joyousness of the aviator amid his desperate risks.

The struggles of war are not essential to the nature of society: at present (1910–15) the major problem is the struggle for existence between the Paleotechnic and Neotechnic orders. Since the Paleotechnic war-obsession stands so definitely in the way of city betterment, let us put the criticism of it in a somewhat different way.

Among lagging peoples agriculture declines; and, with the lowering of the rustic life, its cognate skills and arts, its joys and spirit, its very health decay also. A vicious circle arises and widens; drudgeries, luxurious and servile, mean, even abject, appear and deepen, and replace the old simple fellowship in labour; indulgence or indolence, orgies followed by ennui and apathy, replace rest. Classes become fixed as status through militarism's return; taboos arise and strengthen; and sex, the natural and fundamental spring of the moral life in both sexes, perverts into the dreams and dances of strange sins.

Of all such "progress," such "wealth," such "peace," men weary. The old courage, which in their rustic fathers had faced the chances of life, and mastered them through the courses of Nature, now finds a main outlet in gambling; and this increasingly contaminates legitimate commerce. The ruling class thus becomes increasingly one of wealth, with a corresponding increase of types of populace, submissively ready for any service whatsoever, if only wages be forthcoming, and finding its hope and ecstasy of life in the prospect of also occasionally getting something for nothing, like their betters at that game.

The older rustic castes, high and low, less apt for such modern life, are yet absorbed and enrolled by it, and become guardians and functionaries within, or enter the military caste for external service. Paleotechnic "order" is thus completed, and at the expense of progress; as the history of Russia, of Austria, of Prussia has so often shown us; and, as they tell us, ours is increasingly showing them.

In each such country, and even in its metropolis, though so largely thus created and maintained, the spark of soul which is in every man at length begins to sink within him altogether, or else to flare out into social discontent, it may be with mutterings of revolt. The official orator and bard appear also; as social medicine-men they must at all hazards again arouse manhood, courage, be this even through fear. Thus, fevered with cold and hot, the Paleotects run to and fro; they invent new myths of terror; their guardians new war-dances; these bring forth their treasure, and these build vast and vaster temples to the fear-gods. They carve their clubs, they lengthen and crowd their war-canoes, and one day they sail forth to battle. Be this at the time crowned with victory and glory, with mastery and empire, these have in them no few germs of decay, which also grow towards their ripening.

Is not this, in broadly summarised outline and at its simplest, the anthropology of half the South Seas, even the history of the old pirate and berserk glories of Scandinavia? The only touch of freshness remaining for such an epitome is that this in its fuller outline as above, is what the Scandinavian peoples are now thinking and saying of us, "The Great Powers." For now the Norsemen are in an otherwise evolving frame of mind, with correspondingly different phase of life, different conception of its defence, different practice towards its survival. Saved by their poverty of natural resources, as we used till lately to think, or by good hap, as it now appears, from the modern industrial crowdings, which we, in our terms of mere magnitude, call cities, they are entering upon the development of culture-cities, which already in terms of quality of life and of civilisation alike, are actually and proportionally in advance of ours, even though comparatively favourable examples be taken.

Some years ago it could be said by one Edinburgh man to another: "There is more music and more science in little Bergen than in big Edinburgh." And now Grieg and Nansen are known along the whole chain of villages and townlets whose electric lights twinkle nightly from Tromsø down and round to Christiania itself, and even to us as well. Once, indeed, our Scottish singers and thinkers also were known throughout their land and beyond, but that was in times of comparative poverty, before these days of "business" and "education," now alike so illusory in their numerical estimations.

In summary, then, the struggles of war are not so essential to the nature of society as many nowadays have come to believe; nor even when they occur are they much a matter of big battalions.

Without entering in detail into the social factors of war, which would expand these few paragraphs into a volume, it is enough here to insist upon the thesis of this chapter that our essential struggle for existence at present demands a viewpoint different from, and larger than, that of militarists.

Let us give them every credit for their measure of encouragement to Neotechnic skill and invention, and for the spirit of sacrifice they inculcate towards the social weal; but let us also realise that the present main struggle for existence is not that of fleets and armies, but between the Paleotechnic and the Neotechnic orders. And this not merely as regards our manufacturing productivity, upon which some, to do them justice, insist, but yet more throughout our rural and our urban life.

Most simply stated, as we rebuild our cities as well as our fleets, as we modernise our universities and colleges, our culture-institutes and schools as we have sought to do our fighting ships, there will be far less fear of war, and far more assurance of survival in whatever issue. And conversely, failing this needed uplift of our general level of civilisation, each added weight of armour may go but to keep it down.

1923

Le Corbusier

Engineer's
Aesthetic and
Architecture

Le Corbusier (1887–1965) was the pseudonym of the Swiss architect, urbanist, furniture designer, artist, and writer Charles Edouard Jeanneret-Gris. He began his career in Switzerland, moved to Paris in 1916 where he formed a close artistic partnership with the painter Amédee Ozenfant and together developed the style they called Purism. They also began the journal *l'Esprit Nouveau*, for which Jeanneret developed his architectural pseudonym in 1920. Simultaneously, he began an architectural partnership with his cousin Pierre Jeanneret, which produced a pioneering body of modern architecture over the next fifty years.

Le Corbusier's writings have been every bit as influential as his buildings, and perhaps the most important of his many books was *Vers une Architecture* of 1923, originally written as a series of articles in *l'Esprit Nouveau*, and subsequently translated into English as *Towards a New Architecture* in 1931. In that book, Le Corbusier summarizes the issues facing architecture in the first decades of the twentieth century: filth, disease, pollution, the effects of the car and transit, and the incorporation of mechanical and electrical systems. The section excerpted here is a summary that explains the challenge in a series of oppositions between the work of architects and that of engineers, between technical constraints and visual composition.

The excerpt is important for this collection in the influences understood to shape the work of the engineer, who must follow the "natural law" of economy and of efficiency. The evolutionary forces of technology are seen to operate through the person and profession of the engineer, and to provide a beacon for the architect.

The Engineer's Æsthetic and Architecture – two things that march together and follow one from the other – the one at its full height, the other in an unhappy state of retrogression.

The Engineer, inspired by the law of Economy and governed by mathematical calculation, puts us in accord with universal law. He achieves harmony.

The Architect, by his arrangement of forms, realizes an order which is a pure creation of his spirit; by forms and shapes he affects our senses to an acute degree, and provokes plastic emotions; by the relationships which he creates he wakes in us profound echoes, he gives us the measure of an order which we feel to be in accordance with that of our world, he determines the various movements of our heart and of our understanding; it is then that we experience the sense of beauty.

The Engineer's Æsthetic and Architecture – two things that march together and follow one from the other – the one at its full height, the other in an unhappy state of retrogression.

A QUESTION of morality; lack of truth is intolerable, we perish in untruth.

Architecture is one of the most urgent needs of man, for the house has always been the indispensable and first tool that he has forged for himself. Man's stock of tools marks out the stages of civilization, the stone age, the bronze age, the iron age. Tools are the result of successive improvement; the effort of all generations is embodied in them. The tool is the direct and immediate expression of progress; it gives man essential assistance and essential freedom also. We throw the out-of-date tool on the scrap-heap: the carbine, the culverin, the growler and the old loco-motive. This action is a manifestation of health, of moral health, of *morale* also; it is not right that we should produce bad things because of a bad tool; nor is it right that we should waste our energy, our health and our courage because of a bad tool; it must be thrown away and replaced.

But men live in old houses and they have not yet thought of building houses adapted to themselves. The lair has been dear to their hearts since all time. To such a degree and so strongly that they have established the cult of the home. A *roof!* then other household gods. Religions have established themselves on dogmas, the dogmas do not change; but civilizations change and religions tumble to dust. Houses have not changed. But the cult of the house has remained the same for centuries. The house will also fall to dust.

A man who practises a religion and does not believe in it is a poor wretch; he is to be pitied. We are to be pitied for living in unworthy houses, since they ruin our health and our *morale*. It is our lot to have become sedentary creatures; our houses gnaw at us in our sluggishness, like a consumption. We shall soon need far too many sanatoriums. We are to be pitied. Our houses disgust us; we fly from them and frequent restaurants and night clubs; or we gather together in our houses gloomily and secretly like wretched animals; we are becoming demoralized.

Engineers fabricate the tools of their time. Everything, that is to say, except houses and moth-eaten boudoirs.

There exists in France a great national school of architecture, and there are, in every country, architectural schools of various kinds, to mystify young minds and teach them dissimulation and the obsequiousness of the toady. National schools!

Our engineers are healthy and virile, active and useful, balanced and happy in their work. Our architects are disillusioned and unemployed, boastful or peevish. This is because there will soon be nothing more for them to do. *We no longer have the money* to erect historical souvenirs. At the same time, we have got to wash!

Our engineers provide for these things and they will be our builders.

Nevertheless there does exist this thing called ARCHITECTURE, an admirable thing, the loveliest of all. A product of happy peoples and a thing which in itself produces happy peoples.

The happy towns are those that have an architecture.

Architecture can be found in the telephone and in the Parthenon. How easily could it be at home in our houses! Houses make the street and the street makes the town and the town is a personality which takes to itself a soul, which can feel, suffer and wonder. How at home architecture could be in street and town!

The diagnosis is clear.

Our engineers produce architecture, for they employ a mathematical calculation which derives from natural law, and their works give us the feeling of HARMONY. The engineer therefore has his own aesthetic, for he must, in making his calculations, qualify some of the terms of his equation; and it is here that taste intervenes. Now, in handling a mathematical problem, a man is regarding it from a purely abstract point of view, and in such a state, his taste must follow a sure and certain path.

Architects, emerging from the Schools, those hot-houses where blue hortensias and green chrysanthemums are forced, and where unclean orchids are cultivated, enter into the town in the spirit of a milkman who should, as it were, sell his milk mixed with vitriol or poison.

People still believe here and there in architects, as they believe blindly in all doctors. It is very necessary, of course, that houses should hold together! It is very necessary to have recourse to the man of art! Art, according to Larousse, is the application of knowledge to the realization of a conception. Now, today, it is the engineer who knows, who knows the best way to construct, to heat, to ventilate, to light. Is it not true? Our diagnosis is that, to begin at the beginning, the engineer who proceeds by knowledge shows the way and holds the truth. It is that architecture, which is a matter of plastic emotion, should in its own domain BEGIN AT THE BEGINNING ALSO, AND SHOULD USE THOSE ELEMENTS WHICH ARE CAPABLE OF AFFECTING OUR SENSES, AND OF REWARDING THE DESIRE OF OUR EYES, and should dispose them in such a way THAT THE SIGHT OF THEM AFFECTS US IMMEDIATELY by their delicacy or their brutality, their riot or their serenity, their indifference or their interest; these elements are plastic elements, forms which our eyes see clearly and which our mind can measure. These forms, elementary or subtle, tractable or brutal, work physiologically upon our senses (sphere, cube, cylinder, horizontal, vertical, oblique, etc.), and excite them. Being moved, we are able to get beyond the cruder sensations; certain relationships are thus born which work upon our perceptions and put us into a state of satisfaction (in consonance with the laws of the universe which govern us and to which all our acts are subjected), in which man can employ fully his gifts of memory, of analysis, of reasoning, and of creation.

Architecture today is no longer conscious of its own beginnings.

Architects work in styles "or discuss questions of structure in and out of season; their clients, the public, still think in terms of conventional appearance, and reason on the foundations of an insufficient education. Our external world has been enormously transformed in its outward appearance and in the use made of it, by reason of the machine. We have gained a new perspective and a new social life, but we have not yet adapted the house thereto.

The time has therefore come to put forward the problem of the house, of the street and of the town, and to deal with both the architect and the engineer.

For the *architect* we have written our "THREE REMINDERS:"

MASS which is the element by which our senses perceive and measure and are most fully affected.
SURFACE which is the envelope of the mass and which can diminish or enlarge the sensation the latter gives us.
PLAN which is the generator both of mass and surface and is that by which the whole is irrevocably fixed.

Then, still for the architect, "REGULATING LINES" showing by these one of the means by which architecture achieves that tangible form of mathematics which gives us such a grateful perception of order. We wished to set forth facts of greater value than those in many dissertations on the soul of stones. We have confined ourselves to the natural philosophy of the matter, *to things that can be known*.

We have not forgotten the dweller in the house and the crowd in the town. We are well aware that a great part of the present evil state of architecture is due to the *client*, to the man who gives the order, who makes his choice and alters it and who pays. For him we have written "EYES WHICH DO NOT SEE."

We are all acquainted with too many big business men, bankers and merchants, who tell us: "Ah, but I am merely a man of affairs, I live entirely outside the art world, I am a Philistine." We protest and tell them: "All your energies are directed towards this magnificent end which is the forging of the tools of an epoch, and which is creating throughout the whole world this accumulation of very beautiful things in which economic law reigns supreme, and mathematical exactness is joined to daring and imagination. That is what you do; that, to be exact, is Beauty."

One can see these same business men, bankers and merchants, away from their businesses in their own homes, where everything seems to contradict their real existence – rooms too small, a conglomeration of useless and disparate objects, and a sickening spirit reigning over so many shams – Aubusson, Salon d'Automne, styles of all sorts and absurd bric-à-brac. Our industrial friends seem sheepish and shrivelled like tigers in a cage; it is very clear that they are happier at their factories or in their banks. We claim, in the name of the steamship, of the airplane, and of the motor-car, the right to health, logic, daring, harmony, perfection.

We shall be understood. These are evident truths. It is not foolishness to hasten forward a clearing up of things.

Finally, it will be a delight to talk of ARCHITECTURE after so many grain-stores, workshops, machines, and sky-scrapers. ARCHITECTURE is a thing of art, a phenomenon of the emotions, lying outside questions of construction and beyond them. The purpose of construction is TO MAKE THINGS HOLD TOGETHER; of architecture TO MOVE US. Architectural emotion exists when the work rings within us in tune with a universe whose laws we obey, recognize and respect. When certain harmonies have been attained, the work captures us. Architecture is a matter of "harmonies," it is "a pure creation of the spirit."

Today, painting has outsped the other arts.

It is the first to have attained attunement with the epoch.[1] Modern painting has left on one side wall decoration, tapestry, and the ornamental urn and has sequestered itself in a frame – flourishing, full of matter, far removed from a distracting realism; it lends itself to meditation. Art is no longer anecdotal, it is a source of meditation; after the day's work it is good to meditate.

On the one hand the mass of people look for a decent dwelling, and this question is of burning importance. On the other hand the man of initiative, of action, of thought, the LEADER, demands a shelter for his meditations in a quiet and sure spot; a problem which is indispensable to the health of specialized people. Painters and sculptors, champions of the art of today, you who have to bear so much mockery and who suffer so much indifference, let us purge our houses, give your help that we may reconstruct our towns. Your works will then be able to take their place in the framework of the period and you will everywhere be admitted and understood. Tell yourselves that architecture has indeed need of your attention. Do not forget the problem of architecture.

---

1   I mean, of course, the vital change brought about by cubism and later researches, and not the lamentable fall from grace which has for the last two years seized upon painters, distracted by lack of sales and taken to task by critics as little instructed as insensitive (1921).

Siegfried Giedion (1888–1968) was a Swiss historian of architecture. He was a student of the art historian Heinrich Wölfflin and a close friend of Walter Gropius, Le Corbusier, and others in the modern movement. In 1928 he helped found the Congrès International d'Architecture Moderne (CIAM), serving as its secretary general, and beginning a life-long project of both promoting modern architecture and examining its origins in a commanding series of books and articles. Following the methods of the Swiss school of art history, Giedion sought to identify the *zeitgeist* or spirit of modernism in comparative studies of modern art, modern physics, and modern industrial construction.

The work that secured his reputation as the voice of the modern movement was *Space, Time and Architecture*, initially given as a series of lectures at Harvard in 1938–39, then published as a book in 1941, and still available today in its fifth edition. The book became required reading for nearly two generations of young architects, and described the compelling similarities between the space-time theories of Einstein and the pictoral experiments of modern artists and architects. But the real achievement of the work lay in its compelling account of nineteenth-century experiments with new materials and methods. That book was followed in 1948 by an even deeper investigation titled, *Mechanization Takes Command: A Contribution to Anonymous History*, which dug deep into the archives of patent offices and designers to understand mechanization in all its aspects. That account is often cited by historians of technology as a fundamental work in their field.

The following excerpt is drawn from his first book on modern architecture, *Bauen in Frankreich: Bauen in Eisen, Bauen in Eisenbeton* (Building in France, Building in Iron, Building in Ferro-Concrete) of 1928. In these sections he establishes an analogy between construction and "life processes" to make the argument that nineteenth-century stylistic experiment had missed the real changes occurring in architecture. The dynamic, physiological understanding of history underlay much of the rest of his work, and led to his later interest in anonymous histories. He also used that interpretation to establish an ethical standard for future aesthetic experiments: "Thus, the point is reached where building falls in line with the general life process."

1928

Siegfried Giedion

Construction.
Industry.
Architecture

Construction

Is CONSTRUCTION something EXTERNAL?

We are being driven into an indivisible life process. We see life more and more as a moving yet indivisible whole. The boundaries of individual fields blur. Where does science end, where does art begin, what is applied technology, what belongs to pure knowledge? Fields permeate and fertilize each other as they overlap. It is hardly of interest to us today where the conceptual boundary between art and science is drawn. We value these fields not hierarchically but as equally justified emanations of the highest impulse: LIFE! To grasp life as a totality, to allow no divisions, is among the most important concerns of the age.

Physiologists have shown that a person's body build and nature are inseparably connected. Science traces specific characters back to certain bodily types. The connection between respiration and mental balance has been discovered. The body takes its form internally through breathing, gymnastics, sport. To overdevelop an arm muscle, or to douse the face with cosmetics like an isolated body (as the arteries harden), is no longer acceptable.

Construction is also not mere ratio.[1] The attitude that drove the previous century to expand our knowledge of matter, so much that it resulted in a previously inconceivable command of it, is as much the expression of an instinctive drive as is any artistic symbol.

We say that art anticipates, but when we are convinced of the indivisibility of the life process, we must add: industry, technology, and construction also anticipate.

Let us go further: architecture, which has certainly abused the name of art in many ways, has for a century led us in a circle from one failure to another.

Aside from a certain *haut-goût* charm the artistic drapery of the past century has become musty. What remains unfaded of the architecture is those rare instances when construction breaks through. Construction based entirely on provisional purposes, service, and change is the only part of building that shows an unerringly consistent development. Construction in the nineteenth century plays the role of the sub-conscious. Outwardly, construction still boasts the old pathos; underneath, concealed behind facades, the basis of our present existence is taking shape.

› Industry

Industry completes the transition from handicraft to machine production.

Industry is only part of the problem connected with the transition from individual to collective design.

Machine work means serial design, precision. Handicraft has its own special charm that can never be replaced: the uniqueness of the product.

But without machine work there is no higher technology. By hand one can neither mill sprocket wheels that fit frictionlessly together, nor draw out uniform wire, nor profile iron precisely. The transition from individual to collective design is taking place in all fields, practical as well as spiritual ones.

Now, it is the case that INDUSTRY, which is intensively involved with the life process, displayed this change before other fields – private life or art – took note of it. Industry, big industry, is a result of the French Revolution.[2]

The Assemblee Nationale initiated its development with the *Proclamation of the Liberty of Labor* of 2 March 1791.

With this proclamation of free competition the guild system (les corporations) was at once abolished.[3]

Before the French Revolution articles for everyday use were produced by the guilds. Guild membership was just as limited as the number of workers or helpers each member could take on and the kinds of product each could produce. That meant privilege in favor of a few and an extraordinary burden (gene onereuse) on the consumer. The complex instrument of industry was created through the possibility of a free division of labor.

Like construction, industry is an inner expression of the life process.

Though we are objectively able to create anticipatory designs, old mental "residues" prevent us for a long time from drawing the human consequences:

INDUSTRY anticipates society's inner upheaval just as construction anticipates the future expression of building.

Even before industry existed in its present sense – around 1820 – Henri de Saint-Simon (1760–1825)[4] understood that it was the central concept of the century and that it was destined to turn life inside out:

"The whole of society rests upon industry."

It seems that the force of Saint-Simon's influence on the schools and tendencies of the century lay, above all, in his ability to grasp the emerging reality and to transform it into a utopia. It is the opposite method to the cultural idealism that dominated Germany at the time, which neglected reality in order to pursue emanations of pure spirit.

Saint-Simon foresaw the great concentrations of labor, the urban centers, and the factories with thousands of workers that transferred the results of research directly into action. As a consequence of an industrial economy he foresaw the dawn of a classless society, the end of war, and the end of national borders: a single army of workers spanning the globe. The end of man's exploitation of man (*l'exploitation de l'homme par l'homme*) will have been achieved. The eye of the visionary no doubt simplifies and leaps over intermediary stages:[5] Saint-Simon never reckoned with the century's divided soul, which in architecture as in society imposed the old formal apparatus on the new system.

The anonymous process of production and the interconnected procedures that industry offers only now fully take hold of and reshape our nature.

› Architecture

The concept of architecture is linked to the material of stone. Heaviness and monumentality belong to the nature of this material, just as the clear division between supporting and supported parts does.

The great dimensions that stone requires are for us still habitually connected with each building. It is entirely understandable that, with their unusually modest dimensions, the first buildings executed in tensile materials time and again evoked among contemporaries the concern that the building might collapse.

Architecture is linked to the concept of "monumentality." When the new building materials – iron and ferroconcrete – assume the forms of gravity and "monumentality," they are essentially misused.

It seems doubtful whether the limited concept of "architecture" will indeed endure.

We can hardly answer the question: What belongs to architecture? Where does it begin, where does it end?

Fields overlap: walls no longer rigidly define streets. The street has been transformed into a stream of movement. Rail lines and trains, together with the railroad station, form a single whole. Suspended elevators in glazed shafts belong to it just as much as the insulating filling between the supports. The antenna has coalesced with the structure, just as the limbs of a towering steel frame enter into a relationship with city and harbor. Tall buildings are bisected by rail lines. The fluctuating element becomes a part of building.

Architecture has been drawn into the current from the isolated position it had shared with painting and sculpture.

We are beginning to transform the surface of the earth. We thrust beneath, above, and over the surface. Architecture is only a part of this process, even if a special one. Hence there is no "style," no proper building style. Collective design. A fluid transition of things.

By their design, all buildings today are as open as possible. They blur their arbitrary boundaries. Seek connection and interpenetration.

In the air-flooded stairs of the Eiffel Tower, better yet, in the steel limbs of *a pont transbordeur*, we confront the basic aesthetic experience of today's building: through the delicate iron net suspended in midair stream things, ships, sea, houses, masts, landscape, and harbor. They lose their delimited form: as one descends, they circle into each other and intermingle simultaneously.

One would not wish to carry over into housing this absolute experience that no previous age has known.[6] Yet it remains embryonic in each design of the new architecture: there is only a great, indivisible space in which relations and interpenetrations, rather than boundaries, reign.

The concept of architecture has become too narrow. One can no longer contain, like radium in a bottle, the need to create that which is called art and explain what remains of life devoid of it.

The ponderous movement of human affairs has as its consequence that the new attitude toward life manifests itself much sooner in the objective fields – such as construction, industry – than in those fields that lie close to us.

Only now is the housing form being seized by those hidden forces that a century ago drove man to the constructional and industrial attitude.

Our inner attitude today demands of the house:

Greatest possible overcoming of gravity. Light proportions. Openness, free flow of air: things that were first indicated in an abstract way by the constructional designs of the past century.

Thus, the point is reached where building falls in line with the general life process.

1   We mean here not just the creative intuition that every great constructor must have. It is well known that he fixes the dimensions mostly emotionally and that calculation often comes later only as a test. We mean construction itself, which is not determined by purpose alone, but which seems also to transcend rational values and is expressive. This also challenges the old prejudice that art and construction may be neatly divided, by presenting art as "unintentional" and "purposeless," and construction alone as "purposeful."

2   There were several industrial and joint-stock companies already under the ancien régime. [Charles] Ballot, *L'Introduction du machinisme dans l'industrie francaise* (Paris, 1923), p. 23, discusses the epoch from 1780–92 and another 1792–1815, which, under the effect of the Revolution, introduced the machine to a few areas (cotton, wool). Industry in today's sense was first introduced around 1830.

3   Davioud, "Un discours d'architecte," *Encyclopedie d'architecture*, 1878, p. 27.

4   *Systeme industriel*, 1821. *Catechisme des industriels*, 1823.

5   Saint-Simon, himself rooted in the feudal system, only formulated the elementary contradictions of the military and industrial system. His students quickly drew the consequences of his system.

6   A fascination of Corbusier's houses consists in the fact that he has attempted this as much as possible.

1929

Le Corbusier

Architecture: The
Expression of the
Materials and
Methods of our
Times

Le Corbusier (1887–1965) was the pseudonym of the Swiss architect,
urbanist, furniture designer, artist, and writer Charles Edouard Jeanneret-
Gris (see Le Corbusier, 1923 for full biography).

The following article was published in the *Architectural Record*, in August
1929, as part of a series solicited by the editors. It follows an article by
Frank Lloyd Wright on the nature of materials, which may have suggested
the title. On the one hand, the essay describes the deterministic effects
of new materials and methods that characterize the first generation of
technology studies. On the other hand, he grants the architect great
power in choosing or adapting those influences.

Let us not confuse outward show, however impressive, with an essential truth which is still indistinct in the whirlpool of an epoch in the full tide of evolution.

By "*impressive outward show*," it is implied that the architecture of today appears to be dictated in the eloquence of its form by modern materials and methods. "*Essential truth*" suggests an architecture that results from the state of mind of an epoch and that an architecture exists, *takes* form and is *expressed* only at that *very moment when a general evolution of mind is accomplished.* It is at that moment alone when the mind has recognized and admitted a system of thought which, above all, represents in every field a profound modification of previous states. There is no architecture *during* periods of crisis; architecture comes *after* periods of crisis.

The crisis then has passed? From the consideration of the world about us the opposite seems certain. Perhaps not; a few spirits (not all – far from that, but only those of leaders – and that is enough) have passed through the crisis, and have formulated a new attitude of mind which follows *completed changes.* Only objects – material reality – are in a state of complete disturbance. And why are they? Because precisely at this moment, there breathes a new spirit and the entire world – both man and materials – must inevitably follow the implacable destiny of a new tendency.

Is there then indeed an origin to this profound upheaval? Most certainly. *It has existed for a hundred years.* During the century our brains have escaped from ancient customs. Our life has gone from day to day, changed bit by bit. And thus we scarcely appreciate it. We were unable to know where all this was leading, we could feel only that it *was* leading, powerfully, violently, and ever and ever more rapidly.

Meanwhile, shallow spirits of limited vision cried out: "The world is being wrecked, all is lost." And in desperation, like shipwrecked sailors grasping at floating debris, we clung to the past. Never before had so much archaeology been done as during those heroic times when science was pushing us, each day more insistently, along the adventurous paths that lead towards the unknown.

Is not architecture determined by new materials and new methods? (It is high time I were defining what architecture is.) Indeed to all in America belong the new materials, with you modern methods are in use. But for a hundred years your architecture has not evolved. Alone your programs have changed. And you construct your skyscrapers in the manner of students of the Ecole des Beaux-Arts building a private house. I repeat: a hundred years of new materials and new methods have made no change whatsoever in your architectural viewpoint.

*   *   *   *

It is time, though, to define architecture. Architecture is not building. Architecture is that cast of synthetical thought in response to which the multiple elements of architecture are led synchronically to *express a purpose.* And as this synthetical purpose is absolutely disinterested, having for object neither to make durable, nor to build rapidly, nor to keep warm, nor to promote sanitation, nor to standardize the domestic usefulness of the house, I would say, since it is above any utilitarian objective, it is an elevated purpose. Its object is to bring us benefits of a different nature from those of material usefulness; its aim is to transport us to an inspired state and thus to bring us enjoyment.

Saying this I find myself in accord with the humblest accomplishment of the simplest conscientious laborer, and on the other hand I put myself in agreement with all the great traditions of the past.

*   *   *   *

Nevertheless, there exists in these days, an absorption in definitely practical ideas which is precisely expressed by the subject which was suggested to me, "Architecture, the expression of the materials and methods of our times."

I will even say that it is the clue to the present situation. And here is the reason: A system of thought is imbued with life only when there exists a balance between the results of evolution and the spiritual direction of its progress.

What, then, is the direction of its progress today?

A hundred years of a mechanical era have brought forth an entirely new spectacle. Geometry is supreme. Precision is everywhere. The right angle prevails. There no longer exists any object that does not tend to severity.

Industrialism has stated the postulate of economy: To attain the maximum of result at the minimum of expense.

Science, mathematics, analysis, and hypothesis, have all created an authentic machinery of thought. An imperative need of clarity, the search for the *solution*. It is for that which the mathematicians term the "*elegant solution.*"

Has not this all-pervading precision, exactness, and accuracy definitely annihilated the imperceptible, distance and mystery? Miraculously, *quite the contrary* is the case. This century has officially opened to us gates yawning on the infinite, on majesty, silence, and mystery. More than ever before, man's soul is pathetically brought face to face with itself. Never was there an epoch so powerfully, so unanimously inspired. Poetry is everywhere, constant, immanent.

*   *   *   *

Here, then, is set forth that point of view which constitutes the present era, a veritable magnetic pole towards which swings the compass of *our initiatives, of all our initiatives.*

Let us come to the point. What, in view of the purity and supreme clarity of this new state of thought, are our present architectural forms? Do we concern ourselves with this gleaming liberty of disinterestedness, of courage and poetry? Alas, how timid we are, how firmly we are chained, like slaves. The past has ensnared us, whereas its law is to cry to us, "carry on – why don't you progress and move forward?" We are cowardly and timorous, lazy and without imagination.

Cowardly, timorous, lazy, and without imagination, because, now and invariably, we want our new houses to resemble the old. What a poverty of creative ability!

Meanwhile the means are at hand; science, mathematics, industry, organization.

We still permit our houses to lie close to a damp and unhealthy ground. We are still discussing whether or not our houses are to have roofs, while roof gardens bring health, joy, and an upheaval of plan replete with magnificent liberties. We are still building our houses of stone, with massive walls, while light and slender cars are speeding at sixty miles an hour through snows or under the tropical sun. We are still employing masons and carpenters *on the job*, to work in rain or snow, or fair weather, while factories could turn out to perfection that which we accept poorly executed.

And so forth and so on.

*   *   *   *

Here, now, are my conclusions. In what way are we to allow so many innovations? How are we to select these forms still unknown in the building of houses? How are we to arrange them in such a manner as will bring us anew before an architectural phenomenon as will make us feel once more the vigorous delights of architecture?

A state of new enthusiasm exists; a system of thought has been wrought by a hundred years of investigation and acquired results. We have a *line of conduct*. Instinctively our choice tends towards such constructive systems, towards such materials as possess forces capable of feeding our enthusiasm. In us moderns the new feelings, an instinct, control actions which are in harmony with each other.

The harmony of former centuries is in confusion. The effect continues but the cause has been swept aside by the mechanical revolution. The mechanical revolution is a new cause – immense phenomenon in the history of mankind. Where are the new effects?

Let us be led by this enthusiasm which animates us. Industrialization, standardization, mass production, all are magnificent implements; let us use these implements.

I wish to give you the basis of my reasoning: I am certain that that which at this moment appears most revolutionary in contemporary architectural creations, be it in France, Germany, Russia, or elsewhere – all that is *still nothing more than the old aspect caught in the quicksands of the past*. It is my opinion that as yet we have seen nothing new, done nothing new. That which will come in architecture will survive only when an urbanism, brought face to face with the present social upheaval, will have created cities of which we have as yet not even an idea, of which we have not yet even considered the possibility.

Such is the progress on the one hand (and it is gigantic by comparison with the means at the disposal of the builders of the Romanesque period, or that of Louis XIV) and on the other hand the architects of the contemporary epoch daring at last to state a problem, and to announce the answer, and thus to give to the world an architectural system which is the resultant of the spirit of an era.

The line of action exists – the modern system of thinking.

The Americans, however, are the people who, having done most for progress, remain for the most part timidly chained to dead traditions.

On the other hand, their willingness to progress further strikes me as boundless. And that is a force which, soon, will swing the balance.

1929

Richard Buckminster Fuller

4D Time Lock

Richard Buckminster Fuller (1895–1983) was an American architect, inventor, engineer, designer, cartographer, mathematician, and poet, known mostly for the development of the geodesic dome. Fuller briefly attended Harvard in 1913, then worked in a variety of businesses before joining the Navy in World War I, an experience that was to shape much of his later thinking. After being discharged in 1919, he joined a company making an innovative concrete block invented by his father-in-law, the architect James Monroe Hewlett. In 1927 he was forced out of the company and began developing "Dymaxion" proposals, forming a company called 4D and developing a Ten Deck building, a World Town Plan, and the Minimum Dymaxion House, a hexagonal duralumin unit suspended by cables from a central supporting mast. In the early 1930s he developed a three-wheeled Dymaxion car, and then a pre-fabricated Dymaxion bathroom. During the Second World War he developed further prototypes for manufactured housing, and then in the decade after the war he consolidated his achievements in a fertile period of invention, developing geodesic and tensegrity structures, which were produced by a succession of companies and slowly accepted as an efficient lightweight form of enclosure. In the 1960s he became a international figure, traveling, lecturing, and focusing on the allocation of global resources.

The following essay formed the twelfth chapter of *4D Time Lock*, the text that Fuller prepared to explain the design of the first Dymaxion house in 1928, which had been presented to a group of architects at the American Institute of Architects (AIA) Convention in St. Louis and sent to other individuals in a series of mailings. In a technique adopted from the Navy, and that he used the rest of his life, the text was followed by a "4D Chronofile" recording the responses to these presentations in chronological order. Fuller worked at many scales simultaneously, developing minute production efficiencies at the same time that he sought the largest, simplest principles encompassing the project at hand. In terms of housing, he became famous a decade later with the question, "Madame, do you know how much your house weighs?"

In this collection, the essay introduces the terminology of the fourth dimension, which had been discussed by architects and designers in various ways since the beginning of the century (Giedion, 1941), but which Fuller brings to the question of production and the lightness of its products. Like Kohr's observation about velocity on a human population (1973), Fuller demonstrates how velocities within a system can be understood through measurements of its effects.

› Abstract design, harmony and fourth dimensional control

Repeating somewhat, let us review what has just been said. Europe, as usual, is leading the world tremendously in design, but it is merely designing surfaces. Inspired with much of the new truth, the inspirational, artistic world is doing this fine new design, revolting for a new era of characteristic expression but all they can do, unfortunately, as artists, is to make new three-dimensional geometric combinations. There is no limit to this any more than there is to musical composition. Industry is today centralized. It produces en masse for individualism and, therefore, all that is practical and useful is produced on a gigantic scale, world encompassing, hitherto little thought of and still apparently unperceived by the world in general.

Your greatest artists today are designing for mass production in print, fabric, and even radio, etc. Industry makes possible one more dimension in design, fourth dimension. In all design today we use synthetic materials, or recomposition of elements, to perform best a given function. A material before it reaches its final lodging, passes through many hands, and over much space, and therefore to be efficient and pleasing, must have no unnecessary weight. When it reaches its destiny, how long will it stay there? For the time limit of its existence. The fourth dimension is time. In the composition of synthetic materials, the fourth dimension is the most important. There are no materials which nature has not mixed with others. To use them today the elements which fulfill the function, debunking them, as it were; removing weight, and combining them again with materials whose longevity or fourth dimension is equivalent to their own. Don't mix bronze and wood in design. Wood and paper, yes, brass and glass, yes.

In consideration of the fact that no matter can exist without *time*, else it would not exist; and that *time* dimension is the most important dimension of all matter; and that all our industry is but a *time* saving institution; that all sport is but a *time* controlling demonstration; and that all art is but an harmonic division, composition, and projection of *time*, and that we are fast approaching a *time* standard (men dollar hours) instead of a gold standard; and that inasmuch as everything is balanced, all these time creations are balanced by credit or faith, as opposed to material coin exchange, which is becoming more and more an antiquated practice and confined only to inconsequentials and the lower classes of trade. In full consideration of this new economic law must the new era home be designed and its plans of industrialization evolved.

When it is clearly understood, by a proper study of the fourth dimension, that all time or temporal matter has but one scale applicable to all the various scales which we now know as color, sound, etc., for in reality sounds, colors, etc., are merely registrations through different nervous systems of the same temporal characteristics, be they hard, rough, sharp, round, smooth, high or low, etc., then will it be realized that with proper fourth dimensional consideration of all the discords, that may disturb the senses, these may be reduced to a minimum. On the other hand by the same fourth dimensional consideration is it possible to provide harmony of presentation in all the material design to which the nerves are subjugated; to the end that the abstract spirit, freed from too constant contemplation of material prosaicness, may at last attain harmonization of individualism by virtue of industry or completely segregated and controlled materialism.

Time and faith are both abstract. As people become more individualistic their lives and contacts become more abstract, though ever greater in volume and distance. They more constantly deal by wire, letter, wireless, or multiple letter. Almost any well known speaker today rarely orates to his company without the presence of a microphone, for clarification of his speech to those visibly present. It would seem that we are possibly approaching a time when the distinguished guest might be spared the actual useless meal and address his audience by telephone instead of attending the banquet in Brownville or Greenville. Relative time, distance, or space is constantly reduced.

In due consideration, then, of this time dimension, it is evident that progressive design must be time saving. Time saving is accomplished by segregation of functions. As functions are segregated and individually solved, involves exceedingly light weight materials. This saves in every handling from original source to ultimate disposition. (Incidentally this deweighting process of material things goes hand in hand with the "Debunking" process of the mind.)

As time is saved by progress, and time is in everything, all material products of industry must necessarily become lighter and lighter. It is worthy of note that this will be definitely reflected in the mirror of economic progress, the stock market, provided the time saving progress is balanced by the increase in good faith, and may be taken advantage of by those who intelligently acquaint themselves of this fact. Judge life and industrial progress by their measure of these tokens: GOOD FAITH and TIME OR WEIGHT SAVING. When these two are well balanced, progress may be further measured by the harmony of design as opposed to prosaicness (harmony is service, artistic appeal, etc.).

There are the very definite abstract conceptions: First, that all matter is of globular, radiating form, and that all dimension is fourth dimensional or radiating spheres, which radiate for a certain period of time. The time dimension being the distance from the center of the sphere to the greatest surface attained by radial measurement. There are the radiating spheres whose wave lengths are attuned to our wave length receptivity of consciousness or nervous system of antenae, which is our conscious zone of human vision, hearing, taste, touch, smell, etc. It is through a definite perception of the scientifically recognizable characteristics of these wave lengths within the conscious spheres that scientists have shaped the rules of the truths thereby revealed. This in turn has made possible scientific exploration into the abstract or unconscious spheres, which have made possible abstract discoveries, such as the radio. For example, a human shouts aloud, creating radiating spheres of sound. There is a definite distance away from that human at which the spheres die out, or the temporal matter, which is sound in this case, has ceased. A cross sectional projection of the fourth dimension is provided by the radiating waves in a pool of water caused by the impact of a stone. The fourth dimension can be measured both as the time and space between the contact of the stone with the water and the extreme longevity attained by radial measurement. It will be readily conceived that the intensity of original impact creating the splash as well as the medium in which the splash is made, together with temperatures and other outside conditions, will affect the longevity of the waves.

Without much further discussion of these fourth dimensional truths, be it explained we have exquisite or rapidly moving spheres, and slow or long wave length spheres, depending on the element and on the zone in which it is active (earth, water, air, electricity, ether, fire). It is the variation in the fourth dimension, or time life in individual elements that finally causes the break down of nature's synthetic materials, such as stone by erosion, which is but a "slow movie" form of the effervescence in champagne. In the modern internal combustion engine, we have arranged a group of similar fourth dimensional metallic material in precision relation, and in proper consideration of dynamic truths. We introduce into what we call the cylinder head two groups of fourth dimensional spheres of greatly varying speed and wave length. Due to their discordant wave characteristics they create explosion or repulsion of exquisite effervescence. These rapidly repulsed and swollen spheres, greatly magnified by the electrical "step up" to the next higher plane of activity, (the mathematical increase being figured by spheroidal content increase, as attained by radial or time distance in the next higher zone of each element, there being always geometrical reduction of friction with each higher plane attained) cannot be opposed by the counter dynamic position of the slow material or metal, and the consequence is the transition into motion, as we can perceive of it, within the material or conscious sphere. It is by proper scientific handling of these subjects, that, with synthetic materials, we have devised mechanics and translated fourth dimension into useful motion. Though as is so often true in the first appearance of a truth, completely unconscious of the material law of time dimension, covering the problem, have certain of its solutions been made. That is, those devising the gasoline motor have not conceived of it as a fourth dimensional design and control, though, that it was in effect. Malodor, noise, rupture, disease, fracture, are one and the same, being characteristic perceptions of the different senses of the wasteful protest of inharmonies of time composition. Of such is radio "static"; of such is thunder; of such is rust; of such is earthquake; and of such is stockyard smell.

The basis of denial of the fourth dimension, which has been supported by the theoretical and fallacious plane and cubical geometry, has been the inability to produce an additional or fourth perpendicular to a cube, as the basis of an additional power multiplication, whereas poor little plane arithmetic and algebra, without geometrical reference, being abstract, indicate the perfect ability to do so. Very rightly do they do so, for if the geometrist will go back to his first perpendicular, he will find it perpendicular to a sphere, for did he not assume a dot as the first basis of his geometrical theorem, which if conceded at all must be spheroidal. Matter if existent at all (and we cannot fallaciously assume a truth that is not), must be spheroidal. Surely the "Planeandsolid" geometrist does not claim his "dot" or "point" to be cubical, for then would he have no further cause for his progressive antics. We see that there is no cubism, and that we can have as many perpendiculars to the inside or outside of the sphere as we may wish. Each power raising, or root taking, is on the basis of spheroidal increase or decrease by that many units of its radial or time dimension. The only "straight line" then is the radial or time line, demonstrable by spheroidal dissection on its radial axis. There is also much laughter at the "Planeandsolids".

Thus we come to understand that no two persons standing on the earth's surface, unless one is directly above the other, may occupy the same time or radial perpendicular to the earth's surface, which indeed bespeaks individualism and explains the conflict of flat surface, cubistic life, as at present set forth in our cubistic cities. It will be further seen, that the term "fourth dimension", bespeaking the fallacious three dimensions of cubism, is in itself incorrect and limiting. As long as we have time, may we have as many powers of time, times the whole or angular segmentations of the sphere, as we may wish. So called gravity is but the expansion of this earth sphere keeping up with the units of non-amplified time control. As we control time, so may we fly off the earth. 98.6° F. is the constant friction heat characteristic of the relatively grouped samples of all elemental spheres of the human body. Could synthetic 4D composition be more standard than the material body?

As a non-descriptive reference, 4D being only the enigmatic term for time, do we use these characters as the trade mark of our industrial activity, occasioned by the new or correct basis of figuring of the infinity of time dimensions?

We are about to industrialize the truth that Columbus had the courage to demonstrate four-and-a-half centuries ago. Think of the conviction that was necessary in that "trans oceanic hop", which couldn't possibly be over in thirty hours, one way or the other; and which had to bring along its self-conscious audience to daily, weekly, and monthly provoke it with being "all wrong". Our flights pale beside it, and its very truth has been stubbornly evaded to this day.

Knud Lönberg-Holm (1895–1972) was a Danish architect, engineer, author, and editor. He worked first in Copenhagen, then for Eric Mendelsohn in Berlin, and is sometimes credited with first showing Mendelsohn photographs of the American concrete grain elevators that so engaged the architectural avant-garde. Lönberg-Holm emigrated to the United States in 1923 and began to teach at the University of Michigan. In 1929 he sent this article to the *Architectural Record*, and according to Lönberg-Holm they "rejected it as too controversial, though the magazine put him on its staff," hiring him to edit the newly formed technical department of the magazine. By 1932 he found even that department too caught up in questions of style and moved down the hall to the Sweet's Catalog Service where he became director of research. At the same time, he also became a regular member of Buckminster Fuller's Structural Studies Associates (SSA), helping to convert the Beaux Arts *T-Square Journal* into *Shelter*, possibly the most progressive architectural magazine published in that period.

He remains best known for his work at Sweet's Catalog with the Czech graphic designer Ladislav Sutnar (1897–1976). Together, they designed a new visual identity and communications system for Sweet's. The collaboration resulted in three books: *Catalog Design*, in 1944, *Designing Information*, in 1947, and *Catalog Design Process*, in 1950. As Buckminster Fuller later described it,

> Lönberg-Holm thus became the research design coordinator of all the research departments of the myriad mass-production suppliers of building materials to America . . . To do all this he developed for himself a theory of information-cataloguing which in many ways anticipated the present cybernetics of information storage, retrieval, and question programming.[1]

That work can be viewed as the realization of his call in this article for the engagement with "industrial organizations."

1929
Knud Lönberg-Holm
Architecture in the Industrial Age

---

1   Richard Buckminster Fuller, "The Age of Astro-Architecture," *Saturday Review* (July 13, 1968): 17–42.

Our increased understanding of social morphology and human affinities to time, space, and matter has not yet been methodically applied to the building problem. It is generally assumed that this problem will be able to solve itself, left to the self-interests of business, politics, real estate, and owner.

The result is discouragingly evident. Our cities are impressive only in mere size of amorphous form. We have progressed mechanically and structurally; but our housing is expensive and inadequate, our architecture an escape from reality. Only purely utilitarian structures show unity of purpose, function, and form.

The malady is recognized by the architectural profession, but the true cause is not understood. Consequently the architect resorts to the most immediate expedients and offers superficial remedies in "modernized" architecture and in increased architectural service.

An unsatisfactory solution of a given problem may be caused by an unclear or contradictory program, inadequate instruments and working methods, or both. More architecture cannot change the inorganic structure of our cities. The solution lies in appropriate city-planning; but a new conception of city-planning based on a clearer understanding of the organic functions of a community must lead to a reorganization of the tools and agencies engaged in the building process.

The building activity of a human society is a continuous space-organizing process, determined by the cosmic orientation of the social group – its religion or philosophy, and its space-time conception. The continuous change in the social order is accompanied by a corresponding change of the tools and methods employed. Arts and crafts become science and industry. An organic social structure is possible only when social functions and building process are guided by related fundamental laws.

Science has changed man's relation to nature and to society. The individual and society alike are forced to find a new balance, a new synthesis. Relations to a visible world have become relations to invisible energy. We have discovered the close relations between phenomena apparently unrelated and gained a new understanding of the growth of a civilization. Illusions have been destroyed. New needs exist, particularly the necessity for a reorganization of life and society to deal with the new reality. We enjoy form as a demonstration of function, and have extended and deepened our conception of beauty. We are sensitive to new qualities.

Matter, light, and color we conceive as visible energy that can be measured and harnessed. Ornament and decoration have lost their value as symbols and have become atavistic exhibitions. We have discovered new relations between our physical surroundings and our psychological reactions. Aesthetics has become psychology; time, a new dimension.

The speed of mechanical transportation has been increased; consequently our sense of distance, our spatial scale, has been altered. The illusion of matter as a solid has been destroyed. Our space is an open space, a space we conquer and penetrate – not a space we close off. Instead of cities closed in by fortifications we have the metropolitan region existing as a sum of relations between individual units; instead of solid stone construction, metal tubes and trusses; instead of pressure, tension; instead of steam, electricity.

The architectural ideology based on aesthetics has lost its validity in the industrial society. The conception of architecture as a fine art in contradistinction to the creations of science and technique, and the resulting conception of form as a value in itself, has brought the architect to exhibit an instinctive antipathy toward the industrial society's mass-production and toward its negation of arbitrary and absolute form, mass, gravity, and of buildings as monuments and media for self-expression.

For him the law of economy applied to time, space, and form – types and norms – becomes restrictive instead of creative. Afflicted with this antipathy toward his actual environment and with a related desire to beautify, i.e., escape the new reality, he deals with form instead of space, ignoring the form-creative process. His form is consequently insignificant and amorphous. Design is limited to the surface, and deteriorates to mere decoration in his concessions to the fleeting fads of the hour.

The victim of aesthetic inhibitions; the architect has lost his leadership. From a professional man with a professional ethics he has become a business man subject to the whims of the buyer.

The progressive architect acutely realizes that his problem means ultimately the negation of his profession. He has no power to meet his dilemma through his architectural work. As an individual businessman he cannot afford the research work necessary for the proper execution of his ideas; moreover, he is confronted by the gulf which separates him from a client unsympathetic toward an experiment at his expense. The rare exceptions from this do not alter the general aspect of the situation. And professional organizations have the problem's solution still less within their command since they are primarily interested in the protection of professional interests.

Collective problems require collective thinking and collective work. Industrial organizations are logical instruments for an industrial age. They function rationally in several distinct divisions, namely, scientific research; social contact or sale, dependent upon the establishment of a basis of understanding between the laboratory and the consumer; production based on modern machinery and economy; the striving for types and norms; the constant elimination of superfluous matter and obsolete form, thereby attaining the material achievements of our day and simultaneously creating a new plastic reality. We must learn to apply these modes of an industrial age to the building problem.

Our cities and buildings are organized space, space-machines to facilitate the free function of human and social needs: working, playing, mating, resting, thinking, and creating needs and human relations seen in the light of contemporary knowledge. These spatial structures must be flexible and always conform to the functions of life. They have no independent value in themselves. The plastic elements – material, light, and color – should be organized in accordance with social, physical, and psychological determinants. The utilitarian factory differs from the living quarters and the emotional stage-setting only in the intended function. The creative process is the same.

Acknowledging the full scope of its implications, it must be admitted that this is a complex social problem. Its successful solution must depend upon the collective efforts of:

> research,
> planning,
> Building industries, specialized according to types.

The organization of progressive forces in architecture, engineering, industry, and sociology would be the logical procedure for a conscious transition from the present division of work to the inevitable future. The functions of this organization would be:

To act as a clearing house for individual research,

To create an economically independent research institute.

The research work-analysis of problem, the determination and definition of types and norms, collection, and organization of material – would provide the basic factors for:

The public instruction – the use of contemporary publicity instruments to create a new attitude in the public.

An experimental school – to develop new builders.

Hugo Häring (1882–1958) was a German architect and theorist who took part in the creation of the Congrès International d'Architecture Moderne (CIAM) – alongside Le Corbusier and Siegfried Giedion – at the Chateau La Sarraz in 1928. Although a central figure in the early modern movement, he is now often overlooked in the English-speaking world. This is partly due to his relatively small number of significant built projects, in which a cowshed at the Gut Garkau Farm stands as perhaps his single most influential work. His theoretical position also served to place him on a parallel path to that of mainstream modernism – closer to the free-form "expressionist" architecture of his compatriots Erich Mendelsohn and Hans Scharoun than to the dominant geometric aesthetic of the International Style. He is one of the few important German architects of his era to remain in the country throughout the Second World War. This decision contributed to his isolation from the English-speaking world, and he also failed to recover his influential position in Germany after the war.

Rather than understanding architecture in terms of economy and repeatability, Häring saw the building as a singular event emerging out of the particularities of place, program, material, and culture. Thus, instead of treating the building as a container indifferent to its contents, Häring saw it as a specific response to the individuality of the design conditions. The following extract was originally published in the yearbook *Innendekoration*, in Stuttgart, in 1932. The text embodies many of Häring's key concerns, including a call for "utilitarian objects without adornments," where the will of the individual artist is subordinated to that of the generative force within the work itself. The text also contains significant pre-echoes of current thinking on the use of "genetic algorithms" derived from performance criteria (see van Berkel, 1999; De Landa, 2002).

1932

Hugo Häring

The House as an Organic Structure

It still seems to many people inconceivable that a house too may be evolved entirely as an 'organic structure', that it may be 'bred' out of the 'form arising out of work performance', in other words that the house may be looked upon as 'man's second skin' and hence as a bodily organ. And yet this development seems inescapable. A new technology, working with light constructions, elastic and malleable building materials, will no longer demand a rectangular house, but permit or put into effect all shapes that make the house into a 'housing organ'. The gradual structural shift from the geometrical to the organic, which is taking place throughout our whole spiritual life and to some extent has already taken place, has made the form of work performance mobile as opposed to geometrical. The need to create form constantly leads the artist to experiment with styles, repeatedly leads him, in the interest of expression, to spread shapes over objects – whereas the form arising out of work performance leads to every object receiving and retaining its own essential shape. The artist stands in the most essential contradiction to the form of work performance so long as he refuses to give up his individuality; for in operating with the form arising out of work performance the artist is no longer concerned with the expression of his own individuality but with the expression of the essence of as perfect as possible a utilitarian object. All 'individuals' – and the stronger they are as personalities, and at times the louder they are, the more this applies – are an obstacle in the path of development, and in fact progress takes place in spite of them. But nor does progress take place without them, without individuals, artists, and strong personalities. There remains an essential difference between the architect and the engineer. The work of the engineer has as its goal merely the performance of material work within the limits or in the domain of economic effects. That the result frequently contains other expressive values as well is a side-effect, a subsidiary phenomenon of his work. The architect, on the other hand, creates *a Gestalt*, a total form, a work of spiritual vitality and fulfilment, an object that belongs to and serves an idea, a higher culture.

This work begins where the engineer, the technologist, leaves off; it begins when the work is given life. Life is not given to the work by fashioning the object, the building, according to a viewpoint alien to it, but by awakening, fostering, and cultivating the essential form enclosed within it.

Lewis Mumford (1895–1990) was an American historian of architecture, technology, and the city. He began his career as the architectural critic for the *New Yorker* and over the course of his life wrote many influential books, such as *Sticks and Stones* (1924), *Technics and Civilization* (1934), *The Culture of Cities* (1938), *Art and Technics* (1952), *The City in History* (1961), and the two-volume *The Myth of the Machine* (1967 and 1970). He was a fellow of the American Academy of Arts and Sciences, and a member of the American Philosophical Society. He was knighted by the British Crown in 1943, and received the United States Presidential Medal of Freedom in 1964.

Mumford was greatly influenced by Patrick Geddes, drawing heavily on the latter's phases of technological development, and his optimism about its prospects. After Mumford's early account of modern architecture, *Sticks and Stones*, he felt compelled to trace its genesis to the industrialization of the nineteenth century, leading him on a life-long journey to better understand the "machine."

The following pieces are taken from *Technics and Civilization.* In "Technical Syncretism," Mumford offers a history of industrialization, in which modern technology develops out of a variety of diverse cultures. In "Toward an Organic Ideology," he detects the emergence of a new machine ethic – using Geddes's term biotechnics to describe a kinder, more responsive form of technology. Though he tempered his optimism about those developments in the years after the Second World War, he saw contemporary technology as ever more systematic and organic.

1934

Lewis Mumford

Technical Syncretism and Toward an Organic Ideology

› 1: Technical syncretism

Civilizations are not self-contained organisms. Modern man could not have found his own particular modes of thought or invented his present technical equipment without drawing freely on the cultures that had preceded him or that continued to develop about him.

Each great differentiation in culture seems to be the outcome, in fact, of a process of syncretism. Flinders Petrie, in his discussion of Egyptian civilization, has shown that the admixture which was necessary for its development and fulfillment even had a racial basis; and in the development of Christianity it is plain that the most diverse foreign elements – a Dionysian earth myth, Greek philosophy, Jewish Messianism, Mithraism, Zoroastrianism – all played a part in giving the specific content and even the form to the ultimate collection of myths and offices that became Christianity.

Before this syncretism can take place, the cultures from which the elements are drawn must either be in a state of dissolution, or sufficiently remote in time or space so that single elements can be extracted from the tangled mass of real institutions. Unless this condition existed the elements themselves would not be free, as it were, to move over toward the new pole. Warfare acts as such an agent of dissociation, and in point of time the mechanical renascence of Western Europe was associated with the shock and stir of the Crusades. For what the new civilization picks up is not the complete forms and institutions of a solid culture, but just those fragments that can be transported and transplanted: it uses inventions, patterns, ideas, in the way that the Gothic builders in England used the occasional stones or tiles of the Roman villa in combination with the native flint and in the entirely different forms of a later architecture. If the villa had still been standing and occupied, it could not have been conveniently quarried. It is the death of the original form, or rather, the remaining life in the ruins, that permits the free working over and integration of the elements of other cultures.

One further fact about syncretism must be noted. In the first stages of integration, before a culture has set its own definite mark upon the materials, before invention has crystallized into satisfactory habits and routine, it is free to draw upon the widest sources. The beginning and the end, the first absorption and the final spread and conquest, after the cultural integration has taken place, are over a worldwide realm.

These generalizations apply to the origin of the present-day machine civilization: a creative syncretism of inventions, gathered from the technical debris of other civilizations, made possible the new mechanical body. The waterwheel, in the form of the Noria, had been used by the Egyptians to raise water, and perhaps by the Sumerians for other purposes; certainly in the early part of the Christian era watermills had become fairly common in Rome. The windmill perhaps came from Persia in the eighth century. Paper, the magnetic needle, gunpowder, came from China, the first two by way of the Arabs: algebra came from India through the Arabs, and chemistry and physiology came via the Arabs, too, while geometry and mechanics had their origins in pre-Christian Greece. The steam engine owed its conception to the great inventor and scientist, Hero of Alexandria: it was the translations of his works in the sixteenth century that turned attention to the possibilities of this instrument of power.

In short, most of the important inventions and discoveries that served as the nucleus for further mechanical development, did not arise, as Spengler would have it, out of some mystical inner drive of the Faustian soul: they were wind-blown seeds from other cultures. After the tenth century in Western Europe the ground was, as I have shown, well plowed and harrowed and dragged, ready to receive these seeds; and while the plants themselves were growing, the cultivators of art and science were busy keeping the soil friable. Taking root in medieval culture, in a different climate and soil, these seeds of the machine sported and took on new forms: perhaps, precisely because they had *not* originated in Western Europe and had no natural enemies there, they grew as rapidly and gigantically as the Canada thistle when it made its way onto the South American pampas. But at no point – and this is the important thing to remember – did the machine represent a complete break. So far from being unprepared for in human history, the modern machine age cannot be understood except in terms of a very long and diverse preparation. The notion that a handful of British inventors suddenly made the wheels hum in the eighteenth century is too crude even to dish up as a fairy tale to children.

› 2: Toward an organic ideology

During the first period of mechanical advance, the application of simple mechanical analogies to complex organic phenomena helped the scientist to create a simple framework for experience in general, including manifestations of life. The "real" from this standpoint was that which could be measured and accurately defined; and the notion that reality might in fact be vague, complex, undefinable, perpetually a little obscure and shifty, did not go with the sure click and movement of machines.

Today this whole abstract framework is in process of reconstruction. Provisionally, it is as useful to say in science that a simple element is a limited kind of organism as it once was to say that an organism was a complicated kind of machine. "Newtonian physics," as Professor A. N. Whitehead says in *Adventures of Ideas*,

> is based upon the independent individuality of every bit of matter. Each stone is conceived as fully describable apart from any reference to any other portion of matter. It might be alone in the universe, the sole occupant of uniform space. Also the stone could be adequately described without reference to past or future. It is to be conceived fully and adequately as wholly constituted within the present moment.

These independent solid objects of Newtonian physics might move, touch each other, collide, or even, by a certain stretch of the imagination, act at a distance: but nothing could penetrate them except in the limited way that light penetrated translucent substances.

This world of separate bodies, unaffected by the accidents of history of geographic location, underwent a profound change with elaboration of the new concepts of matter and energy that went forward from Faraday and von Mayer through Clerk-Maxwell and Willard Gibbs and Ernest Mach to Planck and Einstein. The discovery that solids, liquids, and gases were phases of all forms of matter modified the very conception of substance, while the identification of electricity, light, and heat as aspects of a protean energy, and the final break-up of "solid" matter into particles of this same ultimate energy lessened the gap, not merely between various aspects of the physical world, but between the mechanical and the organic. Both matter in the raw and the more organized and internally self-sustaining organisms could be described as systems of energy in more or less stable, more or less complex, states of equilibrium.

In the seventeenth century the world was conceived as a series of independent systems. First, the dead world of physics, the world of matter and motion, subject to accurate mathematical description. Second, and inferior from the standpoint of factual analysis, was the world of living organisms, an ill-defined realm, subject to the intrusion of a mysterious entity, the vital principle. Third, the world man, a strange being who was a mechanical automaton with reference to the world of physics, but an independent being with a destiny in heaven from the standpoint of the theologian. Today, instead of such a series of parallel systems, the world has conceptually become single system: if it still cannot be unified in a single formula, it is even less conceivable without positing an underlying order that reads through all its manifestations. Those parts of reality that can be reduced to patent order, law, quantitative statement are no more real or ultimate than those parts which remain obscure and illusive: indeed, when applied at the wrong moment or in the wrong place or in a false context the exactness of the description may increase the error of interpretation.

All our really primary data are social and vital. One begins with life; and one knows life, not as a fact in the raw, but only as one is conscious of human society and uses the tools and instruments society has developed through history – words, symbols, grammar, logic, in short, the whole technique of communication and funded experience. The most abstract knowledge, the most impersonal method, is a derivative of this world of socially ordered values. And instead of accepting the Victorian myth of a struggle for existence in a blind and meaningless universe, one must, with Professor Lawrence Henderson, replace this with the picture of a partnership in mutual aid, in which the physical structure of matter itself, and the very distribution of elements on the earth's crust, their quantity, their solubility, their specific gravity, their distribution and chemical combination, are life-furthering and life-sustaining. Even the most rigorous scientific description of the physical basis of life indicates it to be internally teleological.

Now changes in our conceptual apparatus are rarely important or influential unless they are accompanied, more or less independently, by parallel changes in personal habits and social institutions. Mechanical time became important because it was re-enforced by the financial accountancy of capitalism: progress became important as a doctrine because visible improvements were being rapidly made in machines. So the organic approach in thought is important today because we have begun, here and there, to act on these terms even when unaware of the conceptual implications. This development has gone on in architecture from Sullivan and Frank Lloyd Wright to the new architects in Europe, and from Owen and Ebenezer Howard and Patrick Geddes in city design to the community planners in Holland, Germany, and Switzerland who have begun to crystallize in a fresh pattern the whole neotechnic environment. The humane arts of the physician and the psychologist and the architect, the hygienist and community planner, have begun during the last few decades to displace the mechanical arts from their hitherto central position in our economy and our life. Form, pattern, configuration, organism, historical filiation, ecological relationship are concepts that work up and down the ladder of the sciences: the esthetic structure and the social relations are as real as the primary physical qualities that the sciences were once content to isolate. This conceptual change, then, is a widespread movement that is going on in every part of society: in part it arises out of the general resurgence of life – the care of children, the culture of sex, the return to wild nature and the renewed worship of the sun – and in turn it gives intellectual re-enforcement to these spontaneous movements and activities. The very structure of machines themselves, as I pointed out in describing the neotechnic phase, reflects these more vital interests. We now realize that machines, at their best, are lame counterfeits of living organisms. Our finest airplanes are crude uncertain approximations compared with a flying duck: our best electric lamps cannot compare in efficiency with the light of the firefly: our most complicated automatic telephone exchange is a childish contraption compared with the nervous system of the human body.

This reawakening of the vital and the organic in every department undermines the authority of the purely mechanical. Life, which has always paid the fiddler, now begins to call the tune. Like The Walker in Robert Frost's poem, who found a nest of turtle eggs near a railroad track, we are armed for war:

> The next machine that has the power to pass
> Will get this plasm on its polished brass.

But instead of being confined to a resentment that destroys life in the act of hurling defiance, we can now act directly upon the nature of the machine itself, and create another race of these creatures, more effectively adapted to the environment and to the uses of life. At this point, one must go beyond Sombart's so far excellent analysis. Sombart pointed out, in a long list of contrasting productions and inventions, that the clue to modern technology was the displacement of the organic and the living by the artificial and the mechanical. Within technology itself this process, in many departments, is being reversed: we are returning to the organic: at all events, we no longer regard the mechanical as all-embracing and all-sufficient.

Once the organic image takes the place of the mechanical one, one may confidently predict a slowing down of the tempo of research, the tempo of mechanical invention, and the tempo of social change, since a coherent and integrated advance must take place more slowly than a one-sided unrelated advance. Whereas the earlier mechanical world could be represented by the game of checkers, in which a similar series of moves is carried out by identical pieces, qualitatively similar, the new world must be represented by chess, a game in which each order of pieces has a different status, a different value, and a different function: a slower and more exacting game. By the same token, however, the results in technology and in society will be of a more solid nature than those upon which paleotechnic science congratulated itself: for the truth is that every aspect of the earlier order, from the slums in which it housed its workers to the towers of abstraction in which it housed its intellectuals, was jerrybuilt – hastily clapped together for the sake of immediate profits, immediate practical success, with no regard for the wider consequences and implications. The emphasis in future must be, not upon speed and immediate practical conquest, but upon exhaustiveness, inter-relationship, and integration. The co-ordination of our technical effort – such co-ordination and adjustment as is pictured for us in the physiology of the living organism – is more important than extravagant advances along special lines, and equally extravagant retardations along other lines, with a disastrous lack of balance and harmony between the various parts.

The fact is then that, partly thanks to the machine, we have now an insight into a larger world and a more comprehensive intellectual synthesis than that which was originally outlined in our mechanical ideology. We can now see plainly that power, work, regularity, are adequate principles of action only when they cooperate with a humane scheme of living: that any mechanical order we can project must fit into the larger order of life itself. Beyond the necessary intellectual reconstruction, which is already going on in both science and technics, we must build up more organic centers of faith and action in the arts of society and in the discipline of the personality: this implies a re-orientation that will take us far beyond the immediate province of technics itself. These are matters – matters touching the building of communities, the conduct of groups, the development of the arts of communication and expression, the education and the hygiene of personality – that I purpose to take up in another book. Here I will confine attention to co-ordinate readjustments which are clearly indicated and already partly formulated and enacted in the realm of technics and industry.

French-born Karel Honzík (1900–66) was a Czech artist, architect, writer, and educator. He graduated from the Czech Technical University, in Prague, in 1925, and was appointed Professor of the Theory of Architecture there in 1947. His career as a practitioner was short, as he started to concentrate on writing from the mid-1930s. During the brief period of his activity in practice, he was involved with three of his school colleagues in forming the Four Purists, and in leftist politics, with the Left Front and the Union of Socialist Architects. His significant buildings include the Starokosírská Street apartments, of 1928, and the General Pensions Institute, of 1934, with Josef Havlícek (1899–1961).

For Honzík, architecture is intrinsically connected to life itself. This position is conveyed in his 1937 article "Biotechnics," which follows. "Biotechnics" was undoubtedly influenced by what was by this time a school of thought, dating back to the 1870s, which was re-examining nature in order to offer the organic mechanism as a foil to the prevailing notion of mechanic organism. Important in this discussion, for architects, were not only the works of Patrick Geddes and Laszlo Moholy-Nagy, but also that of the biologist Raoul Francé, whose work was similar in spirit to that of D'Arcy Wentworth Thompson. Honzík, like Francé, understood form as an ecological process, in which it was the result of the interaction of internal and external forces. Thus form and function were in continuous interaction, rather than being diametrically opposed. Yet, unlike Francé, Honzík's account offered the possibility of local creativity through the notion of shifting performance requirements. This meant that form could outlive its immediate usefulness, and that forms and function were both dynamic in nature.

Honzík's account of technology is unwavering in positioning it as a natural process. Technological forms are driven by the same fundamental objectives, of efficiency and suitability, as natural ones, and form and function are in a continual ebb and flow. Technology is thus inherently ecological and biological in both its techniques and products.

1937

Karel Honzík

Biotechnics: Functional Design and the Vegetable World

From time to time the illustrated papers publish photographs of the *Victoria Regia* which show the neatly rimmed six-foot-wide leaves of this gigantic water-lily floating on a hot-house tank. Few of those who glance at them realize that these thin platter-like discs are rafts strong enough to support a large dog or a young child. But the engineer who examines the underside of one of them is astonished to find it might serve as a scale model of a reinforced-concrete roof-span. For here the monolithic system of transverse beams supporting slab decking was fully embodied aeons before François Hennebique first worked it out, and there are even stiffening haunches to increase the shear resistance of those parts of these leaf-beams where the tension is most pronounced. There are a number of natural organic forms that closely resemble forms devised by man. When Ozenfant said that if Nature had needed to produce a bottle she would have evolved one closely resembling a bottle as we know it, he was probably thinking of the Samura tree which grows in the desert parts of Argentina. The trunk of this tree not only looks like a Chianti flask but actually stores a supply of water to nourish the foliage during periods of drought. Thus we have two branches of technology, one human and the other phytogenical, constantly evolving towards the same ends. May we not presume that in every problem which engages an engineer or architect, natural laws are inexorably informing his designs, even his calculations and detail-drawings? A whole series of phenomena corroborate the assumption that the interaction between natural forces and matter results in the continual recurrence of certain forms.

A surface subject to pressure inevitably needs support in one direction, and where material is economized in the other as well. The resultant form is one of the most constant in Nature. A plant arrives at it by its own tropism; man by intuition, experience, and calculation. So, too, in decay internal and external forces are actively shaping what is being discarded or replaced (for both living and dead matter seem to follow a single impulse of growth or reorganization); and it is not difficult to find a similar process in the work of man. Sand always slides into a slope of 34–37 degrees, a pyramid being the natural formation of its grains in which their conflicting forces of gravitation and friction come to rest. The cone, the pyramid, parallel lines, the plane and the globe are all "constants" in Nature's technique. Nature seeks an ideal state of equilibrium between these forces. The moment she begins to succeed she ceases to be formless; the result being an embodiment of the shapes characteristic of flowers, crystals, and other organisms. We are told that everything on earth is changing, and so moving; but an equally universal law governs its movement: an urge to find a condition of harmony or rest; harmony, maturity, or crystallization. The shape in which matter achieves a balance of strength is its perfection, a solution of its own particular problem in which there is no waste or superfluity. If we add to, or take something from, that perfection the shape loses its equipoise, its characteristic appearance, and has to start in quest of both afresh. In the same way, human inventions arise from the will of man, and move towards the intrinsic perfection of a final form that can only be invalidated by the emergence of new conditions. Thus the best possible shape of chair could be superseded by one arbitrarily invented for the purpose. But that new shape would soon disappear just because it was not the right one. Again, if humanity started sitting in a different position the perfect shape would have to be modified accordingly.

The design of a man who can foresee each of the various forces which his building will be subjected to should be a consummate one, because he may be expected to avoid all merely transient influences. But such an achievement is rare enough to rank as the highest order of creative work. Perfection is far more often attained by the patient groping of generation after generation, intuition, or the cumulative experience and combined reasoning of hundreds of different men. It is for this reason that the masterpieces of architecture are the expression of whole nations, certain epochs or phases of society. Their purely personal significance fades into the background.

Though veiled by a whole sequence of different semblances, this abstract perfection of form exists only as an ideal. We know our own technique is very imperfect compared with Nature's. But even Nature seldom achieves perfect co-ordination of form and purpose. The beams of the *Victoria Regia* and the column of the bamboo are exceptional examples. Some investigators in the field of biotechnics advance arbitrary explanations as to what was Nature's practical intention in evolving this, that, and the other shape. None of these explain the existence of hundreds of variants among plants or animals of a single type. Why should there be 6,000 varieties of the unicellular Diatomacae living under identical conditions? Which of them represents its perfect form? On the contrary, this infinite multiplicity might lead us to doubt a single general law. In *The Intelligence of Flowers*, Maeterlinck describes their dramatic struggle for shape. He gives us examples of their resourcefulness of invention and their striving to achieve new properties by trial and error that read like laboratory experiments. The correlation of form and function, and that isochronous coincidence which so rarely appears in the works either of Nature or of Man, is a field still waiting to be explored. Remy de Goncourt cites the stag-beetle as an instance of how form tends to outlive function. Though its claws are now a useless ornament, they may have been originally provided as a defense against some extinct enemy. Do we not encounter this persistence of form in every walk of life? From time to time we succeed in shaking off old forms that have become so much top-hamper, and yet we revolt against those that supplant them because they announce a new content. It seems as though form precedes function, or anyhow outlives it; and that there is a continual oscillation between them as in the scales of a balance. We feel instinctively that both are for ever striving to attain a fixation of equilibrium in which to fuse their separate identities. Perhaps one day we shall discover the explanation of this separability of form and function which is so disturbing to us in an architect's work. The anticipation or survival of a form is apt to blind us to the object, the mission, of the form itself. This is hard to appraise, and can only be slowly gleaned in what may be read between, rather than from, those (lines) which are the characteristic mold-marks of its final shaping.

1939

Frederick J. Kiesler

On Correalism and Biotechnique: A Definition and Test of a New Approach to Building Design

Frederick Kiesler (1896–1966) was an Austrian architect, artist, set designer, and writer. He began his career in Vienna, studying at the Technische Hochschule and the Akademie der blidenden Künste. He worked briefly in the office of Adolf Loos, and then in Berlin as a set designer, before emigrating to the United States in 1925. During the 1920s and 1930s, he belonged to De Stijl, the American Union of Decorative Artists and Craftsmen, Buckminster Fuller's Structural Studies Associates (SSA), and the theater faculty at Juilliard; he also formed the Laboratory of Correalism at Columbia University, and through his association with Marcel Duchamp and the exiled Parisian art community, became the "official" architect of the surrealists. He completed a few architectural works, notably the Film Guild Cinema, Peggy Guggenheim's Art of This Century Gallery and the Shrine of the Book in Jerusalem, the home of the Dead Sea Scrolls.

Kiesler maintained a position within the avant-garde for most of his career, beginning with the "Raumstadt" exhibition of 1925, which drew him into the De Stijl, and culminating with his Endless House of 1960, whose ideas began with the Space House of 1934. While he constantly adapted his work to his context and period, his fascination with continuity endured, first explained according to notions adopted from physics, then from evolutionary theory, and later from surrealist theories of the image.

The following essay on Correalism was written during Kiesler's association with Buckminster Fuller and the SSA, which began in the early 1930s and was most visible in the 1932 transformation of the *T-Square Journal* into *Shelter*, described by Fuller as a "Correlating Medium." The abbreviated technical language of the SSA is visible throughout Kiesler's essay, as are the themes of technological transformation and evolution, though Kiesler translates them more completely into the concerns of architects than any other contributor to the SSA. The defining point of the essay was his distinction between the building techniques of nature (biotechnics) and of man (biotechniques), who builds by assembly. He meant this to support his claim that continuous construction could solve the problem of human construction, which fails at the joints of assembly, but it also leads to proposals for truly organic architecture (see Katavolos, 1960; De Landa, 2002). His other fundamental observation, that the "needs" which define functionalism are always evolving and so health is the only possible criteria of design, anticipates the framing of ecological architecture and sustainable design. In a succinct statement, he poses one dilemma of the age of systems.

In this paper[1] I propose to show that the perennial crisis in architectural history is due to the perennial lack of a science dealing with the fundamental laws which seem to govern *man as a nucleus of forces*; that until we develop and apply such a science *to* the field of building design, it will continue to exist as a series of disparate, overspecialized, and unevenly distributed products; and that only such a new science can eliminate the arbitrary divisions of architecture into: Art, Technology, and Economy, and make architecture a socially constructive factor in man's daily activities.

Today we face the task of formulating the *general* laws of the foundations that underlie the many specialized sciences, not in terms of metaphysics (such as religion or philosophy) but in terms of work-energies; and the *specific* task of formulating those that govern building design. But the two are intimately related and we in the building field cannot solve our special problems without comprehension of the foundations of such part-sciences, e.g. physics, chemistry, biology, etc. Thus, it would seem imperative that we summarize some of the concepts of modern science and investigate their validity for our specific problem.

› Concepts from science for the building designer

Man is born in evolution of hereditary trends. He is the nucleus of forces which act upon him, and upon which he acts. Forces are energies. We assume, with contemporary science, that they are of an electromagnetic nature. The inter-relation of organic and inorganic matter is a mutual bombardment of energies which have two characteristics: those of integration and those of disintegration.

By means of gravitation, electricity generates energy into solids of visible matter. This is integration. By magnetism and radiation, electricity degenerates energy into tenuous, invisible matter. This is disintegration.

If this general principle of anabolic and catabolic energies were the sole principle of existence, we would have a static, unchanging world. But these two forces (positive and negative) interchange through physico-chemical reactions, one force striving always for a preponderance over the other. In this way *variations* are constantly created; and in this process of creation, new nuclear concepts and new environments are in continual formation.

› Reality and form

The mutual biological interdependence of organisms is, in the final analysis, the result of the primary demands of all creatures: proper food, habitat, reproduction, defense against inimical forces. Life is all expression of the cooperation, jostling, and strife of individual with individual, and of species with species, for these primary needs.

The visible result of these activating forces is called *matter* and constitutes what is commonly understood as reality. The reason for this superficial interpretation of reality lies in the limitation of man's senses in relation to the forces of the universe. For matter is only one of the expressions of Reality, and not reality itself. If matter alone were reality, life would be static.

What we call "forms," whether they are natural or artificial, are only the visible trading posts of integrating and disintegrating forces mutating at low rates of speed. Reality consists of these two categories of forces which inter-act constantly in visible and invisible configurations. *This exchange of inter-acting forces I call CO-REALITY, and the science of its relationships, CORREALISM. The term "correalism" expresses the dynamics of continual interaction between man and his natural and technological environments.*

› Natural, social, and technological heredity

Biology has divided these forces into two main categories: Heredity and Environment. Man had to evolve a method for dealing with the effects of these overwhelming forces upon himself. For this purpose he created technological environment to help him in his physical survival even within the short span of the age-potential of his own species. This is made more difficult because man is biologically unfit to transmit his experiences to his offspring: each child has to begin anew its adaptations to nature. In short: contrary to prevailing belief, acquired traits and habits of parents can not be transmuted into the make-up of body cells and, by way of procreation, given to their children.[2]

By providing unchangeable genes within the germ-cells *Nature* has safeguarded herself from man interfering fundamentally with her aims, whatever they may be. This "sealed order" of the germ cell contains nature's will which man can influence *during his own life-time, but not beyond that.* This places a deep responsibility upon those who "design" technological environment, because the restriction of its application to only *one* life-span makes it so much more needed as part of man's defense-mechanism. It appears, then, that the only human experiences that can be inherited by children are those of customs and habits by way of: training and education, thus "social heredity" is the only tool man can rely upon. Just as all living organisms are generated through their own species from a long chain of generations, so do ideologies or man-made objects generate from a long line of older ideologies or objects of similar functions. Thus a contemporary chair, for instance, is the product of many generations of other tools for man to rest his body in fatigue. This is heredity in technology.

› What is technological environment?

When the biologist speaks of environment, he invariably means the geographical and animal environment. This definition is perhaps accurate for all creatures except man. For man alone has developed a third environment: a *technological* one which has been his steady companion from his very inception. This technological environment, from "shirts to shelter," has become one of the constituent parts of his total

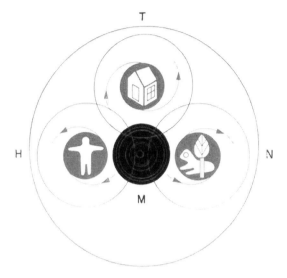

T

H

N

M

Fig 1a
"Man = Heredity + Environment. This diagram expresses the continual interaction of both the total environment on man and the continual interaction of its constituent parts on one another."

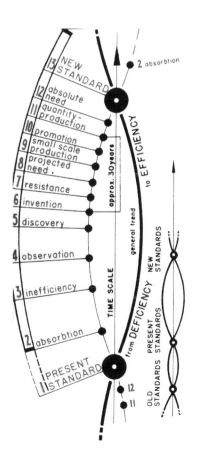

Fig 1b
"Every need follows a characteristic pattern of development. A minimum of twelve progressive stages can be detected in progression from one standard to the next."

environment. Then, the classification of environment becomes three- instead of two-fold:

1 natural environment
2 human environment
3 technological environment

But it is this last factor of technological environment which concerns us here, since it is in this field that the architect works. Man-made, technological tool-objects have been in existence since the Ice Age. *But no branch of science so far has undertaken to investigate, analyze, chart, and measure the direct and indirect, voluntary and involuntary effects of technological environment upon man;* nor has any branch of science charted and formulated the laws which govern the development of technology. We have had numerous accounts of the history of technology but no study of the need-morphology of its growth.

In studying the history of the science of biology one can find with amazement the lack of observation and systematization of natural phenomena: for twenty centuries after the Greeks, no new theory of natural science came until the appearance of Lamarck and Darwin. The scientific theory of evolution is essentially the product of the last hundred years.

An analogous situation exists in technology, and we need not be surprised that no new theory on the phenomena of design has been forthcoming. Just as the scientists of the Middle Ages thought that horses produced wasps; asses, hornets; and cheese, mice, so modern men think that it is industry which produces the technological environment. In reality, the technological environment is produced by *human needs*: absolute needs and simulated needs.

Of what does this technological environment consist? In its simplest terms, it is made up of a whole system of tools, which man has developed for better control of nature. I use the term "tool" advisedly. It is generally agreed that the difference between a machine and a tool is the power by which it is driven, whether manually or by the forces of man's environment – e.g., natural (water) or synthetic (electricity). But this distinction of isolated technological fields must be replaced by an understanding of technological invention as a whole. For the purposes of this analysis, I therefore define "tool" as: *any implement created by man for increased control of nature.* The term "tool" is preferable to the term "machine" because it brings us back to the origin of the machine, and to its ultimate purpose: *enabling man to reach levels of higher productivity.* In this sense, *everything which man uses in his struggle for existence is a tool* and, as such, part of a man-made technological environment, from shirts to shelter, from cannons to poetry, from telephones to painting. No tool exists in isolation. Every technological device is *co-real*: its existence is conditioned by the flux of man's struggle, hence by its relation to his *total environment.*

The persistence of technological environment is marked by constant, if only indirect, infiltration of converted forces embodied in the manufacture of our homes, workshops, transportation shelters, etc. The ratio of fabricated environment to natural environment varies according to the ways in which men make their living. Today, men in urban areas spend about 88 percent of their time indoors; in suburban areas about 70 percent; and in rural areas about 43 percent.

› A qualitative classification of tools

But we must keep in mind that the *technological environment affects man's development, and that technology itself follows laws of heredity in its own development.* We then observe that the principle of heredity also operates in technology. Thus the progressive development of any tool – a knife, a factory, a home – does not follow a straight line any more than does a species of plant or animal. On the contrary, production of any tool in our industrial era seems to develop along three characteristic lines.

The Standard Type, developed by absolute need.
The Variation, evolved from the *Standard Type* for auxiliary purposes.
The Simulated springs directly or indirectly from one of the two foregoing types. This third group of products – and it is by far the largest – distinguishes itself from the *Standard* and the *Variation* chiefly by a lack of material efficiency and insignificant changes in design and materials.

Each of these three types has its special fertilization grounds in which it develops. The *Standard* grows out of scientific knowledge. The *Variations* are a natural adaptation of the *Standard* to specific conditions, and are therefore valid. The *Simulated* product and its temporary survival is only made possible by a lack of knowledge within its social environment.

The *Simulated* are the widest in distribution, the shortest lived, and the most rapidly replaced. The result is a dispersion of energy and a conflict of creative forces whose destructive effect is to slow down the rise of the original *Standard* to higher levels of efficiency.

Adjustments to the basic needs of man require the elimination of the *Simulated* and control of the *Variations*. In the readjustment of industry, the forces (man- and machine-power) which are producing the *Simulated* will be absorbed into the areas of the *Standard* and its *Variations*, thus reinforcing their productivity.

› Evolution of need: from deficiency to efficiency

Since nature demonstrates her will toward *mutative continuity*, man's aim seems also to be: to *sustain and prolong life*. By experience he learned that he was unable to do so with the physical equipment which he inherited. He was therefore compelled to extend the powers of his natural equipment to meet the forces of environment. He had to add to his natural equipment, artificial equipment of defense and offense. Tool-making began. *Man's inherent desire for higher productivity* began to find its material expression.

Man, then, builds tools; and from them arises that man-made complex of relationships which we have called the technological environment. But in order to correct the many obvious maladjustments of this environment, it is necessary to ask: What is the nature of its origin? What is a need? How do needs arise? Are they natural or artificial? Are they static or in evolution? A definition of needs has today become of prime importance to the designer of technological environment. *Investigation on this crucial point cannot be based upon the study of architecture but must be based upon the study of man.* Our duty would therefore be to re-define needs, and upon this basis to re-organize the technological environment. The accompanying chart of the evolution of needs may help to clarify the problem.

We must keep in mind that science in all its branches is based upon man's deficiencies. The direction of man's creation tends constantly from deficiency to efficiency. The main stages in this recurring development are marked by a rise from one standard of living to another. Sociologists speak of "higher" and "lower" standards, but we can only speak of correalist standards, since concepts of higher and lower are entirely relative.[3] *Needs are not static: they evolve.* The intermediary stages of the evolution of needs, as Fig. 1b indicates, seem to develop in the following progression:

1 Present standard
2 Standard is absorbed
3 Absorption demonstrates inefficiency
4 Inefficiency leads to observation
5 Observation leads to discovery
6 Discovery leads to invention
7 Invention meets resistance
8 Resistance leads to "projected need"
9 Projected need leads to small-scale production
10 Small-scale production generates promotion
11 Promotion leads to quantity production
12 Quantity production creates absolute need
13 Absolute need becomes new standard

Fig. 1b shows that actual needs are not the direct incentive to technological and socio-economic changes, as is commonly assumed. Needs evolve, and that evolution is based on the nuclear character of the human structure and its environment.[4]

› Health is man's ultimate need

The failure of an artificial tool to protect man, leads to impaired physical resistance. His health is unbalanced. If by the power of his tools the re-generation of his degenerated physique fails, man's health declines in a progression from fatigue to death. *The fundamental denominator, therefore, to account for the validity of any technological environment, is man's health.* Measured by this crucial, all-embracing criterion of health, technology is one of the most powerful factors for preserving man's energy.

Health appears to be that bodily condition in which the various materials and processes that maintain life-activity are in functional equilibrium.

The resistance-capacity of an individual is the degree in which this equilibrium is able to withstand or absorb the impacts of the environment. There are two sets of these factors: external and internal. The external factors belong to the exigencies of the natural environment. The internal factors are psycho-physiological and are intrinsic to the individual.

Health was originally maintained by organic adaptation to environment. Some of these adaptations are essentially functional (digestion, temperature, blood pressure, etc.), or essentially structural (pigmentation, posture, etc.). There are also adaptations to the human environment, as represented by socio-economic relations (state institutions, industry, trade, marriage, etc.).

The concept of health recognizes fatigue as a part of a continuous natural process. Fatigue is normally produced by the expenditure of energy incident to psycho-physiological action (voluntary and involuntary). This, expended energy, under normal conditions, is replaced by means of physicochemical processes in the body. *When the processes of expending and replacement are in proper balance, we may speak of an optimum efficiency. When this is not the case, we have inefficiency, or waste of energy: de-generation.*[5]

› Environmental control and the maintenance of health

What are the factors which impair the efficiency of the body? Obviously, maladjustments between the body and some parts of its environment, external or internal. Technological environment can be of vital importance in relieving such maladjustments by *protection against fatigue* (preventive) and by *relief of fatigue* (curative).

Unfortunately, history proves that this technological environment has not always been per se beneficial to man's health: on the contrary. Thus, we come to the second factor: In which direction, then, shall technological environment be developed? Development of industry for industry's sake is worse than art for art's sake. Imperative, therefore, is the control of direction of technological production. What is environmental control? Since the means of control are part of the environment, the term would appear to mean simply control of environment by environment. The term becomes clearer, however, when we remember that environment is threefold: natural, human, and technological. Environment control, then, is control of the human and natural environment through *technological environment*.

But control in relation to what? From the correalist viewpoint there can be only one answer: *in relation to man's health*. Control of environment becomes, then, control of health: not control of the environment's health, but control of the health of man and society by environment. The proper term will then read: technological control of environment or *environmental control by technology*.

The maintenance or adequate "management" of technological environment can have only one purpose: to maintain the equilibrium of its health. In turn, the maintenance of the technological environment in proper health can have only one purpose: the maintenance of the equilibrium of man's health.

› Health, the criterion of building design

Hitherto architecture has been judged from four viewpoints: (1) beauty, (2) durability, (3) practicability, and (4) low cost. But these four factors have never altogether coincided in a single work. If a piece of architecture is not beautiful, it is excused on the ground of being cheap; if not cheap, it is excused as being durable; if not practical, it is perhaps beautiful. It would appear, then, that the only way to resolve these age-old contradictions is to find one criterion which will do for all. *This criterion, in my opinion, can only be health.* The rest may be left to personal idiosyncrasies on the part of the consumers and producers, so long as these do not impair the essential criterion.

Thus, architecture, in the future, will not be judged chiefly by its beauty of rhythm, juxtaposition of materials, contemporary style, etc., etc.; it can only be judged by its power to maintain and enhance man's well-being – physical and mental. *Architecture thus becomes a tool for the control of man's health, its degeneration and re-generation.*

› "Form follows function" an obsolete design formula

In the early Twenties, there was again much loose talk about functional design. But when we examine the buildings which were then built, and the drawings which were then presented, we find that no new functions had been invented. All that happened was that, by debunking old *decor* and adding new gadgets, new forms had been wrapped around conventional ways of living. No one could define what function was. Worse still: no new building principles adequate to a new idea of environmental order had been conceived.

The problem was posed in the manner of the Scholastics: should function follow form, or form follow function? Architecture was thus saddled with a new version of an old conundrum: which came first, the hen or the egg? What was overlooked was the very essence of the problem: the inter-relation of form and function with *structure* and the fact that, genetically, all three are contained within the protoplasm of thought.

If we abandon the Scholastic approach, the contemporary designer can learn a valuable lesson from the hen and the egg. In 1912, at the Rockefeller Institute for Medical Research, a hen's egg in process of hatching was opened. The developing chick was removed, and the tiny fleck of its heart was cut out. This bit of living tissue was transferred to a solution in a test tube. There, protected from germs, poisons, heat, and cold and provided with a never-failing supply of oxygen, sugar, and other nutrients, it lived and flourished far better than the heart cells in any living chick ever did.

This experiment confirms the view that, while life comes only from life, it is also dependent on its technological environment. By changing the physical environment, life may be quickened and increased, retarded or destroyed.[6]

What was done for the bit of living tissue at the Rockefeller Institute, experimenters have not yet been able to accomplish for the animal as a whole. But the experiment indicates that a planned chemical environment can be as beneficial for man as for other animals; equally important for man is a properly planned technological environment.

The question investigated in connection with the chick's heart is: at what point and by what means does inanimate matter pass over and become alive? "To find that bridge between nature and man has become the grand quest of science." Similarly, finding the bridge between man and artificial, man-built, technological environment must become the grand quest of future building design.

› New definition of function

We must examine what function has meant, and what function will come to mean in the future, as it concerns the designer. We cannot conceive of function as something static, else growth would cease. The inter-action of environment and man, and the evolution of that inter-action to new possibilities, is not a direct result of environment. It is rather the development by environment of something which was *already inherent* physiologically in the organism.

Function depends not only on natural environment, but also on artificial environment. If functional design depended on the *status quo* of man, it could never develop. It would take care only of man's traditional aspects. But man's evolution has proven that changing environment increases or decreases man's potentialities. Technological environment, being a part of the complex of environmental forces, must consciously contribute to the extraction and development of man's inherent possibilities into a higher order. What these possibilities are depends on the designer's ability to envision and realize them.[7]

Any form is incomplete in itself: it is identified by what it emanates, visibly or invisibly, voluntarily or involuntarily. The new designer will therefore define function as: *a specific nucleus to actions*. It is erroneous to suppose that form follows function. This concept must be replaced by the proper progression of: (1) structure, (2) function, (3) form. All functions and all forms are contained in the structure.

› Defining design and Biotechnique

As in the case of electricity, a polarization creates a nucleus of relationships. These relationships are latent potentialities for further development. In this respect, *all possible needs of man are ever present, but it is only by the demands of the special environmental stimuli that the specific need is brought to the fore.*

Thus it appears that not only is the formula "form follows function" inadequate; the "functional design" based on that formula is likewise inadequate. The term "design" must be re-defined. Since the building designer deals with *forces, not objects*, design is therefore, in my definition, *not the circumscription of a solid but a deliberate polarization of natural forces towards a specific human purpose.*

Such a science of design I have called BIOTECHNIQUE[8] because it is the special skill *of man which he has developed to influence life in a desired direction. Biotechnics*, a term which Sir Patrick Geddes has employed, can be used only in speaking of *nature's* method of building, not of *man's*. There can be no interchange of these two methods, because nature and man build on two different principles: *nature builds by cell division with the aim of continuity*; man can only build by *joining parts together into a unique structure without continuity*. Nevertheless, man-made joinings are ultimately controlled not by man but by nature. The process of disruption through natural forces becomes imminent from the very moment of joining parts. Building design must, therefore, aim at the reduction of joints, making for higher resistance, higher rigidity, easier maintenance, lower costs. Such considerations led me to develop *Continuous Construction.*[9]

The more man recognizes his limitations in building "for a lifetime," the more valid is his structure. As a biologist has said:

> We doubt that an engine might be conceived to which we might bear witness that, after we might have broken it into a hundred pieces, it would reform immediately into a hundred single complete engines. But take that graceful animal, the fresh-water polyp, that is found attached to water lilies in the pond, and cut it into pieces: tomorrow you will find that each piece has become a complete polyp.

The new designer will learn to understand the methods by which nature builds to meet her purposes (biotechnics): but he will not imitate her methods. He will draw the necessary conclusion from the disaster which befell London's Crystal Palace.[10]

The Biotechnical approach tries to develop the possibilities of specific actions contained in any nucleus of human physiology. These potentialities remain at first undiscovered. Only with time are they individually or collectively developed until finally they are consciously demanded. The result will be entirely *new functions* within the old framework of what was considered "human nature," sustained by inventions.

› The objective: minimum biotechnical standards

The two approaches – biotechnical and functional – develop from unlike sources and lead to unlike results. On the one hand, functional design derives from the traditional behavior of any tool; on the other hand, biotechnical design derives from the evolutionary potentialities of man. Functional design develops an object. Biotechnical design develops the human being. Functional design is oscillating. Biotechnical design is inventive. A functional object is inert. A biotechnical object is re-active.

The biotechnician emerges as an important factor in the evolution of society toward a higher standard of living through the control of elements of fatigue and forces of re-generation. This leads to the discovery that no part of the human body is mono-functional; rather, each minute detail is again of nuclear make-up with corollary functions.[11]

Such development can be furthered by the biotechnician who formulates and helps to realize a *biotechnical minimum standard.* Such a biotechnical minimum standard must be based on Correalism and not on mere architectural derivations, which tend to house lower-income groups in dwarf reproductions of giant villas. The *biotechnical minimum standard is that technological environment of home, work-place, and their corollaries which meets the optimum needs of man's health.*

Every object that meets a need is living: it is only dead when it ceases to meet a need or when the need itself disappears. Anything of nature's creation which fulfills a need is a living organism. Similarly, every creation of man's technology is a living organism, whether it be a pillbox, a house, or a motor. Since the criterion of life is activization, we assume that a man no longer active is dead. By analogy we assume that because an object does not express itself in visible activity, it also is dead.

Here our judgment is determined by the limitations of our senses; for, as a matter of fact, when an object moves (a moving locomotive, a flashing electric bulb) we automatically say: it is alive. Conversely, when an object does not move, we automatically assume: it is dead. Our assumption of what is alive or dead is chiefly the result of optical observation. But this nerve center is "short-sighted." With a microscope we can see that a dead piece of cheese is very much alive. The revision of our judgments as to what is "alive" or "dead" must, for the time being, depend solely upon a more profound observation of facts.

› Architecture: generator and de-generator of human energy

The floor on which one walks, the chair on which one sits, the bed on which one rests, the wall that protects, the roof that shelters, and all other units of the man-built environment are significant for what they are: but they also possess *nuclear multiple-force*. It is commonly assumed that these are dead objects; actually they represent an interplay of action with one another and with nature. They are a constant exchange of anabolic and catabolic forces within themselves, and in their coordination with human beings, and through human beings with themselves again, they constitute high potential energy centers.

The modern physicist speaks of constant bombardment of the earth by invisible cosmic rays, of radiation and radio-active elements which cannot be seen or felt, but which, in time, can exert a deadly or beneficent effect upon all life. *This is equally true of the "inter-stellar" organization of a house, a town, or a city.* But here the forces at work are composed not only of animate and inanimate matter, but also of artificial technological bodies.

› Biotechnique as a force of re-generation

The orbit, region, and scope of the activity of technological bodies (be they houses, machinery, or any other tool) are the objectives of the future biotechnician. He will find that any structure he builds is *worth only as much as the ratio of its force of re-generation.*

Despite the imperative need for health-yielding technological tools, obsolete manufacture clutters the market.[12] As far as the building designer is concerned, his contribution to halting such anti-social types of production will be the constant use of the biotechnical approach.

The biotechnical approach has led me to the *evolutionary method of design which, instead of taking its departure from prevailing commodities, employs the study of general physiotechnics.* This enables the biotechnician to avoid giving a mere narrative survey of phenomena, and – on the basis of a genetic account of an unfolding process – to create the necessary need-service. The Mobile-Home-Library shown on the following pages, represents a test of the validity of biotechnical design. The storing of books in the home was chosen as an objective for the first laboratory test because: (1) it is a need in every family's home, and (2) it has become so standardized in the form of a "bookcase" that its re-design seemed at the beginning a wasteful undertaking. The Mobile-Home-Library thus constitutes a documentation for this general statement: *Functionalism shifts the strain from the technological tool to the human being: but, here, biotechnique shifts the strain from the human being to the tool.*

NB Not all figures mentioned in this article have been reproduced.

1   In an earlier manuscript of Mr. Kiesler's ("From Architecture to Life," for Brewer, Warren and Putnam, 1930) the groundwork of this paper was laid; it was first read in approximately its present form at a Symposium on Science and Design held by the Alumnae Association of the Massachusetts Institute of Technology, June 6, 1938; this is its first appearance in print. – Ed.

2   The part of Darwin's theory which stated that "*acquired characteristics are inheritable*" has been disproven (August Weissmann, 1880). Thomas H. Morgan: "the belief in the inheritance of acquired characteristics is not based on scientific evidence but on the very human desire to pass on one's acquisitions to one's children."

3   PROGRESS OF TOOLS RELATIVE TO TIME STRATUM: There is no abstract technological progress. Each stratum of the social development in man's history has produced its own tools to deal with various old and new forces. Each new environment creates new varieties or new standard types of tools which lose their validity if applied backward or forward in history.

4   Examples of nuclear production in industry: corn, subjected to mechanical and chemical treatment, also yields starch, dextrin, glucose, oils, feeds, and other valuable by-products. The hulls of oats yield furfurel, a valuable starting point for chemical synthesis. Waste sugar cane, from which the sugar has been extracted, forms the raw material for making wall-board and insulation. Saw mill wastes are converted into building materials. Similarly, carbon is the nuclear factor for many products: we encounter it in our heating arrangements in the form of coke, charcoal, and coal; we use it in our pencils as graphite, etc.

5   FATIGUE – Fatigue may arise in: (1) the central nervous system, (2) the muscular system, (3) in both combined. "Fatigue may be subjective as experienced by the worker or objective as noted in his actions and output. From a thorough consideration of the literature it is quite evident that a vast amount of emphasis has been laid upon the mechanical or extrinsic factors influencing the working capacity while the multiplicity of original physical and mental states that may limit the working capacity have become almost wholly neglected." From *Waste in Industry*, published by Federated American Engineering Societies, Washington, DC.

6   Jickeli (1902) and Carrel (1912) put forward the hypothesis and finally experimental proof that aging (and death) result from imperfect metabolism within the cell and the subsequent clogging of the cytoplasm with injurious waste. Carrel has shown clearly (tissue-work) a relative potential immortality of the cell, and at the same time its subordination to the fate of the whole organism.

7   In attitudes toward the technological environment, we observe three tendencies as to morphological principle: (a) the functional or synthetic, (b) the formal or transcendental, (c) the mechanical-materialist disintegrative. The mechanical-materialist attitude is not distinctively biological, but is common to nearly all fields of thought. (It dates back to the Greek atomists. The self-deceiving triumph of mechanistic science in the nineteenth century led many to accept mechanical materialism as the only possible scientific method.) Even in biology, but especially in design, it is more akin to the formal than the functional attitude.

8   The term "biotechnique" appeared first in my treatise on "Town Planning"; as "Vitalbau" in "De Stijl" No. 10/11, Paris, 1925, and in America first in "Hound and Horn," May 1934.

9   Not actually formulated until my plans of The Endless House were exhibited in Paris, 1925, and New York, 1933. View of my Space-House (New York, 1933) showing first continuous construction in shelter design and also continuous window framing (right).

10   That structure was built in 1851 by Paxton in imitation of the structural principles of the African water lily's foliage, with its longitudinal and transverse girders. This was an essentially romantic attempt to fashion a man-built structure by *literal application* of nature's design principles. The collapse of the Crystal Palace (1936) was inevitable (Fig. 8). (The fireproofing of buildings – then as now – is far more important than the pursuit of "new forms.")

11   J. R. de le H. Marett: *Race, Sex, and Environment*. Hutchinson's Scientific and Technical Publications, London, 1936.

12   *Waste in Industry*. By the Committee on Elimination of Waste in Industry of the Federated American Engineering Societies, Washington, DC, 1921.

1941

Siegfried Giedion

Industrialization as a Fundamental Event

Siegfried Giedion (1888–1968) was a Swiss historian of architecture (see Giedion, 1928 for full biography).

The following excerpt was drawn from his seminal work, *Space, Time and Architecture: The Growth of a New Tradition* of 1941. Giedion had been invited to give the Charles Eliot Norton lectures at Harvard by Gropius, shortly after the latter arrived there from the Bauhaus. He had written to Giedion, saying the he was one of the only people who could really explain the modern movement, and the success and popularity of the ensuing book would seem to attest to that fact. In an argument familiar within technology studies, modernism is presented as a natural, if not inevitable, outcome of the rapid and radical industrialization of the nineteenth and early twentieth centuries. And like the methods of art historians and psychoanalysts of the period, the emerging sensibility could only be detected in minor, un-selfconsciously executed constructions.

The Industrial Revolution, the abrupt increase in production brought about during the eighteenth century by the introduction of the factory system and the machine, changed the whole appearance of the world, far more so than the social revolution in France. Its effect upon thought and feeling was so profound that even today we cannot estimate how deeply it has penetrated into man's very nature, what great changes it has made there. Certainly there is no one who has escaped these effects, for the Industrial Revolution was not a political upheaval, necessarily limited in its consequences. Rather, it took possession of the whole man and of his whole world. Again, political revolutions subside, after a certain time, into a new social equilibrium, but the equilibrium that went out of human life with the coming of the Industrial Revolution has not been restored to this day. The destruction of man's inner quiet and security has remained the most conspicuous effect of the Industrial Revolution. The individual goes under before the march of production; he is devoured by it.

The heyday of the machine and of unlimited production is heralded in the eighteenth century by the sudden appearance of a widespread urge toward invention. In the England of 1760 this urge had gripped people in all strata of society. Everyone was inventing, from unemployed weavers, small hand workers, farmers' and shepherds' sons like the bridge-builder Telford, to manufacturers like Wedgwood and members of the nobility such as the Duke of Bridgewater (whose tenacious labor was responsible for the creation of the English canal system). Many of these inventors did not even take the trouble to protect their discoveries by taking out patents on them. Many, far from drawing profit from their inventions, were even persecuted because of them. Profit-making and unfair exploitation belong to a later period.

We must, in fact, take care to avoid the delusion that this activity had its source only in material ambitions or in the desire to shine. Its actual source lay much deeper and was one that had for a long time been artificially denied outlet. But at this date the urge to invent could no longer be stemmed. When, as in France, it was kept from entering into important regions of practical activity, it was only diverted, not destroyed. It manifested itself then in the creation of odd mechanical contrivances and of marvelous automatons, lifelike mechanical dolls capable of performing the most amazing feats, from walking to playing musical instruments and drawing pictures. Some of these automatons, in the ingenuity and precision of their workmanship, even succeeded in anticipating the principle of the modern automatic telephone – for example, the "writing doll" made at Neuchatel about 1770 by Pierre Jaquet-Droz. This doll still exists in a perfectly workable condition.

Invention, carried on in this way by men of all nations and all walks of life, led to the industrialization of almost every human pursuit. But this movement which was to give the nineteenth century its essential character is scarcely reflected at all in its official architecture. We should never be able to perceive the real nature of the period from a study of public buildings, state residences, or great monuments. We must turn instead to an examination of humbler structures. It was in routine and entirely practical construction, and not in the Gothic or classical revivals of the early nineteenth century, that the decisive events occurred, the events that led to the evolution of new potentialities.

But life is complex and irrational. When its evolution is blocked in one direction it seeks another (and often an entirely unexpected) outlet. The development of modern industry is essentially material. Nevertheless, in following its material urge, industry unconsciously creates new powers of expression and new possibilities of experience. These possibilities at first remain bound up in quite matter-of-fact enterprises that do not in any way enter into the intimate and personal lives of men. But, slowly and gradually, the new potentialities become a part of private and individual life. Thus a devious line of development leads from innovations in industrial buildings of all kinds – mines, warehouses, railroads, and factories – to the private home and personal life. The history of this metamorphosis is, in large measure, the history of the nineteenth century. Finally these potentialities come to be realized for what they are in themselves, apart from considerations of utility. The architecture of today stands at the end of such a process. Consequently, to understand it, we are obliged to survey in considerable detail developments in regions which seem far removed from aesthetic feeling.

Siegfried Giedion (1888–1968) was a Swiss historian of architecture (see Giedion, 1928 for full biography).

This long excerpt comes from *Mechanization Takes Command: A Contribution to Anonymous History*, published in 1948 and still in print today. As Giedion explained in the introduction, his purpose in this work was to take his investigation of the modern period to a deeper level, investigating the roots of mechanization and its effects on humans. The title of the book is sometimes misinterpreted to suggest that Giedion was simply advocating for mechanization, but the book is in fact a deeply critical study of those effects. He divided the work into seven parts: the first three review the 'anonymous history,' 'springs,' and 'means' of mechanization, then the rest of the book is devoted to 'encounters' between mechanization and the 'organic,' 'human surroundings,' the 'household,' and the 'bath.' It is heavily illustrated (not reproduced here), and uses a variety of arresting images, like the assembly line of hog carcasses, and work from contemporary artists, to bring mechanization to life.

This section is important for this collection in two respects. The first is Giedion's expansion of his earlier investigation of the representations of space-time by artists. In this passage, he looks at the evolving descriptions of motion as a key to understanding mechanization, drawing from the sciences and the arts. Second, the essay charts the emergence of the "factory as an organism," which means that the factory is understood as a series of interconnected processes, and equally that organisms have come to be understood as kinds of factories.

1948

Siegfried Giedion

The Assembly Line and Scientific Management

The assembly line[1] is one of mechanization's most effective tools. It aims at an uninterrupted production process. This is achieved by organizing and integrating the various operations. Its ultimate goal is to mold the manufactory into a single tool wherein all the phases of production, all the machines, become one great unit. The time factor plays an important part; for the machines must be regulated to one another.

More recently the assembly line has been brought under a broader heading: line production. 'Line production is characterized by the continuous regular movement of materials from the stockpile through the necessary stages of fabrication to the finished product. . . . Line production requires a rational layout and frequently, but not necessarily, involves the use of conveyor systems.'[2]

In what follows we shall usually employ the term assembly line, which has become almost a synonym of full mechanization.

Humanly and technically, the problem of the assembly line is solved when the worker no longer has to substitute for any movement of the machine, but simply assists production as a watcher and tester. This was done, quite suddenly, toward the end of the eighteenth century in Oliver Evans' mechanization of the grain-milling process. But in the large-scale manufacture of complicated machinery (automobile chassis), the fully automatic production line was not achieved until 1920.

In the transition phase, still predominant in industry, man acts as a lever of the machine. He must perform certain operations that are not yet carried out by mechanisms. True, the tempo of work is geared to the human organism; but in a deeper sense, the inexorable regularity with which the worker must follow the rhythm of the mechanical system is unnatural to man.

The growth of the assembly line with its labor-saving and production-raising measures is closely bound up with the wish for mass-production. We find it used shortly after 1800 for complicated products, such as the manufacture of biscuit in a victualling office of the British Navy, on a purely handicraft basis, i.e. without the use of machinery. A quite similar process was developed in the 'thirties in the great Cincinnati slaughterhouses, where without mechanical auxiliaries systematic team-work was introduced in the killing and dressing of hogs. The assembly line attitude is present before it can be applied in mechanized form to complicated machine processes.

The assembly line is based upon the speediest, most nearly frictionless transportation from each fabrication process to the next. Conveyor systems are employed to this end. It was Oliver Evans who first incorporated the three basic types of conveyor, as still used today, into a continuous production line.

Toward 1830 a new influence appeared: the introduction of railways. They aroused the imagination of the world. The rail and carriage seemed the most perfect means of transportation. Attempts were soon made to use them in the most diverse branches of industry.

In 1832 the patent was granted in France for a continuous oven 'in the form of a large circular track. The bread was taken around and baked in the course of the trip.'[3] It may be regarded as symptomatic. But in England during that decade, important inventions were made, based on the use of tracks and trolleys.

Among these is the traveling crane – apparently invented by Johann Bodmer in 1833 – which could move weights along a horizontal path high over-head. It was Bodmer who, as we shall see, laid down tracks within a Manchester factory, on which the material was moved in cars directly to the machines.

The horizontally traveling crane forms a step toward the overhead-rail systems that appeared on a large scale in American slaughterhouses of the Middle West during the late 'sixties and finally came into use in the mass fabrication of automobiles (Henry Ford, 1913).

The assembly line in the present-day sense was originally used in food processing when Oliver Evans first applied it in 1783 to grain-milling. In 1833 biscuit manufacture was mechanically performed in an English 'victualling office,' the baking trays being conveyed from machine to machine through the oven and back to their starting point on continuously moving roller-beds. In the late 'fifties the more difficult process of bread baking became mechanized at various places in England and America; and in America, at this time, even fruit was dried in steam chambers with the help of a conveyor by a now forgotten method (Alden process); in the late 'sixties overhead rails, in combination with various machines, are found in the great meat-packing houses of the Middle West.

Every detail of conveyor systems of interest to the engineer or the manufacturer possesses an almost endless literature – but one not very helpful for our purpose. The origin of the assembly line, its almost unnoticed growth through an entire century to virtual dictatorship over everything and everyone, is above all a historical, a human problem. Perhaps that is why we are so ill-informed about its growth. We have no broad survey of the subject, or, apparently, any article outlining the history of this pre-eminent tool of production.

Intimately connected with the assembly line is a problem that has slowly grown to importance since 1900: *scientific management*. Scientific management like the assembly line has much to do with organization. Very early in his experiments during the 'eighties, its founder, Frederick Winslow Taylor, was regulating the speed of various machines by means of individual drive, and became one of the first to use electric motors for this purpose. But of greater significance is scientific management's investigation of the way human work is performed.

Its development has led partly to the alleviation of labor, partly to heedless exploitation of the worker.

Its finest result was the new insight into the nature of work and motion arising from investigations such as Frank B. Gilbreth's. The way Gilbreth made visible the elements as well as the path of human motion is masterly both in the method and the boldness of its application. This aspect of the research, deeply probing the human element, will, we believe, prove the most significant in the long run.

› The continuous production line in the eighteenth century

Oliver Evans

What is most typical of American industry today, production in continuous flow, was a central preoccupation from the first. Before any American industry had come into being, and long before it was building complicated machines, a solitary and prophetic mind set about devising a system wherein mechanical conveyance from one operation to another might eliminate the labor of human hands.

In the last quarter of the eighteenth century, Oliver Evans (1755–1819)[4] built a mill in which the grain passed smoothly and continuously through the various milling processes without the help of human hand. It flowed in a smooth and continuous production line.

Oliver Evans introduced the endless belt and different types of conveyors, regulated to one another in all stages of production. The 'endless belt' (belt conveyor), the 'endless screw' (screw conveyor), and the 'chain of buckets' (bucket conveyor), which he used from the very start, constitute to the present day the three types of conveyor system. Later these three elements became exhaustively technified in their details, but in the method itself there was nothing to change.

In 1783 the model of the automatic mill was complete and in the two following years, 1784–85, the mill itself was built in Redclay Creek valley (figs. 44, 45). This mill could load from either boats or wagons; a scale determined the weight and a screw conveyor (or 'endless Archimedean screw' as Evans calls it) carried the grain inside to the point where it was raised to the top storey by a bucket conveyor (or 'elevator for raising vertically'). It handled three hundred bushels an hour. From this elevator, the grain fell on the mildly inclined 'descender – a broad endless strap of very thin pliant leather, canvas or flannel, revolving over two pulleys.' This belt was set in motion by the weight of the grain and, as Evans adds, 'it moves on the principle of an overshot waterwheel.' A prominent mechanical engineer remarks a century later: 'It is the prototype of belt conveyor of the present day, usually used for horizontal movement.'[5] After intervening operations, the grain was carried down to the millstones and from the millstones back to the top storey. Thus it made its way – which interests us here – through all the floors, from bottom to top and top to bottom, much as the automobile bodies in Henry Ford's plant of 1911.

People refused to believe it would work. How could human hands be thus suddenly superseded? In a rather obscure passage written as a footnote to one of his books twenty years later, Oliver Evans could not keep himself from this comment:

> The human mind seems incapable of believing anything that it cannot conceive and understand ... I speak from experience, for when it was first asserted that merchant flour mills could be constructed to attend themselves, so far as to take meal from the stones, and the wheat from the wagons and raise them to the upper storeys, spreading the meal to cool and gathering it by the same operations into the bolting hopper, etc. until the flour was ready for packing, the projector was answered: You cannot make water run up hill, you cannot make wooden millers.[6]

But the mill Oliver Evans built for himself and his partners on Redclay Creek, 1784–85, did work. The millers of the region came to look and 'they saw that all the operations of milling were going on without the care of any attendant – cleaning, grinding and bolting ... without human intervention.'[7]

On their return home they reported that 'the whole contrivance was a set of rattle traps unworthy the attention of men of common sense.'[8] But soon the economic advantages became clear. The mechanization of milling was soon accepted.

Oliver Evans obtained a patent in 1790 for his 'method of manufacturing flour and meal.' New difficulties arose. We shall return to them presently.

How did this invention come about?

Oliver Evans grew up in the country, in the state of Delaware. The ruins of his father's farmhouse are still standing. When he moved to the city, Philadelphia, then the leading center of culture in America, he was nearly fifty. Evans never went to Europe, and carried on no correspondence with the great scientific personalities of his day. He had to rely solely on his own powers. He dwelt in an agricultural land where farming was carried on by the most primitive methods. His reading was the popular textbooks on the basic laws of mechanics, the mechanics of solids and fluids. These laws, by then long taken for granted, became as new and exciting as they had been in Renaissance time. They took on a fresh vitality, as when an artist infuses with new plastic life objects that had become dull and commonplace.

This is no empty conjecture. Going through his book on the mechanization of the milling process, *The Young Millwright and Miller's Guide*,[9] one finds that about half of it deals with the laws of 'Mechanics and Hydraulics.' The reader can follow almost step by step as the simple theorems, the 'laws of motion and force of falling bodies, of bodies on inclined planes, the laws of the screw and circular motion,' are transformed into mechanical devices, whence is composed the mill that runs by itself, the mill without workers, the automaton.

The paddles of the mill wheel, whose laws of motion under the influence of water Evans studied, are changed into baskets, into buckets on an endless belt, carrying products from a lower to a higher, or from a higher to a lower level. The water on the paddles of the overshot wheel is changed to grain, moving and moving ever onwards; but it does not drive, it is driven.

The difficulties which arose, leading to quarrels and finally to a conflict with the Congress, came from the millers. Aware of the advantages of the mechanized mill, they did not wish to pay royalties to Oliver Evans, and later (1813) attacked the patent in a 'Memorial to Congress.' They asked for 'relief from the oppressive operations' of Oliver Evans' patent.[10] Thomas Jefferson was called in by an expert. His opinion of Oliver Evans' devices was low. He saw only the details, not the thing as a whole. 'The elevator,' he declared, 'is nothing more than the old Persian Wheel of Egypt, and the conveyor is the same thing as the screw of Archimedes.'[11]

If Oliver Evans' invention be split into its simple components, Jefferson is of course right. The chain of pots was used throughout the Ancient World – from Egypt to China – for raising water;[12] and the endless Archimedean screw, the screw conveyor, is found in almost every late Renaissance book dealing with machinery. In the Renaissance it served as a means of 'screwing' water from a lower to a higher level. Thus Agostino Ramelli used a series of Archimedean screws for the raising of water (fig. 43).[13] But Oliver Evans was, so far as we know, the first to use it horizontally, for the transportation of *solids*.

The Renaissance theorists are concerned with simple operations: they aim to raise a heavy load or to transmit force with the aid of lever, gear, or pulley systems. Their work sometimes may assume grandiose form, as when Domenico Fontana (1543–1607), architect, engineer, and town-planner to Sixtus V, lowered the Vatican obelisk on the south side of Saint Peter's, transported it, and raised it on its present site. In contrast to the clumsy proposals of his rivals, Fontana in 1586 used forty sets of windlasses driven by horses to swing the monolith around its center of gravity, while all Rome watched in silence.

All these were tasks of simple lifting and moving, a class to which modern cranes for the handling of coal, minerals, and other goods in harbors, factories, or freight yards also belong.

For Oliver Evans, hoisting and transportation have another meaning. They are but links within the continuous production process: from raw material to finished goods, the human hand shall be replaced by the machine. At a stroke, and without forerunner in this field, Oliver Evans achieved what was to become the pivot of later mechanization.

Evans' method had no analogy in its time. Yet nothing is harder for man than to frame ideas for the barely conceivable future; by nature we tend to approach all things by analogy, be it science and production methods, or emotional phenomena, as in art.

Arthur Schopenhauer once described talent as hitting a mark which ordinary man cannot reach; genius as sighting a point which others cannot even see.

Outwardly Oliver Evans' invention was, as his contemporaries said sneeringly, 'a set of rattle traps.' Moreover, Evans was not, like Benjamin Franklin, a master in dealing with men. Nothing favorable came of his other inventions, among which one at least is of a vision that takes us somewhat aback.[14]

His late successors were far more fortunate in achieving continuous line production. They had an advanced technology to work with and were assisted by an age bent upon nothing more strongly than production.

To the historian it does not matter whether a man joins the successful or the stranded, whether he himself carries his invention from conception to the last cogwheel, or has a staff of thousands of engineers to work out his ideas. What matters is the power of his vision. From this standpoint, Oliver Evans' invention opens a new chapter in the history of mankind.

› The beginnings of the assembly line

Oliver Evans disassembled a complex material (grain), mechanically fabricating a new product therefrom (flour). In the nineteenth century also the problem was to fabricate a product mechanically – a machine for instance. Here the parts are 'assembled' and combined into a new whole. But this is no rule. A whole is often disassembled into its parts as in Evans' mill (in mechanized slaughtering for instance). What marks this period is the unperfected state of the machinery. Men had to be inserted in the mechanisms, as it were, to ensure an uninterrupted production line.

The assembly line forms the backbone of manufacture in our time. The problems involved are no less deeply human than they are organizational and technical. Its slow growth is imperfectly known. In what follows, we shall take at random only a few cross sections from the nineteenth century.

From the nineteenth century on, the assembly line, beyond its labor-saving mechanisms, consists first and foremost in the rationally planned co-operation of groups, teamwork. This is achieved by the division of labor, which Adam Smith in the eighteenth century recognized as the basis of all industry, into tasks regulated to one another in regard to both time and succession.

Systematic beginnings of line production appear in hand methods before the advent of mechanization.

1804

Two decades after Oliver Evans' automatic mill, a human assembly line was established in an English naval arsenal to speed up the production of biscuits. The work was divided into various phases, and the hand operations of the different workers were timed to one another.

A source of 1804, the *Book of Trades*,[15] gives a clear account of this early form of assembly line. A work team of five bakers is to turn out seventy ships' biscuits a minute: twelve ovens; 'each will furnish daily bread for 2,040 men.'

The process of biscuit-making, as practiced at the Victualling Office at Deptford, is curious and interesting.

> The dough, which consists of flour and water only, is worked by a large machine. . . . It is handed over to a second workman, who slices them with a large knife for the bakers, of whom there are five. The first, or the *moulder*, forms the biscuit two at a time, the second, or *marker*, stamps and throws them to the *splitter*, who separates the two pieces and puts them under the hand of the *chucker*, the man that supplies the oven, whose work of throwing the bread on the peel must be so exact that *he cannot look off for a moment*. The fifth, or the *depositor*, receives the biscuits on the peel and arranges them in the oven. The business is to deposit in the oven *seventy biscuits in a minute* and *this is accomplished with the regularity of a clock*, the clacking of the peel, operating like the motion of the pendulum.

This biscuit bakery of the British Navy at Deptford seems to have been well known. More than thirty years later,[16] an observer still found it worthy of detailed description. His account adds nothing essentially new, but does give a more precise picture of the installation which already approximates the idea of the later assembly line: 'The baking establishment consists of two long buildings each divided into two baking offices with six ovens in each, which are ranged back to back. . . . The kneading troughs and kneeling boards are arranged round the outside walls of the building, one opposite each other.'[17]

About this time the 'hand process' came to be replaced by 'a highly ingenious piece of machinery.'

1833

The Superintendent of the Deptford Victualling Office, a Mr. Grant who had devised this 'highly ingenious piece of machinery,' thus brought about what was probably the first assembly line in the food industries. Only one operation, the removing of the dough from the kneader, was performed by hand. All other conveyance from operation to operation took place mechanically, on continuously rotating rollers.

> The arrangement of the several machines should be as near together as possible in order that the hoards may pass from one to the other on rollers. . . . [Doesn't this sound like a doctrine of Henry Ford?] A series of rollers should be fixed against the wall for the purpose of returning the boards to the first table after they have been emptied. At Portsmouth [England] this series of rollers was kept constantly revolving by the steam engine, so that when the empty boards were *placed upon any part of the line* they traveled upon the mixer without further attention.[18]

Several other fields show a like division of the production process into phases, as made famous by Adam Smith's account of the division of labor in a Birmingham needle factory. In the United States, where department stores had slowly been developing since the 'forties, ready-made clothes, in contrast to Europe, were produced from the start. This, before the introduction of sewing machines, led at an early date to a division of labor by teamwork, as in the English biscuit manufactory.

We shall look into one example only: the packing industry, for we shall presently meet with its later development. In Cincinnati, Ohio, where large-scale slaughterhouses originated, travelers as early as the 'thirties were reminded of Adam Smith's division of labor as they observed the slaughtering process and its organization.[19]

By 1837 the point seems to have been reached where, without machinery, a team of twenty men could kill and clean, ready for the cutting-knife, 620 pigs in eight hours.[20]

By mid-century, 'it was found economical to give each workman a special duty . . . one cleaned out the ears; one put off the bristles and hairs, while others scraped the animal more carefully. . . . To show the speed attained at Cincinnati in 1851 the workmen were able to clean three hogs per minute.'[21]

1839

Beginnings of flow work in the building of intricate spinning machinery are discernible in England around 1840. What was taking place in America at this time is still largely unknown. A Swiss inventor, Johann Georg Bodmer (1786–1864),[22] equipped a machine-tool factory which both by its layout and by the construction of its machines was to save movement, labor, and energy in conveyance. The principle given by Henry Ford in *My Life and Work* (1922), to 'place the tools and the man in sequence of operations,' was here followed to a surprising degree.

It was a sort of model workshop, for which almost everything was newly constructed. Nearly every machine was a patent. What improvements were made in the various machine tools are still accurately recorded in Bodmer's patent drawings.[23] Normally a patent runs to only a few pages, here one specification fills fifty-six pages, almost a mechanical catalogue: 'Tools, or Apparatus for Cutting, Planing, Drilling and Rolling Metal' and 'novel arrangements and construction of the various mechanisms.'[24]

Between 1830 and 1850 England was hard at work perfecting these machine tools. It was on this basis that the intensive industrialization proceeded in most branches between 1850 and 1890. What interests us in this connection is how far around 1830 the construction and disposition of machine tools, and means of conveyance, was aimed at a unified production line.

In construction: 'the large lathes being provided overhead with small *travelling cranes* fitted with pulley blocks for the purpose of enabling the workmen more *economically and conveniently to set the articles to be operated on in the lathes* and to remove them after being finished.'

'Small cranes were also erected in sufficient number within easy reach of the planing machines.'

In disposition: 'Gradually nearly all of these *tools* were constructed and were *systematically arranged in rows*, according to a carefully arranged plan.'

In conveyance: 'Several lines of rails traversed the shop from end to end for the easy conveyance on trucks of the parts of machinery to be operated upon. Such arrangements were not common in those days [1839] whatever may be the case now [1868].'[25]

In the first half of the nineteenth century and especially between 1830 and 1850, inventive minds appeared everywhere, measuring themselves against the most diverse problems of industry. Extreme specialization – except in the highly developed spinning machines – was far off. The times still offered the freshness of unfulfilled tasks. Bodmer was one of these versatile inventors; he worked on water wheels, steam engines, locomotives, machine tools, spinning machines; even on the mechanical production of beet sugar. But one problem occupied him from beginning to end: that of *conveyance within production*.

It began as early as 1815 when he built a mill for his brother in Zurich

> with some essential particulars . . . of a hoist of simple construction consisting in fact only of a large and broad flanched strap-pulley and ropedrum . . . the sacks of grain could be made to ascend or descend at pleasure and the operatives could pass from floor to floor by simply tightening or releasing the rope.[26]

Installing a small factory for the manufacture of textile machines in Bolton (1833), Bodmer constructed 'what is now called a *travelling crane*.'[27] It was, as Roe remarks, one of the first, if not the first traveling crane.[28]

Bodmer, like Oliver Evans, was much interested in the endless belt. He used it (1834) to convey heavier materials and to serve new ends. He was the first to employ it for continuous fueling. It was he who invented the traveling grate for boilers and furnaces,[29] 'to obtain the greatest possible degree in economizing pitcoal.'[30] Just as later in the assembly line the speed of the conveyor belt must be regulated to the pace of the worker, here the speed of the traveling grate is regulated to the rate of combustion. 'It was necessary to supply the furnace with fuel at a slow rate and continuously. These considerations led Mr. Bodmer to the adoption of a *traveling grate surface*.'[31] Having divided the rigid grate into movable sections, he goes on to make the most diverse suggestions for his chain grates, for traveling or propelling grates, drum-fire grates, thus broaching the domain of automatic stoking. He tried out a boiler stoked by his traveling fire grates in the Manchester machine-tool factory in 1839. After a while the experiment was dropped. It was still too early. Two decades later, about 1850, the endless belt found a place in American mechanical bakeries, to carry the loaves slowly and continuously through the oven, thus reviving the idea applied by Admiral Coffin in 1810, to which we shall return in the section on the oven and the endless belt.

Bodmer seems to have been still concerned with the problem of the traveling grate. One of his comprehensive patents (1843)[32] brings out further proposals in this field.

Bodmer, as we have mentioned, used his traveling cranes in close co-ordination with large lathes and planing machines in the Manchester machine-tool factory (1839); to this was added a rational arrangement of the machines, and the moving of the material on rails to the machine where needed.

Johann Georg Bodmer was a restless inventor of a type often found in his day. He was spurred from country to country, from invention to invention, as if he might thus bring the times to ripen with the pace of his ideas. He seems to have been sought out for advice,[33] but flourishing success he did not know. He finally died at the place whence he had set out, Zurich. The problem Bodmer attacked time and again, conveyance within production, yielded him true advances as early as 1830, toward the integrated management that was later to find its elaboration in the assembly line.

The 'sixties

The division of labor, which Adam Smith regarded as the pivotal point of industrial-ization after the mid-eighteenth century; Oliver Evans' sudden achievement of the continuous production line, 1783; the manufacture of ships' biscuits as organized in the Victualling Offices, 1804 and 1833; J. G. Bodmer's layout of a Manchester machine-tool factory, with traveling cranes and rails to convey the material to the convenient spot, 1839 – all these were steps toward the assembly line.

Despite scant knowledge of nineteenth-century anonymous history, we can, passing over many facts, tell when, why, and how the specific form of the present-day assembly line first appeared. This is no mere date. It marks the putting into practice of the dominant principle of the twentieth century: industrial production based on efficiency.

The present-day assembly line had its origins in the packing industry. It origin-ated there because many of its devices were invented in the late 'sixties and 'sev-enties, when slaughtering and its manifold operations had to be mechanized.

These inventions – kept in the Patent Office at Washington, and some of which we have chosen to illustrate the industrialization of slaughtering – proved, with few exceptions, unfit for practical use. They did not work; in the slaughtering process the material to be handled is a complex, irregularly shaped object: the hog. Even when dead, the hog largely refuses to submit to the machine. Machine tools for planing iron, undeviating to the millionth of an inch, could be constructed around 1850. Down to the present day, no one has succeeded in inventing a mechanism capable of severing the ham from the carcass. We are dealing here with an organic material, ever changing, ever different, impossible to operate upon by revolving cutters. Hence all the essential operations in the mass production of dressed meat have to be performed by hand. For the speeding of output there was but one solution: to eliminate loss of time between each operation and the next, and to reduce the energy expended by the worker on the manipulation of heavy car-casses. In continuous flow, hanging from an endlessly moving chain at twenty-four-inch intervals, they now move in procession past a row of standing workers each of whom performs a single operation. Here was the birth of the modern assembly line.[34]

This production line in slaughtering comes only in the third act, after the hog has been caught and killed, scalded and scraped.[35] It begins as soon as the carcass, with a gambrel through its hind legs, is switched onto the overhead rail, where, drawn by the endless chain, it is ready to be opened, all but beheaded, to be disemboweled, inspected, split, and stamped. This is the sole phase of slaughtering in which continuous line production could be carried out. The killing and the cleaning could not be done in complete mechanization. Neither, after a sojourn in the chilling room, could the fourth phase, the final dressing and cutting, be mechanically performed.

In the literal sense, Thomas Jefferson, who himself delighted in devices to open doors automatically or to convey bottles from his Monticello wine cellar, was right in stating that Oliver Evans' elevators and conveyors were known from Antiquity, from pre-Roman times. In the literal sense, too, there is nothing in the mechanism of the assembly line, or in the aligned workmen in the packing houses, that could not have been invented in Antiquity: a slaughtered pig hung on a moving chain and in some way aided by wheels or rollers required no new discovery and could equally well have been set up in one of the large Roman slaughterhouses. The devices themselves – influenced, it is true, by suspended or aerial railways – were extremely simple. What was revolutionary and what could not have been invented in earlier periods, in other countries, or even in other industries, was the way they were used to speed into mass production an organic material which defies handling by purely mechanical means.

All that remains, so far as we know, to bear witness for the early period, is a panoramic painting which the Cincinnati packers sent to the Vienna International Exhibition of 1873, and which, with some liberties in the disposition, as observed by *Harper's Magazine* in September of that year,[36] records the hog-slaughtering process through all its stages, from the catching of the pig to the boiling out of the lard. What interests us at the moment (the fuller description will be found in our chapter 'Meat') is a single phase in which lies the genesis of the assembly line. If one defines the assembly line as a work method wherein the object is mechanically conveyed from operation to operation, here is indeed its origin.

Despite careful investigation and the help of Cincinnati's local historians, no other pictorial evidence for the birth period of the assembly line could be found. It was explained, not too convincingly, that Cincinnati was at first ashamed to trace its wealth to pork packing. All the city's activities, its musical life for instance, can be accurately followed. But in tracing the first mechanization of the butcher's trade and the beginnings of the assembly line we have no foothold.

Thus far we can only speculate, as if we were studying some faintly known epoch that has left no documents. The hypothesis is that the assembly line arose in Cincinnati. Devices for use in connection with it, patented in the late 'sixties, stem from Cincinnatian inventors. They indicate that overhead rails fastened to the ceiling were not unusual at this time.

Over forty packing-houses were operated in Cincinnati in the 'fifties. The city remained the center of the industry down to the Civil War, and most of the patents lead back to it.

1869

The overhead rail systems in the great slaughterhouses ultimately led to the conveyor system, which did not reach full development until the following century. The track, high above head level, carries small wheeled trolleys which are either drawn by chains or rolled by their own weight down an incline. Invented by a Cincinnatian in 1869,[37] a hog-weighing device for pork-packing houses shows how overhead rails – as had appeared in J. G. Bodmer's traveling cranes by about 1830 – have now developed into whole railways. 'The hogs are transferred from drying-room to cutting-block by means of an *elevated railway.*'[38] The inventor plainly speaks of improving equipment already current: 'My improvement consists in providing the railway with a detachable section, that is connected to the weighing scale . . . the hog is *suspended from a carriage or truck* which is permitted to run down an inclined portion.'

The patent's well-thought-out overhead tracks, hoveringly suspended from the ceiling, betray that this is no longer a novelty. There had already been experimentation along these lines. In the 'fifties the thought of building an 'Elevated Railroad' over Broadway in New York was played with by engineers, 'A locomotive is to run on the rails and carries a suspended car which will pass between the space of the supporting arches'.[39]

› The appearance of scientific management

Around 1900

The position is clear. Competition is growing. Wage-cutting has proved impractical as a means of lowering production costs. The machine tools are at hand. They will become continually further differentiated and more specialized, but few real improvements seem likely to raise productivity.

The question is narrowing down to: What can be done *within* the plant to lower costs and raise productivity? Before the turn of the century, the attention of industrialists was being claimed not so much by new inventions as by new *organization*. Work in factories was computed by rule of thumb. Scientific methods should take the place of inventions. Hence the question: How is work performed? The work process is investigated, as well as each movement and the manner of its performance. These must be known to the fraction of a second.

In the last decades of the century, a number of men, often independently of one another, took up the problem of rationalizing operations within the factory. Beyond question it was the unremitting effort of Frederick Winslow Taylor (1856–1915) and his circle that, within a quarter of a century, laid the foundations of that ever-growing field they themselves named *scientific management*.

By 1880, when after two years as a worker Taylor became foreman in the Midvale Steel Company (Philadelphia), he resolved to investigate the work process through time studies. He recalled one of his school teachers who had used a stop watch to determine how long different pupils took to finish an exercise. As a youth

Taylor spent several years in Europe with his family; he received a high school education, and served an apprenticeship as a molder and tool maker in a small
Philadelphia factory. In 1878 he started as a worker in the Midvale Steel Company,
where he was promoted to foreman, master, and engineer, until in 1889 he began
reorganizing factories of various types. Meanwhile he had completed his engineering studies at night. His name was already known when, for three years, 1898 to
1901, he was in close collaboration with the Bethlehem Steel Works. This was his
most fertile period, both in production engineering and in invention, for it was at
this time that he made his discovery of high-speed steel. By around 1900, he had
developed his method of scientific management.

Taylor had already appeared in print, but it was not until 1906, some quarter of
a century after his first studies, that he read a paper on 'The Art of Cutting Metal' to
a group of engineers in New York, giving broad insight into his accomplishments in
the field with which he was most familiar.

The problem he deals with is the thorough analysis of a work process. Everything
superfluous must go, for the sake of efficiency and, as Taylor is ever stressing, for the
easing of labor, its functional performance.

Work should be done easily and so far as possible without fatigue. But always
behind this lies the constant goal to which the period was magically drawn: production, greater production at any price. The human body is studied to discover
how far it can be transformed into a mechanism.

Taylor once constructed a great steam hammer, whose parts were so finely calculated that the elasticity of its molecular forces served to heighten its efficiency.
The steam hammer 'was kept in its alignment by the elasticity of its parts which
yielded to the force of a blow and returned to their former positions.'[40]

Similarly does he proceed in the study of human efficiency: He approaches the
limit of elasticity. It has been frequently noted that he picked the best workers for
his experiments and fixed the task accordingly. The human organism is more
complex than the steam hammer, whose inner forces may be included in the reckoning. The body retaliates, though not always in an immediately recognizable
manner, when worked too long near the limit of its capacity.

Taylor's most important invention, high-speed steel, which he made in 1898 in
the Bethlehem Steel Works, also has to do with the exploration of a limit. When
tools were run at their top speed until they became red hot, they showed an 'extraordinary property of retaining what hardness they have. It turned out that at a
certain degree of heat [over 725° Fahrenheit], they kept the sharpness of cutting
steel as well as their "red-hardness," the greatest improvement taking place just
before the melting point.'[41]

The stretching of human capacities and the stretching of the properties of steel
derive from the same roots.

Organization proceeds thus: the managers pool their experience to survey
the field and if possible to recognize already known rules. The most capable
workers are chosen for experiments. By constant observation wrong or slow-
working methods are replaced by rational ones. This, Taylor says,[42] means a division of labor between the management and the operatives. One labor technician

in the planning or distributing office was often necessary for every three workers in the factory.

A methodical system develops, in the beginning at least, which Taylor himself calls the 'military type of organization.' 'As you know,' he said in his Harvard lectures (from 1909 on),

> one of the cardinal principles of the military type of management is that every man in the
> organization shall receive his orders directly through the one superior officer who is over him.
> The general superintendent of the works transmits his orders on tickets or written cardboards
> through the various officers to the workmen in the same way that orders through a general in
> command of a division are transmitted.[43]

Taylor and his successors do not want to command only. They provide for departments through which the worker himself can suggest improvements and share in the economies. The gifted workers may perhaps benefit, but the average man cannot escape automatization.

The hierarchy from general superintendent to worker, the soldierly discipline for efficiency's sake, doubtless offer industrial parallels to military life. But let there be no mistake: Taylorism and military activity are essentially unalike. The soldier indeed has to obey. But when under greatest stress, he faces tasks which demand *personal* initiative. His mechanical weapon becomes useless as soon as there is no moral impulse behind it. In the present situation where the machine is not far enough developed to perform certain operations, Taylorism demands of the mass of workers, not initiative but automatization. Human movements become levers in the machine.

The factory as an organism

Taylor organized industries of the most diverse types: steel mills, arsenals, ferro-concrete constructions, ball-bearing works. He would have his 'fundamental principles of scientific management' worked out in every sphere of life, in 'the management of our homes, farms, of the business of our tradesmen, of our churches, of our governmental departments.'[44]

The significance of his work lies in a further increase of *mechanical efficiency*. He is a specialist of the 1900 type: He conceives the object of his research – the factory – as a closed organism, as a goal in itself. What is manufactured in it and for what purpose are questions beyond his scope.

He owned shares in factories, received income from patents and his organizing work, but he seems never to have been tempted to become a big businessman himself. Taylor was eminently at home in the practical world. But by virtue of his analytical talent, his was one of those laboratory minds bound to the hardships and delights of research. By 1901, already having earned what he judged sufficient for his needs, he retired to devote himself wholly to his investigations.

Freud, the founder of psychoanalysis, by the exceeding penetration of his diagnostic and therapeutic methods, opened new access to the structure of the psyche. That F. W. Taylor was born in the same year as Freud, 1856, is of course coincidental. But a common trait of the scientific and artistic groups around the turn of the century was to employ an unprecedented sharpness of analysis in revealing the inside of processes.[45]

Space-time studies in scientific management

By the weight of all his energy Frederick Taylor opened the way for further elaboration of his method. Refinements soon appeared. There followed an alliance between scientific management. and experimental psychology. Independently of scientific management, psychology had already devised tests to determine the person best suited to certain occupations. The basis of these tests was the time taken to react to a given impression. These techniques had been developed in psychological laboratories. Hugo Muensterberg, a German psychologist who taught at Harvard, was among the first to survey the results of scientific management, then (1912) coming into its own, and to point out that from the psychological standpoint, it was still reckoning by rule of thumb.[46] Testing was also experimented with in America – Stephen Calvin on schoolwork.

Scientific management's approaching of psychology was connected with the giving up of Taylor's stop-watch methods. Frank B. Gilbreth (1868–1924) and his wife, the psychologist Lillian M. Gilbreth, often in collaboration, developed methods which led to a visual representation of the work process. Gilbreth began his studies while working as a large-scale contracting engineer in Boston. He investigated the best way of doing work – in industry and handicrafts indifferently.

The freshness and directness with which the age-old manipulations were observed – Taylor studied the coal-shoveler, Gilbreth the bricklayer – form perhaps the closest parallel to the functional improvement of such traditionary tools as the hammer, saw, spade, or plane in America from around 1830. Gilbreth gives us a step by step account of how without elaborate apparatus he proceeded in rationalizing the most traditional trade of bricklaying.[47] An adjustable scaffold for piling up the bricks was all he used. It did away with the workman's need to 'bend over and raise the weight of his body a thousand times a day,' thus almost tripling a man's daily output, from 1,000 to 2,700 bricks.

The method responsible for this was the study of motion. From the question: 'How long does it take to do a piece of work?' one came to a representation of the path and elements of a movement. Soon the stop watch was eliminated, to be replaced by objective recording apparatus. The Gilbreths were thus led deeper and deeper toward the inside of human motion and its visualization. This was accomplished through time and space studies.

Scientific management and contemporary art

Scientific management, like the assembly line, is deeply concerned with organization. But its most significant achievement is the study of the *human work process*, the way work is performed by the worker.

The purpose of research in scientific management is: 'Analyzing the motions of the workmen in the machine shop . . . all the operations for example which were performed while putting work into or taking work out from the machine.'[48]

This should eliminate unnecessary motions and reduce the time of an operation to a minimum. If we temporarily set aside all technical details and inquire into the essence of the methods employed, we find that they center around space-time studies. Their purpose was to determine the path of a motion through space and its duration in time.

In formulating the laws of mechanics, the physicists of the Renaissance investigated the relation between motion and time. The laws of human work are now investigated in a similar way, so that rough guessing and rule of thumb may yield to precise laws, so far as is possible in the human sphere.

What interests us here is the plunge *inside* the working process.

Frank B. Gilbreth succeeded in extending and refining the study of time and motion. 'time study,' he says in his popular *Primer of Scientific Management*, 'is the art of recording, analyzing, synthesizing the time of the elements of any operation.'[49]

Stop-watch methods were not precise enough for Frederick Taylor's successors. The stop watch is mute and can say nothing about how a motion is performed. The human eye is untrustworthy; reaction time varies with the observer. The form of the movement remains invisible and cannot be investigated. Gilbreth's problem was to portray its elements, to delineate its path.

In his earlier research the goal is not yet clear. His study of ferro-concrete building (1908) lays down some four hundred rules, a sort of military dispatch system as preferred by Frederick Taylor. New conceptions already announce themselves in his large square book, *Concrete System*. It is saturated with pictures illustrating the different phases: 'almost a stenographic report of what a successful contractor said to his workmen.'[50] But in his *Bricklaying System* of the following year, he clearly states what he wishes to inaugurate – an era of motion study. 'The motion study in this book,' he declares, 'is but the beginning of a motion study era.'[51]

The precise recording of movement, c.1912

It is not surprising that Gilbreth made use of the motion-picture camera as soon as it appeared in France. For further insight into the process of movement, he used a black background with a net of co-ordinates to ascertain the various phases.

But this was not a satisfactory solution. It did not make the trajectory of the movement clearly visible, and portrayed it only in conjunction with the body. To accomplish the separation, Gilbreth constructed a device of appealing simplicity. An ordinary camera and a simple electric bulb were all he needed to make visible the absolute path of a movement. He fastened a small electric light to the limb that performed the work, so that the movement left its track on the plate as a luminous white curve. This apparatus he called a 'motion recorder' – Cyclograph. The very form of the movement, invisible to the naked eye, is now captured. The light patterns reveal all hesitation or habits interfering with the worker's dexterity and automaticity. In a word, they embrace the sources of error as well as the perfect performance.

Later Gilbreth translated the image of the movement into models constructed of wire. These wire curves, their windings, their sinuosities, show exactly how the action was carried out. They show where the hand faltered and where it performed its task without hesitation. Thus the workman can be taught which of his gestures are right and which are wrong. For Gilbreth these models were a means of making the worker motion-minded. They revealed the character of the individual's own work. The worker might compare the record of his motions to the wire models, and correct his inefficiencies. Moreover, the gestures captured in wire a life of their own. It is no accident that modern artists sometimes turn to the same material in constructing their airy sculpture.

What followed Gilbreth's Cyclograph was but an elaboration of the method. The principle remains unchanged.

Frank B. Gilbreth investigated the forms of movement. It is not surprising that their trajectories became for him entities with independent laws.

He began to study the similarities of human activities. He believed 'that the skill in trades and in all forms of athletics, and even in such professions as surgery, is based on one common set of fundamental principles.'[52]

He made Cyclographs of champions in widely varying fields – champion fencers, champion bricklayers, expert pitchers, famous surgeons, and the champion oyster opener of Rhode Island – to find 'the points of similarity between their motions.'[53]

The light curves and the wire models reveal the motion in full plasticity. Motion acquires a form of its own and a life of its own. For eyes trained by contemporary art, there is a direct emotional appeal in these shapes, which the eye does not find in nature.

The light curves that visualize the movements of 'a girl folding a handkerchief', showing all the unconscious intricacies, belong to that type of phenomena in which the motion means everything, the object performing it nothing.

We have found no mention of Marey's work in Gilbreth's studies. But it matters little, for our purpose, whether Gilbreth had heard of it or not. Marey had recorded trajectories on a single plate, and mentioned that a Geneva scientist used incandescent lamps for the same purpose. Gilbreth, with his Chronocyclograph, was the first to give us intimate insight into the pure path, as well as the time element, of a movement.

Problems of motion presented themselves to scientists, to production engineers, and to artists. Independently they found similar methods of solving them. Unexpectedly, we encounter the same tendency in art and in scientific management as soon as the latter touches on absolutes and illuminates the structure of manual operations by penetrating the elements and the path of the motion.

The fact that a similarity of methods can be found arising unconsciously in such heterogeneous fields is among the most hopeful symptoms of our period.

This research takes a new starting point. It uses the time factor in making visible the elements of a motion. 'The timing . . . is done on the elements of the process.'[54] Space-time relations form the very basis of the method: Motion is dissected into phases so as to reveal its inner structure.

This characteristic is not limited to scientific management. It is deeply rooted in our epoch. About the same time the dissection of movement appears quite independently as an artistic problem in painting. From the standpoint of motion we can distinguish a close succession of two stages in contemporary art.

First, movement is dissected into separate phases so that the forms appear side by side or overlapping. This occurs around 1910.

The second stage makes the *form* of movement into an object of expression. Scientific management does this for purposes of analysis. In art, calligraphic forms are endowed with the power of symbols. This occurs around 1920.

The development continues into a third stage, of which we know only the beginning. During the 'thirties motion forms increasingly become a pictorial language to express psychic content.

Movement in successive phases, c.1912

The Italian Futurists attempted to represent movement in successive phases – Carlo Carrà with his 'Rattling taxi,' Giacomo Balla with his 'Dog on the leash' (1912).

The boldest handling of phase representation was Marcel Duchamp's 'Nude descending the staircase'. The sequence of movements – which the eye but summarily perceives – forms the starting point of the picture. From their succession, a new synthesis, a new artistic form emerges, giving representation to the heretofore unrepresentable: movement in its phases.

One easily recognizes in this picture the influence of the Futurists, of Archipenko's early sculpture with its hollowed forms, and of Cubism at its peak. Yet the question of influence is overshadowed not only by the masterful rendering, but by the more universal issue: What attempts are made from other sides to solve Marcel Duchamp's problem? What do the scientists have to say about it? Looked at it in this way, Duchamp's problem appears deeply interfused with the period. How early the physiologists showed interest in these problems, we have already seen. In his celebrated studies of the 'seventies on the motions of men and animals, Eadweard Muybridge set up a series of thirty cameras at twelve-inch intervals, releasing their shutters electromagnetically as soon as the moving object passed before the plate. Muybridge attempted – and from several sides simultaneously – to record the phases of simple movements such as rising, sitting, and walking downstairs. He thereby obtained a sequence of motion phases. Each picture showed the object in an isolated phase as arrested by each camera.

Etienne Jules Marey came closer to reconstructing the path of a movement from its phases. He used but a single camera and, in his research of greatest interest today, a single plate. At the beginning his movers wore white garments against a black background. But this resulted in an overlapping series. He therefore clad his model in black, with a bright metal strip extending along the feet and up the torso and arms. This gave a coherent motion sequence in which the forms no longer obliterated one another.[55] A half century later H. E. Edgerton invented the stroboscope, whose highly perfected technical equipment (radio interrupter) could freeze motion to the millionth part of a second. The problem was conceived by both Marey and Edgerton along lines that are methodologically similar.

Although Marcel Duchamp's 'Nude descending the staircase' created a sensation when exhibited at the New York Armory Show of 1913, it lay beyond the public's comprehension – a failure of understanding not limited to one place or one country. It is not enough to say that the American public was here making its first acquaintance of the new trends. The reason must be traced to the deeply ingrown fallacy that problems of feeling have nothing to do with problems of science, notwithstanding the fact that every true culture has taken it for granted that thinking and feeling are interdependent.

Movement in its own right, c.1920

In the second phase *the pure form of movement* becomes an artistic object in Its own right. It does not have to reproduce naturalistically an outside object. Every age has known the impact upon feeling of lines, curves, signs. All good ornament stands witness to this.

And this is no less true of movement in space; it too can be experienced as an absolute, likewise disengaged from the performer.

Is not the endless flow of movement in skating more significant than the body of the skater? As we watch a fireworks display, is it merely the luminous trajectory against the dark background that arrests us? Is it not rather the disembodied movement of the rockets through space that so appeals to our imagination?

What occurs in painting around 1920 is but the artistic extension of this faculty. For a work process to be understandable, it must be made visible; for he who performs it does not know his own movement. And this holds good for our subconscious processes.

These symbols of movement are spontaneous condensations, like the sound-poems of the Dadaists and, later, the Surrealists' quest for an 'automatic writing' (1924). A poet such as Paul Eluard confirms this (1939), as he comments on the 'integral truth' (*vérité totale*) sought by Picasso and every real artist of the time. 'Picasso has created fetishes, but fetishes possessing a life of their own. Not mere intermediary signs but signs in motion. Their motion makes them concrete things.'[56]

Signs in movement, movement in signs. Paul Klee, perhaps the boldest explorer of the subconscious, held that 'pictorial art springs from movement, is in itself interrupted motion and is conceived as motion.'[57]

Klee's *Pedagogical Sketchbook* emerges ever more clearly as a key to contemporary art. This pithily worded notebook summarizes his teaching at the Weimar Bauhaus. Here the master does more than teach; he admits us into the workshop. Klee's elucidation of an artistic problem comes astonishingly close to the frame-work of Oresme's thought. Nothing is static. A line, he begins, 'an activated line, a line moving freely along, is a stroll for strolling's sake. Its performer is a point in transit.'[58] Everything for him is the outcome of motion, even the circle, which, plastically speaking, seems to rest in itself – and which geometry defines as a curve whose points are at an equal distance from the center. For Klee the circle originates in the rotation of a pendulum. And from the circle Klee develops 'The Spiral', 1925, a spiral-head crowns the 'Queen of Hearts', 1921.

Surely it is no accident that in Klee's work first appeared the direction-pointing arrow as a rectangle headed by a triangle, a form that gained international currency. In his *Pedagogical Sketchbook* Klee explains in his way, which is at the same time symbolic and direct, the coming about of this form. Kandinsky's canvas 'Pink Square' (1923) is a cosmic storm, a cosmogony of shooting lines, of arrows, of planetary rings, and the figure '3' expanded in sickle form.

Drawings and lithographs, Paul Klee's favorite expression around 1920, offer the natural medium for rapidly executed and continuous motion. Very soon his motion symbols extend into the organic. A bold step it was in 1921 to form the image of man out of the symbols of movement, as if to portray him by the things he does and thinks.

A third stage announces itself, a development of which we know only the beginning: the form of movement becomes a means of expression in painting just as perspective had formerly been the means of expressing a specific content, an isolated scene. When motion rather than perspective is chosen as the means of expression, it yields instead of a static picture a dynamic one. The titles that Klee gives his pictures – 'Lady in the South,' 'Spinster,' 'Anchored,' 'Park with Birds,' 'Temple Reflected in Water,' 'Aging Couple' – might be titles of the static genre pictures of the ruling taste.

Here the same title stands for something quite different. Just as Gilbreth made visible the form and the true meaning of bodily movement, so Klee was able to give visible form to the innermost processes of the psyche. This perspective cannot do. The search now is for relations that are manifold, fluctuating, and far from static.

The whole picture becomes motion process.

Let us take a picture of Klee's late period – 'Aging Couple' for instance. In truth it hardly needs a title. It lives in its motion-form with a life of its own. As in a good Renaissance painting, its power stems not so much from its content as from distinguished handling of the means of expression. An eye not yet accustomed to the pictorial language based on motion process will at first see no more than this, and perhaps the striking interaction of the colors – brilliant yellow, brown, pink-violet, and green. Whoever has learned the pictorial language based on symbols will see portrayed all that is masklike, antagonistic, and evil in this 'Aging Couple' and how, in a single circuit, the movement embraces and binds the two faces. Without Picasso's pitiless surgical intervention and without his pathos, anatomy is made submissive to expression and movement. This is the year of 'Guernica.'

In less than two decades, art learned to use motion forms to represent psychic processes with lapidary cast and dynamic color. This may be the beginning of a third step, leading toward mastery of symbolic language free of atavistic reference.

In Joan Miro's painting, around 1924, appear signs, numbers, and serpentine curves. Their use is hesitant at first, haphazard and Dadaistic. Toward 1930 they gain in power. The faculty awakens in Miro to endow color, both by the shape it fills, and by its relation to the whole image, with a luminous quality bordering on the magical. Miro's forms, which used to flutter lightly like paper streamers through space, take on weight and definition. What was a boldness in 1921, when Klee lithographed his 'Queen of Hearts,' is now taken for granted. Personages, animals, erotic constellations, turn into signs, motion forms imbued with the force of symbols; and this artist of the post-Kleeian generation seems almost predestined to translate them into murals.

By signs and forms, the artists express the unknown within us, to interpret the winding paths of the mind as really and efficiently as motion-curves serve scientific management.

Both lie equally rooted within us, for movement and the symbols of movement become of one flesh with our being.

Forerunners, successors?

Charles Babbage

Do time and motion studies have any historical forerunners?

It was pointed out (1912)[59] that the early nineteenth and even the eighteenth century had known approaches to Taylor's method. Cited as chief witness was a disciple of Adam Smith, Charles Babbage, professor of mathematics at Cambridge. His book *On the Economy of Machinery and Manufacture* (Cambridge, 1832, and often reprinted) gives tables 'for the cost and time of each operation' in needle manufacture. Babbage quotes the tabulation of the Frenchman Perronet,[60] who in 1760 timed by the clock and computed the cost of each operation in the manufacture of 12,000 needles.

It would be straining the truth to regard these men as precursors of Taylor's method, or to suppose that they anticipated it. The using of a watch is an external. Babbage employs it only to make clear the advantages of the division of labor; it occurs in his chapter on that topic.

Taylor was perfectly right in giving the simple answer: 'Time studies began in the machine shop of the Midvale Steel Company and in 1881.'[61]

Babbage's time measurements were to show the advantages inherent in the division of labor. The time factor in scientific management serves to reveal the very elements of motion.

Charles Bedaux

Is the success of Charles Bedaux, mainly in the 1930s,[62] to be regarded as a further development of scientific management? Doubtless his 'close analysis and systematic observation of industrial operations' were taken over from Taylor and especially from Gilbreth, but the main purpose was to establish more perfect wage systems. Bedaux, who came to New York from France in 1911, said that he applied 'corrections for speed of performance.' To this end, he introduced a unit of human power similar to the 'dyn,' which the physicists use to measure mechanical work. Bedaux calls this unit a 'B.' And he defines it: 'A "B" is a fraction of a minute of work plus a fraction of a minute of rest always aggregating unity but varying in proportion according to the nature of strain.'[63] His 'B' forms the basis of a wage system which has aroused greater hostility among workers than any other measure in scientific management, since it can be used to exploit labor with unusual severity.

The aims have shifted. With Taylor and his successors the stress fell on analysis and organization of operations; with Gilbreth and the elucidation of human work processes by the visualization of movement, the human factor comes to the fore: elimination of waste motion, the reduction of fatigue, the training of the handicapped. With Bedaux, attention centers upon 'labor measurement,' on the wage scale. It stands for a much earlier conception of business enterprise. The suspicion of espionage under which he came, and his inglorious end during the Second World War, show Bedaux's methods in an even more crudely materialistic light.

› The assembly line in the twentieth century

1913–14
This is the time when Henry Ford brings the assembly line into the limelight of success. The assembly line was in full stride at Ford's Highland Park plant by 1915, the year of F. W. Taylor's death. Two methods overlap. Henry Ford does not mention Taylor; he is the self-taught man, who does everything by himself. The results Taylor had attained by decades of perseverance have become common knowledge. The instruction cards on which Taylor set so much value Ford is able to discard. The conveyor belt, the traveling platform, the overhead rails and material conveyors take their place. These are automatic instructions that work more efficiently than Taylor's written cards. Motion analysis has become largely unnecessary, for the task of the assembly-line worker is reduced to a few manipulations. Taylor's stop watch nevertheless remains, measuring the time of operations to the fraction of a second.

When the assembly line was introduced in Cincinnati and then in Chicago, over thirty years before Henry Ford, the stimulus arose in the mechanization of a manual trade, slaughtering. In this period, much experience was gathered regarding the speed at which the moving-line should travel and how the workers conducted themselves toward it. By 1900 conveyor systems were used even in department stores, but without affording continuous flow.

After 1900 the machine industry lapsed into that routine which leads to a crippling of the creative impulse. Its experience seems to have become irrevocably frozen into formulae. This is the period when experts appear with analogies, and argue the impossibility of all that lies beyond their routine. No one has written more amusingly of this than Henry Ford himself.[64] In such periods, every problem seems solved and every path trodden. Nothing remains of the morning freshness of the 'thirties, when a J. G. Bodmer could invent and construct from beginning to end both the machines and the tools with which to fabricate them. A new impulse could spring only from a new product, one that had to be created from the ground up: This around 1900 was the automobile.

Henry Ford's function is to have first recognized democratic possibilities in the vehicle that had always ranked as a privilege. The idea of transforming so complicated a mechanism as the motorcar from a luxury article into one of common use, and of bringing its price within reach of the average man, would have been unthinkable in Europe.

The belief that the automobile could be made an article of mass production, and from this conviction the complete revolutionizing of the manufacture of the product assure Henry Ford his historical position.

Like mass production in butchering, mass production of a new means of transportation, the automobile, became a stimulus for the assembly line, which from there spread to the inflexibly routinized machine factories.

'The Ford shop's assembling practice is to place the most suitable components on elevated ways or rails and to carry it past successive groups of workmen, who fix the various components to the principal component, until the assembling is completed.'[65] How this was carried out in Ford's Highland Park factory at Detroit in 1913–14; how, in April 1913, 'the first experiment of an assembly line on assembling the fly-wheel of a magneto'[66] was attempted; how the motor assembly was split into eighty-four different operations, taking a third of the former time; how the chassis was first placed on rails, operated by a rope and pulley, can be read in Ford's own book or in detailed accounts printed as early as 1915.[67]

To realize his conviction that the automobile must become a people's vehicle Henry Ford employs the means and the ideas of his time. He uses them like building-stones, often with fresh meaning, and simplifying them wherever possible. The assembly line supplants Taylor's motion studies and the yet more complex fatigue studies of his successors. The interchangeability of parts, already known in the field of agricultural machinery in the 'sixties for maintenance of the reaper, takes on another nuance in Ford's hands. He stresses its usefulness for the automobile: 'The machinery of today, especially that which is used in general life away from the machine shop, has to have its parts absolutely interchangeable, so that it can be repaired by non-skilled men.'[68]

He follows Taylor's method, unusual for the time, of so far as possible reducing working hours and raising wages. Here too the foreman retains his function. But when Taylor, in his famous experiments on shoveling, tells his laborers in the yard of the Bethlehem Steel Company: 'Pete and Mike, you fellows understand your job all right, both of you fellows are first class men, but we want to pay you double wages,'[69] he still is set upon raising production within the factory. Henry Ford goes further, and regards low wages as 'the cutting of buying power and the curtailment of the home market.'[70] Indeed Henry Ford views production and sales as a unit and, long before the high-pressure salesmanship of the 1930s, builds a world-wide organization to distribute his products. The efficiency of the sales force is as precisely worked out as the tempo of the assembly line.

A further broadening of the circle might take up the question: How has the automobile affected living habits? In what measure has it stimulated and in what measure has it destroyed? How far, then, is its production to be encouraged and to what extent curbed?

As a phenomenon, Henry Ford crystallizes anew the independent pioneering spirit of 1830 and 1860. In a period of elaborate banking and credit institutions, a period governed by the stock exchange, when the lawyers are needed at every move, Henry Ford trusts none of them and operates without banks.

In an age when anonymous corporations grow to giant proportions, he would exercise patriarchal power over his worker force, like a master over his journeymen. He would be independent of everyone in everything. He gathers in his own hands forests, iron and coal mines, smelting furnaces, rubber plantations, and other raw materials.

But just as great cities become increasingly ungovernable when they overgrow themselves, great industrial concentrations elude the patriarchal hand when they develop to the gigantic.

Ford did not have to spend his life, like Oliver Evans, furthering ideas ungrasped by his contemporaries. He may have had the same indomitable energy; but he also had the advantage of coming not at the start but at the end of the mechanistic phase. Success does not depend on genius or energy alone, but on the extent to which one's contemporaries have been prepared by what has gone before.

The assembly line too, as conceived by Henry Ford, forms in many ways the fruition of a long development.

The automatic assembly line, c.1920
Toward the end of the eighteenth century, Oliver Evans at one stroke achieved continuous line production, an automatic unit in which man acted only as an observer.

More than a century and a half later the curve gradually closes. Again we are approaching the point where a continuous production line, with man serving only as an observer, is the objective. Now it is not for the automatic grinding of grain, but for the building of complicated machinery, involving hundreds of different operations.

It is increasingly clear that the assembly line, as developed from the packing houses, through the automobile industry, and beyond, forms an intermediate stage: Man has still to perform whatever movements the engineer cannot yet delegate to the machine. Quite possibly this form of mechanical work will in some future day be pointed to as a symptom of our barbarism.

The impulse toward a new phase, the automatic assembly line, again has its starting point in the automobile industry. The reason is plain: for the first time an industry was faced with the problem of building a most complex mechanism by the million. A new scale was thus introduced.

After Ford's assembly line was in operation, L. R. Smith, Milwaukee manufacturer, raised the question (1916): 'Can automobile frames be built without men?'

'Its answers rested in the subconscious mind of engineers. We set out,' he says, 'to build automobile frames, without men. We wanted to do this on a scale far beyond that necessary to meet the immediate requirements of the automobile industry.'[71]

Now the question that could not be permanently evaded is raised, from within the industry, not from outside influences: 'It is highly probable that watching our workers do the same thing over and over again, day in and day out, sent us on our quest for the 100% mechanization of frame manufacture.'[72]

It was this often clairvoyant optimism which, at a time when the industry as a whole was producing no more than a million and a half automobiles a year, conceived, and within five years built, a single plant to produce more than a million yearly. 'A completed frame leaves the conveyor end, brushed and cleaned for the paint line, every ten seconds of the production shift. It takes ninety minutes from the strip of steel as received from the mill to the delivery of an enameled automobile frame into storage.'[73]

Here scientific management, so far as it is the analysis of human motions, is replaced by new tools of production. Five hundred engineers transform a factory into an automatic unit, producing more quickly, cheaply, and profitably, and freeing man from mechanical movement.

This automatic assembly line begins with an 'inspection machine,' which 'straightens and checks every piece of strip steel as received from the mill.'[74] The material is worked upon and moves back and forth through the factory on the most varied types of conveyor systems in an uninterrupted process. First in a sub-assembly line, often in parallel operations, the steel bars are cut, punched, and formed. A second group of machines assembles the various parts until they are finally clamped together in the general assembly line. 'Automatic rivet feeding heads swing into position and rivets are shot into the holes waiting to receive them. Air pressure accomplishes this task.'[75] The rivets are pressed into rows of automatic riveters with enormous jaws like the heads of mythical birds. Cleaning and painting follow.

Something of the 1830 spirit of Johann Georg Bodmer lives on in the way manufacturing tools – presses, riveters, conveyor systems – are freshly invented, constructed, and integrated. No longer is the individual machine alone automatized, as is usual in bulk manufacture. Here, extremely precise time charts guide the automatic co-operation of instruments which, like the atom or a planetary system, consist of separate units, yet gravitate about one another in obedience to their inherent laws.

The human aspect of the assembly line

It is not easy to take a historical view of recent periods, especially in so sensitive and ramified an aspect as the inquest into human work.

The assembly line and scientific management are essentially rationalizing measures. Tendencies in this direction extend relatively far back. But it was in the twentieth century that they were elaborated and became a sweeping influence. In the *second decade* (with Frederick Taylor as the central figure), it was scientific management that aroused the greatest attention: the interest of industry, the opposition of workers, public discussion, and governmental enquiries.[76] This is the period of its further refinement and of its joining with experimental psychology (Frank B. Gilbreth, central and most universal figure).

In the third decade (Henry Ford, the central figure), the assembly line moves to a key position in all industry. Its scope is constantly growing. In the time of full mechanization, the production engineer gained sway over manufactures of the most diverse types, seeking every possible opening in which an assembly line might be inserted. The forming of a more comprehensive picture would well reward the effort, for the assembly line becomes almost a symbol of the period between the two world wars.

Looking at the impact of mechanization on man, we must stress those aspects which bear upon man's very nature. We must sharply distinguish the impulse that gave rise to the assembly line and scientific management from the human repercussions. The impulse sprang from the epoch's imperious demand: production, ever-faster production, production at any cost. As soon as evaluation is called for, we find often diametrically opposed views: on the one side, a disgruntled worker; on the other, the enthusiastic promoter of the idea.

Taylor, 1912: 'After a long struggle, worker and employer regard one another as friends.'[77]
Complains the *worker*: 'driven at an inhuman pace by foremen picked for their brutality.'[78]

The *advocate of scientific management*, 1914: 'The speed boss does not drive the men at all. He is their servant. . . . The correct speed is the best speed at which the men can work day after day, year after year, and continuously improve in health.'[79]
Complains the *worker*: 'There was never a moment of leisure or opportunity to turn my head. . . . The men have no rest except for fifteen or twenty minutes at lunch time and can go to the toilet only when substitutes are ready to relieve them.'[80]

These are personal utterances picked at random. The unions were hostile to scientific management. Trade-union organization was late in permeating the United States. In the Bethlehem Steel Works, for example, in which Taylor carried out his famous shovel and high-speed steel experiments, 'not a single employee was a member of a union' even ten years later (1910).[81] The unions saw their tactics endangered 'through building up loyalty to the management,'[82] and above all, they saw scientific management as a new means of exploitation.

Later a change in trade-union policy led to reformulation of the program. 'Labor is fully conscious that the world needs things for use and that the standards of life can improve only as production for use increases. Labor is conscious to work out better methods of industry.'[83]

Not to be overlooked are those aspects which have to do with the class struggle. They, however, lie outside the actual problems of this book, whose task is to describe the impact of a mechanized world on the human organism and on human feeling.

In a Chicago packing house, hogs, hanging head downwards, moved uninterruptedly past a staunch Negro woman at the curve of the conveyor system. Her task was to stamp, with a rubber stamp, the carcasses examined by the inspectors. With a sweeping movement she smacked the rubber stamp on each skin.

Perhaps we start from false premises; but in an outside observer a strange feeling was aroused: a creature of the human race trained to do nothing else but, day after day, and eight hours each day, stamp thousand after thousand of carcasses in four places.

Henry Ford tells (1922)[84] of a worker who had to perform a particularly monotonous task, actually one single motion of the hand. At his request, he was moved to another position, but after a few weeks he asked to be put back in his old job. Here Henry Ford hits on a phenomenon known to every urbanist who has slum-dwellers to resettle: No matter how primitive and unsanitary conditions may be, a certain number will always be found who refuse to leave their slum for new houses, and who prefer by far their old and familiar conditions.

The modern assembly line as it appears, probably for the first time, in the packing houses of Cincinnati, and certain measures of scientific management, which use man as part of an automatic process, are transitional phenomena, prevailing only so long as machinery is unable to perform certain operations of its own accord.

A document that translates the human response to this phase into artistic symbols is Charlie Chaplin's film *Modern Times*.[85] When the picture was first shown in New York, in February 1936, a radical periodical took the attitude: 'What his political views are I don't know and I do not care.'[86] The decisive point in this document is the revolt against subordination to the machine.

It is the story of an individual who, eight hours a day, year in and year out, must perform the same motion, and for whom the whole world is transformed into nuts to be turned by his wrench. The monotony and compulsion of the high-speed conveyor belt destroy his mental balance. 'The mechanized individualist goes mad and proceeds to turn the factory into the madhouse that it really always has been.'[87] He loosens dangerous screws that accelerate the assembly line to a mad pace. In the nose of a foreman, the buttons of an office girl, the breasts of a fat woman, everywhere he sees nuts that have to be tightened. By grotesque exaggeration the human core of the problem is revealed. What is this automatism, this reflex movement of screw-tightening, but the observation which can be made every day upon the workers streaming from the factories, who have the machine in their very gait?

It is the ceaseless mechanizing drive that leads Chaplin to invent the eating machine, which feeds the worker automatically without loss of time; he does not need to stop for lunch, and the assembly line goes on.

All this, though intensified to the grotesque, has a glint of that inner truth which dwells in Shakespeare's comedies.

True, the eating machine is rejected by the manager as too complicated. But a few years later, does not reality begin to approach that symbol of eating in factory tempo? At lunch counters, do not endless belts carry hot plates from kitchen to customer? In drug stores and in the basements of 5- and 10-cent stores do not counter after counter wind like mountain paths to feed as many men as quickly as possible?

The assembly line and scientific management can be put to work within quite opposite economic systems. Their implications, like those of mechanization as a whole, are not unilaterally tied to any one system. They reach into the depths of a basic human problem – labor – and the historical verdict will depend on how far one may expect the human being to become part of an automaton.

Before these methods had come into being, the Reverend William Ellery Channing, one of the great New England preachers of the 1830s, formulated with finality the problem of the assembly line and of any purely mechanical use of man: 'I do not look on a human being as a machine, made to be kept in action by a foreign force, to accomplish an unvarying succession of motions, to do a fixed amount of work, and then to fall to pieces at death. . . .'[88]

1   The term 'assembly line' is of recent date. Only in the supplement of the *Oxford English Dictionary* (1933) was there added this further meaning of assembly: 'The action or method of assembling a machine or its parts' (1897); the *assembly line* is not listed; *assembly-room*, however, is defined as 'a room in a workshop where the parts of some composite articles are assembled.' An American source of 1897 is referred to.

2   As defined in *Wartime Technological Developments*, U.S. Senate, subcommittee monograph No 2, May 1945, p. 348.

3   Aribert's patent.

4   Particulars on the inventor's life and work will be found in the painstakingly documented work of Greville and Dorothy Bathe, *Oliver Evans*, Philadelphia, 1935.

5   Coleman Sellers, Jr., 'Oliver Evans and His Inventions,' in *The Journal of the Franklin Institute*, Philadelphia, vol. xcii (1886), p. 4.

6   In a short history of the steam engine in *Young Steam Engineer's Guide* (Philadelphia, 1804), where he compares himself with the Marquess of Worcester, adding the above passage.

7   Coleman Sellers, Jr., op. cit. p. 2.

8   Ibid.

9   *The Young Millwright and Miller's Guide*, Philadelphia, 1795, with an appendix on business management by his partner Elincott, also translated into French, saw fifteen editions up to 1860. They have been carefully collated by Greville Bathe. The book was used as a standard work for more than half a century.

10 G. Bathe, op. cit. pp. 189–90.

11 Ibid. p. 91.

12 Recent studies have noted from a drawing of Pieter Breughel that the chain of pots 'were used in Holland as dredgers during the digging of a canal in 1561.' Zimmer, 'Early History of Conveying Machines,' in *Transactions of the Newcomen Society* (London, 1924–25), vol. 4, p. 31.

13 Agostino Ramelli, *Le Diverse et Artificiose Machine Del Capitano Agostino Ramelli*, A Parigi, 1588.

14 We are thinking not so much of his 'amphibious digger,' a steam-dredging machine for cleaning the docks of the city (1804, cf. G. Bathe, op. cit. p. 108), or of his high-pressure steam engine, as of the astonishing precision with which he laid down a method for mechanical ice making that remained current during a half century.

15 *The Book of Trades, or Library of the Useful Arts*, London, 1804, pp. 107–8. The first American edition of this source appeared in Philadelphia, 1807.

16 Peter Barlow, *Manufactures and Machinery in Britain*, London, 1836.

17 Ibid. p. 801.

18 Ibid. p. 804.

19 Harriet Martineau, *Retrospect of Western Travels*, New York, 1838, vol. 2, p. 45. Quoted in R. A. Clemen, *The American Livestock and Meat Industry*, New York, 1923.

20 R. A. Clemen, *The American Livestock and Meat Industry*, New York, 1923.

21 Ibid. p. 121.

22 The rediscovery of Johann Georg Bodmer in our time is due to J. W. Roe, who in his book *English and American Toolbuilders*, New Haven, 1916, pp. 75–80, accords Bodmer the place he deserves. He bases his article on the *Minutes of the Institution of Civil Engineers*, London, 1868, xxviii, 573ff., which shortly after Bodmer's death printed a detailed memoir ending with an eight-page list of his patents.

23 British Patent No. 8070, A.D. 1839 – British Patent No. 8912, A.D. 1841.

24 British Patent No. 8070, A.D. 1839, p. 2.

25 The best picture is given by the untouched contemporary account, Institution of Civil Engineers' memoir on Bodmer, op. cit. p. 588.

26 Institution of Civil Engineers' memoir on Bodmer, op. cit. p. 579.

27 Ibid. p. 581.

28   J. W. Roe, op. cit.

29   British Patent No. 6617, A.D. 1834.

30   Institution of Civil Engineers, op. cit. p. 584.

31   Ibid.

32   British Patent No. 9899, A.D. 1843. The specification runs to seventeen pages.

33   'It was about this time [1834] that the formation of a railway between London and Birmingham was contemplated. One of the directors invited Mr. Bodmer to give his views as to the
best system of carriages. On this occasion Mr. Bodmer proposed the construction of carriages since adopted in the USA, in parts of Germany, in Switzerland, and the distinctive
feature there is a longitudinal passage through the middle of each carriage, so that the guard
can pass from one end of the train to the other with greatest ease and security.' Minutes of
the Institution of Civil Engineers, op. cit. p. 585.

34   It matters little that the process here is one of disassembling, not of assembling, as in the
automobile industry. The method of mass production which forms the common denominator
is what counts.

35   Cf. pp. 228–9.

36   *Harper's Magazine*, 6 Sept. 1873, p. 778.

37   T. Morrison, Hog Weighing Apparatus, U.S. Patent No. 92,083, 29 June 1869.

38   Ibid.

39   *The Scientific American* (New York), vol. ix, Pt., 15 Oct. 1853.

40   *Iron Age*, New York, vol. 96, p. 1029.

41   Frank Barklay Copley, *Frederick W. Taylor, Father of Scientific Management*, New York, 1923,
vol. 2, p. 84. The term 'red-hardness' from F. W. Taylor, *The Art of Cutting Metal*, New York,
1906, p. 223.

42   Compare Taylor's basic publication *Shop Management*, 1903, and *Principles of Scientific
Management*, 1911.

43   F. B. Copley, op. cit. vol. 2, p. 213.

44   F. W. Taylor, *The Principle of Scientific Management*, New York, 1911, p. 8.

45   Freud published his studies on hysteria at the same time (1895) as Taylor was delivering his
first lectures to American engineers.

46   Hugo Muensterberg, *Psychology and Industrial Development*, Boston, 1913. This book was
important also for Muensterberg's experiments for the improvement of electric railway and telephone service, the examination of ship's officers not able to meet emergencies, and for his
research in the field, much further developed since, of advertising, display, and salesmanship.

47   Frank B. Gilbreth, *Bricklaying System*, New York, 1909.

48   F. B. Copley, op. cit. vol. 1, p. 223.

49   F. B. Gilbreth, *Primer* of Scientific *Management*, New York, 1914, p. 7.

50   Gilbreth, *Concrete System*, New York, 1908.

51   Gilbreth, *Bricklaying System*, New York, 1909, p. 140.

52   Frank B. and Lillian M. Gilbreth, *Motion Study for the Handicapped*, London, 1920, p. 15.

53   Ibid. p. 16. 'A prominent surgeon,' writes Gilbreth in connection with one of his experiments,

is perfectly willing to be photographed performing a delicate operation but when the fact
is mentioned that this is being done to find the similarity between his actions and other
skilled workers, he becomes scornfully incredulous. How can such a thing be? He, a skillfully trained, highly developed product of long years of study to be likened to a bricklayer!

With the same contemptuous incredulity, a well-known physicist rejected the idea of a relation between the methods of present-day physics and the methods of contemporary art.

54   Frank B. and Lillian M. Gilbreth, *Motion Study for the Handicapped*, London, 1920, p. 7.

55   E. J. Marey, *La Méthode graphique dans les sciences expérimentales*, with appendix:
'Développement de la méthode graphique par l'emploi de la photographie,' Paris, 1885, p. 34.

56   Paul Eluard, *Picasso*, London Bulletin 15, 1939.

57   W. Grohmann, *The Drawings of Paul Klee*, New York, 1944.

58   Paul Klee, *Pedagogical Sketchbook*, first issued as second *Bauhausbuch* under the direction of Walter Gropius and L. Moholy-Nagy, English ed. New York, 1944.

59   By the sub-committee on Administration of the American Society of Engineers.

60   Babbage, op. cit. p. 146.

61   Copley, op. cit. vol. 1, p. 226.

62   The Bedaux Company, *More Production, Better Morale*, A Program for American Industry, New York, 1942. In 1912, 720 corporations with 675,000 workers adopted the Bedaux system.

63   Charles Bedaux, *Labor Management*, a pamphlet, New York, 1928 (many subsequent editions).

64   Henry Ford, *My Life and Work*, New York, 1922, p. 86.

65   Horace Lucien Arnold and Fay Leone Fanrote, *Ford Methods and the Ford Shop*, N. Y., 1915, p. 102.

66   Henry Ford, op. cit. p. 80.

67   Arnold and Fanrote, op. cit.

68   Henry Ford, *Moving Forward*, New York, 1930, p. 128.

69   Copley, op. cit. vol. 2, p. 58.

70   Henry Ford, *My Life and Work*, chapter on wages.

71   L. R. Smith, 'We Build a Plant to Run without Men,' *The Magazine of Business*, New York, February 1929.

72   Ibid.

73   Ibid.

74   Sidney G. Koon, '10,000 Automobile Frames a Day,' in *The Iron Age*, 5 June 1930.

75   Ibid.

76   *Hearings before special committee of the house of Representatives to investigate Taylor's and other systems of Shop Management.* 3 vols. Government Printing Office, 1912.

77   *Bulletin of the Taylor Society*, June–August 1912, p. 103.

78   Robert L. Cruden, *The End of the Ford Myth*, International Pamphlets no. 24, New York, 1932.

79   Gilbreth, op. cit. p. 65.

80   R. L. Cruden, op. cit. p. 4.

81   Drury, *Scientific Management*, New York, 1915, p. 176.

82   Ibid. p. 175.

83   Ibid. p. 27.

84   Henry Ford, *My Life and Work*, in the chapter on 'The Torture of the Machine.'

85   Chaplin worked for five years on this silent film. He began in 1931 at the time when René Clair in *A Nous la Liberté* brought the endless belt and the mechanized man into the film. But a somewhat primitive romanticism and too superficial comparisons – prison life and assembly line – destroy the symbolic force of Clair's satire.

86   *New Masses*, 18 Feb. 1936, vol. 18, no. 6.

87   *Herald Tribune*, New York, 7 Feb. 1936.

88   Rev. William Ellery Channing, *Self Culture*, Introductory address to the Franklin Lectures, delivered at Boston, Sept. 1838.

The German architect Ludwig Mies van der Rohe (1886–1969) was a prolific architect, and a significant educator, both in Germany, and, later, in the United States. After working under Peter Behrens (1868–1940), Mies established his own office, working as a neo-classicist. After the First World War, Mies became taken with the avant-garde, abandoning the ornament of his earlier work in favor of a skin and bones approach. In 1921, Mies produced his most daring proposal with the unrealized glass skyscraper project. This was followed, in 1929, with the German Pavilion for the Barcelona exhibition, and Villa Tugendhat, in 1930. In the 1920s, Mies was also associated with G magazine, and was architectural director of the Deutscher Werkbund, for which he organized the influential Weißenhof Siedlung prototype housing fair. He was also the last director of the Bauhaus, seeing it move to Berlin, and eventually be shut down by the state, precipitating his 1937 move to the USA.

When Mies arrived in the USA, he was already a mature architect with an international reputation. Upon his arrival, he was made director of the Illinois Institute of Technology, on the condition that he design its new campus. While in the United States, Mies would revolutionize architectural technology, designing the first steel and glass curtain-wall building with 860–880 Lake Shore Drive, completed in 1952. Other significant works of the later part of his career include buildings in Chicago – the Farnsworth House, IBM Plaza, the Federal Building – the Seagram Building in New York, the TD Centre in Toronto, Westmount Square in Montréal, and the Neue Nationalgalerie in Berlin.

"Technology and Architecture" was a speech delivered at the Illinois Institute of Technology in 1950. In it, Mies presents technology as both method and thing in itself. As a thing, technology has a history and form, and is itself generative of meaning. Thus technology, given the opportunity, transcends itself to become an expression of the spirit. In doing so, it expresses itself through the components of its own construction.

1950

Ludwig Mies van der Rohe

Technology and Architecture

Technology is rooted in the past. It dominates the present and tends into the future. It is a real historical movement – one of the great movements which shape and represent their epoch. It can be compared only with the Classic discovery of man as a person, the Roman will to power, and the religious movement of the Middle Ages. Technology is far more than a method, it is a world in itself. As a method it is superior in almost every respect. But only where it is left to itself, as in gigantic structures of engineering, there technology reveals its true nature. There it is evident that it is not only a useful means, but that it is something, something in itself, something that has a meaning and a powerful form – so powerful in fact, that it is not easy to name it. Is that still technology or is it architecture? And that may be the reason why some people are convinced that architecture will be outmoded and replaced by technology. Such a conviction is not based on clear thinking. The opposite happens. Wherever technology reaches its real fulfilment, it transcends into architecture. It is true that architecture depends on facts, but its real field of activity is in the realm of significance. I hope you will understand that architecture has nothing to do with the inventions of forms. It is not a playground for children, young or old. Architecture is the real battleground of the spirit. Architecture wrote the history of the epochs and gave them their names. Architecture depends on its time. It is the crystallization of its inner structure, the slow unfolding of its form. That is the reason why technology and architecture are so closely related. Our real hope is that they will grow together, that some day the one will be the expression of the other. Only then will we have an architecture worthy of its name: architecture as a true symbol of our time.

Team 10 was an intentionally loose affiliation of young architects that developed within CIAM (Congrès Internationale d'Architecture Moderne) in the 1950s in rebellion against the institutionalization of that group. Team 10 received its name from the planning committee for CIAM X in Dubrovnik in 1956, which ultimately initiated the dissolution of CIAM. The core group included Jaap Bakema, George Candilis, Giancarlo de Carlo, Aldo van Eyck, Alison and Peter Smithson, and Shadrach Woods, whose common cause was a rejection of the "functional hierarchy of the 'Charte d'Athenes'", the agreement formed at the fourth CIAM meeting in 1933.

The "Team 10 Primer" was a collection of "articles, essays, and diagrams which Team 10 regard as being central to their individual positions." It was first compiled for students by Alison Smithson in 1962, then further enhanced and formally published in 1968. The Doorn Manifesto itself consisted of eight points and was reprinted in the Primer with the elaboration added by Alison Smithson.

The influence of Patrick Geddes is explicit both in the characterization of city-environment interaction and in the sociological categories and techniques of categorization utilized by the group. The group deployed many techniques to make sense of the complex "ecology of the situation," often characterized by Peter Smithson's network diagram of a Brubeck performance. In a section of the Primer drawn from Alison and Peter Smithson's 1960 publication *Uppercase 3*, they cited their goal "to realize the implication of flow and movement in the architecture itself."

1954/ 1962

Team 10

The Doorn
Manifesto

1   It is useless to consider the house except as a part of a community owing to the inter-action of these on each other.
2   We should not waste our time codifying the elements of the house until the other relationship has been crystallized.
3   'Habitat' is concerned with the particular house in the particular type of community.
4   Communities are the same everywhere.
    (1) Detached house-farm.
    (2) Village.
    (3) Towns of various sorts (industrial/admin./special).
    (4) Cities (multi-functional).
5   They can be shown in relationship to their environment (habitat) in the Geddes valley section.
6   Any community must be internally convenient – have ease of circulation; in consequence, whatever type of transport is available, density must increase as population increases, i.e. (1) is least dense, (4) is most dense.
7   We must therefore study the dwelling and the groupings that are necessary to produce convenient communities at various points on the valley section.
8   The appropriateness of any solution may lie in the field of architectural invention rather than social anthropology.

It had become obvious that town building was beyond the scope of purely analytical thinking – that the problem of human relations fell through the net of the 'four functions'. In an attempt to correct this, the Doorn Manifesto proposed: 'To comprehend the pattern of human associations we must consider every community in its particular environment.'

What exactly are the principles from which a town is to develop? The principle of a community's development can be derived from the ecology of the situation, from a study of the human, the natural, and the constructed, and their action on each other.

If the validity of the form of a community rests in the pattern of life, then it follows that the first principle should be continuous objective analysis of the human structure and its change.

Such an analysis would not only include 'what happens', 'the organisms' habits', 'modes of life and relations to their surroundings', such things as living in certain places, going to school, traveling to work and visiting shops, but also 'what motivates' the reasons for going to particular schools, choosing that type of work and visiting those particular shops. In other words, trying to uncover a pattern of reality which includes human aspirations.

The social structure to which the town-planner has to give form is not only different but much more complex than ever before . . .

Richard Neutra (1892–1970) was born in Vienna, studied with Adolph Loos, and then worked for Karl Moser and Erik Mendelsohn. He emigrated to the United States in 1923, and worked for Frank Lloyd Wright in 1925 before moving to Los Angeles to collaborate with Rudolf Schindler. He built the Lovell "Health House" in 1929, which was published in Hitchcock and Johnson's *International Style Exhibition* in 1932 and established his reputation as a modernist architect. A decade later he even won the commission to design a house in Palm Springs for the Kaufmans, the clients of Wright's famous Fallingwater.

Shortly after the completion of the Lovell house, Neutra built his own office and residence, partly funded by the Dutch industrialist C. H. van der Leeuw and called the VDL Research House. It was a cleverly arranged house on a small site, which involved the exploration of Neutra's parallel interests in progressive building technology and "applied biology." Like Kiesler, Neutra understood that function was not a simple concept and required a full assessment of human health. The environmental and biological aspects of his work became more explicit in the years after the Second World War and in 1954 he wrote *Survival Through Design*, followed in 1971 by *Building with Nature*.

This excerpt on Performance Guarantees provides two useful contributions to the changing notion of technology in architecture. The first is an historical review and critique of the complex notion of function, with the development of Neutra's elegant criteria, the "*index of livability*." The second is his consideration of technology and building as a continuation of "organic evolution." That second aspect leads him to the question of "the living environment" and the manner in which building material and process had become industrialized and globalized as exemplified by Sweet's annual building material catalog.

1954

Richard Neutra

Survival Through Design

› The NATURAL ENVIRONMENT IS DOCTORED UP CONTINUOUSLY and warped by the acts of the human brain

Nature has too long been outraged by design of nose rings, corsets, and foul-aired subways. Perhaps our mass-fabricators of today have shown themselves particularly out of touch with nature. But ever since Sodom and Gomorrah, organic normalcy has been raped again and again by man, that super-animal still struggling for its own balance. There have been warners, prophets, great floods, and new beginnings.

What we here may briefly call nature comprises all the requirements and characteristics of live organisms. This entire world of organic phenomena is, in the escapades of our still obvious immaturity, often treated against 'the natural grain' and contrary to the 'supreme plan' – that of biological consistency and requirement. In former ages it was a sin to do this and for such failings the deity threatened to liquidate the sinners. We may now have dropped – perhaps too carelessly – the moral accent. Yet to us, too, the issue is still one of survival by virtue of wholesomeness, or damnation and death through our own default.

In human design, we could conceivably see *organic evolution continued*, and extending into a man-shaped future. At any rate, that phenomenally intensive development in the multi-layered cortex of the human upper brain has not yet with certainty been proved a blind alley or a dismal failure. To be sure, this distinctly human brain harbors trouble, but it also may furnish some as yet untried survival aids. We have been laggards in calling upon all our potential powers and resources to arrange for us in a bearable manner an individual and communal living space. The toxic trash piles of our neglects and misdeeds, old and fresh, surround us in our physical environment. The confused wreckage of centuries, unrelated to any current practical purpose, is mixed in a most disturbing manner with our often feeble, often arbitrary, attempts at creating order.

Organically oriented design could, we hope, combat the chance character of the surrounding scene. Physiology must direct and check the technical advance in constructed environment. This setting of ours is all powerful; it comprises everything man-made to supply man, from the airy storage compartment of our toothbrush to the illumination of a speedway interchange, or of the neighborhood day-care center for toddlers.

A great deal of what has been vaguely called beauty will be involved in this proposed new and watchful scrutiny of man-made environment. It will come into question perhaps far more often than anybody could imagine in our current drab disorder. But the sort of beauty we speak of here will have given up its now too precarious grounds of self-defense. Designers will recognize that gradually but surely they must underbuild their proposals and compositions with more solid physiological foundations rather than with mere speculative conversation or sales talk. An eternal residuum of mystery may always lie deeply buried in this field, and yet the realm of research, testing, and provability increases from day to day.

All our expensive long-term investments in constructed environment will be considered legitimate only if the designs have a high, provable *index of livability*. Such designs must be conceived by a profession brought up in social responsibility, skilled, and intent on aiding the survival of a race that is in grave danger of becoming self-destructive.

Design is the cardinal means by which human beings have long tried to modify their natural environment, piecemeal and wholesale. The physical surroundings had to be made more habitable and more in keeping with rising aspirations. Each design becomes an ancestor to a great number of other designs and engenders a new crop of aspirations.

There were many failures in the past. Cities such as Rome have been called eternal only to become monuments, less of stability than of a continuing need for being remade. Rome and many of its buildings have been cruelly rehandled by inner and outer barbarians. The Eternal City bears striking testimony to the ship-wreck of a multitude of plans and designs that have forever remained frustrated fragments. In the present, things may be different from what they were in the past, perhaps, but certainly not better. The controversial, calamitous character of contemporary towns, from 'modern' Mexico, Milan, Manila, back to Middletown, USA, is known to all of us when we but cross the street from our office building to where we have parked the car.

Through the mental work of design, which is supposed to improve our lives, the race appears generally to stray farther and farther from the natural scene. The paradisical habitat of earliest man is considered a myth today and his natural situation may originally have posed him harsh enough problems. Yet those of our man-designed, man-constructed environment are often more trying and more severe tests to our natural resistance.

Man's own cramped-together creations, anything from underground sewage systems and subways to a badly hemmed-in sky overhead, irritatingly criss-crossed by a maze of electric wires, should not prove as inescapable as fate. Lightning and the plague, once so formidable, have been countered by proper measures; must we then here find ourselves helpless? Must we remain victims, strangled and suffocated by our own design which has surrounded us with man-devouring metropolises, drab small towns manifesting a lack of order devastating to the soul, blighted countrysides along railroad tracks and highways, studded with petty 'mere-utility' structures, shaded by telephone poles and scented by gasoline fumes?

Design, the act of putting constructs in an order, or disorder, seems to be human destiny. It seems to be the way into trouble and it may be the way out. It is the specific responsibility to which our species has matured, and constitutes the only chance of the thinking, foreseeing, and constructing animal, that we are, to preserve life on this shrunken planet and to survive with grace.

Such survival is undoubtedly our grand objective, according to an innate pattern of feeling. It is a matter of urgent concern to everyone – from the loftiest philosopher to the most matter-of-fact businessman. Design to contribute to survival of the race is more than design as a long-hair luxury or as a lubrication of bigger and better trade.

Never have the opportunities for general and integrated design on a world-wide scale been as breathtaking as they are today. The Second World War has left huge areas of destruction in its wake but promptly a clamor rose, from Le Havre, France, to Agana, Guam, that things should be re-built in the 'old way.'

Yet pitiful attempts at resurrection of what is bygone are not the best we can do to honor the past. Also, naïve parochial outlook needs supplementation by global forethought, experience, and contemporary know-how. With all sincere respect for regionalism, there does exist now a cosmopolitan 'joint responsibility' for reconstruction anywhere. Human planning cannot really remain compartmental or sectional in an age of mutually braced security. Vast regions, which were formerly colonial, are awakening to their own contemporary participation with needs and supplies enormously stepped up. Technological progress in advanced centers is spreading and forcing a changed way of life even on the far-away, backward portion of the globe. And under the pressure of this progress if it is to be integrated, conscientious design is needed everywhere.

What sort of design? What are its governing principles and on what objective foundations can it be based? Is there anything to rely on behind all that bewildering multiform activity of ours? Is there anything which eloquent philosophers could put into words?

The writer has long felt tempted to put into words the fact that at this day and age no speculative philosophy, no deductive method alone, no talking-it-out can yield us all the principles of design. In our time new instruments and obligations have come to us from research penetrating into life's performance. Physiology is a pursuit and a science which opens the door to broad and intensive application. We begin to wield tools which will enable us to do the patient spade-work which must be done. It will be fascinating because it is so novel.

With knowledge of the soil and subsoil of human nature and its potentials, we shall raise our heads over the turmoil of daily production and command views over an earth which we shall have to keep green with life if we mean to survive – not cramped full with all the doubtful doings of a too thoroughly commercialized technology. Tangible observation rather than abstract speculation will have to be the proper guide. And drifting will no longer do.

› Performance guarantees versus old 'quality' ideas

Forms around us became dictated by an industrial technology and justified by 'operation'

After naturalism, many movements followed one another in swift, sometimes confusing succession. But a 'scientific' ambition, inaugurated by the naturalists in literature and the impressionists in painting, had become one of the artists' permanent drives. Instead of interpreting romantic subjects, late nineteenth-century painters decided to set these aside and tried to render the natural phenomena of light and color, to paint according to scientific optics. They were selective recorders and most patient experimentalists, like Seurat, the inventor of pointillism. It was only fifty years later that mathematical physicists began wondering whether strictly speaking the observer and his very means of observation do not affect what he sees.

When color and light in nature were no longer a fascinating novelty to the artist with his modern searching mind, his interest shifted to the study of 'pure form and color,' of 'the new media and materials,' or of the artist's 'ego and his sub-conscious.' The results of such inquiry and research were what he painted and carved. In whatever devious currents post-impressionistic painting divided itself, art never again reverted to the bygone innocence. Scientific aspiration persisted. More especially, a scientific-sounding terminology, loosely borrowed from various sciences, now seemed necessary to many artists and critics. Although they were not scientists themselves, they depended on the language of science, almost as much as the medieval artists – who were no saints – had depended on the language of their Christian faith. Thus in the first quarter of the twentieth century, it was, for instance, the psychoanalytic terminology that inspired many artists and their public. Despite their enthusiasm, both had often only a faint idea of what it was all about.

As a young man I was befriended by Professor Freud's sons and had the chance to observe on social visits in his home that the great man himself was indifferent, if not hostile, to the then current expressionistic art, fraught with 'depth psychology.' Sigmund Freud was a connoisseur but kept aloof from consciously revolutionary, controversial, and programmatic novelties. They did not attract him as did Cretan jewelry, Greek statuary, and Hellenistic painting.

Science, fascinating because it was beyond the layman's grasp, or popularized with questionable accuracy, imprinted itself on art manifestos, but often to the annoyance of the scientist.

While the old romantic approach was being shunned for a time by the artists, science itself consolidated its inductive method and preferred operational concepts to the handy package labeled 'eternal truth.' Even philosophy, as far as it survived, began to be permeated by this matter-of-fact attitude. In America, pragmatism and behaviorism attracted wide attention. Following James, Dewey and instrumentalism proclaimed that an idea was true if it worked.

If a thing had truth because it worked, it now also had beauty because it functioned. A hundred years ago the American sculptor Horatio Greenough declared that the structural form created by man must follow function, just as was the case for living organisms, according to the new science of biology. Dr. Giedion has most interestingly recorded less well-known predecessors of these ideas among French designers and writers.

An impressive literary precedent is Gottfried Semper's wise and voluminous book *The Style*. Semper, a contemporary of Greenough, practiced architecture in Dresden, Zurich, and Vienna, and his writing was translated in part into English by John Wellborn Root, the greatest architect of Chicago's 'pre-Columbian' period. Semper's programmatic statements: 'The solution of modern problems must be freely developed from the premises given by modernity' and 'Any technical product is the result of use and material,' were undoubtedly known to Louis H. Sullivan and cherished by him. But however radical the ideas of Gottfried Semper and his French counterpart Viollet-le-Duc may have been, these men never abandoned traditional formalism in practice.

It was Sullivan who in 1892 decided to house Pullman cars and locomotives at the Chicago World's Fair in a Transportation Building of nontraditional form; and it was Otto Wagner who, simultaneously, built two or three dozen stations of the

Vienna subway and elevated rapid transit lines in the new style of the time. The same issue was dealt with on similar terms by one man in Central Europe and another far away in the Middle West of North America, where forty railroad companies had begotten a metropolis which was slowly to emerge from grimy chaos.

I know from my early and frustrated attempts to get Sullivan's writings into print that publishers, only a generation ago, failed to realize the revolutionary significance of Sullivan and the interest his consistent 'Kindergarten Chats' would finally arouse. Perhaps he did not state in so many words the relation between *morphology*, the science of organic shapes, fabrics, and textures on the one hand, and *physiology*, the discipline of life functions on the other. Yet the very idea of this interdependence certainly permeated his profound conversations which inspired and comforted me. Greenough's articles of 1850 probably had remained unknown to Sullivan. At least I do not recall hearing him mention these articles to me.

Assuredly, in every piece of constructed machinery (and why not of building engineering, too?) form seemed to follow function, and perfect functioning seemed to be a criterion of perfect form. Beauty was due for a re-definition by the engineer as well as by the biologist.

The rebirth of aesthetics on a 'scientific-naturalistic' basis seemed to be at hand. A universal solution for all aesthetic problems had all at once been proclaimed, a monopoly of interpretation, and a rule of action seemed established: Investigate the functions of a proposed construction, give it adequate functional form, and it will be a 'beautiful' form – whether or not it fits into our traditional scheme of shapes. Design no longer had to comply with social convention; rather, it was computable through a critical analysis of the available materials and determining requirements. Design-result could be almost automatic. This was a point of view quite unfamiliar to Palladio and Vignola.

What Louis Sullivan, as a saddened and dying man, was kind enough to tell me, a young tyro, about the changed functions of today's building, as well as the need for developing new and fitting formal solutions for them – these were ideas which reflected the general trend of thought of the closing nineteenth century. He was the ingenious recipient of the ideas of his time, destined to formulate these general and fundamental beliefs in specific application to building design.

Possibly in some former periods architects occasionally played the part of pioneers and educators by introducing original ideas of their own. But in the 1890s, the geniuses Wagner and Sullivan distinguished themselves mainly by their relatively higher receptivity to already current thought. Their great merit was to be far ahead of petty-minded colleagues and of their profession, which in general was arrested in its development and impervious to the demands of modern life.

Architecture was now expected to become a real and significant part of current existence instead of remaining the archeological game into which it had degenerated. The straggling architect had finally caught up with his time. An integrated environment seemed really just around the corner. But soon the very biased concept of 'utility' was rashly coupled and popularly confused with the much broader one of function. This led to a distortion of Sullivan's thoughts and paved the way for a reaction.

The ancient idea of Democritus and Lucretius that forms of life developed by an automatic natural selection of suitable elements (while the nonsuitable ones disappear in a cosmic wastebasket) had had its celebrated comeback in the biological philosophy of Darwin. By way of the short formula of the survival of the 'fittest' it had penetrated into the socio-economic neighborhood of the designer. *Pressure of circumstance* which molds a solution was now recognized and honored.

Routine practice in architecture, which throughout the nineteenth century had not fully acknowledged technological progress and indulged in eclectic play with shapely morsels and tidbits from all by-gones and the nooks, islands, and continents of the globe, was in need of a shake-up. The shock came from the new evolutionist doctrine. It was now a credo that everything truly alive at a particular time had to be a *fitting* expression of contemporary needs and means.

To progressive minds in architecture, Greek columns and other symbols of the mystically tinted statics of the past were atavisms. Vestigial organs, such as the vermiform appendix, no longer function and, therefore, it was reasoned, must disappear. It was felt they should vanish by atrophy or else be speedily cut out lest they cause trouble. At an earlier stage of development, such organs might have been fine and useful, but now they were being carried along as a pointless and even harmful burden turned toxic by disuse.

The question arose: Can such dead matter be at all 'beautiful'? According to the newly formulated functional definition of beauty, the answer was no.

Beyond doubt, these Greek columns had lost a good deal of their prime appeal since they had been moved from Sicily or from Cape Sunion – which serenely looks over the wine-colored Mediterranean – to LaSalle Street of the noisy Chicago 'Loop' or Wall Street in Manhattan, crowded with a quite different sort of life and looks. These columns now served to camouflage a new technique foreign to them, and often a whole pile of stories towered above their sorely befuddled epistyles. It all became an arbitrary collection of senseless, accidental props, while originally these forms had been revered as invented by gods to play a noble, exclusive role in their system of structural symbolism.

Greek columns had perhaps been fluted to give them the expression of resilient, strong members of fibrous organic material, and they showed a pronounced swelling at the lower part of their shaft to indicate something like a visible capacity of elastic compression under load. They were carefully 'proportioned' and enriched with symbolic accents, as is the ritual dance that has come a long way from primitive society.

But their careful proportions and symbolic accents did not really fit these ancient paraphernalia into the dry logic of an office building which stands or falls with its concealed modern steel skeleton, whereas the Parthenon actually stood and fell with its exposed truly supporting Doric columns.

Symbolism in structural members, aiming to dramatize their static function, was probably in order at a time when traditional faith and experience, all initiated by a god-teacher, guided the construction crew. The glorified customary proportions of the load-bearing members were sufficient to convince the beholder that the structure was secure, which fact could, after all, then only be guessed and suggested, not mathematically computed. The symbolic detail reminded him of mystical wisdom which, as a protective force, stood behind it all.

However, symbols of strength were now deflated by exact computations of strength which supplanted them. The LaSalle Street bank or office building was thoroughly 'figured' by people with engineering degrees who ascertained the structural capacities of framing members and their fabricated connections. Other accredited engineers as representatives of the public interest checked the computations, and only then did the city building department pass on them. Nothing here was aesthetically proportioned, but dimensions and safety factors for every part were prescribed by regulations and ordinances and chosen without any due mysticism.

Once a steel column was thus computed and dimensioned, nobody could proportion it differently; common sense forbade it and the law was strict. In consequence, the architect divorced the rational engineer and, all by himself, conceived and gave birth – as though by parthenogenesis – to a dream column, quite independent of the structural one. This latter column was to be the *beautiful* one. Apart from the intrinsic steel-skeleton, it was made of false, inflated masonry and faced with conventionally fluted terra cotta. This symbol of an ancient golden age still rises quite casually over the parked cars of the uninitiated – and the very uninterested.

The divorce of 'beauty' from 'utility' can only puzzle the consumer. One must not be surprised that this supernumerary beauty never deeply touched the souls of the people in Cleveland or Buffalo. In such context, it would hardly have touched anyone in Periclean Athens either. For a while it really was enjoyed by the professionals. The man in the street was merely impressed by the historical prestige of these façades and by the luxurious waste of a startling investment in surplus make-up.

This superficial application of beauty, borrowed from the past, turned into an elaborate curse. It was taught by an erudite caste of intellectuals and carried out by humdrum draftsmen, all of whom, as Sullivan felt, lacked confidence in their own age and failed to appreciate its lively possibilities and vital needs.

Although these building designers were officially bound and pledged to historical precedent, on many occasions they indulged in a playful good time, rather like the Marx brothers. They juggled all sorts of historical items and amusingly divested them of any original meaning. Truly the boys of the architectural fraternity were far from tragic or historically serious. If it had not so often been stupid routine, it might have been downright fun to kick the *Petit Trianon* on top of a twenty-three-story hotel and call it a penthouse.

In the new camp, however, which professed the doctrine of an inevitable development determined by environment, there did reign a kind of almost tragic fatalism. The amusing game of making an arbitrary patchwork quilt was superseded by the grave pursuit of integration.

In the eighteenth century, Herder, young Goethe's admired older friend, advanced the theory that the character of the songs or literature of a given people is determined by the living environment. It was another hundred years before men consciously found architecture, too, was part of their environmental destiny. Sullivan detested the flood of architectural old-world imports as a tedious hangover from which American design was to be freed, and posed the question: Does not life itself discard its past forms?

The wide, uninterrupted span of necessity with its tragic flavor was dear to Louis Sullivan. Despite his essential optimism he was fascinated by this same tragic and continuous wide span in the modern music of his beloved Richard Wagner, which had overshadowed the easy coloratura tricks and carefree compilations of a Rossini or Donizetti. No longer were borrowings to be made from old *bel canto*, because its charms, whether in music or in architecture, simply could not be borrowed without badly fading out.

But Sullivan had additional good reasons for opposing the adaptation of old forms. These forms had been inaugurated in the architecture of priestly castes, absolute sovereigns, and feudal aristocracies. The America of the railroad age was very different from the diminutive Greek democracies, half slave, with their very limited class of free full-fledged consumers.

Modern life and production were, on the contrary, determined by the machine and based on a mass consumership. Sullivan was the first architect to see American masses, as Walt Whitman had seen them, a grand, far-flung nation of American men and women. In actual fact, however, modern industry and its consumership were broadening to international dimensions, more international than the Roman Empire or anything that had ever existed.

Once upon a time, the material specifications had been short and simple. For the Parthenon they were marble, quarried in the neighborhood. This was the only material employed from flooring to roofing. Now, the material specifications, not only of a huge monument but even of a little road-side service station could easily fill a heavy tome if they were to be pounded out on a typewriter. There are fire-enameled sheet metal and glazing and structural steel, conduits, wires, pipes, plumbing installations, sash, roofing, plated hardware, and what-have-you. Countless finished products of complex industries which are located in many sections of the country – of the globe – make up the 'raw materials' of even the smallest building.

The glorious 'unity of material' was a thing of the past. The 'raw materials' were no longer raw, but themselves end products of long drawn-out and widely scattered manufacturing processes. The new builder and designer quarried his material from *Sweets*, the great annual building material catalogue. And *Sweets* began to stand on shelves in Mexico City, Shanghai, Melbourne, and Johannesburg. The quarry was anything but local. Just as cars were shipped from Detroit to all points of the compass, so structural steel and sacks of cement found their way from a low-wage industrial country like Belgium to distant Singapore and Rio. American fixtures filtered into many regions of the planet. The building market had become cosmopolitan.

Materials and building supplies, traveling around the earth, were purchased from agents and distributors who knew little about the qualities, composition, or manufacturing processes of their merchandise. Nevertheless, the so-called quality specification still lingered on in now empty phrases such as 'good workmanship and material.' Brunelleschi may have well used this language in fifteenth-century Florence to admonish the dependable craftsmen who built his Segrestia Vecchia. Now it became more sensible to say: 'Everything according to the standards of the American Society for Testing Materials.'

Today, apart from specialists, nobody in the building trade knows much about how billet steel is best made, or what its qualities are. Most material and supply items are innocently purchased over the telephone. Common knowledge of materials in the old sense is gone. Such knowledge has become far too involved to be accessible to the ordinary consumer, or even to his building attorney, the architect. This unavoidable ignorance dims the value of pronouncements on sheer quality.

Also in neighboring fields, quality specifications have been replaced by performance specifications, that is, by a description of the performance capacity and operational objective. These are the criteria according to which a turbo-generator, or a sewage-disposal plant, is actually purchased. Similarly, the buyer of an automobile seldom knows what is inside the engine housing, nor does he hire an expert to find it out. He may come to the showroom for the gloriously advertised style, but what he wants to know or asks about, besides the retail price, is the mileage per gallon of gas and the endurance record of a particular make. And he wants to venture a reasonable guess about when the major repair bills will begin pouring in on him. What is actually given him or what he asks for from the supplier's agent is a performance guarantee. All the incidental talk about quality in itself seems now to be recognized, at least by the enlightened buyer, as vague and unverifiable sales talk.

Qualities can be explained only by a craftsman, not by a salesman; but performance can be guaranteed to the consumer by the manufacturer or his distributor. Thus the functional concept, the pragmatic concept of commercial values, gradually came into being – and the more mystical concept of quality faded away because it was too nebulous to offer security.

Industrial technology had begun to flavor all concepts, from security to beauty.

The German architect Konrad Wachsmann (1901–80) was concerned predominantly with the issues brought about by the industrialization of building construction. Originally apprenticed as a cabinetmaker, he started architectural studies in Berlin in 1922, moving to Dresden in 1923, before studying at the Akademie der Künste, in Berlin, under Hans Pölzig (1869–1936). It was Pölzig who turned Wachsmann to the problems of mass-production, when, in 1926, Pölzig connected him with Christoph und Unmack, a large producer of prefabricated wooden buildings. His years of research formed the basis of a realized 1929 commission for a house for Albert Einstein. In 1938, after struggling through the Depression, he left for Paris, before emigrating to the United States, in 1941. During his first years in the United States, he was associated with Walter Gropius, with whom he worked on the prefabrication of metal construction. His goal was to achieve maximum variation in a limited number of components. The focus of work throughout his career, which included appointments as Professor at the Illinois Institute of Technology, from 1949 to 1964, and at the University of Southern California, from 1964 to 1974, where he founded the Graduate Program in Industrialization, was on the module and the joint, and was documented in *The Turning Point of Building*, in 1961.

Konrad Wachsmann saw the fundamental problem of building as one posed by the industrialization of the building site: standardization. This was primarily an issue of modularization, through the standardization of building components. As buildings have become more technical, they have incorporated more systems, each with its own module. Architecture becomes the synergy of diverse modules. Thus, his account of technology, as expressed in "Seven Theses," is one in which he calls on the architect to engage the complex problems of modularization.

Wachsmann's understanding of technology is another reading of it as a mechanic organism. While similar to his peers in his interest in the interplay of static and dynamic, his position is distinguished from Kiesler's, for instance, in his opposition to a biomorphic expressionism. Rather, he advocated an internal logic of interconnectivity through the methods of building themselves.

1957

Konrad Wachsmann

Seven Theses

Science and technology make possible the establishment of tasks whose solution demands precise study before end results can be formulated.

The machine is the tool of our age. It is the cause of those effects through which the social order manifests itself.

New materials, methods, processes, knowledge in the fields of statics and dynamics, planning techniques and sociological conditions must be accepted.

The building must evolve indirectly, obeying the conditions of industrialization, through the multiplication of cells and elements.

Modular systems of co-ordination, scientific experimental methods, the laws of automation, and precision influence creative thought.

Very complex static and mechanical problems demand the closest possible co-operation with industry and specialists in ideal teams composed of masters.

Human and aesthetic ideas will receive new impulses through the uncompromising application of contemporary knowledge and ability.

Originally from Leeds, England, architectural historian Peter Collins (1920–81) was Professor of Architecture at McGill University, Montreal, from 1956 until 1981. His first book *Concrete: The Vision of a New Architecture* was published in 1959 (reprinted in 2004). His third and final major publication *Architectural Judgement* (1971), was inspired by his study of the legal system and recommended the development of design principles based on the analysis of precedent. Collins' best known work: *Changing Ideals in Modern Architecture: 1750–1950*, in which this extract appeared, was originally published in 1965. It has recently been reissued (1998) with a new introduction by Kenneth Frampton.

Collins' substantial and – for its time – pioneering work in architectural historiography, tried to show the extent to which the roots of modernism reached back into the eighteenth century. The book exerted a significant influence on the course of late-twentieth-century architecture by inviting a critical reassessment of modernism's historical sources – and, perhaps unwittingly, helping to lay the foundations for the emergence of postmodern historical revivalism.

In this essay, which was originally published as an article in the British journal *Architectural Review* in 1959, Collins set out to question the pseudo-scientific reliance on zoological and evolutionary metaphors to justify a reductive and often confused notion of functionalism in architecture. In a subsequent chapter of *Changing Ideals in Modern Architecture* entitled "The Mechanical Analogy," he went on to describe the emergence of machine metaphors in the architectural writings of the nineteenth century. While critical of the tendency to think of buildings as isolated mechanical devices, Collins laments the lack of engagement between many modernist buildings and their immediate surroundings: "One great advantage of the biological analogy was that it laid particular emphasis on the importance of environment, since clearly all living organisms depend on environments for their existence, and constitute in themselves environments which influence other organisms nearby."[1]

1959
Peter Collins
The Biological Analogy

---

1   Peter Collins, *Changing Ideals in Modern Architecture 1750–1950*, second edition (Montreal: McGill-Queen's University Press, 1998), p. 166.

The purpose of analogy is to familiarize us with new ideas by linking them to ideas we already understand. It is not uncommon, however, for new theories to be linked analogously to ideas which are hardly understood by anyone, but which have captured the popular imagination by their progressive appeal. Perhaps also the less an audience understands of a subject used as an analogy, the more impressive does the argument appear. This results, if I may fall into the same trap myself, from some kind of osmosis of profundity. People today talk glibly of 'chain reactions' as if the principles of nuclear physics were obvious to anyone. A century ago, the favourite words were 'germ' and 'evolution'. Now that we are celebrating the centenary of the publication of *The Origin of Species*, it may be useful to examine the influence of biological analogies on architectural theory, and try to assess their usefulness with respect to the architecture of our own day.

The origins of the biological analogy, like so many ideas which have influenced modern architectural doctrines, can be traced to about the year 1750. At that time, two epoch-making scientific books were published: Linnaeus's *Species Plantarum* (1753), in which the entire vegetable kingdom was classified binominally according to the disposition of the female reproductive organs, or 'styles', and Buffon's *Histoire Naturelle* (1749), a vast compendium which attempted to incorporate all biological phenomena into a general interpretation of the laws governing the universe. Linnaeus's work does not immediately concern this present inquiry. Buffon, however, is of considerable relevance, since he disagreed both with Linnaeus's immutable species, and with his whole doctrine of classification by arbitrarily chosen characteristics. On the contrary, he believed that this kind of compilation obscured the fact that all species must have derived from a single type, and, supporting his views on this subject both by the evidence of fossil shells, and by reference to mammoths recently discovered in Siberia he put forward a philosophy of creation in which the idea of evolution was expressed clearly for the first time.

In so far as his system relates to biological ideas used later by architectural theorists, there are two features which deserve mention. The first is that, in hitting upon the idea of evolution, he saw it as essentially a process of degeneration, not of improvement, since his religious beliefs (or his respect for those held by his contemporaries) prevented him from assigning the evolutionary process to any but the lower animals. On the other hand he was the first scientist to distinguish correctly between the 'vegetative' and specifically 'animal' parts of animals, whereby an animal may be regarded simply as a vegetable organism endowed with the power of moving from place to place. Thus 'organic life' has come to mean, for architectural theorists at least, the sum of the functions of the 'vegetative' class, for all living organisms, whether plants or animals, possess them to a more or less marked degree.

The scientist who first gave classical expression to this meaning of 'organic' was Xavier Bichat, whose *Physiological Researches on Life and Death* was published in 1800. Until then it was normal, especially in view of the humanistic culture of the age, for the biological analogy to refer to animals rather than plants. Lord Kames, for example, who disliked symmetry in gardens, contended nevertheless that 'in organized bodies comprehended under one view, nature studies regularity, which for the same reason, ought to be studied in architecture'. At the beginning of the nineteenth century, however, 'organic' came to be regarded less as a quality of 'life which moves'. It was thus the asymmetry of plants and viscera, rather than the symmetry of animal skeletons, which came to be accepted as characteristic of organic structures, whereby biology could still be adduced to support the architectural fashions of the age.

The most important enunciations of evolutionary theory at this time were those published by Lamarck. Lamarck was essentially a botanist of the school of Buffon, but when, at the age of fifty, he was appointed professor of Zoology by the National Convention without any previous experience at all, he was obliged to transfer his attention to the study of anatomy. As a result of this combination of disciplines, he was eventually led to conclude that living forms had not evolved retrogressively as Buffon had believed, but progressively. This change of attitude was only to be expected. Buffon, living in the age of Rousseau, and at a time when the Book of Genesis was literally accepted, naturally favoured a hypothesis implying a Fall from perfection. Lamarck, in the age of Revolution, and at a time when the idea of Progress was literally accepted, naturally favoured a contrary view.

Similarly, it was not entirely strange that Lamarck should suggest that evolution was due to environment. The importance of this influence on art, law and society had already been emphasized by Winckelmann, Montesquieu and de Goguet respectively, although they did not, as far as I know, go so far as to say that it actually caused evolution direct. This, however, was the essence of Lamarck's revolutionary argument. 'It is not', he wrote, 'the organs – that is to say, the form and character of the animal's bodily parts – which have given rise to its habits and peculiar properties, but, on the contrary, it is its habits and manner of life and the conditions in which its ancestors lived that has in the course of time fashioned its bodily form, its organs and its qualities.'

The word 'biology', or science of life, was invented by Lamarck in about 1800; at the same time, the word 'morphology', or science of form, was invented by Goethe, who in his own day was as famous as a scientist as he was as a poet. Being a poet, however, he understood the term morphology in a much wider sense than we do today (when the subjects of its study are confined to the comparison and relationships of living structures and their development), and included nonliving forms such as rocks. This, as we shall see, was to be another element of confusion in the biological analogy in that, from its inception, there was uncertainty as to whether morphology was concerned with structures which live, or with structures which grow. Félix Vicq d'Azyr, for example, at the end of the eighteenth century, had rejected the old comparison between the growth of organisms and the growth of crystals, contending that crystals are mathematically regular in shape and homogeneous in structure, whereas organisms are of rounded shapes and complex composition. On the other hand Jacob Schleiden, fifty years later, considered that life was nothing more or less than a 'form-building force', and he considered the growth of crystals and organisms to belong to the same category of phenomena. As late as 1898, Herbert Spencer could still assert that the growth of crystals and organisms was 'an essentially similar process'. Since it was Spencer's biological works which mainly influenced Frank Lloyd Wright, the possible effects of this ambiguity will be obvious.

Moreover, as soon as the new science of morphology was established, and pursued methodically by the study of comparative anatomy, two dilemmas in the interpretation of the facts at once made themselves apparent: does form follow function, or does function follow form? To the layman, the conundrum might appear futile and insoluble, but to those familiar with the history of modern architectural theories its importance will need no justification. Amongst biologists, the distinction was considered sufficiently important to perpetuate a bitter quarrel for half a century, the leader of the 'form follows function' school being Georges Cuvier, the leader of the opposing faction being Geoffroy Saint-Hilaire. Cuvier (who was incidentally a friend of the architect A. T. Brogniart, and obtained his assistance in examining fossilized building stones) stated that every modification of a function entailed the modification of an organ. Geoffroy Saint-Hilaire protested against arguing from function to structure as an 'abuse of final causes'. The controversy might well have continued indefinitely had it not been that advances in cell-theory distracted attention from morphology, by causing organisms to be seen no longer as cleverly constructed mechanisms but simply as an aggregate of cells.

In the event, when the biological analogy was first seriously applied to art theory, the delicate topic of 'form versus function' was avoided completely, since interest was concentrated on the way forms grow, rather than on the way they work. From the time aesthetics became associated with psychology in the middle of the eighteenth century, philosophers had been trying to explain how inspiration (or 'genius' as it was sometimes called) grew in the human mind. Buffon himself, in his speech on *Style* to the Academie Française (1753) was perhaps the first to hint at a biological analogy when he remarked that 'the human mind can create nothing, and only produces after having been fertilized by experience and meditation, in that its perceptions are the germs of its products'. Later Young, in his *Conjectures on Original Composition* (1759), stated that 'an original may be said to be of a vegetable nature; it rises spontaneously from the vital root of genius; it grows, it is not made'. But it was left to Samuel Taylor Coleridge to express the idea as a complete artistic theory.

There seems little doubt that Coleridge derived his ideas from Germany, where he had studied in his youth and where such ideas had long been in circulation. Young's *Conjectures*, though virtually ignored in England, had been twice translated into German within two years of its publication, and had become an important part of the gospel of Storm and Stress. J. G. Herder, in his essay *On the Knowing and Feeling of the Human Soul* (1778), had used plants as an analogy for the development of art forms from the soil of their own time and place. Goethe, in his famous early essay on German architecture, had described Gothic as the organic product of growth in the mind of genius. But Coleridge, who was himself an amateur biologist, not merely translated these views into English; he organized the attack against the whole 'Mechanico-Corpuscular' philosophy of creation. 'The form is mechanic', he wrote, 'when on any given material we impress a predetermined form, as when to a mass of wet clay we give whatever shape we wish it to retain when hardened. The organic form, on the other hand, is innate, it shapes as it develops itself from within, and the fullness of its development is one and the same with the perfection of its outward form.'

Several criticisms relevant to the present enquiry may be made concerning Coleridge's views. One is that the process of artistic creation is explained by him as virtually an unwilled and unconscious process of mind. The second is that however violently he might attack the 'mechanical' theory, it has been frequently used by biologists to explain how living organisms actually work. It was not only early philosophers such as Descartes who regarded the animal body as a machine. One of the most famous of Cuvier's disciples, Henri Milne-Edwards, stated that he had 'tried to grasp the manner in which organic forms might have been invented by comparing and studying living things as if they were machines created by the industry of man'. Finally, it is worth noting that no explanation of morphological development was more mechanistic than Darwin's 'Natural Selection'.

It has already been pointed out that by 1859 there was nothing novel in the idea of evolution as applied to the theory of life, even though the term 'evolution' was not used in this sense until 1831. This is equally true with regard to the theory of architecture. The classical architects of the early eighteenth century believed implicitly in evolution, since they believed that the moderns had improved on the Romans, just as the Romans improved on the Greeks. Even mid-nineteenth-century writers on architecture such as Fergusson, who specifically criticized Lamarck's theories, believed in architectural evolution because they believed in Progress. For biologists the novelty of Darwin's theory was that it attributed evolution to a selection of *existing* forms (or, to put it another way, the elimination of obsolescent forms) by Nature herself. It thus inevitably weighed the balance in favour of the 'function follows form' school by presupposing 'that the forms existed in the first place. Lamarck had claimed that a change in environment actually modifies the form of animals, and that these changes are transmitted by heredity. Darwin claimed, on the contrary, that the changes were arbitrary and accidental, and that species changed only because the unfunctional forms never survived. He compared the action of natural selection to that of a man building a house from field-stones of various shapes. The shapes of these stones, he said, would be due to definite causes, but the uses to which the stones were put in the building would not be explicable by those causes. Yet as Charles Singer has pointed out, when a man builds a house, there is the intervention of a definite purpose, directed towards a fixed end and governed by a clearly conceived idea.

> The builder in the proper sense of the word *selects*. But the acts of selection – mental events in the builder's mind – have no relation to the 'causes' which produced the stones. They cannot therefore be compared with the action of Natural Selection.

Architectural theorists who are guilty of similarly inexact analogies between building and botany may find consolation in the thought that a classic precedent was furnished by the Master himself.

If in fact we look at those phenomena which scientists consider as biological, we shall see that the number of exact parallels which can be drawn are slight. Vicq d'Azyr classified organic functions into nine categories: digestion, nutrition, circulation, respiration, secretion, ossification, generation, irritability, and sensibility, and of these only circulation would seem to have any analogy with the function of buildings. Similarly, if we examine morphological systems of classification, whether it be the Linnaean system (based on one selected feature), Cuvier's system (based on total structure related to inner parts), or the system of von Baer (based on what he called the 'spatial relationship' of organic elements, i.e. radial, longitudinal, massive, and vertebrate), there seems little even remotely suggestive of buildings and the way they are designed. It would seem as if the analogy must always be general and poetic, and in fact the features held in common seem limited to four: the relationship of organisms to their environment, the correlation between organs, the relationship of form to function, and the principle of vitality itself.

The most comprehensible analogy concerns the influence of environment on design, an idea which undoubtedly derived its main stimulus from Darwin, although it first emerged in the work of Alexander von Humboldt, who opposed the academic methods of Linnaeus and suggested that plants should be classified according to the climates in which they were found, rather than according to inherent characters determinable in a museum. Being of a romantic and aesthetic disposition, he sought a system of classification through the impression made by landscapes when simply looked at by the ordinary observer. He was very interested in architecture and described in detail the pre-Columbian buildings he found in Central America. He nowhere seems to have suggested, however, that the design of buildings had much relationship with topography and vegetation, although he thought that pyramids were best suited to mountainous ground. Only in the sphere of engineering did he exert any influence on construction, in that his description of Peruvian suspension bridges is known to have suggested modern experiments in this field.

Darwin naturally took von Humboldt's doctrine considerably further by contending that Nature had *selected* those forms which were most suitable for the environments in which they were situated, but he offered no suggestion as to how Nature created such forms in the first place. He had in fact no training and probably little interest in pure morphology, and in so far as his work affected morphological studies, it was to cause the public to regard organisms *historically.* In his first draft of *The Origin of Species*, written in 1842, he remarked that 'we must look at every complicated mechanism and instinct as the summary of a long history of useful contrivances much like a work of art'. Whether or not he actually regarded the history of architecture as analogous with natural selection, I do not know. But there can be little doubt that, so far as his biological theory of the relationship of form to environment is concerned, the relevance of Darwinism to architecture has tended to decrease. Improvements in air-conditioning equipment are making architectural form increasingly independent of climatic considerations. Only in districts where distinctive local materials can be used for domestic architecture is there any likelihood of regional characteristics influencing form, and even in newly developed areas where the example has been set, such as Arizona, there seems little evidence of a desire to carry the movement very far.

As regards the 'correlation between organs' (which one might perhaps compare with the relationship between the parts of a building), the fact was first enunciated as a biological principle by Vicq d'Azyr, who pointed out that a certain shape of tooth presupposes a certain type of structure in the extremities and the digestive canal, because the animal's bodily parts are adapted to its way of living.

This idea was taken even further by Cuvier, who, from small fossil fragments, showed how one could reconstruct extinct animals by a sequence of deductions based on the interdependence of each organic part. Yet in so far as this discovery relates to architectural theory, it suggests merely a curious parallel with the Renaissance theory of modular proportions, whereby, as the Humanists had observed, the proportions of the human body are so standardized that if one were to find the finger of an antique statue, it would be possible, theoretically, to reconstruct the whole (a fact enthusiastically seized upon by the great forgers of the age). However, the only use to which Cuvier's discovery was put by nineteenth-century theorists was in proving that the 'imitation of styles' was morally wrong, since it left false evidence for future historians. After describing zoological reconstructions of prehistoric animals in his *True Principles of Beauty in Art* (1849), Fergusson added: 'With the same facility with which a fossil impress or a bone does this for the geologist, does any true style of art enable the archaeologist to tell from a few fragments in what century the building to which it belonged was erected.'

In general, the only major biological fact which seems directly analogous to modern architecture concerns the relationship of form to function, but as we have seen, the theory that form follows function was hotly contested by those who believed that function follows form. It is curious to note that this dilemma was specifically pointed out by Herbert Spencer, from whose writings (so Frank Lloyd Wright tells us) Louis Sullivan derived many, if not all, his biological ideas. However, since nobody has ever denied the obvious fact that form and function are in some way related, it is worth considering how this relationship does fit in with a theory of design.

In case it should be objected that such a topic is not part of the 'Organic' theory at all, but of the 'Functional' theory, it is opportune to suggest that whereas in the functional analogy, the relationship between form and function is considered as necessary to *beauty*, in the biological analogy, it is considered as necessary to *life*. Historians are generally agreed that credit for this new interpretation must be given, as far as architectural theory is concerned, to Louis Sullivan, although it may be noted that he never expressed it or applied it until after he had met Wright. It had been foreshadowed by Greenough and Baudelaire, who, perhaps with von Humboldt in mind, suggested that the best critics were those who had travelled alone through forests and prairies, contemplating, dissecting, and writing. 'They know', he wrote, 'the admirable, inevitable relationship between form and function.' Similarly, Viollet-le-Duc, like Ruskin before him, drew attention to the way mediaeval sculptors had studied the morphology of vegetation, and how they understood that the contours of plants 'always express a function, or submit themselves to the necessities of the organisms'. He did not, however, draw any major philosophical conclusions from this observation, except to say that the masons 'sought to bring out in the structures of their buildings those qualities they found in vegetation'. The French Rationalists were in fact more interested in the idea that form follows structure (which they found quite intelligible without the use of elaborate analogies), so that there can be little doubt that it was Sullivan who first made biological analogies the foundation of a total architectural doctrine.

Sullivan seems to have derived little inspiration from Viollet-le-Duc's theories, since his main interest was in composition rather than in construction (which he left to Adler). Yet following the anti-academic fashion of his age, he objected to the term 'composition', although in the circumstances it is difficult to see why. Since 'decomposition' is the chief characteristic of organisms which are dead, it might reasonably be inferred that 'composition' is the chief characteristic of organisms which are living. But, like so many theorists who have found the biological analogy stimulating, he never really pursued it very deeply, and made little distinction as to whether it referred to the object created or the process of design. Whilst some of his writings suggest a Lamarckian interpretation of evolution (as when he wrote that 'it was not simply a matter of form expressing function; the vital idea was that the function *created* or organized its form'), most of them suggest the Coleridgean analogy between biology and poetic vision. It is perhaps significant that his first enunciation of an architectural doctrine – the address on *Inspiration* given to the Western Association of Architects' Convention – was in the form of a long poem intelligible only to three other people in the room.

In the present century the biological analogy has been associated primarily with Frank Lloyd Wright, into whose young hands Sullivan enthusiastically transmitted his copy of Spencer's biological works. What Wright has meant by 'Organic Architecture' has never been clear; the difficulty is that for Wright it meant so much: crystalline plan forms, the possibility of growth by asymmetrical addition, the relationship of composition to site and client, the use of local materials, the individuality of every created thing, the need for every artist to endow his work with the integrity of his innermost being, and so on. But primarily it meant for him *a living* architecture; an architecture in which useless forms were sloughed off as part of the process of a nation's growth, and in which every composition, every element and every detail was deliberately shaped for the job it had to perform. To this interpretation no one can take exception, and perhaps the safest thing to say of the Biological Analogy is that it is simply a more poetic expression of the ideal of *L'Architecture Vivante*.

It is now a century since the *Revue Generale de l'Architecture* launched the slogan 'Organic Architecture' in this sense, although at the time it proved premature. 'We have named it Organic', wrote the editor in 1863,* 'because it is, in relationship to the Historic and Eclectic Schools, what the organized life of animals and vegetables is in relationship to the unorganized existence of the rocks which form the substratum of the world.' Since then, many developments have occurred in biological theory, and many in architectural practice. Occasionally some of the former can be paralleled with some of the latter. Claude Bernard's discoveries concerning the way the body adapts itself to changing conditions (or vaso-motor mechanism) suggest clear parallels with the flexibility of modern planning. Similarly Milne-Edwards's law of economy, which states that nature does not always create a new organ for a new function, but often adapts undifferentiated parts to special functions, or even converts to other uses organs already specialized, suggests many interesting parallels in this present age of standardized forms. Most important of all, Wilhelm Roux's discovery that the blood-vascular system is largely determined by direct adaptation to functional requirements demonstrates that form does occasionally follow function after all. But in general, detailed analogies are as dangerous now as when the slogan was first formulated, and apart from holding that architecture must be a living art, we cannot go much deeper into the mystery of life than when *The Origin of Species* was first given to an astonished and excited world.

Within the last few years, however, one surprising change has occurred in the philosophy of architecture which provides a curiously apposite termination to a study of the influence of Darwin. The nineteenth century's naïve faith in evolutionary progress is now being seriously challenged, and a suspicion has arisen that Buffon's approach may not have been entirely wrong. This does not of course mean that optimism has given place to pessimism, but simply that we no longer accept, like the followers of Darwin, the idea that every change is for the best. Recently however it has become clear, in both Europe and America, that the leading architectural periodicals are no longer content merely to divide all new buildings into the two categories: 'evolutionary' and 'vestigial', and leave it at that; they are subjecting contemporary architecture to systematic criticism in order to determine how improvements can best be brought about.

This, of course, is the very opposite of natural selection, but it has become necessary because we can no longer afford to regard every new 'contemporary' building as automatically an advance on the rest. In the early years of the International Style, there was much to be said for accepting every manifestation of the new spirit uncritically, since premature disparagement might have stunted its early growth. Today, when the functional forms evolved by the leading modern architects are so widely accepted, there is obvious danger of their misuse, and nothing can better serve the advancement of architecture than that examples of this should be publicly singled out.

An even more cogent reason for the new critical attitude is that, just as biologists have become very conscious of 'biotic' environment (i.e. the influence of free organisms on each other), so we are becoming much more aware that 'environment' does not only comprise natural scenery, but also the accumulated legacy of the buildings in our towns. The urban scene, especially in America, is in many districts predominantly 'contemporary', so that modern architecture has no longer an excuse for ignoring its neighbours. On the other hand, with the general acceptance of functionalism, there is no need to perpetuate the early revolutionaries' aggressive disdain for the so-called 'beaux-arts' styles. Such buildings, when juxtaposed against our own, bear gratifying testimony to the victory of the fittest, but they also carry the awful warning that, in architecture, it is not necessarily only the Fittest which Survive.

---

* The earliest use I have found of the word 'organic' with specific reference to a 'living' architecture occurs in Lamennais' beautiful eulogy of Gothic buildings in *De l'Art et du Beau* (1841): 'Ce qui les caractérise, c'est le travail organique qui de tant d'éléments divers a fait une seule forme, dont les innombrables parties . . . se fondent en un corps unique et vivant.'

138

1960

Peter Reyner Banham

Functionalism and Technology

Peter Reyner Banham (1922–88) was an English architectural historian, critic, teacher, writer, and journalist. After reading art history at the Courtauld Institute, he joined the editorial staff of the *Architectural Review* in 1952, developing his racy style of writing and a particular sympathy for the Italian Futurists and their enthusiastic embrace of technology. That same year Banham convened the first full meeting of the *Independent Group*, a collection of artists and designers dissatisfied with orthodox modernism and interested in consumerism, mass culture, fashion, and styling. In 1958 he completed a PhD under Nikolaus Pevsner, and then in 1960 published his very influential book *Theory and Design in the First Machine Age*, a revised version of his dissertation, followed by *The New Brutalism* in 1966. At that time he also began lecturing at the Bartlett School of Architecture, and by 1969 was appointed Chair of Architectural History at the Bartlett, just as he published *The Architecture of the Well-Tempered Environment*. In 1977 he moved to the United States to become the Chairman of Design Studies at the University of New York, Buffalo, then assumed the Chair of Art History at the University of California, Santa Cruz in 1980.

This excerpt on "Functionalism and Technology" formed the final chapter of *Theory and Design in the First Machine Age*, and characterizes Banham's broad scope and particular insights. Following in the footsteps of Pevsner and also Siegfried Giedion, he continued to develop the broad thesis that modern architecture resulted from the encounter with new means of production, but also saw it as a cultural problem in which architecture was itself a discrete culture distinct from that of industry and technology. And, like Le Corbusier (1923), he believed that the engineers and inventors were the more dynamic and innovative cultural group. As he concluded the book, the architect who proposes to run with technology knows now that he will be in fast company, and that, in order to keep up, he may have to emulate the Futurists and discard his whole cultural load, including the professional garments by which he is recognized as an architect.

The essay is useful for this collection as a review of functionalism as a design credo among designers in their encounter with new technologies. At this point, Banham remained engaged with the ideal of the Futurists, simply committing himself to the heady process of "constantly accelerating change," and in that respect maintained the belief in a true or purely technological form of building.

By the middle of the Thirties it was already common practice to use the word *Functionalism*, as a blanket term for the progressive architecture of the Twenties and its canon of approved forerunners that had been set up by writers like Siegfried Giedion. Yet, leaving the shortlived *G* episode in Berlin on one side, it is doubtful if the ideas implicit in Functionalism – let alone the word itself – were ever significantly present in the minds of any of the influential architects of the period. Scholiasts may care to dispute the exact date on which this misleading word was first used as the label for the International Style, but there is little doubt that the first consequential use was in Alberto Sartoris's book *Gli Elementi dell'architettura Funzionale*, which appeared in Milan in 1932. Responsibility for the term is laid on Le Corbusier's shoulders – the work was originally to have been called *Architettura Razionale*, or something similar, but, in a letter which is reprinted as a preface to the book, Le Corbusier wrote

> The title of your book is limited: it is a real fault to be constrained to put the word *Rational* on one side of the barricade, and leave only the word *Academe* to be put on the other. Instead of Rational say *Functional*.

Most critics of the Thirties were perfectly happy to make this substitution of words, but not of ideas, and *Functional* has, almost without exception been interpreted in the limited sense that Le Corbusier attributed to, *Rational*, a tendency which culminated in the revival of a nineteenth-century determinism such as both Le Corbusier and Gropius had rejected summed up in Louis Sullivan's empty jingle,

> Form follows function.

Functionalism, as a creed or programme, may have a certain austere nobility, but it is poverty-stricken symbolically. The architecture of the Twenties, though capable of its own austerity and nobility, was heavily, and designedly, loaded with symbolic meanings that were discarded or ignored by its apologists in the Thirties. Two main reasons emerge for this decision to fight on a narrowed front. First, most of those apologists came from outside the countries – Holland, Germany, and France – that had done most to create the new style, and came to it late. They thus failed to participate in those exchanges of ideas, collisions of men and movements, congresses and polemics, in which the main lines of thought and practice were roughed out before 1925, and they were strangers to the local conditions that coloured them. Thus, Siegfried Giedion, Swiss, caught only the tail end of this process in 1923; Sartoris, Italian, missed it almost completely; Lewis Mumford, American, in spite of his sociological perceptiveness, was too remotely placed to have any real sense of the aesthetic issues involved – hence his largely irrelevant tergiversations on the problem of monumentality.

The second reason for deciding to fight on the narrowed front was that there was no longer any choice of whether or not to fight. With the International Style outlawed politically in Germany and Russia, and crippled economically in France, the style and its friends were fighting for a toehold in politically-suspicious Fascist Italy, aesthetically-indifferent England, and depression-stunned America. Under these circumstances it was better to advocate or defend the new architecture on logical and economic grounds than on grounds of aesthetics or symbolisms that might stir nothing but hostility. This may have been good tactics – the point remains arguable – but it was certainly misrepresentation. Emotion had played a much larger part than logic in the creation of the style; inexpensive buildings had been clothed in it, but it was no more an inherently economical style than any other. The true aim of the style had clearly been, to quote Gropius's words about the Bauhaus and its relation to the world of the Machine Age

> . . . to invent and create forms symbolising that world,

and it is in respect of such symbolic forms that its historical justification must lie.

How far it had succeeded in its own terms in creating such terms, and in carrying such symbolism, can best be judged by examining two buildings, widely held to be masterpieces, and both designed in 1928. One of them is the German Pavilion at the Barcelona Exhibition of 1929, a work of Mies van der Rohe, so purely symbolic in intention that the concept of Functionalism would need to be stretched to the point of unrecognisability before it be made to fit it – the more so since it is not easy to formulate in Rational terms precisely what it was intended to symbolise. A loose background, rather than a precise exposition, of the probable intentions can be established from Mies's pronouncements on exhibitions in 1928:

> The era of monumental expositions that make money is past. Today we judge exposition by what it accomplishes in the cultural field.
>
> Economic, technical, and cultural conditions have changed radically. Both technology and industry face entirely new problems. It is very important for our culture and our society, as well as for technology and industry, to find good solutions. German industry, and indeed European industry as a whole, must understand and solve these specific tasks. The path must lead from quantity towards quality – from the extensive to the intensive.
>
> Along this path industry and technology will join with the forces of thought and culture.
>
> We are in a period of transition – a transition that will change the world.
>
> To explain and help along this transition will be the responsibility of future expositions. . . .

The ambiguities of these statements were resolved in the Pavilion by architectural usages that tapped many sources of symbolism – or, at least sources of architectural prestige. Attention has been drawn to echoes of Wright, of *de Stijl* and *Schinkelschüler* tradition, in the Pavilion, but its full richness is only apparent when these references are rendered precise. All three of these echoes are, in practice, summed up in a mode of occupying space which is strictly Elementarist. Its horizontal planes, which have been likened to Wright, and its scattered vertical surfaces, whose distribution on plan has been referred to van Doesburg, mark out one of Moholy's 'pieces of space' in such a way that a 'full penetration with outer space' is effectively achieved. Further, the distribution of the columns which support the roof slab without assistance from the vertical planes, is completely regular and their spacing suggests the Elementarist concept of space as a measurable continuum, irrespective of the objects it contains. And again, the podium on which the whole structure stands, in which Philip Johnson has found 'a touch of Schinkel', extending on one side a good way beyond the area covered by the roof slab, is also a composition in its own right in plan because of the two pools let into it, and thus resembles the patterned base-boards which form an active part in those Abstract studies of volumetric relations that came from the Ladowski–Lissitsky circle, and, like them, appears to symbolise 'infinite space' as an active component of the whole design.

To this last effect the materials also contribute, since the marble floor of the podium, everywhere visible, or at least appreciable even where covered by carpeting, emphasises the spatial continuity of the complete scheme. But this marble, and the marbling of the walls, has another level of meaning – the feeling of luxury it imparts sustains the idea of transition from quantity to quality of which Mies had spoken, and introduces further paradoxical echoes of both Berlage and Loos. These walls are space-creators, in Berlage's sense and have been 'let alone from floor to cornice' in the manner that Berlage admired in Wright; yet, if it be objected that the sheets of marble or onyx with which they are faced are 'decoration hung on them' such as Berlage disapproved, one could properly counter that Adolf Loos, the enemy of decoration, was prepared to admit large areas of strongly patterned marble as wall-cladding in his interiors.

The continuity of the space is further demonstrated by the transparency of the glass walls that occur in various parts of the scheme, so that a visitor's eye might pass from space to space even where his foot could not. On the other hand the glass was tinted so that its materiality could also be appreciated, in the manner of Artur Korn's *There and Not There* paradox. The glass of these walls is carried in chromium glazing bars, and the chromium ace is repeated on the coverings of the cruciform columns. This confrontation of rich modern materials with the rich ancient material of the marble is a manifestation of that tradition of the parity of artistic and anti-artistic materials that runs back through Dadaism and Futurism to the *papiers collés* of the Cubists.

One can also distinguish something faintly Dadaist and even anti-Rationalist in the-non-structural parts of the Pavilion. A Mondriaanesque Abstract logical consistency, for instance, would have dictated something other than the naturalistic nude statue by Kolbe that stands in the smaller pool – in this architecture it has something of the incongruity of Duchamp's 'Bottle-rack' in an art exhibition, though it lives happily enough with the marble wall that serves as a background to it. Again, the movable furniture, and particularly the massive steel-framed chairs flout, consciously, one suspects, the canons of economy inherent in that Rationalism that del Marle had proposed as the motive force behind the employment of steel in chairs; they are rhetorically over-size, immensely heavy, and do not use the material in such a way as to extract maximum performance from it.

It is clear that even if it were profitable to apply strict standards of Rationalist efficiency or Functionalist formal determinism to such a structure, most of what makes it architecturally effective would go unnoted in such an analysis. The same is true of the designs of Le Corbusier, whose work, while often extremely practical, does not yield up its secrets to logical analysis alone. In his *Dom-ino* project for instance, he postulated a structure whose only given elements were the floor slabs and the columns that supported them. The disposition of the walls was thus left at liberty, but some critics have logically extrapolated also that this left Le Corbusier at the mercy of his floor slabs. Nothing could be farther from the truth as far as his completed buildings are concerned which, from the villa at Chaux-de-Fonds onwards, have their floor slabs treated in a most cavalier fashion, and much of their internal architecture created by breaking through from one storey to another. Conversely, if there is a building in which the horizontal slabs are absolute, it is Mies's Barcelona Pavilion – the pools merely diversify the surface of the podium, nothing breaks through the roof slab and nothing rises above it; the whole building is designed almost in two dimensions, and this is true of much of his later work as well.

In the case of the other building of 1928 which it is proposed to study here, Le Corbusier's house, *Les Heures Claires*, built for the Savoye family at Poissy-sur-Seine and completed in 1930, the vertical penetrations are of crucial importance to the whole design. They are not large in plan but, since they are effected by a pedestrian ramp, whose balustrades make bold diagonals across many internal views, they are very conspicuous to a person using the house. Furthermore, this ramp was designed as the preferred route of what the architect calls the *promenade architecturale* through the various spaces of the building – a concept which appears to lie close to that almost mystical meaning of the word 'axis' that he had employed in *Vers une Architecture*. The floors connected by this ramp are strongly characterised functionally – *on vit par étage* – the ground floor being taken up with services and servants, transport and entrance facilities, and a guest room; the first floor given over to the main living accommodation, virtually a week-end bungalow complete with patio; and the highest floor a roof garden with sun-bathing deck and viewing platform, surrounded by a windscreen wall.

This, of course, is only the functional breakdown; what makes the building architecture by Le Corbusier's standards and enables it to touch the heart, is the way these three floors have been handled visually. The house as a whole is white – *le couleur-type* – and square – one of *les plus belles formes* – set down in a sea of uninterrupted grass – *le terrain idéal* – which the architect has called a Virgilian Landscape. Upon this traditional ground he erected one of the least traditional buildings of his career, rich in the imagery of the Twenties. The ground floor is set back a considerable distance on three sides from the perimeter of the block, and the consequent shadow into which it is plunged was deepened by dark paint and light-absorbent areas of fenestration. When the house is viewed from the grounds, this floor hardly registers visually, and the whole upper part of the house appears to be delicately poised in space, supported only by the row of slender pilotis under the edge of the first floor – precisely that species of material–immaterial illusionism that Oud had prophesied, but that Le Corbusier more often practised.

However, the setting back of the ground floor has further meaning. It leaves room for a motor-car to pass between the wall and the pilotis supporting the floor above; the curve of this wall on the side away from the road was, Le Corbusier claims, dictated by the minimum turning circle of a car. A car, having set down its passengers at the main entrance on the apex of this curve, could pass down the other side of the building, still under the cover of the floor above, and return to the main road along a drive parallel to that on which it had approached the house. This appears to be nothing less than a typically Corbusian 'inversion' of the test-track on the roof of Matté-Trucco's Fiat factory, tucked under the building instead of laid on top of it, creating a suitably emotive approach to the home of a fully motorised post-Futurist family. Inside this floor, the entrance hall has an irregular plan, but is given a business-like and ship-shape appearance by narrow-paned industrial glazing, by the plain balustrades of the ramp and the spiral staircase leading to the floor above, and by the washbasin, light fittings, etc. which, as in the *Pavilion de l'Esprit Nouveau*, appear to be of industrial or nautical extraction. On the main living floor above, the planning shows less of that *Beaux-Arts* formality that had appeared in the slightly earlier house at Garches, but is composed much as an Abstract painting might have been composed, by jig-sawing together a number of rectangles to fit into a given square plan. The feeling of the arrangement of parts within a pre-determined frame is heightened by the continuous and unvaried window-strip – the ultimate *fenêtre en longueur* – that runs right round this floor, irrespective of the needs of the rooms or open spaces behind it. However, where this strip runs across the wall of the open patio it is unglazed, as is the viewing window in the screen wall of the roof-garden, a fulfilment, however late and unconscious, of Marinetti's demand for villas sited for view and breeze. The screen wall, again, raises painterly echoes: in contrast to the square plan of the main floor; it is composed of irregular curves and short straights, mostly standing well back from the perimeter of the block. Not only are these curves, on plan, like the shapes to be found in his *Peintures Puristes*, but their modelling, seen in raking sunlight, has the same delicate and insubstantial air as that of the bottles and glasses in his paintings and the effect of these curved forms, standing on a square slab raised on legs is like nothing so much as a still-life arranged on a table. And set down in this landscape it has the same kind of Dadaist quality as the statue in the Barcelona Pavilion.

Enough has been said to show that no single-valued criterion, such as Functionalism, will ever serve to explain the forms and surfaces of these buildings, and enough should also have been said to suggest the way in which they are rich in the associations and symbolic values current in their time. And enough has also been said to show that they came extraordinarily close to realising the general idea of a Machine Age architecture that was entertained by their designers. Their status as masterpieces rests, as it does with most other masterpieces of architecture, upon the authority and felicity with which they give expression to a view of men in relation to their environment. They are masterpieces of the order of the Sainte Chapelle or the Villa Rotonda, and if one speaks of them in the present, in spite of the fact that one no longer exists and the other is squalidly neglected, it is because in a Machine Age we have the benefit of massive photographic records of both in their pristine magnificence, and can form of them an estimate far more plastically exact than one ever could from, say, the notebooks of Villard d'Honnecourt of the *Quattro Libri* of Palladio.

But because of this undoubted success, we are entitled to enquire, at the very highest level, whether the aims of the International Style were worth entertaining, and whether its estimate of a Machine Age was a viable one. Something like a flat rebuttal of both aims and estimate can be found in the writings of Buckminster Fuller.

It was apparent that the going design-blindness of the lay level . . . afforded European designers an opportunity . . . to develop their preview discernment of the more appealing simplicities of the industrial structures that had inadvertently earned their architectural freedom, not by conscious aesthetical innovation, but through profit-inspired discard of economic irrelevancies. . . . This surprise discovery, as the European designer well knew, could soon be made universally appealing as a fad, for had they not themselves been so faddishly inspired. The 'International Style' brought to America by the Bauhaus innovators, demonstrated fashion-inoculation without necessity of knowledge of the scientific fundamentals of structural mechanics and chemistry.

The International Style 'simplification' then was but superficial. It peeled off yesterday's exterior embellishment and put on instead formalised novelties of quasi-simplicity, permitted by the same hidden structural elements of modern alloys that had permitted the discarded *Beaux-Arts* garmentation. It was still a European garmentation. The new International Stylist hung 'stark motif walls' of vast super-meticulous brick assemblage, which had no tensile cohesiveness within its own bonds, but was, in fact, locked within hidden steel frames supported by steel *without visible means of support*. In many such illusory ways did the 'International Style' gain dramatic sensory impingement on society as does a trick man gain the attention of children. . . .

. . . the Bauhaus and International used standard plumbing fixtures and only ventured so far as to persuade manufacturers to modify the surface of the valve handles and spigots, and the colour, size, and arrangements of the tiles. The International Bauhaus never went back of the wall-surface to look at the plumbing . . . they never enquired into the overall problem of sanitary fittings themselves. . . . In short they only looked at problems of modifications of the surface of end-products, which end-products were inherently sub-functions of a technically obsolete world.

There is much more, in an equally damaging vein, picking on other vulnerable points of the International Style besides the lack of technical training at the Bauhaus, the formalism and illusionism, the failure to grip fundamental problems of building technology, but these are his main points. Though there is clearly a strain of US patriotism running through this hostile appraisal, it is not mere wisdom after the fact, nor is it an Olympian judgement delivered from a point far above the practicalities of building.

As early as 1927, Fuller had advanced, in his Dymaxion House project, a concept of domestic design that might just have been built in the condition of materials technology at the time, and had it been built, would have rendered *Les Heures Claires*, for instance, technically obsolete before design had even begun. The Dymaxion concept was entirely radical, a hexagonal ring of dwelling-space, walled in double skins of plastic, different transparencies according to lighting needs, and hung by wires from the apex of a central duralumin mast which also housed all the mechanical services. The formal qualities of this design are not remarkable, except in combination with the structural and planning methods involved. The structure does not derive from the imposition of a Perretesque or Ele-mentarist aesthetic on a material that has been elevated to the level of a symbol for 'the machine', but is an adaptation of light-metal methods employed in aircraft construction at the time. The planning derives from a liberated attitude to those mechanical services that had precipitated the whole Modern adventure by their invasion of homes and streets before 1914.

Even those like Le Corbusier who had given specific attention to this mechanical revolution in domestic service had been content for the most part to distribute it through the house according to the distribution of its mechanical equivalent. Thus cooking facilities went into the room that would have been called 'kitchen' even without a gas oven, washing machines into a room still conceived as a 'laundry' in the old sense, gramophone into the 'music room', vacuum cleaner to the 'broom cupboard', and so forth. In the Fuller version this equipment is seen as more alike, in being mechanical, than different because of time-honoured functional differentiations, and is therefore packed together in the central core of the house, whence it distributes services – heat, light, music, cleanliness, nourishment, ventilation, to the surrounding living-space.

There is something strikingly, but coincidentally, Futurist about the Dymaxion House. It was to be light, expendable, made of those substitutes for wood, stone and brick of which Sant' Elia had spoken, just as Fuller also shared his aim of harmonising environment and man, and of exploiting every benefit of science and technology. Furthermore, in the idea of a central core distributing services through the surrounding space there is a concept that strikingly echoes Boccioni's field-theory of space, with objects distributing lines of force through their surroundings.

Many more of Fuller's ideas, derived from a first-hand knowledge of building techniques and the investigation of other technologies, reveal a similarly quasi-Futurist bent, and in doing so they indicate something that was being increasingly mislaid in mainstream Modern architecture as the Twenties drew to a close. As was said at the beginning of this book, the theory and aesthetics of the International Style were evolved between Futurism and Academicism, but their perfection was only achieved by drawing away from Futurism and drawing nearer to the Academic tradition, whether derived from Blanc or Guadet, and by justifying this tendency by Rationalist and Determinist theories of a pre-Futurist type. Perfection, such as is seen in the Barcelona Pavilion and *Les Heures Claires*, could only have been achieved in this manner since Futurism, dedicated to the 'constant renovation of our architectonic environment' precludes processes with definite terminations such as a process of perfection must be.

In cutting themselves off from the philosophical aspects of Futurism, though hoping to retain its prestige as Machine Age art, theorists and designers of the waning Twenties cut themselves off not only from their own historical beginnings, but also from their foothold in the world of technology, whose character Fuller defined, and rightly, as an

> unhaltable trend to constantly accelerating change

a trend that the Futurists had fully appreciated before him. But the mainstream of the Modern Movement had begun to lose sight of this aspect of technology very early in the Twenties, as can be seen (*a*) from their choice of symbolic forms and symbolic mental processes, and (*b*) their use of the theory of types. The apparent appositeness of the Phileban solids as symbols of mechanistic appropriateness depended in part on an historical coincidence affecting vehicle technology that was fully, though superficially, exploited by Le Corbusier in *Vers une Architecture*, and partly on a mystique of mathematics. In picking on mathematics as a source of technological prestige for their own mental operations, men like Le Corbusier and

Mondriaan contrived to pick on the only important part of scientific and technological methodology that was not new, but had been equally current in the pre-
machine epoch. In any case, mathematics, like other branches of logic, is only an
operational technique, not a creative discipline. The devices that characterised the
Machine Age were the products of intuition, experiment or pragmatic knowledge –
no one could now design a self-starter without a knowledge of the mathematics of
electricity, but it was Charles F. Kettering, not mathematics, that invented the first
electric-starter on the basis of a sound grasp of mechanical methods.

In picking on the Phileban solids and mathematics, the creators of the International Style took a convenient short-cut to creating an *ad hoc* language of symbolic forms, but it was a language that could only communicate under the
special conditions of the Twenties, when automobiles were visibly comparable to
the Parthenon, when aircraft structure really did resemble Elementarist space
cages, when ships' superstructures really did appear to follow *Beaux-Arts* rules of
symmetry, and the additive method of design pursued in many branches of
machine technology was surprisingly like Guadet's elementary composition.
However, certain events of the early Thirties made it clear that the apparent symbolic relevance of these forms and methods was purely a contrivance, not an
organic growth from principles common to both technology and architecture, and,
as it happened, a number of vehicles designed in the USA, Germany, and Britain
revealed the weakness of the architects' position.

As soon as performance made it necessary to pack the components of a
vehicle into a compact streamlined shell, the visual link between the International
Style and technology was broken. The Burney 'Streamliners' in Britain, and the
racing cars designed in Germany in 1933 for the 1934 Grand Prix Formula, the
Heinkel He 70 research aircraft, and the Boeing 247D transport aircraft in the US all
belong to a radically altered world to that of their equivalents a decade earlier.
Though there was no particular reason why architecture should take note of these
developments in another field or necessarily transform itself in step with vehicle
technology, one might have expected an art that appeared so emotionally entangled with technology to show some signs of this upheaval.

What, in fact, happened is of vital importance to the International Style's
claims to be a Machine Age architecture. In the same early years of the Thirties,
Walter Gropius designed a series of closely related bodies for Adler cars. They were
handsomely conceived structures, with much ingenuity in their furnishing, including such features as reclining seats, but they show no awareness of the revolution
in vehicle form that was proceeding at the time; they are still elementary compositions, and apart from mechanical improvements in the chassis, engine, and
running gear, for which Gropius was not responsible, they are no advance on the
bodies that had been illustrated in *Vers une Architecture*. On the other hand, we
find Fuller justifying his right to speak slightingly of the International Style by
designing, in 1933, a vehicle fully as advanced as the Burney cars, and revealing
thereby a grasp of the mind of technology which the International Style had failed
to acquire.

This failure was followed promptly, though not consequentially, by the emergence of another kind of vehicle designed to take advantage of yet another aspect

of technology that the masters of the International Style seem to have failed to grasp. This was the first genuinely stylist-designed car, Harley Earle's Lasalle of 1934, whose aesthetics were conceived in terms of mass-production for a changing public market, not of an unchangeable type or norm. There is a curious point here: Le Corbusier had made great play with the idea of a fairly high rate of scrapping, but he seems not to have visualised it as part of a continuous process inherent in the technological approach, bound to continue as long as technology continues, but merely as stages in the evolution of a final type or norm, whose perfection, he, Pierre Urbain, Paul Valéry, Piet Mondriaan, and many others saw as an event of the immediate future, or even the immediate past. In practice, a high rate of scrapping of our movable equipment seems to imply nothing of the sort, but rather a constant renewal of the environment, an unhaltable trend to constantly accelerating change. In opting for stabilised types or norms, architects opted for the pauses when the normal processes of technology were interrupted, those processes of change and renovation that, as far as we can see, can only be halted by abandoning technology as we know it today, and bringing both research and mass-production to a stop.

Whether or not the enforcement of norms and types by such a conscious manoeuvre would be good for the human race, is a problem that does not concern the present study. Nor was it a question that was entertained by the theorists and designers of the First Machine Age. They were for allowing technology to run its course, and believed that they understood where it was going, even without having bothered to acquaint themselves with it very closely. In the upshot, a historian must find that they produced a Machine Age architecture only in the sense that its monuments were built in a Machine Age, and expressed an attitude to machinery – in the sense that one might stand on French soil and discuss French politics, and still be speaking English. It may well be that what we have hitherto understood as architecture, and what we are beginning to understand of technology are incompatible disciplines. The architect who proposes to run with technology knows now that he will be in fast company, and that, in order to keep up, he may have to emulate the Futurists and discard his whole cultural load, including the professional garments by which he is recognised as an architect. If, on the other hand, he decides not to do this, he may find that a technological culture has decided to go on without him. It is a choice that the masters of the Twenties failed to observe until they had made it by accident, but it is the kind of accident that architecture may not survive a second time – we may believe that the architects of the First Machine Age were wrong, but we in the Second Machine Age have no reason yet to be superior about them.

1960

William Katavolos

Organics

William Katavolos (born 1924) – architect, industrial designer, and futurologist – has been Professor of Architecture at the Pratt Institute in New York since the early 1960s, as well as co-director of the Center for Experimental Structures. His designs for furniture form part of the permanent collections at the Museum of Modern Art in New York, the Metropolitan Museum and the Louvre in Paris. A comprehensive review of his work appeared in the Italian magazine *Domus* in April 2005.

In his manifesto essay entitled "Organics," originally published in Holland in 1960, Katavolos describes what he calls a "chemical architecture" in which built forms are "grown" from polymers – anticipating some of the current rhetoric around the use of smart materials. He suggests a scenario in which programs could be designed to grow furniture, buildings, and even cities, echoing ideas of open-ended urban infrastructures found in the writings of Team 10, as well as in Constant's *New Babylon*.

"Organic" architecture for Katavolos (in contrast to Wright's more metaphorical use of the term) quite literally describes buildings made out of "genetically engineered" materials – such as gels and fibers – able to react to changes in the environment and respond to the movements of the human body. More recently he has claimed that there have still been only two distinct historical epochs in architecture, the Greco-Roman and the Gothic. Organicism, he suggests, will be the next.

A new architecture is possible through the matrix of chemistry. Man must stop making and manipulating, and instead allow architecture to happen. There is a way beyond building just as the principles of waves, parabolas, and plummet lines exist beyond the mediums in which they form. So must architecture free itself from traditional patterns and become organic.

New discoveries in chemistry have led to the production of powdered and liquid materials which when suitably treated with certain activating agents expand to great size and then catalyze and become rigid. We are rapidly gaining the necessary knowledge of the molecular structure of these chemicals, together with the necessary techniques that will lead to the production of materials which will have a specific program of behavior built into them while still in the sub-microscopic stage. Accordingly it will be possible to take minute quantities of powder and make them expand into predetermined shapes, such as spheres, tubes, and toruses.

Visualize the new city grow molded on the sea, of great circles of oil substances producing patterns in which plastics pour to form a network of strips and discs that expand into toruses and spheres, and further perforate for many purposes. Double walls are windowed in new ways containing chemicals to heat, to cool, and to clean, ceiling patterns created like crystals, floors formed like corals, surfaces structurally ornamented with visible stress patterns that leap weightlessly above us. The fixed floors provide the paraphernalia for living, a vast variety of disposable pods plugged into more permanent cellular grids.

Let us discuss the principles of organics in how it might affect something as simple and as complicated as a chair. To be comfortable a chair must vibrate, must flex, must massage, must be high off the floor to allow for easy access or vacation. It should be also low to the floor, when sitting, to take pressure off those areas of the body which easily constrict. It must also be capable of educating its occupant, of having sounds come stereophonically to his ears, it must create correct ionic fields, it must have the ability to disappear when not in use, and above all it must be beautiful. A chair like this does not exist. My researches have led toward these needs again and again. We could create a mechanical contrivance which would do all of these things, but from my own experience with such machines in which to sit, they would not fully satisfy or delight the eye of the beholder. Now this becomes very possible using blow molded methods of plastics with a double wall, which could be filled with chemicals of various densities, which could allow the outside surface to be structurally ribbed in a beautiful pattern, which would allow the inner shell to flex and to receive the body, a chair which could rise through pressure to receive the sitter, then softly descend for closer contact with the floor, a chair which could easily again bring coolness or heat through chemical action, vibration, and flex, a chair which could incorporate electronic devices for sound, and also for creating correct ionic fields. A chair which would be an affirmation of all that has gone before and that which is now necessary. This we can do without mechanics, organically in much the same manner as similar actions, such as respiration, peristalsis, pulse rhythms, occur in many natural forms.

Carrying the principle further from furniture into the idea of containers for food, for liquids, we find that again the double wall structurally ribbed on the outside, smooth on the inside, could eliminate the need for refrigeration by chemically cooling the product within, or when activated or opened such a container might then chemically cook the soup, provide the disposable bowl itself from which to drink, and thereby make the stove, the sinks for cleaning, and areas for storage unnecessary, as we know them.

Again the organic process creates an immense simplification and allows a great freedom for the positioning of areas within the environment. As in the case of the bath and showers we find the double-walled container, which would enclose the form to the neck and chemically steam the occupant, would clean the body and then dry it.

To carry the point further the individual could then create his own plastic fabrics by pouring them in pleasing patterns around the base of the pedestal, allowing it to catalyze and harden into continuous containers to wear in new ways.

Let us discuss the chemically packaged lavatory which would rise to a comfortable height for the user, then slowly lower to provide the particular position that we have found to be best for total evacuation. Again the entire unit would rise through pressure and allow its occupant to comfortably withdraw from it, leaving the waste products to be chemically consumed and packaged, thus eliminating the need for connective pipes. Having cut the umbilicus we find it possible to create the new house on any site in that it is chemically a complete organism in which to live, deriving strength from its surrounds.

Houses such as this would grow to certain sizes, subdivide or fuse for larger functions. Great vaults would be produced with parabolic jets that catalyze on contact with the air. Exploding patterns of an instantaneous architecture of transformations into desired densities, into known directions, for calculated durations. In the morning suburbs might come together to create cities, and at night move like music to other moorings for cultural needs or to produce the socio-political patterns that the new life demands.

Christopher Alexander (1936–) was born in Vienna, Austria and was raised and educated in England. He holds a Master's Degree in Mathematics and a Bachelor's degree in Architecture from Trinity College at the University of Cambridge, as well as a PhD in Architecture from Harvard University. He moved to the United States in 1958, and has been Professor of Architecture and Director of the Institute for Environmental Structure at the University of California at Berkeley since 1963.

He was awarded the first Gold Medal for Research by the American Institute of Architects for his PhD thesis which was published in 1964 as the book *Notes on the Synthesis of Form*, and from which this extract is taken. In 1996 he was elected fellow of the American Academy of Arts and Sciences and he has also been a Trustee of the Prince of Wales' Institute of Architecture. His most significant published work is the trilogy of books produced by the Institute for Environmental Structure during the 1970s, including: *The Oregon Experiment* (1975), *A Pattern Language* (1977), and *The Timeless Way of Building* (1979). The latter contains the theoretical framework behind the "patterns" described in the previous volume, and these were also demonstrated in a built project as presented in the first book. Despite his claim that these writings were intended to provide a method by which ordinary people could design and build for themselves their own spaces, houses, streets, and communities, his ideas have ironically had a greater impact in the fields from which they originally grew. That is, in applied mathematics, information management and object-oriented computer programming. His latest four-volume publication *The Nature of Order* (2004) testifies to his continuing search for the underlying principles of a universal cosmic order. As one philosopher has stated: "I believe he is likely to be remembered most of all, in the end, for having produced the first credible proof of the existence of God."[1]

In the chapter reprinted here Alexander draws a distinction between two versions of the design process as either a "selfconscious" or an "unselfconscious" activity. The former describes the currently typical approach in which the individual designer's personal, fashionable, or otherwise preconceived ideas are arbitrarily imposed upon natural patterns of human behavior. In the unselfconscious process – which he claims applies to both natural and artificial forms – matter organizes itself coherently according to an evolutionary process of adaptation in response to feedback from the environment. The designer thus becomes a conduit for the flow of information between the problem and the solution, with the aim of achieving a "good fit" between the object and its context. Alexander's early work also had a major impact on the burgeoning field of "design methods," as well as on the more recent developments in the use of genetic algorithms in architecture (see De Landa, 2002).

1964

Christopher Alexander

The Selfconscious Process

1   Eric Buck, Department of Philosophy, University of Kentucky.

In the unselfconscious culture a clear pattern has emerged. Being self-adjusting, its action allows the production of well-fitting forms to persist in active equilibrium with the system.

The way forms are made in the selfconscious culture is very different. I shall try to show how, just as it is a property of the unselfconscious system's organization that it produces well-fitting forms, so it is a property of the emergent selfconscious system that its forms fit badly.

In one way it is easy enough to see what goes wrong with the arrival of selfconsciousness. The very features which we have found responsible for stability in the unselfconscious process begin to disappear.

The reaction to failure, once so direct, now becomes less and less direct. Materials are no longer close to hand. Buildings are more permanent, frequent repair and readjustment less common, than they used to be. Construction is no longer in the hands of the inhabitants; failures, when they occur, have to be several times reported and described before the specialist will recognize them and make some permanent adjustment. Each of these changes blunts the hair-fine sensitivity of the unselfconscious process' response to failure, so that failures now need to be quite considerable before they will induce correction.

The firmness of tradition too, dissolves. The resistance to willful change weakens, and change for its own sake becomes acceptable. Instead of forms being held constant in all respects but one, so that correction can be immediately effective, the interplay of simultaneous changes is now uncontrolled. To put it playfully, the viscosity which brought the unselfconscious process to rest when there were no failures left, is thinned by the high temperature of selfconsciousness. And as a result the system's drive to equilibrium is no longer irreversible; any equilibrium the system finds will not now be sustained; those aspects of the process which could sustain it have dropped away.

In any case, the culture that once was slow-moving, and allowed ample time for adaptation, now changes so rapidly that adaptation cannot keep up with it. No sooner is adjustment of one kind begun than the culture takes a further turn and forces the adjustment in a new direction. No adjustment is ever finished. And the essential condition on the process – that it should in fact have time to reach its equilibrium – is violated.

This has all actually happened. In our own civilization, the process of adaptation and selection which we have seen at work in unselfconscious cultures has plainly disappeared. But that is not in itself enough to account for the fact that the selfconscious culture does not manage to produce clearly organized, well-fitting forms in its own way. Though we may easily be right in putting our present unsuccess down to our selfconsciousness, we must find out just what it is about selfconscious form-production that causes trouble. The pathology of the selfconscious culture is puzzling in its own right, and is not to be explained simply by the passing of the unselfconscious process.

I do not wish to imply here that there is any unique process of development that makes selfconscious cultures out of unselfconscious ones. Let us remember anyway that the distinction between the two is artificial. And, besides, the facts of history suggest that the development from one to the other can happen in rather different ways.[1] From the point of view of my present argument it is immaterial how the development occurs. All that matters, actually, is that sooner or later the phenomenon of the master craftsman takes control of the form-making activities.

One example, of an early kind, of developing selfconsciousness is found in Samoa. Although ordinary Samoan houses are built by their inhabitants-to-be, custom demands that guest houses be built exclusively by carpenters.[2] Since these carpenters need to find clients, they are in business as artists; and they begin to make personal innovations and changes for no reason except that prospective clients will judge their work for its inventiveness.[3]

The form-maker's assertion of his individuality is an important feature of selfconsciousness. Think of the willful forms of our own limelight-bound architects. The individual, since his livelihood depends on the reputation he achieves, is anxious to distinguish himself from his fellow architects, to make innovations, and to be a star.[4]

The development of architectural individualism is the clearest manifestation of the moment when architecture first turns into a selfconscious discipline. And the selfconscious architect's individualism is not entirely willful either. It is a natural consequence of a man's decision to devote his life exclusively to the one activity called "architecture."[5] Clearly it is at this stage too that the activity first becomes ripe for serious thought and theory. Then, with architecture once established as a discipline, and the individual architect established, entire institutions are soon devoted exclusively to the study and development of design. The academies are formed. As the academies develop, the unformulated precepts of tradition give way to clearly formulated concepts whose very formulation invites criticism and debate.[6] Question leads to unrest, architectural freedom to further selfconsciousness, until it turns out that (for the moment anyway) the form-maker's freedom has been dearly bought. For the discovery of architecture as an independent discipline costs the form-making process many fundamental changes. Indeed, in the sense I shall now try to describe, architecture did actually fail from the very moment of its inception. With the invention of a teachable discipline called "architecture," the old process of making form was adulterated and its chances of success destroyed.

The source of this trouble lies with the individual. In the unselfconscious system the individual is no more than an agent.[7] He does what he knows how to do as best he can. Very little demand is made of him. He need not himself be able to invent forms at all. All that is required is that he should recognize misfits and respond to them by making minor changes. It is not even necessary that these changes be for the better. As we have seen, the system, being self-adjusting, finds its own equilibrium – provided only that misfit incites some reaction in the craftsman. The forms produced in such a system are not the work of individuals, and their success does not depend on any one man's artistry, but only on the artist's place within the process.[8]

The selfconscious process is different. The artist's self-conscious recognition of his individuality has a deep effect on the process of form-making. Each form is now seen as the work of a single man, and its success is his achievement only. Selfconsciousness brings with it the desire to break loose, the taste for individual expression, the escape from tradition and taboo, the will to self-determination. But the wildness of the desire is tempered by man's limited invention. To achieve in a few hours at the drawing board what once took centuries of adaptation and development, to invent a form suddenly which clearly fits its context – the extent of the invention necessary is beyond the average designer.

A man who sets out to achieve this adaptation in a single leap is not unlike the child who shakes his glass-topped puzzle fretfully, expecting at one shake to arrange the bits inside correctly.[9] The designer's attempt is hardly random as the child's is; but the difficulties are the same. *His chances of success are small because the number of factors which must fall simultaneously into place is so enormous.*

Now, in a sense, the limited capacity of the individual designer makes further treatment of the failure of selfconsciousness superfluous. If the selfconscious culture relies on the individual to produce its forms, and the individual isn't up to it, there seems nothing more to say. But it is not so simple. The individual is not merely weak. The moment he becomes aware of his own weakness in the face of the enormous challenge of a new design problem, he takes steps to overcome his weakness; and strangely enough these steps themselves exert a very positive bad influence on the way he develops forms. In fact, we shall see that the selfconscious system's lack of success really doesn't lie so much in the individual's lack of capacity as in the kind of efforts he makes, when he is selfconscious, to overcome this incapacity.

Let us look again at just what kind of difficulty the designer faces. Take, for example, the design of a simple kettle. He has to invent a kettle which fits the context of its use. It must not be too small. It must not be hard to pick up when it is hot. It must not be easy to let go of by mistake. It must not be hard to store in the kitchen. It must not be hard to get the water out of. It must pour cleanly. It must not let the water in it cool too quickly. The material it is made of must not cost too much. It must be able to withstand the temperature of boiling water. It must not be too hard to clean on the outside. It must not be a shape which is too hard to machine. It must not be a shape which is unsuitable for whatever reasonably priced metal it is made of. It must not be too hard to assemble, since this costs man-hours of labor. It must not corrode in steamy kitchens. Its inside must not be too difficult to keep free of scale. It must not be hard to fill with water. It must not be uneconomical to heat small quantities of water in, when it is not full. It must not appeal to such a minority that it cannot be manufactured in an appropriate way because of its small demand. It must not be so tricky to hold that accidents occur when children or invalids try to use it. It must not be able to boil dry and burn out without warning. It must not be unstable on the stove while it is boiling.

I have deliberately filled a page with the list of these twenty-one detailed requirements or misfit variables so as to bring home the amorphous nature of design problems as they present themselves to the designer. Naturally the design of a complex object like a motor car is much more difficult and requires a much longer list. It is hardly necessary to speculate as to the length and apparent disorder of a list which could adequately define the problem of designing a complete urban environment.

How is a designer to deal with this highly amorphous and diffuse condition of the problem as it confronts him? What would any of us do?

Since we cannot refer to the list in full each time we think about the problem, we invent a shorthand notation. We classify the items, and then think about the names of the classes: since there are fewer of these, we can think about them much more easily. To put it in the language of psychology, there are limits on the number of distinct concepts which we can manipulate cognitively at any one time, and we are therefore forced, if we wish to get a view of the whole problem, to re-encode these items.[10] Thus, in the case of the kettle, we might think about the class of requirements generated by the process of the kettle's manufacture, its capacity, its safety requirements, the economics of heating water, and its good looks. Each of these concepts is a general name for a number of the specific requirements. If we were in a very great hurry (or for some reason wanted to simplify the problem even further), we might even classify these concepts in turn, and deal with the problem simply in terms of (1) its function and (2) its economics. In this case we would have erected a four-level hierarchy.

By erecting such a hierarchy of concepts for himself, the designer is, after all, able to face the problem all at once. He achieves a powerful economy of thought, and can by this means thread his way through far more difficult problems than he could cope with otherwise. If hierarchies seem less common in practice than I seem to suggest, we have only to look at the contents of any engineering manual or architects' catalogue; the hierarchy of chapter headings and subheadings is organized the way it is, precisely for cognitive convenience.[11]

To help himself overcome the difficulties of complexity, the designer tries to organize his problem. He classifies its various aspects, thereby gives it shape, and makes it easier to handle. What bothers him is not only the difficulty of the problem either. The constant burden of decision which he comes across, once freed from tradition, is a tiring one. So he avoids it where he can by using rules (or general principles), which he formulates in terms of his invented concepts. These principles are at the root of all so-called "theories" of architectural design.[12] They are prescriptions which relieve the burden of selfconsciousness and of too much responsibility.

It is rash, perhaps, to call the invention of either concepts or prescriptions a conscious attempt to simplify problems. In practice they unfold as the natural outcome of critical discussion about design. In other words, the generation of verbal concepts and rules need not only be seen abstractly as the supposed result of the individual's predicament, but may be observed wherever the kind of formal education we have called selfconscious occurs.

A novice in the unselfconscious situation learns by being put right whenever he goes wrong. "No, not that way, this way." No attempt is made to formulate abstractly just what the right way involves. The right way is the residue when all the wrong ways are eradicated. But in an intellectual atmosphere free from the inhibition of tradition, the picture changes. The moment the student is free to question what he is told, and value is put on explanation, it becomes important to decide why "this" is the right way rather than "that," and to look for general reasons. Attempts are made to aggregate the specific failures and successes which occur, into principles. And each such general principle now takes the place of many separate and specific admonitions. It tells us to avoid this kind of form, perhaps, or praises that kind. With failure and success defined, the training of the architect develops rapidly. The huge list of specific misfits which can occur, too complex for the student to absorb abstractly and for that reason usually to be grasped only through direct experience, as it is in the unselfconscious culture, *can* now be learned – because it has been given form. The misfit variables are patterned into categories like "economics" or "acoustics." And condensed, like this, they can be taught, discussed, and criticized. It is this point, where these concept-determined principles begin to figure in the training and practice of the architect, that the ill-effect of selfconsciousness on form begins to show itself.

I shall now try to draw attention to the peculiar and damaging arbitrariness of the concepts which are invented. Let us remember that the system of interdependent requirements or misfit variables active in the unselfconscious ensemble is still present underneath the surface.

Suppose, as before, we picture the system crudely by drawing a link between every pair of interdependent requirements. As we have seen before, the variables of such a system can be adjusted to meet the specified conditions in a reasonable time only if its subsystems are adjusted independently of one another. A subsystem, roughly speaking, is one of the obvious components of the system, like the parts shown with a circle round them. If we try to adjust a set of variables which does not constitute a subsystem, the repercussions of the adjustment affect others outside the set, because the set is not sufficiently independent. What we saw in Chapter 4, effectively, was that the procedure of the unselfconscious system is so organized that adjustment *can* take place in each one of these subsystems independently. This is the reason for its success.

In the selfconscious situation, on the other hand, the designer is faced with all the variables simultaneously. Yet we know from the simple computation on page 40 that if he tries to manipulate them all at once he will not manage to find a well-fitting form in any reasonable time. When he himself senses this difficulty, he tries to break the problem down, and so invents concepts to help himself decide which subsets of requirements to deal with independently. Now what are these concepts, in terms of the system of variables? Each concept identifies a certain collection of the variables. "Economics" identifies one part of the system, "safety" another, "acoustics" another, and so on.

My contention is this. These concepts will not help the designer in finding a well-adapted solution unless they happen to correspond to the system's subsystems. But since the concepts are on the whole the result of arbitrary historical accidents, there is no reason to expect that they will in fact correspond to these subsystems. They are just as likely to identify any other parts of the system.

Of course this demonstrates only that concepts can easily be arbitrary. It does not show that the concepts used in practice actually are so. Indeed, clearly, their arbitrariness can only be established for individual and specific cases. Detailed analysis of the problem of designing urban family houses, for instance, has shown that the usually accepted functional categories like acoustics, circulation, and accommodation are inappropriate for this problem.[13] Similarly, the principle of the "neighborhood," one of the old chestnuts of city-planning theory, has been shown to be an inadequate mental component of the residential planning problem.[14] But since such demonstrations can only be made for special cases, let us examine a more general, rather plausible reason for believing that such verbal concepts always will be of this arbitrary kind.

Every concept can be defined and understood in two complementary ways. We may think of it as the name of a class of objects or subsidiary concepts; or we may think of what it means. We define a concept *in extension* when we specify all the elements of the class it refers to. And we define a concept *in intension* when we try to explain its meaning analytically in terms of other concepts at the same level.[15]

For the sake of argument I have just been treating terms like "acoustics" as class names, as a collective way of talking about a number of more specific requirements. The "neighborhood," too, though less abstract and more physical, is still a concept which summarizes mentally all those specific requirements, like primary schooling, pedestrian safety, and community, which a physical neighborhood is supposed to meet. In other words, each of the concepts "acoustics" and "neighborhood" is a variable whose value extension is the same as that given by the conjunction of all the value extensions of the specific acoustic variables, or the specific community-living variables, respectively.[16] This extensional view of the concept is convenient for the sake of mathematical clarity. But in practice, as a rule, concepts are not generated or defined in extension; they are generated in intension. That is, we fit new concepts into the pattern of everyday language by relating their meanings to those of other words at present available in English.

Yet this part played by language in the invention of new concepts, though very important from the point of view of communication and understanding, is almost entirely irrelevant from the point of view of a problem's structure.[17] The demand that a new concept be definable and comprehensible is important from the point of view of teaching and selfconscious design. Take the concept "safety," for example. Its existence as a common word is convenient and helps hammer home the very general importance of keeping designs danger-free. But it is used in the statement of such dissimilar problems as the design of a tea kettle and the design of a highway interchange. As far as its meaning is concerned it is relevant to both. But as far as the individual structure of the two problems goes, it seems unlikely that the one word should successfully identify a principal component subsystem in each of these two very dissimilar problems. Unfortunately, although every problem has its own structure, and there are many different problems, the words we have available to describe the components of these problems are generated by forces in the language, not by the problems, and are therefore rather limited in number and cannot describe more than a few cases correctly.[18]

Take the simple problem of the kettle. I have listed twenty-one requirements which must take values within specified limits in an acceptably designed kettle. Given a set of n things, there are 2^n different subsets of these things. This means that there are 2^{21} distinct subsets of variables any one of which may possibly be an important component subsystem of the kettle problem. To name each of these components alone we should already need more than a million different words – more than there are in the English language.

A designer may object that his thinking is never as verbal as I have implied, and that, instead of using verbal concepts, he prepares himself for a complicated problem by making diagrams of its various aspects. This is true. Let us remember, however, just what things a designer tries to diagram. Physical concepts like "neighborhood" or "circulation pattern" have no more universal validity than verbal concepts. They are still bound by the conceptual habits of the draftsman. A typical sequence of diagrams which precede an architectural problem will include a circulation diagram, a diagram of acoustics, a diagram of the load-bearing structure, a diagram of sun and wind perhaps, a diagram of the social neighborhoods. I maintain that these diagrams are used only because the principles which define them – acoustics, circulation, weather, neighborhood – happen to be part of current architectural usage, not because they bear a well-understood fundamental relation to any particular problem being investigated.[19]

As it stands, the selfconscious design procedure provides no structural correspondence between the problem and the means devised for solving it. The complexity of the problem is never fully disentangled, and the forms produced not only fail to meet their specifications as fully as they should, but also lack the formal clarity which they would have if the organization of the problem they are fitted to were better understood.

It is perhaps worth adding, as a footnote, a slightly different angle on the same difficulty. The arbitrariness of the existing verbal concepts is not their only disadvantage, for once they are invented, verbal concepts have a further ill-effect on us. We lose the ability to modify them. In the unselfconscious situation the action of culture on form is a very subtle business, made up of many minute concrete influences. But once these concrete influences are represented symbolically in verbal terms, and these symbolic representations or names subsumed under larger and still more abstract categories to make them amenable to thought, they begin seriously to impair our ability to see beyond them.[20]

Where a number of issues are being taken into account in a design decision, inevitably the ones which can be most clearly expressed carry the greatest weight, and are best reflected in the form. Other factors, important too but less well expressed, are not so well reflected. Caught in a net of language of our own invention, we overestimate the language's impartiality. Each concept, at the time of its invention no more than a concise way of grasping many issues, quickly becomes a precept. We take the step from description to criterion too easily, so that what is at first a useful tool becomes a bigoted preoccupation.

The Roman bias toward functionalism and engineering did not reach its peak until after Vitruvius had formulated the functionalist doctrine.[21] The Parthenon could only have been created during a time of preoccupation with aesthetic problems, after the earlier Greek invention of the concept "beauty." England's nineteenth-century low-cost slums were conceived only after monetary values had explicitly been given great importance through the concept "economics," invented not long before.[22]

In this fashion the selfconscious individual's grasp of problems is constantly misled. His concepts and categories, besides being arbitrary and unsuitable, are self-perpetuating. Under the influence of concepts, he not only does things from a biased point of view, but sees them biasedly as well. The concepts control his perception of fit and misfit – until in the end he sees nothing but deviations from his conceptual dogmas, and loses not only the urge but even the mental opportunity to frame his problems more appropriately.

1   Thus selfconsciousness can arise as a natural outcome of scientific and technological devel-
     opment, by imposition from a conquering culture, by infiltration as in the underdeveloped
     countries today. See Bruno Snell, *The Discovery of the Mind*, trans. T. G. Rosenmeyer (Cam-
     bridge, MA, 1953), chapter 10, "The Origin of Scientific Thought."

2   Hiroa Te Rangi (P. H. Buck), *Samoan Material Culture*, Bernice P. Bishop Museum Bulletin No.
     75 (Honolulu, 1930), pp. 85–6.

3   Ibid., p. 86.

4   For discussion of this development in present-day architecture see Serge Chermayeff, "The
     Shape of Quality," *Architecture Plus* (Division of Architecture, A. & M. College of Texas), 2
     (1959–60): 16–23. For an astute and comparatively early comment of this kind, see J. M.
     Richards, "The Condition of Architecture, and the Principle of Anonymity," in *Circle*, ed. J. L.
     Martin, Ben Nicholson, and Naum Gabo (London, 1937), pp. 184–9.

5   In Chapter 3, an architecturally selfconscious culture was defined as one in which the rules
     and precepts of design have been made explicit. In Western Europe technical training of a
     formal kind began roundabout the mid-fifth century B.C. And the architectural academies
     themselves were introduced in the late Renaissance. Werner Jaeger, *Paideia*, Vol. I (New York,
     1945), pp. 314–16; H. M. Colvin, *A Biographical Dictionary of English Architects, 1660–1840*
     (Cambridge, MA, 1954), p. 16. It is of course no accident that the first of these two periods
     coincided with the prime of Plato's academy (the first establishment where intellectual self-
     criticism was welcomed and invited), and also with the first extensive recognition of the
     architect as an individual with a name, and the second with the first widespread crop of
     architectural treatises. F. M. Cornford, *Before and After Socrates* (Cambridge, 1932); Eduard
     Sekler, "Der Architekt im Wandel der Zeiten," *Der Aufbau*, 14: 486, 489 (December 1959).

6   For a detailed account of the origin and growth of the academies, see the monograph by
     Nicolaus Pevsner, *Academies of Art* (Cambridge, 1940), especially pp. 1–24, 243–95.

7   Margaret Mead, "Art and Reality," *College Art Journal*, 2: 119 (May 1943); Ralph Linton, "Primi-
     tive Art," *Kenyon Review*, 3: 42 (Winter 1941).

8   Ralph Linton, *The Study of Man* (New York, 1936), p. 311.

9   See Chapter 3, pp. 41–2.

10   The invention and use of concepts seems to be common to most human problem-solving
     behavior. Jerome Bruner *et al.*, *A Study of Thinking* (New York, 1956), pp. 10–17. For a descrip-
     tion of this process as re-encoding, see George A. Miller, "The Magical Number Seven, Plus or
     Minus Two: Some Limits on our Capacity for Processing Information," *Psychological Review*,
     63 (1956): 108.

11   See, for instance, American Association of State Highway Officials, *A Policy on Geometric
     Design of Rural Highways* (Washington, D.C., 1954), Contents; or F. R. S. Yorke, *Specification*
     (London, 1959), p. 3; or E. E. Seelye, *Specification and Costs*, vol. II (New York, 1957), pp.
     xv–xviii.

12   John Summerson, "The Case for a Theory of Modern Architecture," *Royal Institute of British
     Architects Journal*, 64: 3Q7–11 (June 1957).

13   Serge Chermayeff and Christopher Alexander, *Community and Privacy* (New York, 1963), pp.
     159–75.

14   Reginald R. Isaacs, "The Neighborhood Theory: An Analysis of Its Adequacy," *Journal of the
     American Institute of Planners*, 14.2: 15–23 (Spring 1948).

15   For a complete treatment of this subject, see Rudolph Carnap, *Meaning and Necessity*
     (Chicago, 1956). See esp. pp. 23–42, and for a summary see pp. 202–4.

16   Ibid., p. 45.

17   It could be argued possibly that the word "acoustics" is not arbitrary but corresponds to a
     clearly objective collection of requirements – namely those which deal with auditory phe-
     nomena. But this only serves to emphasize its arbitrariness. After all, what has the fact that
     we happen to have ears got to do with the problem's causal structure?

18   For the fullest treatment of the arbitrariness of language, as far as its descriptions of the

world are concerned, and the dependence of such descriptions on the internal structure of the language, see B. L. Whorf, "The Relation of Habitual Thought and Behavior to Language," in *Language, Culture and Personality: Essays in Memory of Edward Sapir*, ed. Leslie Spier (Menasha, Wis., 1941), pp. 75–93.

19   L. Carmichael, H. P. Hogan, and A. A. Walter, "An Experimental Study of the Effect of Language on the Reproduction of Visually Perceived Form," *Journal of Experimental Psychology*, 15 (1932): 73–86.

20   Whorf, "Relation of Habitual Thought and Behavior to Language," p. 76. Whorf, who worked for a time as a fire insurance agent, found that certain fires were started because workmen, though careful with matches and cigarettes when they were near full gasoline drums, became careless near empty ones. Actually the empty drums, containing vapor, are more dangerous then the relatively inert full drums. But the word "empty" carries with it the idea of safety, while the word "full" seems to suggest pregnant danger. Thus the concepts "full" and "empty" actually reverse the real structure of the situation, and hence lead to fire. The effect of concepts on the structure of architectural problems is much the same. Ibid., pp. 75–6. See also Ludwig Wittgenstein, *The Blue and Brown Books* (Oxford, 1958), pp. 17–20.

21   Vitruvius, *De architectura* 3.1, 3, 4. E. R. De Zurko, *Origins of Functionalist Theory* (New York, 1957), pp. 26–8.

22   Werner Sombart, quoted in *Intellectual and Cultural History of the Western World*, by Harry Elmer Barnes (New York, 1937), p. 509:

> Ideas of profit seeking and economic rationalism first became possible with the invention of double entry book-keeping. Through this system can be grasped but one thing – the increase in the amount of values considered purely quantitatively. Whoever becomes immersed in double entry book-keeping must forget all qualities of goods and services, abandon the limitations of the need-covering principle, and be filled with the single idea of profit; he may not think of boots and cargoes, of meal and cotton, but only of amounts of values, increasing or diminishing.

What is more, these concepts even shut out requirements very close to the center of the intended meaning! Thus in the case of "economics" even such obvious misfit variables as the cost of maintenance and depreciation have only recently been made the subject of architectural consideration. See J. C. Weston, "Economics of Building," *Royal Institute of British Architects Journal*, 62: 256–7 (April 1955), 63: 268–78 (May 1956), 63: 316–29 (June 1956). As for the cost of social overheads – the milkman's rounds; the laundries and TB sanatoria which have to cope with the effects of smoke from open fireplaces – even the economists are only just beginning to consider these. See Benjamin Higgins, *Economic Development* (New York, 1959), pp. 254–6, 660–1. Yet the cost of the form is found in all these things. The true cost of a form is much more complicated than the concept "economics" at first suggests.

Marshall McLuhan (1911–80) was a Canadian educator, philosopher, scholar, and critic, who is best known for his work in communications and media studies. His ground-breaking work in media studies has taken on increased salience in the decade since the emergence of the world-wide-web and the plethora of visualization technologies unleashed by the personal computer. The strong resurgence in interest in McLuhan's work has been fueled by *Wired* magazine, which claimed him as their patron saint, and which republished his key texts in the mid-1990s.

The essay on housing is taken from McLuhan's seminal work *Understanding Media*, which proposed that the medium of communication is more important than the content. He argued that media themselves embody certain values which are necessarily biased and enable particular ways of understanding, at the expense of others. In labeling media as *hot* and *cool*, he categorized them in terms of the degree of definition offered by the particular medium. High definition (*hot*) media fill one sense with data to the deprivation of others. Low definition (*cool*) media, on the other hand, require audience participation to make up for missing information. For McLuhan, books were *hot*, and television was *cool*.

McLuhan's description of housing as an extension of the skin or as an article of clothing is part of the larger argument he shared with Buckminster Fuller that technology in general could be understood as an extension of human capacities. It draws on developments in sociology and anthropology, and through his participation in *Ekistics*, entered into the broad discourse about city growth and planning in the most radical application of system theory.[1]

1964

Marshall McLuhan

Housing: New Look and New Outlook

1   Mark Wigley, "Network Fever," *Grey Room* 4 (Summer 2001): 82–122.

If clothing is an extension of our private skins to store and channel our own heat and energy, housing is a collective means of achieving the same end for the family or the group. Housing as shelter is an extension of our bodily heat-control mechanisms – a collective skin or garment. Cities are an even further extension of bodily organs to accommodate the needs of large groups. Many readers are familiar with the way in which James Joyce organized *Ulysses* by assigning the various city forms of walls, streets, civic buildings, and media to the various bodily organs. Such a parallel between the city and the human body enabled Joyce to establish a further parallel between ancient Ithaca and modern Dublin, creating a sense of human unity in depth, transcending history.

Baudelaire originally intended to call his *Fleurs du Mal, Les Limbes*, having in mind the city as corporate extensions of our physical organs. Our letting-go of ourselves, self-alienations as it were, in order to amplify or increase the power of various functions, Baudelaire considered to be flowers of growths of evil. The city as amplification of human lusts and sensual striving had for him an entire organic and psychic unity.

Literate man, civilized man, tends to restrict and enclose space and to separate functions, whereas tribal man had freely extended the form of his body to include the universe. Acting as an organ of the cosmos, tribal man accepted his bodily functions as modes of participation in the divine energies. The human body in Indian religious thought was ritually related to the cosmic image, and this in turn was assimilated into the form of house. Housing was an image of both the body and the universe for tribal and nonliterate societies. The building of the house with its hearth as fire-altar was ritually associated with the act of creation. This same ritual was even more deeply embedded in the building of the ancient cities, their shape and process having been deliberately modeled as an act of divine praise. The city and the home in the tribal world (as in China and India today) can be accepted as iconic embodiments of the *word*, the divine *mythos*, the universal aspiration. Even in our present electric age, many people yearn for this inclusive strategy of acquiring significance for their own private and isolated beings.

Literate man, once having accepted an analytic technology of fragmentation, is not nearly so accessible to cosmic patterns as tribal man. He prefers separateness and compartmented spaces, rather than the open cosmos. He becomes less inclined to accept his body as a model of the universe, or to see his house – or any other of the media of communication, for that matter – as a ritual extension of his body. Once men have adopted the visual dynamic of the phonetic alphabet, they begin to lose the tribal man's obsession with cosmic order and ritual as recurrent in the physical organs and their social extension. Indifference to the cosmic, however, fosters intense concentration on minute segments and specialist tasks, which is the unique strength of Western man. For the specialist is one who never makes small mistakes while moving toward the grand fallacy.

Men live in round houses until they become sedentary and specialized in their work organization. Anthropologists have often noted this change from round to square without knowing its cause. The media analyst can help the anthropologist in this matter, although the explanation will not be obvious to people of visual culture. The visual man, likewise, cannot see much difference between the motion picture and TV, or between a Corvair and a Volkswagen, for this difference is not between two visual spaces, but between tactile and visual ones. A tent or a wigwam is not an enclosed or visual space. Neither is a cave nor a hole in the ground. These kinds of space – the tent, the wigwam, the igloo, the cave – are not "enclosed" in the visual sense because they follow dynamic lines of force, like a triangle. When enclosed, or translated into visual space, architecture tends to lose its tactile kinetic pressure. A square is the enclosure of a visual space; that is, it consists of space properties abstracted from manifest tensions. A triangle follows lines of force, this being the most economical way of anchoring a vertical object. A square moves beyond such kinetic pressures to enclose visual space relations, while depending upon diagonal anchors. This separation of the visual from direct tactile and kinetic pressure, and its translation into new dwelling spaces, occurs only when men have learned to practice specialization of their senses, and fragmentation of their work skills. The square room or house speaks the language of the sedentary specialist, while the round hut or igloo, like the conical wigwam, tells of the integral nomadic ways of food-gathering communities.

This entire discussion is offered at considerable risk of misapprehension because these are, spatially, highly technical matters. Nevertheless, when such spaces are understood, they offer the key to a great many enigmas, past and present. They explain the change from circular-dome architecture to gothic forms, a change occasioned by alteration in the ratio or proportion of the sense lives in the members of a society. Such a shift occurs with the extension of the body in new social technology and invention. A new extension sets up a new equilibrium among all of the senses and faculties leading, as we say, to a "new outlook" – new attitudes and preferences in many areas.

In the simplest terms, as already noted, housing is an effort to extend the body's heat-control mechanism. Clothing tackles the problem more directly but less fundamentally, and privately rather than socially. Both clothing and housing store warmth and energy and make these readily accessible for the execution of many tasks otherwise impossible. In making heat and energy accessible socially, to the family or the group, housing fosters new skills and new learning, performing the basic functions of all other media. Heat control is the key factor in housing, as well as in clothing. The Eskimo's dwelling is a good example. The Eskimo can go for days without food at 50 degrees below zero. The unclad native, deprived of nourishment, dies in a few hours.

It may surprise many to learn that the primitive shape of the igloo is, nonetheless, traceable to the primus stove. Eskimos have lived for ages in round stone houses, and, for the most part, still do. The igloo, made of snow blocks, is a fairly recent development in the life of this stone-age people. To live in such structures became possible with the coming of the white man and his portable stove. The igloo is an ephemeral shelter, devised for temporary use by trappers. The Eskimo became a trapper only after he had made contact with the white man; up until then he had been simply a food-gatherer. Let the igloo serve as an example of the way in which a new pattern is introduced into an ancient way of life by the intensification of a single factor – in this instance, artificial heat. In the same way, the intensification of a single factor in our complex lives leads naturally to a new balance among our technologically extended faculties, resulting in a new look and a new "outlook" with new motivations and inventions.

In the twentieth century we are familiar with the changes in housing and architecture that are the result of electric energy made available to elevators. The same energy devoted to lighting has altered our living and working spaces even more radically. Electric light abolished the divisions of night and day, of inner and outer, and of the subterranean and the terrestrial. It altered every consideration of space for work and production as much as the other electric media had altered the space-time experience of society. All this is reasonably familiar. Less familiar is the architectural revolution made possible by improvements in heating centuries ago. With the mining of coal on a large scale in the Renaissance, inhabitants in the colder climates discovered great new resources of personal energy. New means of heating permitted the manufacture of glass and the enlargement of living quarters and the raising of ceilings. The Burgher house of the Renaissance became at once bedroom, kitchen, workshop, and sale outlet.

Once housing is seen as group (or corporate) clothing and heat control, the new means of heating can be understood as causing change in spatial form. Lighting, however, is almost as decisive as heating in causing these changes in architectural and city spaces. That is the reason why the story of glass is so closely related to the history of housing. The story of the mirror is a main chapter in the history of dress and manners and the sense of the self.

Recently an imaginative school principal in a slum area provided each student in the school with a photograph of himself. The classrooms of the school were abundantly supplied with large mirrors. The result was an astounding increase in the learning rate. The slum child has ordinarily very little visual orientation. He does not see himself as becoming something. He does not envisage distant goals and objectives. He is deeply involved in his own world from day to day, and can establish no beachhead in the highly specialized sense life of visual man. The plight of the slum child, via the TV image, is increasingly extended to the entire population.

Clothing and housing, as extensions of skin and heat-control mechanisms, are media of communication, first of all, in the sense that they shape and rearrange the patterns of human association and community. Varied techniques of lighting and heating would seem only to give new flexibility and scope to what is the basic principle of these media of clothing and housing; namely, their extension of our bodily heat-control mechanisms in a way that enables us to attain some degree of equilibrium in a changing environment.

Modern engineering provides means of housing that range from the space capsule to walls created by air jets. Some firms now specialize in providing large buildings with inside walls and floors that can be moved at will. Such flexibility naturally tends toward the organic. Human sensitivity seems once more to be attuned to the universal currents that made of tribal man a cosmic skin-diver.

It is not only the *Ulysses* of James Joyce that testifies to this trend. Recent studies of the Gothic churches have stressed the organic aims of their builders. The saints took the body seriously as the symbolic vesture of the spirit, and they regarded the Church as a second body, viewing its every detail with great completeness. Before James Joyce provided his detailed image of the metropolis as a second body, Baudelaire had provided a similar "dialogue" between the parts of the body extended to form the metropolis, in his *Fleurs du Mal*.

Electric lighting has brought into the cultural complex of the extensions of man in housing and city, an organic flexibility unknown to any other age. If color photography has created "museums without walls," how much more has electric lighting created space without walls, and day without night. Whether in the night city, the night highway, or the night ball game, sketching and writing with light have moved from the domain of the pictorial photograph to the live, dynamic spaces created by out-of-door lighting.

Not many ages ago, glass windows were unknown luxuries. With light control by glass came also a means of controlling the regularity of domestic routine, and steady application to crafts and trade without regard to cold or rain. The world was put in a frame. With electric light not only can we carry out the most precise operations with no regard for time or place or climate, but we can photograph the submicroscopic as easily as we can enter the subterranean world of the mine and of the cave-painters.

Lighting as an extension of our powers affords the clearest-cut example of how such extensions alter our perceptions. If people are inclined to doubt whether the wheel or typography or the plane could change our habits of sense perception, their doubts end with electric lighting. In this domain, the medium is the message, and when the light is on there is a world of sense that disappears when the light is off.

"Painting with light" is jargon from the world of stage-electricity. The uses of light in the world of motion, whether in the motorcar or the movie or the microscope, are as diverse as the uses of electricity in the world of power. Light is information without "content," much as the missile is a vehicle without the additions of wheel or highway. As the missile is a self-contained transportation system that consumes not only its fuel but its engine, so light is a self-contained communication system in which the medium is the message.

The recent development of the laser ray has introduced new possibilities for light. The laser ray is an amplification of light by intensified radiation. Concentration of radiant energy has made available some new properties in light. The laser ray – by thickening light, as it were – enables it to be modulated to carry information as do radio waves. But because of its greater intensity, a single laser beam can carry as much information as all the combined radio and TV channels in the United States. Such beams are not within the range of vision, and may well have a military future as lethal agents.

From the air at night, the seeming chaos of the urban area manifests itself as a delicate embroidery on a dark velvet ground. Gyorgy Kepes has developed these aerial effects of the city at night as a new art form of "landscape by light through" rather than "light on." His new electric landscapes have complete congruity with the TV image, which also exists by light *through* rather than by light *on*.

The French painter André Girard began painting directly on film before the photographic movies became popular. In that early phase it was easy to speculate about "painting with light" and about introducing movement into the art of painting. Said Girard: "I would not be surprised if, fifty years from now, almost no one would pay attention to paintings whose subjects remain *still* in their always too-narrow frames."

The coming of TV inspired him anew:

> Once I saw suddenly, in a control room, the sensitive eye of the camera presenting to me, one after another, the faces, the landscapes, the expressions of a big painting of mine in an order which I had never thought of. I had the feeling of a composer listening to one of his operas, all scenes mixed up in an order different from the one he wrote. It was like seeing a building from a fast elevator that showed you the roof before the basement, and made quick stops at some floors but not others.

Since that phase, Girard has worked out new techniques of control for painting with light in association with CBS and NBC technicians. The relevance of his work for housing is that it enables us to conceive of totally new possibilities for architectural and artistic modulation of space. Painting with light is a kind of housing-without-walls. The same electric technology, extended to the job of providing global thermostatic controls, points to the obsolescence of housing as an extension of the heat-control mechanisms of the body. It is equally conceivable that the electric extension of the process of collective consciousness, in making consciousness-without-walls, might render language walls obsolescent. Languages are stuttering extensions of our five senses, in varying ratios and wavelengths. An immediate simulation of consciousness would by-pass speech in a kind of massive extrasensory perception, just as global thermostats could by-pass those extensions of skin and body that we call houses. Such an extension of the process of consciousness by electric simulation may easily occur in the 1960s.

Peter Reyner Banham (1922–88) was an English architectural historian, critic, teacher, writer, and journalist (see Banham, 1960 for full biography).

The following essay is characteristic of Banham's provocative, polemical style of writing, influenced by the New Journalism of Tom Wolfe, and also by his long, close attention to popular and consumer culture. It was written while he was on a research fellowship in Chicago (1964–66), and displays his particular enthusiasm for post-war American culture. The essay lays the groundwork for his subsequent book on environmental technologies, *The Architecture of the Well-Tempered Environment* (1969), and establishes the opposition of building-as-structure versus building-as-power-consuming-device that he turns into a foundation myth in this book. But for all his criticism of architecture as monument, he strongly objected when the *Well-Tempered Environment* was cataloged under technology. Like the Futurists, he saw technology as the redemption of design.

The essay was reprinted at least three times in other magazines and collections, twice paired with Martin Pawley's "Time-House" essay, which opposed the apparent functional determinism of Banham's article.[1] It was the reduction of house to its environmental services that gives the article its polemical force, and which also taps into the powerful, popular notion that technologies evolve according to their own logic and efficiencies (see Superstudio, 1972 for a similar reduction to systems). But his argument is in no way naïve, he locates the tendency toward reduction in both the logic of the systems and the desires of mobile Americans, presenting it as both determinism and enthusiasm.

1965

Peter Reyner Banham

A Home is not a House

1   "A Home is Not a House," *Architectural Design* (January, 1969): 45–8; *Meaning in Architecture*, Charles Jencks and George Baird (eds) (New York: G. Braziller, 1970; *Architecture Culture 1943–1968: A Documentary Anthology*, Joan Ockman (ed.) (New York: Columbia Books of Architecture/Rizzoli, 1993).

When your house contains such a complex of piping, flues, ducts, wires, lights, inlets, outlets, ovens, sinks, refuse disposers, hi-fi reverberators, antennae, conduits, freezers, heaters – when it contains so many services that the hard-ware could stand up by itself without any assistance from the house, why have a house to hold it up? When the cost of all this tackle is half of the total outlay (or more, as it often is) what is the house doing except concealing your mechanical pudenda from the stares of folks on the sidewalk? Once or twice recently there have been buildings where the public was genuinely confused about what was mechanical services, what was structure – many visitors to Philadelphia take quite a time to work out that the floors of Louis Kahn's laboratory towers are not supported by the flanking brick duct boxes, and when they have worked it out, they are inclined to wonder if it was worth all the trouble of giving them an independent supporting structure.

No doubt about it, a great deal of the attention captured by those labs derives from Kahn's attempt to put the drama of mechanical services on show – and if, in the end, it fails to do that convincingly, the psychological importance of the gesture remains, at least in the eyes of his fellow architects. Services are a topic on which architectural practice has alternated capriciously between the brazen and the coy – there was the grand old Let-it-dangle period, when every ceiling was a mess of gaily painted entrails, as in the council chambers of the UN building, and there have been fits of pudicity when even the most innocent anatomical details have been hurriedly veiled with a suspended ceiling.

Basically, there are two reasons for all this blowing hot and cold (if you will excuse the air-conditioning industry's oldest working pun). The first is that mechanical services are too new to have been absorbed into the proverbial wisdom of the profession: none of the great slogans – Form Follows Function, *accusez la structure*, Firmness Commodity and Delight, Truth to Materials, *Wenig ist Mehr* – is much use in coping with the mechanical invasion. The nearest thing, in a significantly negative way, is Le Corbusier's "*Pour Ledoux, c'était facile – pas de tubes*," which seems to be gaining proverbial-type currency as the expression of a profound nostalgia for the golden age before piping set in.

The second reason is that the mechanical invasion is a fact, and architects – especially American architects – sense that it is a cultural threat to their position in the world. American architects are certainly right to feel this, because their professional specialty, the art of creating monumental spaces, has never been securely established on this continent. It remains a transplant from an older culture and architects in America are constantly harking back to that culture. The generation of Stanford White and Louis Sullivan were prone to behave like *émigrés* from France, Frank Lloyd Wright was apt to take cover behind sentimental Teutonicisms like *Lieber Meister*, the big boys of the Thirties and Forties came from Aachen and Berlin anyhow, the pacemakers of the Fifties and Sixties are men of international culture like Charles Eames and Philip Johnson, and so too, in many ways, are the coming men of today, like Myron Goldsmith.

Left to their own devices, Americans do not monumentalize or make architecture. From the Cape Cod cottage, through the balloon frame to the perfection of permanently pleated aluminum siding with embossed wood-graining, they have tended to build a brick chimney and lean a collection of shacks against it. When Groff Conklin wrote (in "The Weather-Conditioned House") that "A house is nothing but a hollow shell . . . a shell is all a house or any structure in which human beings live and work, really is. And most shells in nature are extraordinarily inefficient barriers to cold and heat" he was expressing an extremely American view, backed by a long-established grass-roots tradition.

And since that tradition agrees with him that the American hollow shell is such an inefficient heat barrier, Americans have always been prepared to pump more heat, light, and power into their shelters than have other peoples. America's monumental space is, I suppose, the great outdoors – the porch, the terrace, Whitman's rail-traced plains, Kerouac's infinite road, and now, the Great Up There. Even within the house, Americans rapidly learned to dispense with the partitions that Europeans need to keep space architectural and within bounds, and long before Wright began blundering through the walls that subdivided polite architecture into living room, games room, card room, gun room, etc., humbler Americans had been slipping into a way of life adapted to informally planned interiors that were, effectively, large single spaces.

Now, large single volumes wrapped in flimsy shells have to be lighted and heated in a manner quite different and more generous than the cubicular interiors of the European tradition around which the concept of domestic architecture first crystallized. Right from the start, from the Franklin stove and the kerosene lamp, the American interior has had to be better serviced if it was to support a civilized culture, and this is one of the reasons that the US has been the forcing ground of mechanical services in buildings – so if services are to be felt anywhere as a threat to architecture, it should be in America.

"The plumber is the quartermaster of American culture," wrote Adolf Loos, father of all European platitudes about the superiority of US plumbing. He knew what he was talking about; his brief visit to the States in the Nineties convinced him that the outstanding virtues of the American way of life were its informality (no need to wear a top hat to call on local officials) and its cleanliness – which was bound to be noticed by a Viennese with as highly developed a set of Freudian compulsions as he had. That obsession with clean (which can become one of the higher absurdities of America's lysol-breathing Kleenex-culture) was another psychological motive that drove the nation toward mechanical services. The early justifications of air-conditioning were not just that people had to breathe: Konrad Meier ("Reflections on Heating and Ventilating," 1904) wrote fastidiously of "excessive amounts of water vapor, sickly odors from respiratory organs, unclean teeth, perspiration, untidy clothing, the presence of microbes due to various conditions, stuffy air from dusty carpets and draperies . . . cause greater discomfort and greater ill health."

(Have a wash, and come back for the next paragraph.)

Most pioneer air-conditioning men seem to have been nose-obsessed in this way: best friends could just about force themselves to tell America of her national B.O. – and then, compulsive salesmen to a man, promptly prescribed their own patent improved panacea for ventilating the hell out of her. Somewhere among

these clustering concepts – cleanliness, the lightweight shell, the mechanical services, the informality and indifference to monumental architectural values, the
passion for the outdoors – always seemed to me to lurk some elusive master
concept that would never quite come into focus. It finally came clear and legible to
me in June 1964, in the most highly appropriate and symptomatic circumstances.

I was standing up to my chest-hair in water, making home movies (I get that
NASA kick from taking expensive hardware into hostile environments) at the
campus beach at Southern Illinois. This beach combines the outdoor and the clean
in a highly American manner – scenically it is the ole swimmin' hole of Huckleberry
Finn tradition, but it is properly policed (by sophomore lifeguards sitting on Eames
chairs on poles in the water) and it's *chlorinated* too. From where I stood, I could
see not only immensely elaborate family barbecues and picnics in progress on the
sterilized sand, but also, through and above the trees, the basketry interlaces of
one of Buckminster Fuller's experimental domes. And it hit me then, that if dirty old
Nature could be kept under the proper degree of control (sex left in, streptococci
taken out) by other means, the United States would be happy to dispense with
architecture and buildings altogether.

Bucky Fuller, of course, is very big on this proposition: his famous non-rhet-
orical question, "Madam, do you know what your house weighs?" articulates a sub-
versive suspicion of the monumental. This suspicion is inarticulately shared by the
untold thousands of Americans who have already shed the deadweight of domestic
architecture and live in mobile homes which, though they may never actually be
moved, still deliver rather better performance as shelter than do ground-anchored
structures costing at least three times as much and weighing ten times more. If
someone could devise a package that would effectively disconnect the mobile
home from the dangling wires of the town electricity supply, the bottled gas containers insecurely perched on a packing case and the semi-unspeakable sanitary
arrangements that stem from not being connected to the main sewer – then we
should really see some changes. It may not be so far away either; defense cutbacks
may send aerospace spin-off spinning in some new directions quite soon, and that
kind of miniaturization-talent applied to a genuinely self-contained and regenera-
tive standard-of-living package that could be towed behind a trailer home or
clipped to it, could produce a sort of U-haul unit that might be picked up or
dropped off at depots across the face of the nation. Avis might still become the first
in U-Tility, even if they have to go on being a trying second in car hire.

Out of this might come a domestic revolution beside which modern architecture would look like Kiddibrix, because you might be able to dispense with the
trailer home as well. A standard-of-living package (the phrase and the concept are
both Bucky Fuller's) that really worked might, like so many sophisticated inventions, return Man nearer to a natural state in spite of his complex culture (much as
the supersession of the Morse telegraph by the Bell Telephone restored his power
of speech nationwide). Man started with two basic ways of controlling environment: one by avoiding the issue and hiding under a rock, tree, tent, or roof (this led
ultimately to architecture as we know it) and the other by actually interfering with
the local meteorology, usually by means of a campfire, which, in a more polished
form, might lead to the kind of situation now under discussion. Unlike the living

space trapped with our forebears under a rock or roof, the space around a campfire has many unique qualities which architecture cannot hope to equal, above all, its freedom and variability.

The direction and strength of the wind will decide the main shape and dimensions of that space, stretching the area of tolerable warmth into a long oval, but the output of light will not be affected by the wind, and the area of tolerable illumination will be a circle overlapping the oval of warmth. There will thus be a variety of environmental choices balancing light against warmth according to need and interest. If you want to do close work, like shrinking a human head, you sit in one place, but if you want to sleep you curl up somewhere different; the floating knuckle-bones game would come to rest somewhere quite different to the environment that suited the meeting of the initiation-rites steering committee . . . and all this would be jim dandy if camp-fires were not so perishing inefficient, unreliable, smoky, and the rest of it.

But a properly set up standard-of-living package, breathing out warm air along the ground (instead of sucking in cold along the ground like a campfire), radiating soft light and Dionne Warwick in heart-warming stereo, with well-aged protein turning in an infra-red glow in the rotisserie, and the ice-maker discreetly coughing cubes into glasses on the swing-out bar – this could do something for a woodland glade or creek-side rock that Playboy could never do for its penthouse. But how are you going to manhandle this hunk of technology down to the creek? It doesn't have to be that massive; aerospace needs, for instance, have done wild things to solid-state technology, producing even tiny refrigerating transistors. They don't as yet mop up any great quantity of heat, but what are you going to do in this glade anyhow; put a whole steer in deep-freeze? Nor do you have to manhandle it – it could ride on a cushion of air (its own air-conditioning output, for instance) like a hovercraft or domestic vacuum cleaner.

All this will eat up quite a lot of power, transistors notwithstanding. But one should remember that few Americans are ever far from a source of between 100 and 400 horsepower – the automobile. Beefed-up car batteries and a self-reeling cable drum could probably get this package breathing warm bourbon fumes o'er Eden long before microwave power transmission or miniaturized atomic power plants come in. The ear is already one of the strongest arms in America's environmental weaponry, and an essential component in one non-architectural anti-building that is already familiar to most of the nation – the drive-in movie house. Only, the word *house* is a manifest misnomer – just a flat piece of ground where the operating company provides visual images and piped sound, and the rest of the situation comes on wheels. You bring your own seat, heat, and shelter as part of the car. You also bring Coke, cookies, Kleenex, Chesterfields, spare clothes, shoes, the Pill, and god-wot else they don't provide at Radio City.

The car, in short, is already doing quite a lot of the standard-of-living package's job – the smoochy couple dancing to the music of the radio in their parked convertible have created a ballroom in the wilderness (dance floor by courtesy of the Highway Dept. of course) and all this is paradisal till it starts to rain. Even then, you're not licked – it takes very little air pressure to inflate a transparent Mylar airdome, the conditioned-air output of your mobile package might be able to

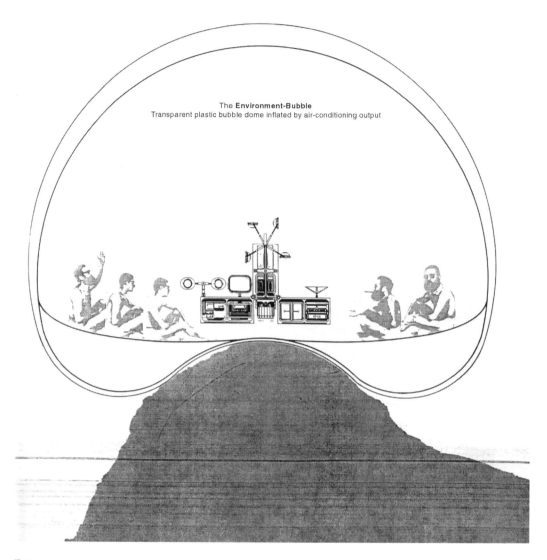

The **Environment-Bubble**
Transparent plastic bubble dome inflated by air-conditioning output

Fig 2
Francois Dallegret and Reyner Banham, The Environment Bubble, 1969 © Francois Dallegret. Licensed by DACS 2006

do it, with or without a little boosting, and the dome itself, folded into a parachute pack, might be part of the package. From within your thirty-foot hemisphere of warm dry *lebensraum* you could have spectacular ringside views of the wind felling trees, snow swirling through the glade, the forest fire coming over the hill or Constance Chatterley running swiftly to you know whom through the downpour.

But ... surely this is not a home, you can't bring up a family in a polythene bag? This can never replace the time-honored ranch-style tri-level standing proudly in a landscape of five defeated shrubs, flanked on one side by a ranch-style tri-level with six shrubs and on the other by a ranch-style tri-level with four small boys and a private dust bowl. If the countless Americans who are successfully raising nice children in trailers will excuse me for a moment, I have a few suggestions to make to the even more countless Americans who are so insecure that they have to hide inside fake monuments of Permastone and instant roofing. There are, admittedly, very sound day-to-day advantages to having warm broadloom on a firm floor underfoot, rather than pine needles and poison ivy. America's pioneer house builders recognized this by commonly building their brick chimneys on a brick floor slab. A transparent airdome could be anchored to such a slab just as easily as could a balloon frame, and the standard-of-living-package could hover busily in a sort of glorified barbecue pit in the middle of the slab. But an airdome is not the sort of thing that the kids, or a distracted Pumpkin-eater could run in and out of when the fit took them – believe me, fighting your way out of an airdome can be worse than trying to get out of a collapsed rain-soaked tent if you make the wrong first move.

But the relationship of the services-kit to the floor slab could be re-arranged to get over this difficulty; all the standard-of-living tackle (or most of it) could be redeployed on the upper side of a sheltering membrane floating above the floor, radiating heat, light and what-not downwards and leaving the whole perimeter wide-open for random egress – and equally casual ingress, too, I guess. That crazy modern-movement dream of the interpenetration of indoors and outdoors could become real at last by abolishing the doors. Technically, of course, it would be just about possible to make the power-membrane literally float, hovercraft style. Anyone who has had to stand in the ground-effect of a helicopter will know that this solution has little to recommend it apart from the instant disposal of waste paper. The noise, power consumption, and physical discomfort would be really something wild. But if the power-membrane could be carried on a column or two, here and there, or even on a brick-built bathroom unit, then we are almost in sight of what might be technically possible before the Great Society is much older.

The basic proposition is simply that the power-membrane should blow down a curtain of warmed/cooled/conditioned air around the perimeter of the windward side of the un-house, and leave the surrounding weather to waft it through the living space, whose relationship in plan to the membrane above need not be a one-to-one relationship. The membrane would probably have to go beyond the limits of the floor slab, anyhow, in order to prevent rain blow-in, though the air-curtain will be active on precisely the side on which the rain is blowing and, being conditioned, will tend to mop up the moisture as it falls. The distribution of the air-curtain will be governed by various electronic light and weather sensors, and by that radical new invention, the weathervane. For really foul weather automatic storm shutters would be required, but in all but the most wildly inconstant climates, it should be possible to design the conditioning kit to deal with most of the weather most of the time, without the power consumption becoming ridiculously greater than for an ordinary inefficient monumental type house.

Obviously, it would still be appreciably greater, but this whole argument hinges on the observation that it is the American Way to spend money on services and upkeep rather than on permanent structure as do the peasant cultures of the Old World. In any case, we don't know where we shall be with things like solar power in the next decade, and to anyone who wants to entertain an almost-possible vision of air-conditioning for absolutely free, let me recommend *Shortstack* (another smart trick with a polythene tube) in the December 1964 issue of *Analog*. In fact, quite a number of the obvious common-sense objections to the un-house may prove to be self-evaporating: for instance, noise may be no problem because there would be no surrounding wall to reflect it back into the living space, and, in any case, the constant whisper of the air-curtain would provide a fair threshold of loudness that sounds would have to beat before they began to be comprehensible and therefore disturbing. Bugs? Wild life? In summer they should be no worse than with the doors and windows of an ordinary house open; in winter all right-thinking creatures either migrate or hibernate; but, in any case, why not encourage the normal processes of Darwinian competition to tidy up the situation for you? All that is needed is to trigger the process by means of a general purpose lure; this would radiate mating calls and sexy scents and thus attract all sorts of mutually incompatible predators and prey into a compact pool of unspeakable carnage. A closed-circuit television camera could relay the state of play to a screen inside the dwelling and provide a twenty-four-hour program that would make the ratings for Bonanza look like chicken feed.

And privacy? This seems to be such a nominal concept in American life as factually lived that it is difficult to believe that anyone is seriously worried. The answer, under the suburban conditions that this whole argument implies, is the same as for the glass houses architects were designing so busily a decade ago – more sophisticated landscaping. This, after all, is the homeland of the bulldozer and the transplantation of grown trees – why let the Parks Commissioner have all the fun'?

As was said above, this argument implies suburbia which, for better or worse, is where America wants to live. It has nothing to say about the city, which, like architecture, is an insecure foreign growth on the continent. What is under discussion here is an extension of the Jeffersonian dream beyond the agrarian sentimentality of Frank Lloyd Wright's Usonian/Broadacre version – the dream of the good life in the clean countryside, power-point homesteading in a paradise garden of appliances. This dream of the un-house may sound very anti-architectural but it is so only in degree, and architecture deprived of its European roots but trying to strike new ones in an alien soil has come close to the anti-house once or twice already. Wright was not joking when he talked of the "destruction of the box," even though the spatial promise of the phrase is rarely realized to the full in the all-too-solid fact. Grass-roots architects of the plains like Bruce Goff and Herb Greene have produced houses whose supposed monumental form is clearly of little consequence to the functional business of living in and around them.

But it is in one building that seems at first sight nothing but monumental form that the threat or promise of the un-house has been most clearly demonstrated – the Johnson House at New Canaan. So much has been misleadingly said (by Philip Johnson himself, as well as others) to prove this a work of architecture in the European tradition, that its many intensely American aspects are usually missed. Yet when you have dug through all the erudition about Ledoux and Malevitsch and Palladio and stuff that has been published, one very suggestive source or prototype remains less easily explained away – the admitted persistence in Johnson's mind of the visual image of a burned-out New England township, the insubstantial shells of the houses consumed by the fire, leaving the brick floor slabs and standing chimneys. The New Canaan glasshouse consists essentially of just these two elements, a heated brick floor slab, and a standing unit which is a chimney/fireplace on one side and a bathroom on the other.

Around this has been draped precisely the kind of insubstantial shell that Conklin was discussing, only even less substantial than that. The roof, certainly, is solid, but psychologically it is dominated by the absence of visual enclosure all around. As many pilgrims to this site have noticed, the house does not stop at the glass, and the terrace, and even the trees beyond, are visually part of the living space in winter, physically and operationally so in summer when the four doors are open. The "house" is little more than a service core set in infinite space, or alternatively, a detached porch looking out in all directions at the Great Out There. In summer, indeed, the glass would be a bit of a nonsense if the trees did not shade it, and in the recent scorching fall, the sun reaching in through the bare trees created such a greenhouse effect that parts of the interior were acutely uncomfortable – the house would have been better off without its glass walls.

When Philip Johnson says that the place is not a controlled environment, however, it is not these aspects of undisciplined glazing he has in mind, but that "when it gets cold I have to move toward the fire, and when it gets too hot I just move away." In fact, he is simply exploiting the campfire phenomenon (he is also pretending that the floor-heating does not make the whole area habitable, which it does) and in any case, what does he mean by a controlled environment? It is not the same thing as a uniform environment, it is simply an environment suited to what you are going to do next, and whether you build a stone monument, move away from the fire or turn on the air-conditioning, it is the same basic human gesture you are making.

Only, the monument is such a ponderous solution that it astounds me that Americans are still prepared to employ it, except out of some profound sense of insecurity, a persistent inability to rid themselves of those habits of mind they left Europe to escape. In the open-fronted society, with its social and personal mobility, its interchangeability of components and personnel, its gadgetry and almost universal expendability, the persistence of architecture-as-monumental-space must appear as evidence of the sentimentality of the tough.

1969

Richard Buckminster Fuller

Comprehensive Propensities

Richard Buckminster Fuller (1895–1983) was an American architect, inventor, engineer, designer, cartographer, mathematician, and poet, known primarily for the invention of the geodesic dome (see Fuller, 1929 for full biography).

Fuller was also a cosmologist, whose view of an integrated, orderly, stable universe was reflected in an overarching interest in structure, efficiency, and energy. "Comprehensive Propensities" is the introductory chapter of the ambitiously titled *Operating Manual for Spaceship Earth*, in which Fuller criticizes poor design, global imbalances, and land-use. These he attributes to a tendency toward short-term solutions, and increasing specialization. What is needed, for Fuller, is a global view. This is the propensity for comprehensive knowledge of the child, before it has been mitigated by the specialization of adulthood.

"Comprehensive Propensities" can thus seen alongside a series of critiques of specialization, echoing the concerns of Frank Lloyd Wright, Siegfried Giedion, and others. Similarly, in his invocation of the nomad, who is able to command vast areas using a general knowledge, his view is shared by Marshall McLuhan. In his understanding of the world's problems as being inherently problems of design, solved by globally deployed technological solutions, his work stands in for a subset of ecological modernist thought.

I am enthusiastic over humanity's extraordinary and sometimes very timely ingenuities. If you are in a shipwreck and all the boats are gone, a piano top buoyant enough to keep you afloat that comes along makes a fortuitous life preserver. But this is not to say that the best way to design a life preserver is in the form of a piano top. I think that we are clinging to a great many piano tops in accepting yesterday's fortuitous contrivings as constituting the only means for solving a given problem. Our brains deal exclusively with special-case experiences. Only our minds are able to discover the generalized principles operating without exception in each and every special-experience case which if detected and mastered will give knowledgeable advantage in all instances.

Because our spontaneous initiative has been frustrated, too often inadvertently, in earliest childhood we do not tend, customarily, to dare to think competently regarding our potentials. We find it socially easier to go on with our narrow, shortsighted specializations and leave it to others – primarily to the politicians – to find some way of resolving our common dilemmas. Countering that spontaneous grown-up trend to narrowness I will do my, hopefully "childish," best to confront as many of our problems as possible by employing the longest-distance thinking of which I am capable – though that may not take us very far into the future.

Having been trained at the US Naval Academy and practically experienced in the powerfully effective forecasting arts of celestial navigation, pilotage, ballistics, and logistics, and in the long-range, anticipatory, design science governing yesterday's naval mastery of the world from which our present day's general systems theory has been derived, I recall that in 1927 I set about deliberately exploring to see how far ahead we could make competent forecasts regarding the direction in which all humanity is trending and to see how effectively we could interpret the physical details of what comprehensive evolution might be portending as disclosed by the available data. I came to the conclusion that it is possible to make a fairly reasonable forecast of about twenty-five years. That seems to be about one industrial "tooling" generation. On the average, all inventions seem to get melted up about every twenty-five years, after which the metals come back into recirculation in new and usually more effective uses. At any rate, in 1927 I evolved a forecast. Most of my 1927's prognosticating went only to 1952 – that is, for a quarter-century, but some of it went on for a half-century, to 1977.

In 1927 when people had occasion to ask me about my prognostications and I told them what I thought it would be appropriate to do about what I could see ahead for the 1950s, 1960s, and 1970s people used to say to me, "Very amusing – you are a thousand years ahead of your time." Having myself studied the increments in which we can think forwardly I was amazed at the ease with which the rest of society seemed to be able to see a thousand years ahead while I could see only one-fortieth of that time distance. As time went on people began to tell me that I was a hundred years ahead, and now they tell me that I'm a little behind the times. But I have learned about public reaction to the unfamiliar and also about the ease and speed with which the transformed reality becomes so "natural" as misseemingly to have been always obvious. So I knew that their last observations were made only because the evolutionary events I had foreseen have occurred on schedule.

However, all that experience gives me confidence in discussing the next quarter-century's events. First, I'd like to explore a few thoughts about the vital data confronting us right now – such as the fact that more than half of humanity as yet exists in miserable poverty, prematurely doomed, unless we alter our comprehensive physical circumstances. It is certainly no solution to evict the poor, replacing their squalid housing with much more expensive buildings which the original tenants can't afford to reoccupy. Our society adopts many such superficial palliatives. Because yesterday's negatives are moved out of sight from their familiar locations many persons are willing to pretend to themselves that the problems have been solved. I feel that one of the reasons why we are struggling inadequately today is that we reckon our costs on too shortsighted a basis and are later overwhelmed with the unexpected costs brought about by our shortsightedness.

Of course, our failures are a consequence of many factors, but possibly one of the most important is the fact that society operates on the theory that specialization is the key to success, not realizing that specialization precludes comprehensive thinking. This means that the potentially-integratable–techno-economic advantages accruing to society from the myriad specializations are not comprehended integratively and therefore are not realized, or they are realized only in negative ways, in new weaponry or the industrial support only of warfaring.

All universities have been progressively organized for ever finer specialization. Society assumes that specialization is natural, inevitable, and desirable. Yet in observing a little child, we find it is interested in everything and spontaneously apprehends, comprehends, and co-ordinates an ever-expanding inventory of experiences. Children are enthusiastic planetarium audiences. Nothing seems to be more prominent about human life than its wanting to understand all and put everything together.

One of humanity's prime drives is to understand and be understood. All other living creatures are designed for highly specialized tasks. Man seems unique as the comprehensive comprehender and co-ordinator of local universe affairs. If the total scheme of nature required man to be a specialist she would have made him so by having him born with one eye and a microscope attached to it.

What nature needed man to be was adaptive in many if not any direction; wherefore she gave man a mind as well as a co-ordinating switchboard brain. Mind apprehends and comprehends the general principles governing flight and deep sea diving, and man puts on his wings or his lungs, then takes them off when not using them. The specialist bird is greatly impeded by its wings when trying to walk. The fish cannot come out of the sea and walk upon land, for birds and fish are specialists.

Of course, we are beginning to learn a little in the behavioral sciences regarding how little we know about children and the educational processes. We had assumed the child to be an empty brain receptacle into which we could inject our methodically gained wisdom until that child, too, became educated. In the light of modern behavioral science experiments that was not a good working assumption.

Inasmuch as the new life always manifests comprehensive propensities I would like to know why it is that we have disregarded all children's significantly spontaneous and comprehensive curiosity and in our formal education have deliberately instituted processes leading only to narrow specialization. We do not have to go very far back in history for the answer. We get back to great, powerful men of the sword, exploiting their prowess fortuitously and ambitiously, surrounded by the abysmal ignorance of world society. We find early society struggling under economic conditions wherein less than 1 percent of humanity seemed able to live its full span of years. This forlorn economic prospect resulted from the seeming inadequacy of vital resources and from an illiterate society's inability to cope successfully with the environment, while saddled also with preconditioned instincts which inadvertently produced many new human babies. Amongst the strugglers we had cunning leaders who said, "Follow me, and we'll make out better than the others." It was the most powerful and shrewd of these leaders who, as we shall see, invented and developed specialization.

Looking at the total historical pattern of man around the Earth and observing that three-quarters of the Earth is water, it seems obvious why men, unaware that they would some day contrive to fly and penetrate the ocean in submarines, thought of themselves exclusively as pedestrians – as dry land specialists. Confined to the quarter of the Earth's surface which is dry land it is easy to see how they came to specialize further as farmers or hunters – or, commanded by their leader, became specialized as soldiers. Less than half of the dry 25 percent of the Earth's surface was immediately favorable to the support of human life. Thus, throughout history 99.9 percent of humanity has occupied only 10 percent of the total Earth surface, dwelling only where life support was visibly obvious. The favorable land was not in one piece, but consisted of a myriad of relatively small parcels widely dispersed over the surface of the enormous Earth sphere. The small isolated groups of humanity were utterly unaware of one another's existence. They were everywhere ignorant of the vast variety of very different environments and resource patterns occurring other than where they dwelt.

But there were a few human beings who gradually, through the process of invention and experiment, built and operated, first, local river and bay, next, alongshore, then off-shore rafts, dugouts, grass boats, and outrigger sailing canoes. Finally, they developed voluminous rib-bellied fishing vessels, and thereby ventured out to sea for progressively longer periods. Developing ever larger and more capable ships, the seafarers eventually were able to remain for months on the high seas. Thus, these venturers came to live normally at sea. This led them inevitably into world-around, swift, fortune-producing enterprise. Thus they became the first world men.

The men who were able to establish themselves on the oceans had also to be extraordinarily effective with the sword upon both land and sea. They had also to have great anticipatory vision, great ship designing capability, and original scientific conceptioning, mathematical skill in navigation and exploration techniques for coping in fog, night, and storm with the invisible hazards of rocks, shoals, and currents. The great sea venturers had to be able to command all the people in their dry land realm in order to commandeer the adequate metalworking, woodworking, weaving, and other skills necessary to produce their large, complex ships. They had to establish and maintain their authority in order that they themselves and the craftsmen preoccupied in producing the ship be adequately fed by the food-producing hunters and farmers of their realm. Here we see the specialization being greatly amplified under the supreme authority of the comprehensively visionary and brilliantly co-ordinated top swordsman, sea venturer. If his "ship came in" – that is, returned safely from its years' long venturing – all the people in his realm prospered and their leader's power was vastly amplified.

There were very few of these top power men. But as they went on their sea ventures they gradually found that the waters interconnected all the world's people and lands. They learned this unbeknownst to their illiterate sailors, who, often as not, having been hit over the head in a saloon and dragged aboard to wake up at sea, saw only a lot of water and, without navigational knowledge, had no idea where they had traveled.

The sea masters soon found that the people in each of the different places visited knew nothing of people in other places. The great venturers found the resources of Earth very unevenly distributed, and discovered that by bringing together various resources occurring remotely from one another one complemented the other in producing tools, services, and consumables of high advantage and value. Thus resources in one place which previously had seemed to be absolutely worthless suddenly became highly valued. Enormous wealth was generated by what the sea venturers could do in the way of integrating resources and distributing the products to the, everywhere around the world, amazed and eager customers. The ship-owning captains found that they could carry fantastically large cargoes in their ships, due to nature's floatability – cargoes so large they could not possibly be carried on the backs of animals or the backs of men. Furthermore, the ships could sail across a bay or sea, traveling shorter distances in much less time than it took to go around the shores and over the intervening mountains. So these very few masters of the water world became incalculably rich and powerful.

To understand the development of *intellectual specialization*, which is our first objective, we must study further the comprehensive intellectual capabilities of the sea leaders in contradistinction to the myriad of physical, muscle, and craft-skill specializations which their intellect and their skillful swordplay commanded. The great sea venturers thought always in terms of the world, because the world's waters are continuous and cover three-quarters of the Earth planet. This meant that before the invention and use of cables and wireless 99.9 percent of humanity thought only in the terms of their own local terrain. Despite our recently developed communications intimacy and popular awareness of total Earth we, too, in 1969 are as yet politically organized entirely in the terms of exclusive and utterly obsolete sovereign separateness.

This "sovereign" – meaning top-weapons enforced – "national" claim upon humans born in various lands leads to ever more severely specialized servitude and highly personalized identity classification. As a consequence of the slavish "categoryitis" the scientifically illogical, and as we shall see, often meaningless questions "Where do you live?" "What are you?" "What religion?" "What race?" "What nationality?" are all thought of today as logical questions. By the twenty-first century it either will have become evident to humanity that these questions are absurd and anti-evolutionary or men will no longer be living on Earth. If you don't comprehend why that is so, listen to me closely.

In 1969, the magazine *Progressive Architecture* devoted an issue to "space planning," meaning the layout of the "billions of square feet of office space" completed in the twenty-five years after the Second World War. Of interest to the editors was the scale of the subject and the rapidly evolving "science" of designing "anonymous space." While most of the magazine involved fairly straightforward reporting on the subject, the essay reproduced here was effectively a summary of an issue published in 1967 on "Performance Design," which was the editor's term for operations research and general systems theory in planning practices.

As the editors had explained in the earlier issue, operations research and systems analysis had emerged from technological advances by the military, especially in the development of radar and missile targeting, though they cited the computer and television as two products that had already altered civilian life. As the pace of technological change was felt to quicken, efforts to forecast the next change become more common, and more urgent. What is interesting for the question of technology and architecture is the application of management theories and system thinking to the profession and design process themselves. With that step, design and planning are not only subject to the increased efficiencies of the assembly line, but to the evolutionary forces of the organism. The ultimate stage – "On-line Planning" with no designers – refers not only to Boyce's theories, but to Robert Probst's work at the Herman Miller Furniture company. Probst had imagined a new cadre of distributed design agents located within companies and helped develop a remarkably effective array of furniture to accommodate the rapid changes of configuration. The result was the evolution of that most ubiquitous and robust element in the billions of square feet of office: the cubicle.

1969

James R. Boyce

What is the Systems Approach?

The systems approach is more than a technological flash in the pan; it is an attitude toward planning which may change the fundamental beliefs of architects, designers, and planners everywhere.

The 1960s will be remembered as a time when America experienced a fundamental division in attitudes. The tremor began gradually but, at the close of the decade, increased toward a major cultural polarization. One pole, dominated by conservative industrial institutions, might be termed the technocratic. The other, the humanistic pole, is represented by a radical, liberal group that is anti-establishment. In architecture the technocratic pole is represented by conglomerates, most large architectural firms, and the industrialized builders. The humanist pole includes new left student movements and social activist architects dealing with community problems, often more involved with politics than plans.

The technocrats have been so manipulated by the industrial machine that when we squint our eyes the great movement of modern science appears to shrink to a mere puppet show run by cigar-smoking entrepreneurs.

On the other hand, it must be said with equal vigor, that the humanists' ranks have been dominated by negativistic leaders who chose to abandon reason, science, and everything else that has been touched by established institutions.

During the past two decades a third body of thought has been forming. Felt only as a ripple in the midst of the technocratic–humanist conflict, this third force is now entering into the arena with its own ideas on how to solve growing world problems. Called *the systems approach*,[1] this force calls for a return to the use of rationalist-based principles for solving large-scale planning and design problems.

But let us step back briefly and view this approach historically, as well as in light of our understanding of today's human needs. Renaissance man found it advantageous to subdivide his bodies of knowledge, at first through crafts, and later through fields of science and the arts. This process has often been labeled instrumentalization. Without this method for repeatedly breaking down and expanding knowledge, we never would have accumulated the wealth of information which we now possess about man and his environments. This accumulation of information was requisite to detailed analyses yet it yielded few insights into problems. This is because the problems crossed over into a number of disciplines. Recent history thus suggested a strangling paradox: while we were increasing our detailed understanding of man and environments, we were apparently decreasing our capability to cope with the problems that man and environment inflicted upon each other. We could build a comfortable air conditioning system for a building but failed to air condition our cities.

Modes of government quickly adopted the scientific divisions of knowledge, making it difficult to legislate collective actions. It once was thought that problems were always with us, and turned up only at a given point in time by a given culture's point of view. Today the problems are literally enveloping us beyond all cultural dynamics. The natural environment is rapidly being polluted and destroyed. Even the most cautious conservationists agree that man's very existence is threatened unless the patterns are reversed quickly. World economies and governments are geared toward population growth, with starvation and territorial conflicts constant problems. The ecological problems are now so enormous that they are essentially cross-disciplinary.

The systems approach provides a method of dealing with large-scale problems. Counter to popular thought this approach is more than computers, methods for design, and PERT charts. The systems approach requires a revitalized planning purpose to drive our actions. Rationalist based methods are used, then, to collectively manipulate the instruments of knowledge toward that planning purpose. Men have just begun to realize the power of man's image of himself, and the world's image of itself. Moreover, the ability of men and the world to fulfill their images can be recognized within the systems approach. Man's worldly endeavors are shifting from the artisan and his artifacts to problems and performance.

This brief introduction to the systems approach leads to the three major design management processes: sequential, cyclic, and evolutionary.

› Sequential-design processes

As projects grow in scale, their success or failure depends more on the role of project management and design methods. Sequential design is the most common of the processes. It can be found in most architectural offices. These processes are characterized by a rather abrupt beginning and end. This reflects architecture's current propensity not to involve itself before or after the building process. The linear sequential process must be familiar to all architects; services are begun with schematic design (SD), then design development (DD), construction documents (CD), and finally construction administration (CA).[2]

A refinement of the linear process is the linear overlapping sequential process.[3] It is utilized primarily to reduce the planning period to fit early construction scheduling demands.

Another nearly identical sequential process is the parallel alternative. The difference is that during the SD and DD periods alternate design solutions are simultaneously but independently developed and then evaluated one against the other by an individual or a review board.[4]

The final sequential process popularized by the aero-space industries, is known as the collapsible time-frame.[5] This process relies on careful pre-planning and rigid control to integrate many design and development functions toward some exact completion date. This technique entails a more detailed breakdown and control of sub-tasks within each of SD, DD, CD, and CA.

If a project is small and well defined, with design priorities clearly established at the outset, (e.g., design a low cost office building, a single detached dwelling, or a small bank) then sequential processes are satisfactory.

However, these processes fail as an exploratory approach to solving non-simplistic problems. On projects with a diversity of client-users and activities, such as housing, commercial, or educational complexes, we usually glean knowledge about hidden design priorities only after we have made tentative design decisions and developed these decisions toward formal schemes and evaluations. This process may have to be repeated several times before the final problem definition and solution appear. With these solutions implementary decisions can be reached.

The design result from a sequential process is almost predetermined due to the absence of formal evaluation procedures. Any evaluation after the building is completed is usually left to the mercies of architectural critics and historians.

It is also difficult to bring about cooperation among the planning participants of a cross-disciplinary project when sequential processes are employed. This is because the job is usually performed by fragmented departments of large offices and the client's opportunity to evaluate the issues and solutions is too brief. Communication, then, rests essentially with project managers, who must convey messages from isolated designers to questioning clients and tired out production men. The success of such a project can usually be measured by the project manager's ability to keep everyone calm and happy, rather than on more substantive planning concerns.

Sequential design management processes were developed in a period when more emphasis was placed on the design act and the artifacts that accompanied a job than on the significance and intricacies of the problems being acted upon. It is questionable whether this problem-solving attitude will survive the shift in emphasis to problems rather than designs.

› Cyclic-design processes

Cyclic-design is the second major design management process. The term "feedback" has become a buzz word for progressive sounding architects. What does it mean when immersed in the design management process? Each of the three cyclic processes employ feedback. The simplest of the three is linear feedback.[6] This process implies the presence of performance criteria against which successive design alternatives are measured until one solution satisfies the criteria. This type of design management suits projects with rigid, well defined, and measurable performance standards (e.g., a radiology laboratory, an operating room, or an aerospace rocket launch and test center).

Developmental planning is a cyclic process that requires any design to be considered as part of a cycle of events over time.[7] Some buildings must be evaluated after a period of use, others require updating that calls for a new plan, and other situations require additional plans after a project is complete. Each new incremental plan differs from the previous one in its recognition of new needs.

The third cyclic process is termed empirical evaluation.[8] By this design management process architects predict how future buildings will work to satisfy user needs. However, a serious question is whether documenting how people presently use buildings is a satisfactory measure of their present or future needs and preferences.

Cyclic processes are, in principle, good "solution-corroboration devices" since they impose the discipline of recycling, or redesigning until a solution meets the criteria.

Cyclic processes are primarily aimed at fulfilling specific performance criteria rather than improving or gaining experience in working with them. More than with any other problem solving method, feedback subjects the success of a project to the correct initial problem definition and delineation of performance criteria.

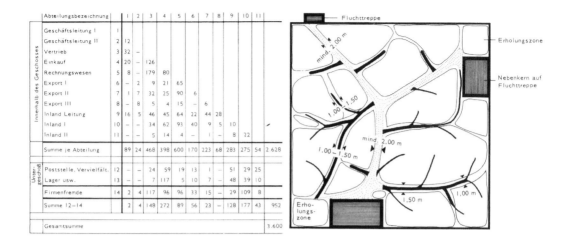

Abteilungsbezeichnung		1	2	3	4	5	6	7	8	9	10	11	
Geschäftsleitung I	1												
Geschäftsleitung II	2	12											
Vertrieb	3	32	–										
Einkauf	4	20	–	126									
Rechnungswesen	5	8	–	179	80								
Export I	6	–	2	9	21	65							
Export II	7	1	7	32	25	90	6						
Export III	8	–	8	5	4	15	–	6					
Inland Leitung	9	16	5	46	45	64	22	44	28				
Inland I	10	–	–	34	62	93	40	9	5	10			
Inland II	11	–	–	5	14	4	–	1	–	8	22		
Summe je Abteilung		89	24	468	398	600	170	223	68	283	275	54	2.628
Poststelle, Vervielfält.	12	–	–	24	59	19	13	1	–	51	29	25	
Lager usw.	13	–	–	7	117	5	10	7	–	48	39	10	
Firmenfremde	14	2	4	117	96	96	33	15	–	29	109	8	
Summe 12–14		2	4	148	272	89	56	23	–	128	177	43	952
Gesamtsumme													3.600

Fig 3
Office layout as a self-organizing system

Cyclic processes are very good when used as a planning method to integrate design and use. By employing feedback from the use-evaluation of a building complex, we can program and define problems better for the next related project design. As planning tools, cyclic processes aid in the communications, coordination, and cooperation between users and designers.

In summary, the chief weakness of cyclic processes in design and planning lies in the fact that they have little concern for the problem itself; performance criteria are tacitly fixed in the design and evaluation process. Nor do cyclic processes provide for systematically evaluating and improving upon the description of the problems involved.

› Evolutionary design processes

As architects move from the design of single isolated buildings to large-scale projects, they must adopt new management attitudes and processes to fuse today's urgent social, cultural, and economic problems. Evolutionary processes are helpful in the necessary fusion.

To visualize an evolutionary process, imagine for a moment a problem-describing/problem-solving machine. Then assume that it is possible to continually make a better machine by finding any faults in the previous machine's ability to describe and solve a given problem. After this, the machine is operated by management processes at time – t – on a pre-designated problem so as to evolve a better machine at time – t + 1. The new machine will be more capable of centralizing the design issues and of solving the problem. Not until the point in time has been reached when a sufficiently good machine has been developed will actual design decisions be made.

Cyclic processes provide feedback to fixed performance criteria. Evolutionary processes are concerned with feedback to new performance criteria which came out of previous tentative definitions and solutions. This approach has been described metaphorically as a parent–sibling relationship. The offspring management-machine itself becomes a parent when it reaches the point that it is a better problem describer/solver than *its* parent.

Now to proceed it is necessary to replace the ideated machine with a design team. Then we can proceed to the first of three evolutionary processes, *developmental designs.*[9] This method was originated for cross-disciplinary architecture. With this process instead of choosing the best of alternate solutions, it is possible to find a series of better solutions each time a design solution has been generated.

The fresh thing about developmental design is that the designer or team proclaims total uncertainty of the "problem field" from the outset. The goal of the process is to minimize this uncertainty. Success is measured by how well a designer or team can both define and solve the relevant problems. With developmental design the search is for those problem formulations that best engender the felt necessities – the intuitive, human understanding of the needs of the time – that justify all action in developmental design.

When the interval between design iterations is reduced to a very small period of time the process becomes on-line design.[10] This process is possible when computers are employed to speed the analysis and presentation aspects of architecture. On-line design depends on how fast the designer can: 1) simulate a design action and receive a sufficiently comprehensive evaluation, 2) revise and generate another solution, and 3) sort and record learned behavior. It is possible for a single designer to change both the criteria and the design actions through the computer console, or the same procedures can be followed in an equally effective manner by a team of designer, client, and consultants to more highly corroborated design solutions.

The last of the three evolutionary-type processes is on-line planning.[11] For this type of process, the professional designer is not required. The users of facilities and building products become the designers and planners. On-line planning will soon dominate office facilities planning, where totally industrialized products can be manipulated by users and their computers. As the larger building market becomes more industrialized, the same will take place in architecture and city planning. It appears that to the degree to which products can be standardized to interfit, easily adjust, or systematically decompose to changing needs, to that degree the problems of assembling and maintaining these products can be given over to computers and the users' direct manipulation.

Evolutionary management design processes will be increasingly important as the splintered planning disciplines seek out larger and more user-participatory systems of planning.

---

1   To my knowledge, first called "the systems approach" by members of the Systems Research Center, Case Institute of Technology, whose roster included C. W. Churchman, Hilary Putnam and Kenneth Boulding.

2   1A: Linear. *Architects Handbook of Professional Practice* (Washington, D.C.: American Institute of Architects, 1963) Project Procedures, Chapter 1.

3   1B: Linear Overlapping. The Chicago Office of Skidmore Owings and Merrill, Architects. *Production Management Systems and Synthesis*, Martin Kenneth Starr (Englewood Cliffs, NJ: Prentice Hall, 1964) Production Management Models, pp. 12 and 139.

4   1C: Parallel Alternatives. The San Francisco Office of John Carl Warnecke Architects.

5   1D: Collapsible Time Frame. Aero Space Industries Project Control (Lockheed Missile and Space Division, Sunnyvale, California).

6   2A: Linear Feedback. *An Introduction to Cybernetics*, Ross W. Ashby (New York: John Wiley & Sons, 1967).

7   2B: Developmental Planning. *Developmental Planning*, Richard Meier (New York: McGraw-Hill, 1965).

8   2C: *Empirical Evaluation. Dorms at Berkeley, An Environmental Analysis*, Sim Van der Ryn and Murray Silverstein (New York: Educational Facilities Laboratory, 1967). "Techniques for the Measurement of Performance," Horst Rittel, Robert Fessendon, and Henry Sanoff (Berkeley: Department of Architecture, University of California, 1965).

9   3A: Developmental Design. The New York Offices of Design Futures. The New York Office of Caudill Rowlett Scott, Architects.

10   3B: On-Line Design. ADEPT, a comprehensive computer software and hardware system (including graphic displays) designed and developed by the Boston Office of Design Systems, Inc.

11   3C: On Line Planning. "Totipotentology: The theory of Planning for Alternative Futures," the author first presented at the Parsons School of Design, New York, March, 1969. *The Office – A Facility Based on Change*, Robert Propst (Elmhurst, Illinois: The Business Press, 1968).

1970

Peter Cook

Experiment is an Inevitable

Peter Cook was born in Southend-on-Sea, England, in 1936 and studied architecture at the Bournemouth College of Art from 1953 to 1958 and also at the Architectural Association in London from 1958 to 1960 under the guidance of Peter Smithson (see Team 10, 1954/1962). He is best known as a founder member of the experimental "anti-architectural" practice Archigram, launched in 1961 while Cook was working in the offices of James Cubitt and Partners in London. Archigram actually began as a broadsheet newsletter ("architectural-telegram") and served as a vehicle to promote the group's futuristic ideas for high-technology housing and urban planning schemes through a seductive language of colorful and cartoon-like collages. In recognition of their influence Archigram was awarded the prestigious Gold Medal of the Royal Institute of British Architects in 2002. Recently Cook has also carried out several innovative building projects including the blob-like Kunsthaus art gallery in Graz, Austria, designed in collaboration with Colin Fournier. He is currently Professor of Architecture and Chair of the Bartlett School of Architecture, University College London.

Peter Cook has published ten books in total, the earliest of which – *Architecture: Action and Plan* (1967) – is possibly his most influential. His third book *Experimental Architecture* (1970), from which this extract is taken, continues the "manifesto" approach of the earlier Archigram work. The text celebrates examples of the deployment of new materials and systems, and explores the potential of mass production and prefabrication in the construction industry (see Buckminster Fuller, 1929). This work is also inspired by the writings of Reyner Banham (see Banham, 1960, 1965) who subsequently came to be seen as the major "mouthpiece" of the Archigram group. Much of Cook's writing deals with "the struggle of architecture with technology that is the love–hate situation in today's second machine age."[1]

1   Peter Cook, *Architecture: Action and Plan* (London: Studio Vista, 1967), p. 96.

› The force of ideas and technologies

In this chapter we shall look at a series of substantive and emotive forces which are characteristic of the basic ambiguity of architecture, standing as it does between the practical and the idealistic. In this century there have been many motivations which it has seemed necessary to explode. A succession of logical steps have arisen from the aftermath of wars or from new attitudes about the need for certain types of building. By considering five rather different motivations, which in many ways challenge one another and certainly challenge the attention of young architects, we can see that whichever one we follow we are gradually led to the point where they necessarily suggest far-reaching experiment; and though each experiment is of a different nature, each has been a necessary outcome of the situation.

We can look at the logic of production and the strong connection between industrial processes and architecture, originating in Victorian pragmatism. (Also in the Victorian period we become aware of the beginnings of a much more open attitude towards the rightness of using new components to make up a building.) We are familiar with the cast-iron crockets and railings and the notion of repetition, but it is not until the 1920s that production-line building becomes a really serious proposition, and only then does it become an integral part of the philosophy of a new architecture. Either we can see this, cynically, as a theoretical alignment (i.e. in order to make a new architecture one goes straight to the most up-to-date process, deliberately assuming an anti-historical stand-point) or we can decide that, with a closer and necessary involvement with the economics of building, it becomes inevitable.

Against this materialist corner of the modern movement there is always the strong urge to find new philosophic value in any piece of space or design which is made. Those who propose to erect such stages of values are still taken seriously, even though they may be unable to qualify these values to a general public.

Another similar attitude wishes to evaluate architecture according to preferred constituent elements. This viewpoint suggests that architecture, though an artifact, should arise from a series of basic physical consistencies. Though the imposition of such evaluation is similar to that of the previous group, its train of thought is different, as we shall see when we trace it through to the position of questioning.

Another very strong thread running through the architecture of the last forty years is that where the material itself has provided an incentive for the discovery of new things. And finally we can see the most frequent aspiration that has been overlaid: that looking towards technology as a great force for a new architecture.

› The logic of production

The force and logic of *production* has a practicality which appeals to certain designers. They turn towards an area which has to exist by its very rationale, feeling that if the *process* itself can be fed through this same rationale the resulting product will avoid many idiosyncrasies of peasant building. Once again there is a strong moralist thread to this approach, which would be vulnerable if it were not for the fact that the procedure comes so very close to success. Slowly the building process is being brought closer to something industrial and there are many arguments which suggest that this may be the only way in which building methods can survive.

More frequently, too, the process has been looked at in greater and greater depth, so that one can no longer just regard a serious piece of production building from the point where the factory process starts; one has to go back into the area of consumer surveys. Gradually there has been the bleeding-in of these outside methods. Sequences have been looked at which did not arise in the traditional building industries. New technologies have been overlaid on the older ones, and the notion of prefabrication emerges as an inevitable technique.

Prefabrication has been put forward by each successive generation as the solution to building illogicality and has gradually become one of the central approaches. But it was not until 1927 that it became philosophically respectable, when Walter Gropius evolved a system[1] of panel construction which was not only sophisticated but contained a degree of flexibility with ubiquitousness which suggested that (whether you liked the style or not) previous ways of making houses were by definition archaic. Even at this time the whole business was tightly allied to the notion of modulation, which probably resulted from the desire to make clear the *demonstration* of a prefabricated building. It is interesting to see that later designers have felt the necessity to keep prefabricated sets of parts completely prefabricated, although from time to time there have been strong arguments for combining these elements with others that are in situ or additive or perhaps made by some other system of industrialization. The modular intention has had to come through in spirit and the parts have had to at least appear to be consistent with one another.

The prefabricators have rapidly become preoccupied with their own kind of delicacy: the notions of the magic part and the magic joint and a constant search for the *universal* joint have almost fetishist overtones. The suspicion that there must be some ultimate purity in the putting together of mechanical parts has its own rigorous appeal. It was in the 1940s, when Konrad Wachsmann started to produce beautiful prototypes,[2] joints, and working parts, that production architecture reached its maturity. Wachsmann had his early training in the Germany of carpenters and industrial designers; and it is in Germany too that production architecture has consistently been closest to real industrial design.

America's position in the development of prefabricated architecture has probably been more relaxed because of its ready use of timber in house building. The balloon frame itself is a very rational way of using this material and by the beginning of the twentieth century American 'ready-made' homes were displaying all but the most sophisticated prefabrication series (although, of course, they were wooden systems).

Almost echoing the old joke among architects that even the floor joist and the brick have to be modular, there is a gradual toughening up of the whole process. The real experimental work will probably now be done in the whole design approach to prefabrication and not just in the evolution of the fabric. As Chris Abel suggests in his article 'Ditching the Dinosaur Sanctuary',[3] there is now the need to tune the machine to the consumer rather than rely on some formal straitjacket for prefabricated parts.

We then find that system building, which is the definition of the kind of pre-fabrication with a consistent set of parts, is also sharing this shifting relationship to the consumer market. Philosophically it suggests a way of building which is much closer to the world of car or utensil production, implying that the house, or the large building, can similarly respond to the changing tastes and requirements of succes-sive generations. Yet there is still a strong link with the building industry itself and more often industrialized building has operated some kind of rationalization of earlier, much less sequential methods of making buildings.

The situation which now can be called experimental will be strategic as well as operational; it will involve the design of the process, its economics and its market-ing potential as much as the beauty of its detailing. It is curious then that, in looking for examples to study, we have to fall back on prefabricated parts as illus-tration of the rigorousness which is the main involvement. The clever assembly tends to involve the clever multi-directional joint and every investigation and exper-iment in this area reiterates this problem.

Wachsmann, in his early work, has probably brought the idea of fabric prefab-rication further than anyone else: from its more primitive aspects through to the sophisticated package house system which he designed in 1942 with Walter Gropius, where there was the interface of a very rational panel system and a bril-liant joint. He then moved on to experiments with the topology of constructional steelwork and subsequently to notions of minimal structure supporting maximal space. He is currently working on a building which has no immediately apparent support structure, which in fact beds its tension membrane (the roof) into the ground horizontally either side of the structure. While this, the Town Hall for Califor-nia City, is not strictly concerned with prefabrication, it illustrates Wachsmann's movement towards an ultimate constructional gesture. Significantly, he has worked his way through the middle ground of the abstractly rational jointing theory and the putting-together of parts to a heroic gesture which is the whole building.

Jean Prouvé is perhaps a more typical experimentalist in the field of prefabri-cated parts. His work, mostly in France, has often been in association with famous architects, as a developer-engineer. He has made panels, usually of steel, into intrinsically beautiful buildings by virtue of the finesse with which he is able to resolve the structural potential of pressed metal, its production and its jointing. Some of his buildings may at first appear undistinguished (and it is certainly very difficult for non-architects to appreciate their superiority over any other more normal panel-built buildings), but his experiments are significant for most practi-tioners. His work has been a continual ironing out of the problems of sheet mater-ial, of joint, and of the inherent problem of the exposure of the joint. Most architects in northern countries have to spend much time and will-power on the problem of weathering and system-building has always had the problem of finding a material which does not absorb water or cannot easily be fractured.

In technique, however, Prouvé's work can most clearly be read against car pro-duction and seen as a sophisticated working through of the idea of the component. Other metal buildings (from as early as the 1920s in Germany, through to the 1960s) have usually fallen into the category of panel and post construction. Either they attempt to produce a very small number of basic components with a resulting inadequacy in their jointing; or they admit that a larger number of basic components can be more specific, more subtle and more effective as a piece of practical structure. But some suffer a philosophical loss of face because of the admittance of non-purity. In his exhibition building at Grenoble,[4] Prouvé presents a very cool skin which demonstrates that his panel system has almost reached the ubiquitousness of brick.

The developers of metal cast systems (such as the IBIS [Industrialized Building in Steel] project in the mid-1960s in England and the various projects for the international competition for steel houses, held in 1967) display a gradual constipation of ideas. Perhaps only Herbert Ohl has evolved something as fundamental (in his garage enclosure system, which he has subsequently developed for other building types) as a component which could make almost any enclosure rather than something which is limited by the specific problems of a local condition.

It was, characteristically, Buckminster Fuller who, as far back as 1927, pointed a natural direction in which the production run could be significant to house building. His Wichita house is, in effect, a simple piece of corrugated sheet metal tacked around a central pole structure. This principle is developed through to his Dymaxion house where the components become more sophisticated and can be put together in very much the same way as current furniture kits, i.e., sequentially, but are not made up of parts which are all exactly the same. The Dymaxion bathroom, which failed only as a result of the politics of the construction industry, is the famous example of large-scale building components being produced and marketed in the way that a car is. Fuller's actual technique by-passed the trap of the universal part and, consequently, the implications of his structures are much wider. They imply the possibility of treating buildings as durables which can be bought (and expended) because of their features. Still more important in this context is the fact that the components require each other to be produced industrially because it is actually more efficient and not just a nice idea.

Another side of the commitment to the production run is the philosophy of the shed. Its origins clearly lie in one tradition of architecture which attempts to create large spaces, but its development is much more specifically allied to the development of steel. The Victorian railway stations and the need for really large uninterrupted spaces to meet industrial and military requirements forced the consideration and (naturally enough for heroic reasons) the notion of the totally uninterrupted space, where the structure was only necessary to support the total envelope. The development of the space frame extended this idea and Wachsmann, in the 1940s and 1950s, made projects for space frame structures of gigantic dimensions and sophisticated profile. Particularly as a result of industrial needs and the necessary incorporation of services, good lighting and the alternative profiles of roofing (industrial north lights and ventilation), the idea of the space frame roof as the parent structure emerged as a very strong notion. In the last few years this has seen its most sophisticated and influential application in the work done by Ezra Ehrenkrantz and his team on the SCSD project.[5] They have produced schools where a highly sophisticated steel roof system can carry on top of it all the air conditioning, lighting, electrics, and other services for an equally sophisticated type of school underneath. The implication of its use is that the school itself can be very freely planned and can be repeatedly changed and reorganized. The servicing hangs down and both it and the sub-structures of partitions (or for that matter, anything else) are located between the roof and the floor, always with reference to the top system. Reyner Banham has championed this as a significant step towards the totally ubiquitous structure for the totally free changing building.

The two notions of the totally rational building component and the totally ubiquitous building must at some time come together. Even in the United States there is still most often a reliance upon the normal constructional system; and so far there has not really been a structure incorporating both ideas. Buckminster Fuller has clearly suggested the fusion in his notion of the dome made from equal parts (whether geodesic or not) and his extension of the housing ideas through to his standard of living package. And it is in Reyner Banham's 1965 article entitled 'A Home is not a House' that the notion of the envelope with the autonomous servicing package as the only internal feature is expressed.

Clearly the inspiration of production is central to the mid-twentieth century. Very few experimental architects will now be able to ignore the suggestion of ideals in production. Once again it is a question of definition, and a question of heroics, and we can see that production in the recent past has had too simplistic an aspiration. But its other implication, that of the ability of a rational product to give a member of the public precisely what he wants more quickly, cheaply, and successfully, is more interesting.

1   Experimental panel house, Walter Gropius 1927 at Dessau. See also project for steel panel house, Marcel Breuer, 1925.
2   'Packaged House System', Walter Gropius and Konrad Wachsmann. See *The Turning Point of Building*, Konrad Wachsmann, New York, 1961.
3   'Ditching the Dinosaur Sanctuary', Chris Abel, *Architectural Design*, August 1969.
4   Palais des Expositions, Jean Prouvé, *Architecture d'Aujourd'hui*, 135.
5   'School Construction Systems Development', *Architectural Design*, July 1965 and November 1967.

Superstudio was a radical Italian architecture group formed in Florence in 1966. The first two members were Adolfo Natalini and Cristiano Toraldo di Francia – two young architecture graduates from Florence University who collaborated on an exhibition called 'Superarchitecture' – and they were joined within two years by Gian Piero Frassinelli and Alessandro and Roberto Magris. As they described the trajectory of their work in 1973: "In the beginning we designed objects for production, designs to be turned into wood and steel, glass and brick or plastic – then we produced neutral and usable designs, then finally negative utopias, forewarning images of the horrors which architecture was laying in store for us with its scientific methods for the perpetuation of existing models." Like many of the other radical groups of the 1960s, Superstudio dissolved in the late 1970s as its members left to follow their own paths.

The following essay was prepared to explain their installation for an exhibition curated by Emilio Ambasz at the Museum of Modern Art in 1972, called *Italy: The New Domestic Landscape, Achievements and Problems in Italian Design*. It marked a decisive moment in the group's work, a turn away from architectural production toward "a life without objects." The exhibition came after their most dystopic designs to that point: twelve ideal cities, each a more terrifying object lesson in technology taken to its extremes. The principle element in the MOMA exhibition, supported by collages and a short film, was a mirror box with an outlet, plugs, wires, and a TV in the corner, configured so that it showed the items reflected in an infinite grid disappearing to the horizon. The network of infrastructure was a direct extension of the infinite grid used in the group's earlier projects, reduced to its minimum.

In an eerily evocative description of contemporary infrastructures and mobility, the installation offered "a network of energy and information extending to every properly inhabitable area" in which "nomadism becomes the permanent condition." The film clip displayed on the TV was the first of five short films conceived to explore five *Fundamental Acts* – Life, Education, Ceremony, Love, Death – that might redeem architecture from its supportive role in the technological system. Their dilemma is legible throughout the project: can design be purified sufficiently to achieve that redemption or is technology-as-a-system itself the problem?

1972

Superstudio

Microevent/
Microenvironment

Fig 4
'All you have to do is stop and connect a plug: the desired microclimate is immediately created (temperature, humidity, etc.); you plug in to the network of information, you switch on the food and water blenders. . . .' Superstudio, 1972

› Description of the Microevent/Microenvironment

The proposed microevent is a critical reappraisal of the possibilities of life without objects. It is a reconsideration of the relations between the process of design and the environment through an alternative model of existence, rendered visible by a series of symbolic images. The microenvironment is like a room with walls; the floor and ceiling are covered with black felt; thin luminescent lines make the corner angles stand out clearly.

A cube about six feet wide is placed in the center on a platform about sixteen inches high. All the walls of the cube, except the one facing the entrance, are made of polarized mirrors, so that the model inside becomes clearer and clearer as we move to the end of the room. This model, repeated to infinity by the mirrors, is a square plate of chequered laminated plastic, with a little 'machine' out of which come various terminals. One of the terminals is connected to a TV screen, which transmits a three-minute movie, a documentary on the model seen in various natural and work situations. The sound-track gives information about the original concepts for the model. Meteorological events will be projected on the ceiling: sunrise, sun, clouds, storm, sunset, night.

The lighting of the cube varies according to the phenomena projected. The rest of the room is permanently plunged in darkness.

Specific considerations

In this exhibition, we present the model of a mental attitude. This is not a three-dimensional model of a reality that can be given concrete form by a mere transposition of scale, but a visual rendition of a critical attitude toward (or a hope for) the activity of designing, understood as philosophical speculation, as a means to knowledge, as critical existence.

Design should be considered as a 'cross-discipline,' for it no longer has the function of rendering our requirements more complex through creating a new artificial panorama between man and environment. By finding a connection between data taken from the various humanistic and scientific disciplines (from the technique of body control to philosophy, the disciplines of logic and medicine, to bionomics, geography, etc.), we can visualize an image-guide: the final attempt of design to act as the 'projection' of a society no longer based on work (and on power and violence, which are connected with this), but an unalienated human relationship.

In this exhibition, we present an alternative model for life on earth.

We can imagine a network of energy and information extending to every properly inhabitable area. Life without work and a new 'potentialized' humanity are made possible by such a network. (In the model, this network is represented by a Cartesian 'squared' surface, which is of course to be understood not only in the physical sense, but as a visual-verbal metaphor for an ordered and rational distribution of resources.)

The network of energy can assume different forms.

The first is a linear development.

The others include different planimetrical developments, with the possibility of covering different, and gradually increasing, parts of the habitable areas. The configuration (typology) of the environment depends solely on the percentage of area covered, analogous to the way in which we distinguish a street from a town, a town from a city.

Some of the types
10 percent covered: The network is developed like a continuous ribbon extending over the territory.

50 percent covered: The network is developed like a checkerboard, with areas measuring one square kilometer alternating with squares of open land.

100 percent covered: The network is transformed into a continuous development, the natural confines of which are formed by mountains, coasts, rivers.

It is an image of humanity wandering, playing, sleeping, etc., on this platform. Naked humanity, walking along the highway with banners, magic objects, archeological objects, in fancy dress . . .

The distances between man and man (modified); these generate the ways in which people gather, and therefore 'the places': if a person is alone, the place is a small room; if there are two together, it is a larger room; if there are ten, it is a school; if a hundred, a theater; if a thousand, an assembly hall; if ten thousand, a city; if a million, a metropolis . . .

Nomadism becomes the permanent condition: the movements of individuals interact, thereby creating continual currents. The movements and migrations of the individual can be considered as regulated by precise norms, the distances between man and man, attractions/reactions – love/hate. As with fluids, the movement of one part affects the movements of the whole.

The diminished possibility of physical movement results in an increase in conceptual activities (communications). The model constitutes the logical selection of these developing tendencies: the elimination of all formal structures, the transfer of all designing activity to the conceptual sphere. In substance, the rejection of production and consumption, the rejection of work, are visualized as a physical metaphor: the whole city as a network of energy and communications.

The places where humanity is concentrated in great numbers have always been based on the city network of energy and information, with three-dimensional structures representing the values of the system. In their free time, large crowds on the beaches or in the country are in fact a concentrated mass of people 'served' by mechanical, mobile miniservices (car, radio, portable refrigerator). Concentrations such as the Isle of Wight or Woodstock indicate the possibility of an 'urban' life without the emergence of three-dimensional structures as a basis. The tendency to the spontaneous gathering and dispersing of large crowds becomes more and more detached from the existence of three-dimensional structures.

Free gathering and dispersal, permanent nomadism, the choice of interpersonal relationships beyond any preestablished hierarchy, are characteristics that become increasingly evident in a work-free society.

The types of movement can be considered as the manifestations of the intellectual processes: the logical structure of thought continually compared (or contrasted) to our unconscious motivations.

Our elementary requirements can be satisfied by highly sophisticated (miniaturized) techniques. A greater ability to think, and the integral use of our psychic potential, will then be the foundations and reasons for a life free from want.

Bidonvilles, drop-out city, camping sites, slums, tendopoles, or geodetic domes are all different expressions of an analogous desire to attempt to control the environment by the most economical means.

The membrane dividing exterior and interior becomes increasingly tenuous: the next step will be the disappearance of this membrane and the control of the environment through energy (air-cushions, artificial air currents, barriers of hot or cold air, heat-radiating plates, radiation surfaces, etc.).

Through an examination of the statistics of population growth, an analysis of the relationship between population and the territory that can be exploited for living purposes, new techniques for agricultural production, and ecological theories, we can arrive at a formulation of various hypotheses for survival strategies:

a)   hypothesis for the creation and development of servoskin: personal control of the environment through thermoregulation, techniques for breathing, cyborgs ... mental expansion, full development of senses, techniques of body control (and initially, chemistry and medicine).

b)   hypothesis for total system of communications, software, central memories, personal terminals, etc.

c)   hypothesis for network of energy distribution, acclimatization without protective walls.

d)   mathematical models of the cyclic use of territory, shifting of the population, functioning and non-functioning of the networks.

General considerations

If we look closely, we can see how all the changes in society and culture in this century (or since 1920) have been generated by one force only – the elimination of formal structures as a tendency toward a state of nature free from work.

The destruction of objects, the elimination of the city, and the disappearance of work are closely connected events. By the destruction of objects, we mean the destruction of their attributes of 'status' and the connotations imposed by those in power, so that we live *with* objects (reduced to the condition of neutral and disposable elements) and not *for* objects.

By the elimination of the city, we mean the elimination of the accumulation of the formal structures of power, the elimination of the city as hierarchy and social model, in search of a new free egalitarian state in which everyone can reach different levels in the development of his possibilities, beginning from equal starting points.

By the end of work, we mean the end of specialized and repetitive work, seen as an alienating activity, foreign to the nature of man; the logical consequence will be a new, revolutionary society in which everyone should find the full development of his possibilities, and in which the principle of 'from everyone according to his capacities, to everyone according to his needs' should be put into practice. The construction of a revolutionary society is passing through the phase of radical, concrete criticism of present society, of its way of producing, consuming, living.

Merchandise, according to Guy Debord, in bourgeois society (which acts and perpetuates itself through its products – including political parties and trade unions, which are essential parts of the spectacle) becomes the contemplation of itself.

The production machine produces a second poverty (Galbraith), perpetuating itself even after the fulfilment of its goals, or beyond its essential ends (the satisfaction of primary needs), constantly inducing new needs.

Once clarified that:

a) design is merely an inducement to consume;
b) objects are status symbols, the expressions of models proposed by the ruling class. Their progressive accessibility to the proletariat is part of a 'leveling' strategy intended to avoid the conflagration of the class struggle;
c) the possession of objects is the expression of unconscious motivation: through analysis, the removal of the motivation underlying their desirability may be reached;

. . . then it becomes urgent to proceed to destroy them . . . or does it?

Metamorphoses become frequent when a culture does not have sufficient courage to commit suicide (to eliminate itself) and has no clear alternatives to offer, either.

The theory of intermediate states is the book of changes?

Thus, while the merchandise-form continues on toward its absolute realization, we reduce operations to a minimum. Reducing operations to a minimum, in all fields, is part of a general process of 'reduction.' Only through this reduction process can the field be cleared of false problems and induced needs. Through reduction, we proceed toward a mental state of concentration and knowledge, a condition essential for a truly human existence.

Earlier, we defined the destruction of the syntactical ties that bind the object to the system, the destruction of its significance as superimposed by the ruling classes, as 'destruction of the object.' We have formulated an hypothesis of the reduction of objects to neutral, disposable elements. To this, we can add the hypothesis of the construction of the object through its metamorphosis. The present process of 'overloading' meanings onto an object is part of that strategy of disgust to which we have already referred.

Through the psychological rethinking of an object, we can try for its 'reconstruction.' And this through discontinuous and alogical action, refusing guarantees of value (licenses issued by the system), aspiring to identify with life and total reality.

Objects thus cease to be the vehicles of social communication to become a form of reality and the direct experience of reality.

The metamorphoses which the object has to go through are those during which it is reloaded with the values of myth, of sacredness, of magic, through the reconstruction of relationships between production and use, beyond the abolition of the fictitious ties of production–consumption.

When design as an inducement to consume ceases to exist, an empty area is created, in which, slowly, as on the surface of a mirror, such things as the need to act, mold, transform, give, conserve, modify, come to light.

The alternative image (which is, really, the hope of an image) is a more serene, distended world, in which actions can find their complete sense and life is possible with few, more or less magical, utensils.

Objects, that is, such as mirrors – reflection and measure.

The objects we will need will be only flags or talismans, signals for an existence that continues, or simple utensils for simple operations. Thus, on the one hand, there will remain utensils (with less chrome and decorations); on the other, such symbolic objects as monuments or badges. Objects perhaps created for eternity from marble and mirrors, or for the present from paper and flowers – objects made to die at their appointed hours, and which even have this sense of death among their characteristics. Objects that can easily be carried about, if we should decide to become nomads, or heavy and immovable, if we decide to stay in one place forever.

A journey from A to B
There will be no further need for cities or castles.
There will be no further reason for roads or squares.
Every point will be the same as any other
(excluding a few deserts or mountains which are in no wise inhabitable).
So, having chosen a random point on the map, we'll be able to say my house will be here for three days two months or ten years. And we'll set off that way (let's call it B) without provisions, carrying only objects we're fond of.
The journey from A to B can be long or short, in
any case it will be a constant migration,
with the actions of living at every point along the ideal line
between A (departure) and B (arrival).
It won't, you see, be just the transportation of matter.
These are the objects we'll carry with us:
some strange pressed flowers,
a few videotapes, some family photos,
a drawing on crumpled paper,
an enormous banner of grass and reeds interwoven with
old pieces of material which once were clothes,
a fine suit, a bad book . . .
These will be the objects.
Someone will take with him
only a herd of animals for friends. For instance:
a quartet of Bremermusikanten,
or a horse, two dogs and two doves
or twelve cats, five dogs and a goat.

Yet others will take with them only memory,
become so sharp and bright as to be a visible object.
Others will hold one arm raised, fist clenched.
Someone will have learnt a magic word and will take it with him
as a suitcase or a standard: CALM, COMPREHENSION, CONFIDENCE,
COURAGE, ENERGY, ENTHUSIASM, GOODNESS, GRATITUDE, HARMONY, JOY, LOVE,
PATIENCE, SERENITY, SIMPLICITY, WILL, WISDOM (dark blue).

(This is the complete set of cards in the 'Technique of Evocative Words' by Roberto
Assagioli, M.D.) But almost everybody will take only himself from A to B, a single
visible object, like a complete catalogue
as an enormous Mail Order Catalogue

What we'll do.
We'll keep silence to listen to our own bodies,
we'll hear the sound of blood in our ears,
the slight crackings of our joints or teeth,
we'll examine the texture of our skins, the patterns made by the hairs on our bodies
and heads.
We'll listen to our hearts and our breathing.

We'll watch ourselves living.
We'll do very complicated muscular acrobatics.
We'll do very complicated mental acrobatics.

The mind will fall back on itself to read its own history.
We'll carry out astonishing mental operations.
Perhaps we'll be able to transmit thoughts and images,
then one happy day our minds will be in communication with that of the whole
world.

That which was called philosophy will be the natural physical activity of our minds,
and will at the same time be philosophy, religion, love, politics, science . . .
Perhaps we'll lose the names of these disciplines (and it will be no great loss) when
everybody will be present in essence in our minds.
We'll be able to create and transmit visions and images, perhaps even make little
objects move for fun.

We'll play wonderful games, games of ability and love.
We'll talk a lot, to ourselves and to everybody.
We'll look at the sun, the clouds, the stars.
We'll go to faraway places, just to look at them and hear them.
Some people will become great story-tellers: many will move
to go and listen to them. Some will sing and play.

Stories, songs, music, dancing will be the words we speak and tell
ourselves.
Life will be the only environmental art.

The happy island.

A lady of our acquaintance became hysterical at hearing all this story and said: I certainly have no intention of doing without my vacuum-cleaner and the mowing machine, and the electric iron and the washing machine and refrigerator, and the vase full of flowers, the books, my costume jewellery, doll and clothes!
Whatever you say madam!
Just take whatever you like, or rather equip a happy island for yourself with all your goods.
The only problem is that the sea has receded all round and the island is sticking up in the middle of a plain without any messages in bottles.

The distant mountain.
Look at that distant mountain . . . what can you see?
is that the place to go to? or is it only the limit of the habitable? It's the one and the other, since contradiction no longer exists, it's only a case of being complementary. Thus thought a fairly adult Alice skipping over her rope, very slowly, though without feeling either heat or effort.

The encampment.
You can be where you like, taking with you the tribe or family. There's no need for shelters, since the climatic conditions and the body mechanisms of thermoregulation have been modified to guarantee total comfort.
At the most we can play at making a shelter, or rather at the home, at architecture.

The invisible dome.
All you have to do is stop and connect a plug: the desired microclimate is immediately created (temperature, humidity, etc.); you plug in to the network of information, you switch on the food and water blenders . . .

A short moral tale on design, which is disappearing.
Design, become perfect and rational, proceeds to synthesize different realities by syncretism and finally transforms itself, not coming out of itself, but rather withdrawing into itself, in its final essence of natural philosophy.
Thus designing coincides more and more with existence: no longer existence under the protection of design objects, but existence as a design.
The times being over when utensils generated ideas, and when ideas generated utensils, now ideas are utensils. It is with these new utensils that life forms freely in a cosmic consciousness.

If the instruments of design have become as sharp as lancets and as sensitive as sounding lines, we can use them for a delicate lobotomy.
Thus beyond the convulsions of overproduction a state can be born of calm in which a world takes shape without products and refuse, a zone in which the mind is energy and raw material and is also the final product, the only intangible object for consumption.
The designing of a region free from the pollution of design is very similar to a design for a terrestrial paradise . . .
This is the definitive product – this is the only one of the projects of a marvelous metamorphosis.

The Austrian Leopold Kohr (1909–94) was an economist, political scientist, philosopher, and a leading thinker of the ecology movement. He was also an anarchist who understood anarchy as a state of cooperative living, without the need of external rules and disciplinary power. He left Austria during the Second World War for Paris, and then for Puerto Rico.

He developed the slogan "small is beautiful," made famous by his student Friedrich Schumacher. For Kohr, once a system gets beyond a certain threshold (which, for states, he set at between twelve and fifteen million people), problems start to increase at a much quicker rate than they can be solved. His solution was decentralization. Just as cells divide once they reach a certain size, rather than combine with others to become ever larger, Kohr favors division over amalgamation. In relation to the increasing speed of mobility, the subject of "Velocity Population," taken from his book *The Inner City: From Mud to Marble*, he finds that speed has qualitative effects on our everyday lives, creating a need for new conditions to accommodate an increasing rate of flow. To this, he proposes a solution of polycentral regeneration, whereby regions divide to create small-scale, autonomous areas.

Kohr's work has a new resonance today. His understanding of the city and region was informed by the behavior of systems. His interest in threshold conditions at which systems exhibit new characteristics was based on an understanding of biological systems, which Ivan Illich has traced back to the work of D'Arcy Thompson and his studies on biological growth and form. As Kohr put it, the dinosaurs could not survive, because they were just too big. At a point in which our thinking about systems has moved onto those which are adaptive, complex, and self-regulating, and we are coming to value flexibility, suppleness, and agility in themselves, Kohr's attention to proper fit takes on a new relevance.

1973

Leopold Kohr

Velocity Population

Broadus Mitchell, biographer of Alexander Hamilton and one of America's outstanding economic historians, tells the delightful story of a somewhat mystified
physician who attended the birth of an unusual number of illegitimate children in a
large area in the South of the United States.

What mystified the doctor was, in the first place, that all the girls pointed to
the same person as the father of their babies. But what floored him completely was
that the father, when the doctor finally met him, turned out to be a man in his eighties. "How on earth," asked the doctor, "did you manage to sire all these children?"
"Well," answered the astonishing octogenarian with the raspy voice of old age, "I
admit, I couldn't have done it if I hadn't had a motor bike."

In other words, the velocity of modern transportation permitted the old fellow
to perform over a considerable area what would have been possible only over a
square mile or so had he been restricted to walking. And in a single square mile
there would, of course, not have been that many girls available for such random
motherhood.

However, the velocity with which one nowadays gets around has a more
significant effect than that of increasing the population quantitatively, by adding to
the number of people. The real bombshell is that velocity increases population also
qualitatively, by adding to its mass, just as higher speed has the effect of increasing the mass of atomic particles, or as faster circulation increases the "quantity" of
money, as every student of economics is taught.

This explains why theatres must have emergency exits in addition to ordinary
ones – in case an audience should get panic-stricken and try to leave at a faster
than normal pace. For, as every theatre owner knows, a faster crowd has the same
material effect as a larger crowd. The number of exits he must make available must
therefore be adjusted not to the numerical but to the effective (or velocity) size of
an audience, which is numerical size multiplied by speed.

Now, what applies to people moving within the enclosure of a theatre, applies
also to populations moving within the enclosed space of cities or of nations. The
faster they move as a result of modern means of transportation, ranging from motor
cycles to jet planes, the larger becomes their effective size. In fact, aside from a few
exceptions such as in India, the really terrifying aspect of the world's overpopulation
problem is as yet primarily due not so much to the excessive numbers of human
beings as to the excessive speed with which they have started to move around.

This being the case, one way of solving the problem would be to create, as in a
theatre, "emergency" space to cope with the periods when populations try to move
faster than ordinary pace, as happens in every city during rush hours. This is, in
fact, what planners are doing anyway by constantly adding new roads and widening
the old ones.

The only trouble here is that, in contrast to the fixed space of a theatre, the
addition of "emergency" space within stretchable limits of a city does not relieve
the effect of overcrowding; it actually intensifies it by encouraging a given population to disperse beyond the "walls" over ever wider areas. But the further an
integrated population spreads, the greater is the distance it must cover to attend
its daily chores. And the greater the distance, the faster it must move. And the
faster it moves, the larger becomes effective (or velocity) size.

In the case of a city the size of San Francisco, Bristol, or Puerto Rico's San Juan, this means that a numerical population of let us say 600,000 is blown up to an effective population of perhaps 2,000,000, while its emergency road network is adjusted at best to perhaps 1,000,000. And nothing can ever be done about this gap. For each time that new roads are added at an *arithmetic* ratio, a city's effective or velocity population is on this very ground increased at a *geometric* ratio. This is why the New Jersey Turnpike, opened in 1948, reached the traffic density predicted for 1975 a week after opening day or why, to the surprise of Inspector Martin West of the Surrey police's road-safety department (*The Times* August 18, 1988), the Surrey section of "the motorway (M25) is carrying the levels of traffic predicted for the 1990s" already in the 1980s. "The volume of traffic is causing chaos" not in spite but because of the new motorways.

This leaves, as the only practical solution, the second method by which theatres try to cope with the mass increasing effect of velocity when they exhort their audience: "In case of fire, **WALK, DO NOT RUN.**" For just as increased speed increases the pressure and mass of a crowd, reduced speed reduces it. But as every theatre owner also knows, the only way of really reducing the effective or velocity size of an audience is not by warning it of the disastrous *consequences* of running but by depriving it of the *motive* for running. This he does by making sure that there will be no fire. His real answer to the problem caused by the size increasing effect of accelerated pace lies not so much in emergency exits as in a fireproof structure.

And so it is with the answer to our urban and national problems at least as long as the excessive pressure on space which we associate with the concept of overpopulation is still mainly caused by an increase in motorized velocity rather than an increase in the number of people. Hence also our planners must create the condition which takes away not the *means* of high-velocity travel but the *motive* that forces people to move at an ever faster pace to begin with. In other words, what they must study is not *locomotion* but *motivation*; not the types of vehicles and roads that make Sammy run, but the reason that makes Sammy running – and then deprive him of his reason for running.

Nationally, this is achieved by returning to a high degree of long-distance traffic and speed discouraging regional self-sufficiency, *as is envisaged by the devolutionists*; and, municipally, by a high degree of urban decentralization *or, as it should better be called, polycentral regeneration.* This means: instead of dispersing the central offices of a metropolitan area over its various districts, to turn the boroughs once again into autonomous communities of their own in which the citizen finds everything he needs for his daily existence in locations which are central, but small and close by. Hence the answer is not really decentralization but *centralization on a small scale.*

This is the only way by which the rising traffic pressure of our motorized velocity overpopulation can be reduced by means other than killing or pilling: not by regionalizing central institutions (car licences in Swansea, a London art gallery with an annexe in dockland Manchester) but by centralizing the regions through granting them a high measure of autonomy; and, in cities, not by suburbanizing the slums but by urbanizing the suburbs; not by turning districts of the poor, for the poor, by the poor, into districts of the yuppies, for the yuppies, by the yuppies, both depending on work fifteen miles away from their one-class locations, but by turning each area into a little all-class city of such a different identity, convivial self-sufficiency and aesthetic charm that few would either need or wish to leave it.

Before the end of the twentieth century, when the abysmal spectre of numerical overpopulation will take over, this is all that is at present needed to turn the unmanageable *velocity* overpopulation of cities such as Cardiff, San Francisco, or Bristol from anything up to 2,000,000 back to their manageable numerical dimension of 600,000. And, as for London, why not turn the greater part of it by the end of this century into the federation of villages which the gentle anarchist William Morris envisaged for it. Or write it off.

The Italian-American architect Paolo Soleri was born in 1919 in Turin where he studied architecture at the Turin Polytechnic. After graduation, he came to apprentice with Frank Lloyd Wright at Taliesen in Wisconsin and Arizona. After spending the first half of the 1950s in Italy, he returned to the United States, settling in Arizona where he founded the Cosanti Foundation based on a set of ecological principles. In 1970 he started to develop the planned community, Arcosanti, which was meant to embody the principles of arcology, his particular combination of architecture and ecology. Arcosanti is intended to maximize social interaction and sharing of facilities, minimize energy use and environmental pollution, and allow for ready contact with the natural environment. This excerpt is drawn from his manifesto, *The Bridge Between Matter and Spirit is Matter Becoming Spirit*.

In "Function Follows Form," Soleri inverts the slogan coined by Louis Sullivan, and argues instead that structures engender events. The earth provides such a model: a large structure without particular function that engendered a variety of habitations. Though he notes, as have many others, that when human society replaces biological evolution, the process becomes conscious, requiring "intelligent" decisions to save us from "natural" selection. The centralized organization of the arcology was meant to bring genetics, technology, and creation together for the betterment of spiritual and physical well-being.

The contribution this essay provides to this collection is the role of esthetics in the shift from natural to cultural evolution and equally the importance of scale or scope in providing enough context and interactions for the "esthetogenetic" processes to operate.

1973

Paolo Soleri

Function Follows Form (Structure Before Performance)

The way biological life develops into new forms and consciousness seems to confirm that the instrument has a chronological precedence over the performance. An organism does not willfully construct for himself a new organ so as to attain a certain goal, but stumbles by mutant chance onto a certain characterization given to it fractionally and in infinitesimal doses. It is the fixation of such mutant novelty – the new instrument – that makes the organism able to perform a certain act which could not have been performed before, no matter how great its usefulness or how intense the desire had been for it (see the urban parallel). As such changes happen to increase his chances for survival, they increase also the number of offspring carrying the mutant alteration. The neck of the giraffe does not grow out of the will of the animal, but out of a sequence of genetic variations accidentally useful for his well-being. The function does not originate the form. It maintains it. The long neck of the giraffe is incorporated into the species in as much as it has found a function useful for the animal. At the same time, the chain of offspring of the tennis player will not show any disproportion between right and left arm though he himself might exhibit a right arm more muscular and heavy than the left (acquired specialization in a "fringe" function, not genetic alteration working at the core of survival). If a tyrant dynasty was to demand superchampions from a restricted caste, then the genetic machine would start working (as in the case of selected cattle breeds), or else the genealogical trees of many tennis players' families would be truncated.

There is an element of pure novelty in the progression of links where the new, if accidental, instrument (mutant character) almost forces upon the organism a new kind of relationship with the environment (and with himself in the case of consciousness-reflection). It is the reversal of the abuse "form follows function" formula. It is instead "function follows form," or one can say, "structure precedes performance." The form is there in the genetic emergent mutation, a structure that appears and becomes filled with events. The earth itself is the most comprehensive case in point. It is hard to believe (unacceptable) that its "function" was the generation and sustenance of life.

› The esthetic and the extravagant

Can what is done haphazardly in the genetic world be done consciously and willfully in the mental universe? The esthetic "process" does this as it is that kind of event which justifies itself by its own advent, and its usefulness is not to be found in an a priori functional demand. And as it is for the mutant gene, it is more than often, preponderantly, a regressive mutant, the "extravagant." But now and then the extravagant is pushed aside and creation comes to be and with it the emergence within the species of a new consciousness, a new fragment of etherealization. Thus, in a way in the human world, the mutant accident is substituted by the esthetic event moving the evolutionary thrust from the genetic-biological to the cultural-personal. Of all other activities of man, none seem to be so clearly connecting the past to the future without passing through the purgatory of instrumentalization.

And as it happens in the biological world, often such instrumentalization gives results that are the equivalent of a (non-accidental) bad mutation making the organism less well equipped for survival. A classic example is the automobile. The survival of the race is jeopardized by the mutant incorporating the car into the structure of society. Nor is the analogy too farfetched if one considers (1) the species is in many ways an organism composed not of cells but of persons; and that (2) this organism has transferred most of its evolutionary thrust from biogenetic devices to technological tools. What remains to be seen is if the will or the discriminative power belonging to the single components – the person – has in itself retained or incorporated the ability to distinguish the good from the bad tool (mutant) which is, in human terms, the choice of life or death for the species. Only "intelligent" decisions can disengage the species from the pressure and the fatality of "natural" selection.

In the formula "form follows function" is buried the frightful fact that a high percentage of the functions are the equivalent of behavior originating from bad genetic mutations, those for instance of the "can do, thus must do" ethic (see the automobile). When the feasible becomes the desirable in as much as it swells the gross national or personal product, the mutation it stands for is not a logical or constructive mutation. Chances are that it is one of those fatally bad mutations crowded with dark forebodings.

› Functional coherence

What is then the difference between the invention of the car and the invention of the spoon or of the arcology? To find it one must carry the conditions they advocate to their ultimate consequences. What appears in the automobile context is that those consequences are the negation of the premises on which the invention (mutation) was incorporated into the body-social. The communication-information aim, original to the invention of the automobile, is stunted and finally disappearing as the automobile takes over and paralyzes society. (If the giraffe's neck doesn't stop growing, it will make the animal unfit.) The mystique which is at the driver's seat of this paradox can hardly be found in the spoon, which will remain more or less always a humble device for feeding ourselves.

The arcology? In its most naked purpose arcology is an attempt as much as and far more than the automobile at giving the person the swiftest way to communication, information, and action. It wants to enlarge the personal universe of each individual by centering him in the thick of things. At its most "absurd" limit, the arcology becomes punctiform; that is to say, it contains in "no space" the whole of itself and the organisms it is meant to serve. It resolves itself and its own content into pure spirit. This transformation, as paradoxical as it might appear, is none but the dreamed omnipotence and omnipresence of God. Civitate Dei becoming Godlike ... God himself. The infinite complexity of a being utterly centered upon itself infinitely powerful and infinitely wise (by definition as it is spirit); a Point whose next metamorphosis could be the advent of the explosion known as the "big bang," initiating a new cosmos, the spiritual universe. A universe, liberated from the "slavery" of mass-energy-speed, can perform explosively instead of necessitating the implosive performance (the necessary liberating form).

So what we see on one side is the car, scattering evermore, to a uniform dull-
ness and dumbness, the species of man into isolated, segregated, electronically
plugged-in cells; and on the other, the arcology that puts each city dweller at the
"center" of the city, the ideal position for a person to be conscious and to be part of
the information-communication-action-participation world, to which he belongs as
a social-cultural individual.

It is to be pointed out that at the service level, smallness, co-ordination and
efficiency induce self-effacement. If a Cleveland needs a police force of four to five
thousand, an arcological Cleveland can do with 20 percent of that number. So it is
for the other servo-systems keeping the city in shape: delivery and retrieval of goods
and wastes, delivery and performance of utilities and services – whether manual,
mechanical, or both. To seek the opposite, with the not so original contention that if
the police can become so efficient the city will become a police city, is that amount
of saying that it is better to make everyone into a policeman, a garbage collector, a
postal employee, a truck driver, a telephone maintenance man, a doctor, a nurse, a
sewer supervisor, a bus driver (a car driver?) etc. That is to say, loose man in the
maze of the amorphous where life becomes random and, lastly, God's litter. Delegation of responsibilities effectively controlled by the community is the goal and the
smaller the supervised body and the machinery it uses, the better off the community. This points at more than physical miniaturization. It states also that for the
"same" aims, a lesser number of service men are needed. The percentage of service
and maintenance people is cut to a fraction of its former number, the bureaucratic
machinery shrivels to a shadow of its former self and its inertia all but disappears.
Then society ripens the fruits of swiftness, response, sensitization. It becomes
dynamic, without necessarily becoming mobile. (I say this here because of the mis-
taken identification of mobility with dynamism.) It becomes dynamic essentially
because toil, drudgery and senselessness are drastically cut to size. At this point
miniaturization leaves opportunism to become an imperative – the ethical imperative that demands the conscious and willful use of the universe of matter and energy
in the only way that can sustain the survival and evolution of life.

The other difference between the car, the spoon, and the arcology is in the
possible presence or not of the esthetic in the car, the spoon, and the arcology. The
automobile or the spoon are or were given a very limited slot into reality; their main
purpose is to serve well, to "perform honestly." To such a category belongs a
"beauty" of directness and readability, a proto-esthetic firmness, untroubled,
correct, rational, logical, clinical (all present in the best of our tools and equipment). For the arcology, whose scope is almost inseparable from the context of life
itself if it is true that the environment is meshed in with life, the esthetic potential
is boundless; that is to say, in one and the same more promising and more dangerous. The correct, the rational, the logical, the clinical, ought to give way to the corresponding transcendental twins – the "more than" (more than rational, more than
logical, more than clinical), even at the risk of collapsing into the "less than." In the
most pragmatic light, the worth of man is indissolubly bound to the esthetogenetic
process as it is only by it (for it) that man will find himself to be the witness of a
compassionate environment, worthy of reverence – an environment made of
things, of live things, and of persons.

One can carry on the parallel, in a garbled fashion undoubtedly, mindful that man evolves on the three fronts of genetics, technology, and creation, with the genetic statistically adopting the formula "function follows form"; and the creative working out consciously, that is, non-statistically, "forms" which define their own function. One then sees the man-giraffe whose neck is stretching (is it?); the man-technologist whose instruments – automobiles, typewriters – accumulate his power for transformism; and the man-creator on the shoulders of the two preceding, reaching up to the branches of the tree of creation whose leaves will possibly, if not improbably, bud and grow at the radiance of the three-headed creature.

In a fully responsive arcology, all three of them must be present and working. The portent is to make the whole animal (the city), with its physiological and technological instruments and with its leaves growing amply up there, into a pandemonium, a quivering, of grace, serenity, fire, joy, reverence, excitation, consciousness, expectation. To make this possible for man it is not sufficient to shelter him from pain and punishment with the help of behavioral conditioning. On the left of zero the most we can hope for is the smallest of the negative numbers: as the world of the positive numbers is blocked off by zero, itself, the edge of the two watersheds of adaptations and creation. As adaptative animals, we are purely the creatures of our environment. As creators of the real world we are the agent of a trust that cannot be completely accounted for by the pristine nature of the planet supporting us.

The question arises: Is this the right time for launching into chancy endeavors in the hope that, among the many, one might turn out to be fruitful? If the attempt were a purely blind date with the future, the answer should be no, as the suffering and the deprivations are global and intense, and the stakes are frightfully central to the hypothesis of man as an animal with a future. But if (1) this last is the situation, and (2) the date is not a blind one but indeed is radically joined to the dynamics of evolution, then the unpredictability itself of the resulting "formula" might contain the best ferments for the germination of a novel reality. It might well be the best if not the only hopeful fissure that the cleavage of our problem might have to pray in, for the invention of a more human future.

We in part shape answers by better shaping instruments of which we have experience (adaptation, improvement), but at the same time we must risk a "longer neck" even if the leaves that such neck will afford were not much more than ontologically probable and auspicable. It is then not a pure case of structure before performance – "form before function" – of the genetic world, but a well-rooted start on a background of past experiences and errors, in a journey which chartering in detail would not only be nearsighted but plainly incoherent as the land to be sought is not there to be discovered but has to be invented and created along the way. Therefore, the journey is not into things existing but toward the future of which we are, ourselves, the makers, inventors, and creators. At stake is a more intense physico-spiritual becoming and stronger fulfillments for both the person and the species. And by the way, if this is not clear, the person is also, if not more so, the ghetto-bound and the hunger-stricken, for whom immediate rescue – philanthropic piety – is often less than compassionate and too often mystifying. "We" must "immediately" feed the hungry. Upon each of us is the weight of our personal hypocritical forgetfulness and the direct responsibility for the sufferance and dehumanization of the ill fed. Beyond that, our practical solutions are sugar-coated fraudulence as they only transfer within society the same "original sins" without trying to expel them from the context of human life.

Ruth Schwartz Cowan (born 1941) is an American historian of science, technology, and medicine, with degrees from Barnard College, the University of California at Berkeley and the Johns Hopkins University. She taught at the State University of New York at Stony Brook from 1967 to 2002, attaining the rank of Professor in 1984. From 1992 to 1994 she served as President of the Society for the History of Technology and since 2002 has been Professor of the History and Sociology of Science at the University of Pennsylvania.

Cowan is the author of four books including *The Social History of American Technology* (1997), in which she uses the notion of technological systems as a unifying theme to argue that technological change is neither sudden nor discontinuous, but is always closely related to social developments that determine both the kinds of tools developed and the ways in which they are utilized. Cowan demonstrates that the way in which Americans have viewed technology has been at least as important as the scientific developments themselves. The text reprinted here was originally published in 1976 in the journal *Technology and Culture*, and was subsequently included in the anthology entitled *The Social Shaping of Technology*, in 1985 and 1999. The central thesis of the essay – that new technological appliances have increased rather than reduced domestic drudgery – was also further elaborated in the book from 1983 entitled *More Work for Mother: The Ironies of Household Technology from the Open Hearth to the Microwave*. Confirming the complexity of such a claim, Cowan wrote a subsequent article which observed that the one exception to that rule was the professionalization of health care, which substantially contributed to women's ability to work outside the home.[1]

The essay exemplifies the social constructivist approach to technology, demonstrating the degree to which every tool is part of a larger social and economic system.

1976

Ruth Schwartz Cowan

The "Industrial Revolution" in the Home: Household Technology and Social Change in the Twentieth Century

---

[1]   Ruth Schwartz Cowan, 'Less Work for Mother,' *American Heritage of Invention & Technology* 2 (Spring 1987): 57–63.

When we think about the interaction between technology and society, we tend to think in fairly grandiose terms: massive computers invading the workplace, railroad tracks cutting through vast wildernesses, armies of women and children toiling in the mills. These grand visions have blinded us to an important and rather peculiar technological revolution which has been going on right under our noses: the technological revolution in the home. This revolution has transformed the conduct of our daily lives, but in somewhat unexpected ways. The industrialization of the home was a process very different from the industrialization of other means of production, and the impact of that process was neither what we have been led to believe it was nor what students of the other industrial revolutions would have been led to predict.

Some years ago sociologists of the functionalist school formulated an explanation of the impact of industrial technology on the modern family. Although that explanation was not empirically verified, it has become almost universally accepted.[1] Despite some differences in emphasis, the basic tenets of the traditional interpretation can be roughly summarized as follows:

Before industrialization the family was the basic social unit. Most families were rural, large, and self-sustaining; they produced and processed almost everything that was needed for their own support and for trading in the marketplace, while at the same time performing a host of other functions ranging from mutual protection to entertainment. In these preindustrial families women (adult women, that is) had a lot to do, and their time was almost entirely absorbed by household tasks. Under industrialization the family is much less important. The household is no longer the focus of production; production for the marketplace and production for sustenance have been removed to other locations. Families are smaller and they are urban rather than rural. The number of social functions they perform is much reduced, until almost all that remains is consumption, socialization of small children, and tension management. As their functions diminished, families became atomized; the social bonds that had held them together were loosened. In these postindustrial families women have very little to do, and the tasks with which they fill their time have lost the social utility that they once possessed. Modern women are in trouble, the analysis goes, because modern families are in trouble; and modern families are in trouble because industrial technology has either eliminated or eased almost all their former functions, but modern ideologies have not kept pace with the change. The results of this time lag are several: some women suffer from role anxiety, others land in the divorce courts, some enter the labor market, and others take to burning their brassieres and demanding liberation.

This sociological analysis is a cultural artifact of vast importance. Many Americans believe that it is true and act upon that belief in various ways: some hope to reestablish family solidarity by relearning lost productive crafts – baking bread, tending a vegetable garden – others dismiss the women's liberation movement as 'simply a bunch of affluent housewives who have nothing better to do with their time.' As disparate as they may seem, these reactions have a common ideological source – the standard sociological analysis of the impact of technological change on family life.

As a theory this functionalist approach has much to recommend it, but at present we have very little evidence to back it up. Family history is an infant disciplineline, and what evidence it has produced in recent years does not lend credence to the standard view.[2] Phillippe Ariès has shown, for example, that in France the ideal of the small nuclear family predates industrialization by more than a century.[3] Historical demographers working on data from English and French families have been surprised to find that most families were quite small and that several generations did not ordinarily reside together; the extended family, which is supposed to have been the rule in preindustrial societies, did not occur in colonial New England either.[4] Rural English families routinely employed domestic servants, and even very small English villages had their butchers and bakers and candlestick makers; all these persons must have eased some of the chores that would otherwise have been the housewife's burden.[5] Preindustrial housewives no doubt had much with which to occupy their time, but we may have reason to wonder whether there was quite as much pressure on them as sociological orthodoxy has led us to suppose. The large rural family that was sufficient unto itself back there on the prairies may have been limited to the prairies – or it may never have existed at all (except, that is, in the reveries of the sociologists).

Even if all the empirical evidence were to mesh with the functionalist theory, the theory would still have problems, because its logical structure is rather weak. Comparing the average farm family in 1750 (assuming that you knew what that family was like) with the average urban family in 1950 in order to discover the significant social changes that had occurred is an exercise rather like comparing apples with oranges; the differences between the fruits may have nothing to do with the differences in their evolution. Transferring the analogy to the case at hand, what we really need to know is the difference, say, between an urban laboring family of 1750 and an urban laboring family 100 and then 200 years later, or the difference between the rural non-farm middle classes in all three centuries, or the difference between the urban rich yesterday and today. Surely in each of these cases the analyses will look very different from what we have been led to expect. As a guess we might find that for the urban laboring families the changes have been precisely the opposite of what the model predicted; that is, that their family structure is much firmer today than it was in centuries past. Similarly, for the rural non-farm middle class the results might be equally surprising; we might find that married women of that class rarely did any housework at all in 1890 because they had farm girls as servants, whereas in 1950 they bore the full brunt of the work themselves. I could go on, but the point is, I hope, clear: in order to verify or falsify the functionalist theory, it will be necessary to know more than we presently do about the impact of industrialization on families of similar classes and geographical locations.

With this problem in mind I have, for the purposes of this initial study, deliberately limited myself to one kind of technological change affecting one aspect of family life in only one of the many social classes of families that might have been considered. What happened, I asked, to middle-class American women when the implements with which they did their everyday household work changed? Did the technological change in household appliances have any effect upon the structure of American households, or upon the ideologies that governed the behavior of

American women, or upon the functions that families needed to perform? Middle-class American women were defined as actual or potential readers of the better-quality women's magazines, such as the *Ladies' Home Journal, American Home, Parents' Magazine, Good Housekeeping*, and *McCall's*.[6] Nonfictional material (articles and advertisements) in those magazines was used as a partial indicator of some of the technological and social changes that were occurring.

The *Ladies' Home Journal* has been in continuous publication since 1886. A casual survey of the nonfiction in the *Journal* yields the immediate impression that that decade between the end of World War I and the beginning of the depression witnessed the most drastic changes in patterns of household work. Statistical data bear out this impression. Before 1918, for example, illustrations of homes lit by gaslight could still be found in the *Journal*; by 1928 gaslight had disappeared. In 1917 only one-quarter (24.3 percent) of the dwellings in the United States had been electrified, but by 1920 this figure had doubled (47.4 percent – for rural non-farm and urban dwellings), and by 1930 it had risen to four-fifths (80 percent).[7] If electrification had meant simply the change from gas or oil lamps to electric lights, the changes in the housewife's routines might not have been very great (except for eliminating the chore of cleaning and filling oil lamps); but changes in lighting were the least of the changes that electrification implied. Small electric appliances followed quickly on the heels of the electric light, and some of those augured much more profound changes in the housewife's routine.

Ironing, for example, had traditionally been one of the most dreadful household chores, especially in warm weather when the kitchen stove had to be kept hot for the better part of the day; irons were heavy and they had to be returned to the stove frequently to be reheated. Electric irons eased a good part of this burden.[8] They were relatively inexpensive and very quickly replaced their predecessors; advertisements for electric irons first began to appear in the ladies' magazines after the war, and by the end of the decade the old flatiron had disappeared; by 1929 a survey of 100 Ford employees revealed that ninety-eight of them had the new electric irons in their homes.[9]

Data on the diffusion of electric washing machines are somewhat harder to come by; but it is clear from the advertisements in the magazines, particularly advertisements for laundry soap, that by the middle of the 1920s those machines could be found in a significant number of homes. The washing machine is depicted just about as frequently as the laundry tub by the middle of the 1920s; in 1929, forty-nine out of those 100 Ford workers had the machines in their homes. The washing machines did not drastically reduce the time that had to be spent on household laundry, as they did not go through their cycles automatically and did not spin dry; the housewife had to stand guard, stopping and starting the machine at appropriate times, adding soap, sometimes attaching the drain pipes, and putting the clothes through the wringer manually. The machines did, however, reduce a good part of the drudgery that once had been associated with washday, and this was a matter of no small consequence.[10] Soap powders appeared on the market in the early 1920s, thus eliminating the need to scrape and boil bars of laundry soap.[11] By the end of the 1920s Blue Monday must have been considerably less blue for some housewives – and probably considerably less 'Monday,' for with an electric iron, a washing machine, and a hot water heater, there was no reason to limit the washing to just one day of the week.

Like the routines of washing the laundry, the routines of personal hygiene must have been transformed for many households during the 1920s – the years of the bathroom mania.[12] More and more bathrooms were built in older homes, and new homes began to include them as a matter of course. Before the war most bathroom fixtures (tubs, sinks, and toilets) were made out of porcelain by hand; each bathroom was custom-made for the house in which it was installed. After the war industrialization descended upon the bathroom industry; cast iron enamelware went into mass production and fittings were standardized. In 1921 the dollar value of the production of enameled sanitary fixtures was $2.4 million, the same as it had been in 1915. By 1923, just two years later, that figure had doubled to $4.8 million; it rose again, to $5.1 million, in 1925.[13] The first recessed, double-shell cast iron enameled bathtub was put on the market in the early 1920s. A decade later the standard American bathroom had achieved its standard American form: the recessed tub, plus tiled floors and walls, brass plumbing, a single-unit toilet, an enameled sink, and a medicine chest, all set into a small room which was very often five feet square.[14] The bathroom evolved more quickly than any other room of the house; its standardized form was accomplished in just over a decade.

Along with bathrooms came modernized systems for heating hot water: 61 percent of the homes in Zanesville, Ohio, had indoor plumbing with centrally heated water by 1926, and 33 percent of the homes valued over $2,000 in Muncie, Indiana, had hot and cold running water by 1935.[15] These figures may not be typical of small American cities (or even large American cities) at those times, but they do jibe with the impression that one gets from the magazines: after 1918 references to hot water heated on the kitchen range, either for laundering or for bathing, become increasingly difficult to find.

Similarly, during the 1920s many homes were outfitted with central heating; in Muncie most of the homes of the business class had basement heating in 1924; by 1935 Federal Emergency Relief Administration data for the city indicated that only 22.4 percent of the dwellings valued over $2,000 were still heated by a kitchen stove.[16] What all these changes meant in terms of new habits for the average housewife is somewhat hard to calculate; changes there must have been, but it is difficult to know whether those changes produced an overall saving of labor and/or time. Some chores were eliminated – hauling water, heating water on the stove, maintaining the kitchen fire – but other chores were added – most notably the chore of keeping yet another room scrupulously clean.

It is not, however, difficult to be certain about the changing habits that were associated with the new American kitchen – a kitchen from which the coal stove had disappeared. In Muncie in 1924, cooking with gas was done in two out of three homes; in 1935 only 5 percent of the homes valued over $2,000 still had coal or wood stoves for cooking.[17] After 1918 advertisements for coal and wood stoves disappeared from the *Ladies' Home Journal*; stove manufacturers purveyed only their gas, oil, or electric models. Articles giving advice to homemakers on how to deal with the trials and tribulations of starting, stoking, and maintaining a coal or a wood fire also disappeared. Thus it seems a safe assumption that most middle-class homes had switched to the new method of cooking by the time the depression began. The change in routine that was predicated on the change from coal or wood to gas or oil was profound; aside from the elimination of such chores as loading the fuel and removing the ashes, the new stoves were much easier to light, maintain, and regulate (even when they did not have thermostats, as the earliest models did not).[18] Kitchens were, in addition, much easier to clean when they did not have coal dust regularly tracked through them; one writer in the *Ladies' Home Journal* estimated that kitchen cleaning was reduced by one-half when coal stoves were eliminated.[19]

Along with new stoves came new foodstuffs and new dietary habits. Canned foods had been on the market since the middle of the nineteenth century, but they did not become an appreciable part of the standard middle-class diet until the 1920s – if the recipes given in cookbooks and in women's magazines are a reliable guide. By 1918 the variety of foods available in cans had been considerably expanded from the peas, corn, and succotash of the nineteenth century; an American housewife with sufficient means could have purchased almost any fruit or vegetable and quite a surprising array of ready-made meals in a can – from Heinz's spaghetti in meat sauce to Purity Cross's lobster à la Newburg. By the middle of the 1920s home canning was becoming a lost art. Canning recipes were relegated to the back pages of the women's magazines; the business-class wives of Muncie reported that, while their mothers had once spent the better part of the summer and fall canning, they themselves rarely put up anything, except an occasional jelly or batch of tomatoes.[20] In part this was also due to changes in the technology of marketing food; increased use of refrigerated railroad cars during this period meant that fresh fruits and vegetables were in the markets all year round at reasonable prices.[21] By the early 1920s convenience foods were also appearing on American tables: cold breakfast cereals, pancake mixes, bouillon cubes, and packaged desserts could be found. Wartime shortages accustomed Americans to eating much lighter meals than they had previously been wont to do; and as fewer family members were taking all their meals at home (businessmen started to eat lunch in restaurants downtown, and factories and schools began installing cafeterias), there was simply less cooking to be done, and what there was of it was easier to do.[22]

Many of the changes just described – from hand power to electric power, from coal and wood to gas and oil as fuels for cooking, from one-room heating to central heating, from pumping water to running water – are enormous technological changes. Changes of a similar dimension, either in the fundamental technology of an industry, in the diffusion of that technology, or in the routines of workers, would have long since been labeled an 'industrial revolution.' The change from the laundry tub to the washing machine is no less profound than the change from the hand loom to the power loom; the change from pumping water to turning on a water faucet is no less destructive of traditional habits than the change from manual to electric calculating. It seems odd to speak of an 'industrial revolution' connected with housework, odd because we are talking about the technology of such homely things, and odd because we are not accustomed to thinking of housewives as a labor force or of housework as an economic commodity – but despite this oddity, I think the term is altogether appropriate.

In this case other questions come immediately to mind, questions that we do not hesitate to ask, say, about textile workers in Britain in the early nineteenth century, but we have never thought to ask about housewives in America in the twentieth century. What happened to this particular work force when the technology of its work was revolutionized? Did structural changes occur? Were new jobs created for which new skills were required? Can we discern new ideologies that influenced the behavior of the workers?

The answer to all of these questions, surprisingly enough, seems to be yes. There were marked structural changes in the work force, changes that increased the work load and the job description of the workers that remained. New jobs were created for which new skills were required; these jobs were not physically burdensome, but they may have taken up as much time as the jobs they had replaced. New ideologies were also created, ideologies which reinforced new behavioral patterns, patterns that we might not have been led to expect if we had followed the sociologists' model to the letter. Middle-class housewives, the women who must have first felt the impact of the new household technology, were not flocking into the divorce courts or the labor market or the forums of political protest in the years immediately after the revolution in their work. What they were doing was sterilizing baby bottles, shepherding their children to dancing classes and music lessons, planning nutritious meals, shopping for new clothes, studying child psychology, and hand stitching color-coordinated curtains – all of which chores (and others like them) the standard sociological model has apparently not provided for.

The significant change in the structure of the household labor force was the disappearance of paid and unpaid servants (unmarried daughters, maiden aunts, and grandparents fall in the latter category) as household workers – and the imposition of the entire job on the housewife herself. Leaving aside for a moment the question of which was cause and which effect (did the disappearance of the servant create a demand for the new technology, or did the new technology make the servant obsolete?), the phenomenon itself is relatively easy to document. Before World War I, when illustrators in the women's magazines depicted women doing housework, the women were very often servants. When the lady of the house was drawn, she was often the person being served, or she was supervising the serving, or she was adding an elegant finishing touch to the work. Nurse-maids diapered babies, seamstresses pinned up hems, waitresses served meals, laundresses did the wash, and cooks did the cooking. By the end of the 1920s the servants had disappeared from those illustrations; all those jobs were being done by housewives – elegantly manicured and coiffed, to be sure, but housewives nonetheless.

If we are tempted to suppose that illustrations in advertisements are not a reliable indicator of structural changes of this sort, we can corroborate the changes in other ways. Apparently, the illustrators really did know whereof they drew. Statistically the number of persons throughout the country employed in household service dropped from 1,851,000 in 1910 to 1,411,000 in 1920, while the number of households enumerated in the census rose from 20.3 million to 24.4 million.[23] In Indiana the ratio of households to servants increased from 13.5/1 in 1890 to 30.5/1 in 1920, and in the country as a whole the number of paid domestic servants per 1,000 population dropped from 98.9 in 1900 to 58.0 in 1920.[24] The business-class house-wives of Muncie reported that they employed approximately one-half as many woman-hours of domestic service as their mothers had done.[25]

In case we are tempted to doubt these statistics (and indeed statistics about household labor are particularly unreliable, as the labor is often transient, part-time, or simply unreported), we can turn to articles on the servant problem, the dis-appearance of unpaid family workers, the design of kitchens, or to architectural drawings for houses. All of this evidence reiterates the same point: qualified ser-vants were difficult to find; their wages had risen and their numbers fallen; houses were being designed without maid's rooms; daughters and unmarried aunts were finding jobs downtown; kitchens were being designed for housewives, not for ser-vants.[26] The first home with a kitchen that was not an entirely separate room was designed by Frank Lloyd Wright in 1934.[27] In 1937 Emily Post invented a new charac-ter for her etiquette books: Mrs. Three-in-One, the woman who is her own cook, waitress, and hostess.[28] There must have been many new Mrs. Three-in-Ones abroad in the land during the 1920s.

As the number of household assistants declined, the number of household tasks increased. The middle-class housewife was expected to demonstrate compe-tence at several tasks that previously had not been in her purview or had not existed at all. Child care is the most obvious example. The average housewife had fewer children than her mother had had, but she was expected to do things for her children that her mother would never have dreamed of doing: to prepare their special infant formulas, sterilize their bottles, weigh them every day, see to it that they ate nutritionally balanced meals, keep them isolated and confined when they had even the slightest illness, consult with their teachers frequently, and chauffeur them to dancing lessons, music lessons, and evening parties.[29] There was very little Freudianism in this new attitude toward child care: mothers were not spending more time and effort on their children because they feared the psychological trauma of separation, but because competent nursemaids could not be found, and the new theories of child care required constant attention from well-informed persons – persons who were willing and able to read about the latest discoveries in nutrition, in the control of contagious diseases, or in the techniques of behavioral psychology. These persons simply had to be their mothers.

Consumption of economic goods provides another example of the housewife's expanded job description; like child care, the new tasks associated with consump-tion were not necessarily physically burdensome, but they were time consuming, and they required the acquisition of new skills.[30] Home economists and the editors of women's magazines tried to teach housewives to spend their money wisely. The present generation of housewives, it was argued, had been reared by mothers who did not ordinarily shop for things like clothing, bed linens, or towels; consequently modern housewives did not know how to shop and would have to be taught. Fur-thermore, their mothers had not been accustomed to the wide variety of goods that were now available in the modern marketplace; the new housewives had to be taught not just to be consumers, but to be informed consumers.[31] Several contemporary observers believed that shopping and shopping wisely were occupy-ing increasing amounts of housewives' time.[32]

Several of these contemporary observers also believed that standards of household care changed during the decade of the 1920s.[33] The discovery of the 'household germ' led to almost fetishistic concern about the cleanliness of the home. The amount and frequency of laundering probably increased, as bed linen and underwear were changed more often, children's clothes were made increasingly out of washable fabrics, and men's shirts no longer had replaceable collars and cuffs.[34] Unfortunately all these changes in standards are difficult to document, being changes in the things that people regard as so insignificant as to be unworthy of comment; the improvement in standards seems a likely possibility, but not something that can be proved.

In any event we do have various time studies which demonstrate somewhat surprisingly that housewives with conveniences were spending just as much time on household duties as were housewives without them – or, to put it another way, housework, like so many other types of work, expands to fill the time available.[35] A study comparing the time spent per week in housework by 288 farm families and 154 town families in Oregon in 1928 revealed 61 hours spent by farm wives and 63.4 hours by town wives; in 1929 a US Department of Agriculture study of families in various states produced almost identical results.[36] Surely if the standard sociological model were valid, housewives in towns, where presumably the benefits of specialization and electrification were most likely to be available, should have been spending far less time at their work than their rural sisters. However, just after World War II economists at Bryn Mawr College reported the same phenomenon: 60.55 hours spent by farm housewives, 78.35 hours by women in small cities, 80.57 hours by women in large ones – precisely the reverse of the results that were expected.[37] A recent survey of time studies conducted between 1920 and 1970 concludes that the time spent on housework by non-employed housewives has remained remarkably constant throughout the period.[38] All these results point in the same direction: mechanization of the household meant that time expended on some jobs decreased, but also that new jobs were substituted, and in some cases – notably laundering – time expenditures for old jobs increased because of higher standards. The advantages of mechanization may be somewhat more dubious than they seem at first glance.

As the job of the housewife changed, the connected ideologies also changed; there was a clearly perceptible difference in the attitudes that women brought to housework before and after World War I.[39] Before the war the trials of doing housework in a servantless home were discussed and they were regarded as just that – trials, necessary chores that had to be got through until a qualified servant could be found. After the war, housework changed: it was no longer a trial and a chore, but something quite different – an emotional 'trip.' Laundering was not just laundering, but an expression of love; the housewife who truly loved her family would protect them from the embarrassment of tattletale gray. Feeding the family was not just feeding the family, but a way to express the housewife's artistic inclinations and a way to encourage feelings of family loyalty and affection. Diapering the baby was not just diapering, but a time to build the baby's sense of security and love for the mother. Cleaning the bathroom sink was not just cleaning, but an exercise of protective maternal instincts, providing a way for the housewife to keep her family safe from disease. Tasks of this emotional magnitude could not possibly be delegated to servants, even assuming that qualified servants could be found.

Women who failed at these new household tasks were bound to feel guilt about their failure. If I had to choose one word to characterize the temper of the women's magazines during the 1920s, it would be 'guilt.' Readers of the better-quality women's magazines are portrayed as feeling guilty a good lot of the time, and when they are not guilty they are embarrassed: guilty if their infants have not gained enough weight, embarrassed if their drains are clogged, guilty if their children go to school in soiled clothes, guilty if all the germs behind the bathroom sink are not eradicated, guilty if they fail to notice the first signs of an oncoming cold, embarrassed if accused of having body odor, guilty if their sons go to school without good breakfasts, guilty if their daughters are unpopular because of old-fashioned, or unironed, or – heaven forbid – dirty dresses. In earlier times women were made to feel guilty if they abandoned their children or were too free with their affections. In the years after World War I, American women were made to feel guilty about sending their children to school in scuffed shoes. Between the two kinds of guilt there is a world of difference.

Let us return for a moment to the sociological model with which this essay began. The model predicts that changing patterns of household work will be correlated with at least two striking indicators of social change: the divorce rate and the rate of married women's labor force participation. That correlation may indeed exist, but it certainly is not reflected in the women's magazines of the 1920s and 1930s: divorce and full-time paid employment were not part of the life-style or the life pattern of the middle-class housewife as she was idealized in her magazines.

There were social changes attendant upon the introduction of modern technology into the home, but they were not the changes that the traditional functionalist model predicts; on this point a close analysis of the statistical data corroborates the impression conveyed in the magazines. The divorce rate was indeed rising during the years between the wars, but it was not rising nearly so fast for the middle and upper classes (who had, presumably, easier access to the new technology) as it was for the lower classes. By almost every gauge of socioeconomic status – income, prestige of husband's work, education – the divorce rate is higher for persons lower on the socioeconomic scale – and this is a phenomenon that has been constant over time.[40]

The supposed connection between improved household technology and married women's labor force participation seems just as dubious, and on the same grounds. The single socioeconomic factor which correlates most strongly (in cross-sectional studies) with married women's employment is husband's income, and the correlation is strongly negative; the higher his income, the less likely it will be that she is working.[41] Women's labor force participation increased during the 1920s but this increase was due to the influx of single women into the force. Married women's participation increased slightly during those years, but that increase was largely in factory labor – precisely the kind of work that middle-class women (who were, again, much more likely to have labor-saving devices at home) were least likely to do.[42] If there were a necessary connection between the improvement of household technology and either of these two social indicators, we would expect the data to be precisely the reverse of what in fact has occurred: women in the higher social classes should have fewer functions at home and should therefore be more (rather than less) likely to seek paid employment or divorce.

Thus for middle-class American housewives between the wars, the social changes that we can document are not the social changes that the functionalist model predicts; rather than changes in divorce or patterns of paid employment, we find changes in the structure of the work force, in its skills, and in its ideology. These social changes were concomitant with a series of technological changes in the equipment that was used to do the work. What is the relationship between these two series of phenomena? Is it possible to demonstrate causality or the direction of that causality? Was the decline in the number of households employing servants a cause or an effect of the mechanization of those households? Both are, after all, equally possible. The declining supply of household servants, as well as their rising wages, may have stimulated a demand for new appliances at the same time that the acquisition of new appliances may have made householders less inclined to employ the laborers who were on the market. Are there any techniques available to the historian to help us answer these questions?

In order to establish causality, we need to find a connecting link between the two sets of phenomena, a mechanism that, in real life, could have made the causality work. In this case a connecting link, an intervening agent between the social and the technological changes, comes immediately to mind: the advertiser – by which term I mean a combination of the manufacturer of the new goods, the advertising agent who promoted the goods, and the periodical that published the promotion. All the new devices and new foodstuffs that were being offered to American households were being manufactured and marketed by large companies which had considerable amounts of capital invested in their production: General Electric, Procter & Gamble, General Foods, Lever Brothers, Frigidaire, Campbell's, Del Monte, American Can, Atlantic & Pacific Tea – these were all well-established firms by the time the household revolution began, and they were all in a position to pay for national advertising campaigns to promote their new products and services. And pay they did; one reason for the expanding size and number of women's magazines in the 1920s was, no doubt, the expansion in revenues from available advertisers.[43]

Those national advertising campaigns were likely to have been powerful stimulators of the social changes that occurred in the household labor force; the advertisers probably did not initiate the changes, but they certainly encouraged them. Most of the advertising campaigns manifestly worked, so they must have touched upon areas of real concern for American housewives. Appliance ads specifically suggested that the acquisition of one gadget or another would make it possible to fire the maid, spend more time with the children, or have the afternoon free for shopping.[44] Similarly, many advertisements played upon the embarrassment and guilt which were now associated with household work. Ralston, Cream of Wheat, and Ovaltine were not themselves responsible for the compulsive practice of weighing infants and children repeatedly (after every meal for newborns, every day in infancy, every week later on), but the manufacturers certainly did not stint on capitalizing upon the guilt that women apparently felt if their offspring did not gain the required amounts of weight.[45] And yet again, many of the earliest attempts to spread 'wise' consumer practices were undertaken by large corporations and the magazines that desired their advertising: mail-order shopping guides, 'product-testing' services, pseudoinformative pamphlets, and other such promotional devices were all techniques for urging the housewife to buy new things under the guise of training her in her role as skilled consumer.[46]

Thus the advertisers could well be called the 'ideologues' of the 1920s, encouraging certain very specific social changes – as ideologues are wont to do. Not surprisingly, the changes that occurred were precisely the ones that would gladden the hearts and fatten the purses of the advertisers; fewer household servants meant a greater demand for labor and timesaving devices; more household tasks for women meant more and more specialized products that they would need to buy; more guilt and embarrassment about their failure to succeed at their work meant a greater likelihood that they would buy the products that were intended to minimize that failure. Happy, full-time housewives in intact families spend a lot of money to maintain their households; divorced women and working women do not. The advertisers may not have created the image of the ideal American housewife that dominated the 1920s – the woman who cheerfully and skillfully set about making everyone in her family perfectly happy and perfectly healthy – but they certainly helped to perpetuate it.

The role of the advertiser as connecting link between social change and technological change is at this juncture simply a hypothesis, with nothing much more to recommend it than an argument from plausibility. Further research may serve to test the hypothesis, but testing it may not settle the question of which was cause and which effect – if that question can ever be settled definitely in historical work. What seems most likely in this case, as in so many others, is that cause and effect are not separable, that there is a dynamic interaction between the social changes that married women were experiencing and the technological changes that were occurring in their homes. Viewed this way, the disappearance of competent servants becomes one of the factors that stimulated the mechanization of homes, and this mechanization of homes becomes a factor (though by no means the only one) in the disappearance of servants. Similarly, the emotionalization of housework becomes both cause and effect of the mechanization of that work; and the expansion of time spent on new tasks becomes both cause and effect of the introduction of time-saving devices. For example the social pressure to spend more time in child care may have led to a decision to purchase the devices; once purchased, the devices could indeed have been used to save time – although often they were not.

If one holds the question of causality in abeyance, the example of household work still has some useful lessons to teach about the general problem of technology and social change. The standard sociological model for the impact of modern technology on family life clearly needs some revision: at least for middle-class non-rural American families in the twentieth century, the social changes were not the ones that the standard model predicts. In these families the functions of at least one member, the housewife, have increased rather than decreased; and the dissolution of family life has not in fact occurred.

Our standard notions about what happens to a work force under the pressure of technological change may also need revision. When industries become mechanized and rationalized, we expect certain general changes in the work force to occur: its structure becomes more highly differentiated, individual workers become more specialized, managerial functions increase, and the emotional context of the work disappears. On all four counts our expectations are reversed with regard to household work. The work force became less rather than more differentiated as domestic servants, unmarried daughters, maiden aunts, and grandparents left the household and as chores which had once been performed by commercial agencies (laundries, delivery services, milkmen) were delegated to the housewife. The individual workers also became less specialized; the new housewife was now responsible for every aspect of life in her household, from scrubbing the bathroom floor to keeping abreast of the latest literature in child psychology.

The housewife is just about the only unspecialized worker left in America – a veritable jane-of-all-trades at a time when the jacks-of-all-trades have disappeared. As her work became generalized the housewife was also proletarianized: formerly she was ideally the manager of several other subordinate workers; now she was idealized as the manager and the worker combined. Her managerial functions have not entirely disappeared, but they have certainly diminished and have been replaced by simple manual labor; the middle-class, fairly well educated housewife ceased to be a personnel manager and became, instead, a chauffeur, charwoman, and short-order cook. The implications of this phenomenon, the proletarianization of a work force that had previously seen itself as predominantly managerial, deserve to be explored at greater length than is possible here, because I suspect that they will explain certain aspects of the women's liberation movement of the 1960s and 1970s which have previously eluded explanation: why, for example, the movement's greatest strength lies in social and economic groups who seem, on the surface at least, to need it least – women who are white, well-educated, and middle-class.

Finally, instead of desensitizing the emotions that were connected with household work, the industrial revolution in the home seems to have heightened the emotional context of the work, until a woman's sense of self-worth became a function of her success at arranging bits of fruit to form a clown's face in a gelatin salad. That pervasive social illness, which Betty Friedan characterized as 'the problem that has no name,' arose not among workers who found that their labor brought no emotional satisfaction, but among workers who found that their work was invested with emotional weight far out of proportion to its own inherent value: 'How long,' a friend of mine is fond of asking, 'can we continue to believe that we will have orgasms while waxing the kitchen floor?'

1   For some classic statements of the standard view, see W. F. Ogburn and M. F. Nimkoff, *Technology and the Changing Family* (Cambridge, Mass., 1955); Robert F. Winch, *The Modern Family* (New York, 1952); and William J. Goode, *The Family* (Englewood Cliffs, N.J., 1964).

2   This point is made by Peter Laslett in 'The Comparative History of Household and Family,' in *The American Family in Social Historical Perspective*, ed. Michael Gordon (New York, 1973), pp. 28–9.

3   Phillippe Ariès, *Centuries of Childhood: A Social History of Family Life* (New York, 1960).

4   See Laslett, pp. 20–4; and Philip. J. Greven, 'Family Structure in Seventeenth Century Andover, Massachusetts,' *William and Mary Quarterly* 23 (1966): 234–56.

5   Peter Laslett, *The World We Have Lost* (New York, 1965), passim.

6   For purposes of historical inquiry, this definition of middle-class status corresponds to a sociological reality, although it is not, admittedly, very rigorous. Our contemporary experience confirms that there are class differences reflected in magazines, and this situation seems to have existed in the past as well. On this issue see Robert S. Lynd and Helen M. Lynd, *Middletown: A Study in Contemporary American Culture* (New York, 1929), pp. 240–4, where the marked difference in magazines subscribed to by the business-class wives as opposed to the working-class wives is discussed; Salme Steinberg, 'Reformer in the Marketplace: E. W. Bok and *The Ladies Home Journal*' (PhD diss., Johns Hopkins University, 1973), where the conscious attempt of the publisher to attract a middle-class audience is discussed; and Lee Rainwater *et al.*, *Workingman's Wife* (New York, 1959), which was commissioned by the publisher of working-class women's magazines in an attempt to understand the attitudinal differences between working-class and middle-class women.

7   *Historical Statistics of the United States, Colonial Times to 1957* (Washington, D.C., 1960), p. 510.

8   The gas iron, which was available to women whose homes were supplied with natural gas, was an earlier improvement on the old-fashioned flatiron, but this kind of iron is so rarely mentioned in the sources that I used for this survey that I am unable to determine the extent of its diffusion.

9   Hazel Kyrk, *Economic Problems of the Family* (New York, 1933), p. 368, reporting a study in *Monthly Labor Review* 30 (1930): 1209–52.

10 Although this point seems intuitively obvious, there is some evidence that it may not be true. Studies of energy expenditure during housework have indicated that by far the greatest effort is expended in hauling and lifting the wet wash, tasks which were not eliminated by the introduction of washing machines. In addition, if the introduction of the machines served to increase the total amount of wash that was done by the housewife, this would tend to cancel the energy-saving effects of the machines themselves.

11 Rinso was the first granulated soap; it came on the market in 1918. Lux Flakes had been available since 1906; however it was not intended to be a general laundry product but rather one for laundering delicate fabrics. 'Lever Brothers,' *Fortune* 26 (November 1940): 95.

12 I take this account, and the term, from Lynd and Lynd, p. 97. Obviously, there were many American homes that had bathrooms before the 1920s, particularly urban row houses, and I have found no way of determining whether the increases of the 1920s were more marked than in previous decades. The rural situation was quite different from the urban; the President's Conference on Home Building and Home Ownership reported that in the late 1920s, 71 percent of the urban families surveyed had bathrooms, but only 33 percent of the rural families did (John M. Gries and James Ford, eds, *Homemaking, Home Furnishing and Information Services*, President's Conference on Home Building and Home Ownership, vol. 10 [Washington, D.C., 1932], p. 13).

13 The data above come from Siegfried Giedion, *Mechanization Takes Command* (New York, 1948), pp. 685–703.

14 For a description of the standard bathroom see Helen Sprackling, 'The Modern Bathroom,' *Parents' Magazine* 8 (February 1933): 25.

15 *Zanesville, Ohio and Thirty-six Other American Cities* (New York, 1927), p. 65. Also see Robert S. Lynd and Helen M. Lynd, *Middletown in Transition* (New York, 1936), p. 537. Middletown is Muncie, Indiana.

16 Lynd and Lynd, *Middletown*, p. 96, and *Middletown in Transition*, p. 539.

17 Lynd and Lynd, *Middletown*, p. 98, and *Middletown in Transition*, p. 562.

18 On the advantages of the new stoves, see *Boston Cooking School Cookbook* (Boston, 1916), pp. 15–20; and Russell Lynes, *The Domesticated Americans* (New York, 1957), pp. 119–20.

19 'How to Save Coal While Cooking,' *Ladies' Home Journal* 25 (January 1908): 44.

20 Lynd and Lynd, *Middletown*, p. 156.

21 Ibid.; see also 'Safeway Stores,' *Fortune* 26 (October 1940): 60.

22 Lynd and Lynd, *Middletown*, pp. 134–5 and 153–4.

23 *Historical Statistics*, pp. 16 and 77.

24 For Indiana data, see Lynd and Lynd, *Middletown*, p. 169. For national data, see D. L. Kaplan and M. Claire Casey, *Occupational Trends in the United States, 1900–1950*, US Bureau of the Census Working Paper no. 5 (Washington, D.C., 1958), table 6. The extreme drop in numbers of servants between 1910 and 1920 also lends credence to the notion that this demographic factor stimulated the industrial revolution in housework.

25 Lynd and Lynd, *Middletown*, p. 169.

26 On the disappearance of maiden aunts, unmarried daughters, and grandparents, see Lynd and Lynd, *Middletown*, pp. 25, 99, and 110; Edward Bok, 'Editorial,' *American Home* 1 (October, 1928): 15; 'How to Buy Life Insurance,' *Ladies' Home Journal* 45 (March 1928): 35. The house plans appeared every month in *American Home*, which began publication in 1928. On kitchen design, see Giedion, pp. 603–21; 'Editorial,' *Ladies' Home Journal* 45 (April 1928): 36; advertisement for Hoosier kitchen cabinets, *Ladies' Home Journal* 45 (April 1928): 117. Articles on servant problems include 'The Vanishing Servant Girl,' *Ladies' Home Journal* 35 (May 1918): 48; 'Housework, Then and Now,' *American Home* 8 (June 1932): 128; 'The Servant Problem,' *Fortune* 24 (March 1938): 80–4; and *Report of the YWCA Commission on Domestic Service* (Los Angeles, 1915).

27 Giedion, p. 619. Wright's new kitchen was installed in the Malcolm Willey House, Minneapolis.

28 Emily Post, *Etiquette: The Blue Book of Social Usage*, 5th ed. rev. (New York, 1937), p. 823.

29 This analysis is based upon various child care articles that appeared during the period in the *Ladies' Home Journal, American Home*, and *Parents' Magazine*. See also Lynd and Lynd, *Middletown*, chap. 11.

30 John Kenneth Galbraith has remarked upon the advent of women as consumers in *Economic and the Public Purpose* (Boston, 1973), pp. 29–37.

31 There was a sharp reduction in the number of patterns for home sewing offered by the women's magazines during the 1920s; the patterns were replaced by articles on 'what is available in the shops this season.' On consumer education see, for example, 'How to Buy Towels,' *Ladies' Home Journal* 45 (February 1928): 134; 'Buying Table Linen,' *Ladies' Home Journal* 45 (March 1928): 43; and 'When the Bride Goes Shopping,' *American Home* 1 (January 1928): 370.

32 See, for example, Lynd and Lynd, *Middletown*, pp. 176 and 196; and Margaret G. Reid, *Economics of Household Production* (New York, 1934), chap. 13.

33 See Reid, pp. 64–8; and Kyrk, p. 98.

34 See advertisement for Cleanliness Institute – 'Self-respect thrives on soap and water,' *Ladies' Home Journal* 45 (February 1928): 107. On changing bed linen, see 'When the Bride Goes Shopping,' *American Home* 1 (January 1928): 370. On laundering children's clothes, see 'Making a Layette,' *Ladies' Home Journal* 45 (January 1928): 20; and Josephine Baker, 'The Youngest Generation,' *Ladies' Home Journal* 45 (March 1928): 185.

35 This point is also discussed at length in my paper 'What Did Labor-saving Devices Really Save?' (unpublished).

36 As reported in Kyrk, p. 51.

37 Bryn Mawr College Department of Social Economy, *Women During the War and After* (Philadelphia, 1945); and Ethel Goldwater, 'Woman's Place,' *Commentary* 4 (December 1947): 578–85.

38 JoAnn Vanek, 'Keeping Busy: Time Spent in Housework, United States, 1920–1970' (PhD thesis, University of Michigan, 1973). Vanek reports an average of fifty-three hours per week over the whole period. This figure is significantly lower than the figures reported above,

because each time study of housework has been done on a different basis, including different activities under the aegis of housework, and using different methods of reporting time expenditures; the Bryn Mawr and Oregon studies are useful for the comparative figures that they report internally, but they cannot easily be compared with each other.

39   This analysis is based upon my reading of the middle-class women's magazines between 1918 and 1930. For detailed documentation see my paper 'Two Washes in the Morning and a Bridge Party at Night: The American Housewife between the Wars,' *Women's Studies* (in press). It is quite possible that the appearance of guilt as a strong element in advertising is more the result of new techniques developed by the advertising industry than the result of attitudinal changes in the audience – a possibility that I had not considered when doing the initial research for this paper. See A. Michael McMahon, 'An American Courtship: Psychologists and Advertising Theory in the Progressive Era,' *American Studies* 13 (1972): 5–18.

40   For a summary of the literature on differential divorce rates, see Winch, p. 706; and William J. Goode, *After Divorce* (New York, 1956) p. 44. The earliest papers demonstrating this differential rate appeared in 1927, 1935, and 1939.

41   For a summary of the literature on married women's labor force participation, see Juanita Kreps, *Sex in the Marketplace: American Women at Work* (Baltimore, 1971), pp. 19–24.

42   Valerie Kincaid Oppenheimer, *The Female Labor Force in the United States*, Population Monograph Series, no. 5 (Berkeley, 1970), pp. 1–15; and Lynd and Lynd, *Middletown*, pp. 124–7.

43   On the expanding size, number, and influence of women's magazines during the 1920s, see Lynd and Lynd, *Middletown*, pp. 150 and 240–4.

44   See, for example, the advertising campaigns of General Electric and Hotpoint from 1918 through the rest of the decade of the 1920s; both campaigns stressed the likelihood that electric appliances would become a thrifty replacement for domestic servants.

45   The practice of carefully observing children's weight was initiated by medical authorities, national and local governments, and social welfare agencies, as part of the campaign to improve child health which began about the time of World War I.

46   These practices were ubiquitous, *American Home*, for example, which was published by Doubleday, assisted its advertisers by publishing a list of informative pamphlets that readers could obtain; devoting half a page to an index of its advertisers; specifically naming manufacturers and list prices in articles about products and services; allotting almost one-quarter of the magazine to a mail-order shopping guide which was not (at least ostensibly) paid advertisement; and as part of its editorial policy, urging its readers to buy new goods.

1977

**Kisho
Kurokawa**

The Philosophy of
Metabolism

Kisho Kurokawa (born 1934) is a Japanese architect, writer, and urbanist. In 1960, he was one of the founders of the Metabolist group along with Kiyonori Kikutake (born 1928), Fumihiko Maki (born 1928), Masato Otaka (born 1923), Noboru Kawazoe (born 1926), and Kenji Ekuan (born 1929). He had earned his doctoral degree at the University of Tokyo under Kenzo Tange, whose work inspired Metabolism. The Metabolists sought to apply theories of organic growth to the problem of the expanding metropolitan city. In response to the fast urbanization of Japan, the buildings and projects took the form of large, flexible, expandable structures. Since the disbanding of the Metabolists, Kurokawa has been prolific, designing buildings in Japan and around the world. Notable among these are the extension to the Van Gogh Museum, in Amsterdam, the Kuala Lumpur International Airport, and the master plan and international airport of Astana, Kazakhstan.

The following selection is the first chapter to the book *Metabolism in Architecture*, in which Kurokawa outlines the rise of Metabolism and the role of technology in Japanese life. Metabolism views human society within a continuum of nature, including plant and animal life, and sees technology as an extension of the human condition, and thus as a natural process in itself. Technology is not opposed to nature, but an outgrowth and embodiment of the flows of nature. In consequence, buildings are ephemera, changing over time through the steady replacement of parts and functions.

Kurokawa addresses his writing specifically to the Japanese context, arguing that growth is always locally conditioned, but his broader views about technology and nature add an important element to the collection. In particular the idea that replacement parts might have their own metabolism.

War helped me discover Japanese culture. As I stood amidst the ruins of Nagoya, the third largest city in Japan, there was nothing but scorched earth for as far as I could see. In contrast to the desolate surroundings, the blue of the mountain range on the horizon was dazzling to the eyes.

My father was an architect. After graduating from the Nagoya Industrial College he went to work for the architectural department of the Aichi Prefectural Government and later became chief architect in the architectural department of a private firm. At the end of World War II, when I was only eleven and still in primary school, I already felt myself strongly drawn to architecture. I remember that my father's library contained works on classical Greek and Roman architecture and many volumes by writers such as John Ruskin and William Morris. Reading books of this kind formed in my mind an image of architecture and of cities as entities which are eternal and do not lose their eternal quality even if they are destroyed.

Very little was left of the Japanese cities destroyed by the air raids of World War II. Much in cities in the West is built of brick and stone, which remain as heaps of rubble after the buildings themselves have been destroyed. In Japan, on the other hand, building is mostly of wood (today 80 per cent of the buildings in Tokyo are wooden) and consequently destruction usually levels Japanese cities to the ground. But even then the buildings and cities persist as vivid images in the minds and imaginations of the people. And it was in this sense that I first came into contact with several major characteristics of Japanese culture, after I had lost my home town in the war.

Taking 1867, the year of the Meiji Restoration in response to pressures to open Japan up to the West, as the start of Japan's modern age, we may divide the country's history up to the present into four generations. The first generation was that of the founders of the modernization and Westernization movement. These were the leaders of Meiji Period society, and they founded a large number of private universities, enterprises, and industrial organizations. In architecture this generation introduced a modern educational system to replace the old apprentice system of carpenters and other construction workers, by establishing a national engineering school – called the National Kobu Daigakko. Among the first graduates of this school were Kingo Tatsuno, Yorichika Tsumaki, Tokuma Katayama, Yuzuru Watanabe, and Tatsuzo Sone. These men copied Baroque, Renaissance, and other European architectural styles in designing banks and government buildings. To their way of thinking, the modernization of Japanese architecture meant the introduction of Western styles without any modification at all. They advocated without hesitation that the true modern Japanese architectural style was the Renaissance, the Baroque, or whatever other style they were copying.

The second generation saw the development from the successful achievement of an industrial revolution in Japan, to the Old Liberalism often referred to as Taisho Democracy (the period from 1912 to 1926 is called the Taisho Period), the growth of jingoism, and finally to the war and its conclusion in defeat. Among the architects of this generation were men like Shin'ichiro Okada who tried to incorporate traditional Japanese styles into Western ones. For example, they would cap reinforced concrete buildings with old-fashioned tile roofs. Some architects, such as Sutemi Horiguchi and Mamoru Yamada, came under the direct influence of contemporary

European art movements including Art Nouveau and the Vienna Sezession. Establishment of the greater East Asia Co-prosperity Sphere during World War II was regarded as an opportunity to export ultra-nationalistic Japanese architectural styles, which would be a sign of Japanese authority. The document of surrender, however, was the death notice for this ultra-nationalistic architecture.

In addition to reconsidering the validity of blindly copying imported styles, architects at this time also witnessed the downfall of ultra-nationalism and the destruction of national self-confidence brought on by defeat in the war. The third generation of Kenzo Tange and Kunio Maekawa have created a special world for themselves because they began their truly creative work only after the end of World War II, and were able to respond to the changes wrought in architectural style.

I belong to the fourth generation, whose point of origin is the defeat and destruction in the war. For this reason we are sometimes called the Charred Ruins School. In the hearts of all the members of this generation are traumatic images of events that took place when we were in our formative childhood years: the sudden, tragic destruction of Hiroshima and Nagasaki by atomic bombs and the virtually total reduction of cities and buildings to ashes.

Ours was the first generation to be educated in the totally new post-war system. Indeed, for a period shortly after the end of the war most of the pages in our old textbooks were inked out because their contents were no longer considered suitable.

The architectural models in our minds no longer included the classical European architecture which had interested me in my father's books. Nor were the ultra-nationalistic designs of my father, including the Aichi Prefectural Government which had been spared the flames of war, of any great significance to me. One idea which may well have taken their place in my training is Buddhist thought, since my parents were believers in Jodo Buddhism and were supporting members of a temple of the sect. Furthermore, the head of the Tokai Gakuen, where I attended junior and senior high school, was the distinguished Buddhist priest Benkyo Shio, who later was the chief abbot of the Zojoji in Tokyo. Buddhism later exerted great influence on the architectural philosophy I developed.

When the time came to select a university there were good reasons for choosing Kyoto University: I had been strongly impressed by the interpretation of the social significance of architecture that I read in a book, *Housing Problems in the Future*, written by Uzo Nishiyama, who was a professor at that school. Nishiyama was attempting to employ sociological methods to derive a new, scientific, planning theory to clarify the social meaning of architecture. His philosophy and methods were extremely fresh and exciting. During my four years in college I studied many subjects outside the architectural curriculum and, under the guidance of Professor Nishiyama, participated in social studies of slum areas. But when I learned that the architecture which results from scientific design theories based on such social investigations does not invariably produce fine works of art I made up my mind to leave Kyoto University. I decided to attend graduate school at Tokyo University, where Kenzo Tange (a practising architect, unlike Nishiyama) was teaching. I spent seven years there studying for the master's and doctor's degrees.

It was not until the 1950s that modern architecture made its real debut in

Japan. Maekawa and Tange produced no works before World War II, but became very active during the latter half of the 1940s. It was especially interesting to me that Tange's initial work in this period was the Peace Centre in Hiroshima. I found it meaningless to attempt to revive an already destroyed city by means of a monument; I felt that it was important to let the destroyed be and to create a new Japan.

The architects of the first of the four generations were confident that they were right in attempting to introduce and to copy Western architectural styles. The generation of Maekawa and Tange was equally confident that they were right in their efforts to introduce the philosophy of CIAM and Le Corbusier. In my view, what these men were doing led to confusion and conflict. But architects of the fourth generation, including Kiyonori Kikutake, Arata Isozaki, and myself, refused to take part. Instead, we elected to observe the proceedings from the stalls. We did not take action until the second half of the 1950s.

It was decided that in 1960 the World Design Conference would be held in Japan. Because we were involved in planning the conference, Takashi Asada, general secretary of the conference, architectural critic Noboru Kawazoe, architect Kiyonori Kikutake, and I, found it necessary to meet and hold discussions almost daily. At this time we began to think about the significance of the fourth generation.

During the first half of the 1960s Kenzo Tange designed the Yoyogi National Gymnasium, one of his masterpieces. At about the same time Japan entered a period of astounding economic development that was to last for more than a decade. Not only did this growth strengthen Japan economically and politically, it also, for the first time in history, upset the old Japanese social institutions and gave birth to a mass-oriented society. New people became prominent in all fields; new art movements that refused to be bound to the established orders appeared. For us the confusion produced by those changes provided an excellent opportunity to think about and act upon the cities and buildings that had been destroyed in the war. Confusion in the city made it impossible for monuments or symbols to control or dominate urban spaces. Elements that mutually contradict, oppose, or operate in parallel make the character of cities and architecture multivalent and ambiguous and render visual comprehension of the whole urban landscape impossible.

The Metabolic movement began against this kind of social background and came into being through the preparations for the World Design Conference. These preparations lasted for two years, beginning in 1958; and during the conference the Metabolist group made its first declaration: *Metabolism 1960 – a Proposal for a New Urbanism.* The people who collaborated on this book were architects Kiyonori Kikutake, Fumihiko Maki, Masato Otaka, and myself, and graphic designer Kiyoshi Awazu.

A key passage in this declaration reads: We regard human society as a vital process, a continuous development from atom to nebula. The reason why we use the biological word metabolism is that we believe design and technology should denote human vitality. We do not believe that metabolism indicates only acceptance of a natural, historical process, but we are trying to encourage the active metabolic development of our society through our proposals. This is an important element in our declaration for two reasons. First, it reflects our feelings that human society must be regarded as one part of a continuous natural entity that includes all, animals and plants. Second, it expresses our belief that technology is an extension of humanity. This belief contrasts with the Western belief that modernization is a repetition of a conflict between technology and humanity.

The Japanese physicist Hideki Yukawa, a 1960 Nobel Prize winner, wrote in a newspaper article that religion and science ought not to be thought of as different worlds but that they should be recognized as being connected in a single cycle. This idea should not be new to the Japanese, who have become familiar with it through the long history of Buddhist culture.

Our group hoped to focus discussion on new issues by introducing elements peculiar to and characteristic of Japanese culture into the history of functionally and rationally organized modernization. At the conference I presented a report entitled 'Character in design stems from the universality of new quality' in which I argued that the way to create a character for Japanese design arises from a good use of the Japanese understanding of the continuity and harmony of technology, humanity, and nature in modern society. In addition, such a character gives Japanese design universal validity. By seeking international styles and standards it is not possible to create a style both Japanese and of universal appeal. (This topic is discussed in my book on the conference, *WodeCo*, published in 1960.)

To understand the philosophy of the Metabolist group better, it is necessary to give a brief outline of pertinent characteristics of Japanese social history.

Specifically, we must look at population trends, mobility, and the technological orientation of modernization. First, the population of Japan in 1721 was thirty-one million. At that time the power of the Tokugawa shogunate, which ruled the nation throughout the Edo period (1615–1867), was at its height, with Yoshiyasu as the current shogun. For the next century and a half, the population varied between thirty-one and thirty-three million. After the modernization that began in full force with the Meiji Restoration in 1867 the population began to increase sharply. Between 1875 and 1975, it tripled, to reach the present level of 105 million. According to a projection made by the Institute of Social Engineering, by the year 2000 Japan will have a population of about 135 million. Although increases in the birth rate contribute to this growth, the most significant factor will be the increase of elderly people as a percentage of the total population. Indeed, modernization of the environment, advances in medical science, improvements in health and medical care systems, all supported by and fostered by growth and development of the economy, are so lengthening the average life span that soon Japan will probably have the largest elderly population, relative to the total population, of any country in the world. No developed Western nation shows a population growth pattern – an ageing – of this kind.

Obviously growth of the population and change in the age structure of the population greatly influence the nature of cities, types of residence and nature of architectural spaces. Furthermore, the speed of population growth has made it impossible to satisfy housing demand through the ordinary construction methods of the past. Consequently, the residential environment has deteriorated. The need to provide an annual minimum of 1.6 million dwellings at minimal cost has become a pressing social issue. In Japan, therefore, there has been a strong demand for the industrialization of residential building which has then become a vital subject to the architect. It is to be wondered whether even a shift from high-level economic growth to a long period of low-level growth will affect the dynamic conditions of change in Japanese society. As long as these conditions persist the architect must not accept them passively, as the inevitable results of technological progress. Instead he must help people to master technology and strive to produce a system whereby changes occur as the result of human judgement. The architect's job is not to propose ideal models for society, but to devise spatial equipment that the citizens themselves can operate.

The second Japanese social characteristic of importance here is the mobility of the Japanese people. Since antiquity, the capital of Japan has been moved on a number of occasions, for political or religious reasons: Nara, Kyoto, Kamakura and Tokyo (then Edo) have each in turn been the capital. Another example of the way the Japanese accepted mobility was the requirement which the Tokugawa shogunate made of the clan lords (*daimyo*), who were required to spend each alternate year in the capital city. This entailed a great deal of coming and going on the part of the lords and the members of their households. Still another example of the readiness with which the Japanese people move about is the long-established custom for agricultural workers to migrate to cities to seek employment for periods when farm labour is not needed in the rural areas. Even today, when the income differential between the town and country has practically disappeared, this seasonal shift of the labour force continues all over the nation. Traditionally, religious pilgrimage and travel have occupied an important position in the way the Japanese people spend their leisure time. Pilgrims' associations were even set up, organized on a regional or village basis, to manage funds saved for travel expenses. Travel and pilgrimages to the various temples and shrines in many parts of Japan led to new friendships, study, and cultural exchange among various regions, and even to the formation of a kind of sex industry. Nor was travel only for the masses: intellectuals found travel a way to free themselves from the opposition imposed by authorities and to lead a life of refinement and taste. One of the best works of the famous haiku poet Matsuo Bashō (1644–94) is a kind of poetic travel diary, *The Narrow Road of Oku*. He often said that travel was his home.

Today two types of mobility are still features of Japanese life. First, 10 per cent of the population changes its place of residence each year. Second, mobility is of great importance in the daily lives of wage-earners and students who must travel from home and back, twice daily, and the lives of those who are active in clubs, associations, and recreational organizations and the like. So inherent is the love of mobility to the Japanese character that until only recently most Japanese preferred to take their families to public bath houses instead of building a bathroom or installing a tub in the home. Even today some families drive to local public baths. This readiness to move contrasts with the more settled system of Europe, where life is focused on the home, extending out to a settled local community. In a book called *Homo Movens* (1969) I contrasted this fluid society with the closed community of the West and suggested that the Japanese way suggested possibilities for a new kind of living space.

The history of Japanese modernization is at the same time a history of urbanization. Nowhere in the world is the population more concentrated in cities than in

Japan: 50 per cent (fifty million) of the population lives in and around Tokyo, Osaka, and Nagoya; 10 per cent live in Tokyo alone. The rapid economic growth of the nation since 1960 has intensified the concentration of the population in cities and has consequently aggravated pollution problems. It is forecast, however, that in future the drift into the major urban centres will slacken and that there will be a movement to cities of moderate size. By the year 2000 the population shift to the major cities will have halted, and thirty million people will have moved to regional cities. Without doubt, such a violent ebb and flow will work immense changes in Japanese society. Even should population flow into the large cities and into the regional cities cease, the Japanese fondness for mobility will continue to demand dynamic spatial compositions.

In 1970 the Japanese government had a development plan for the entire nation prepared. I participated in the production of the plan, which took the mobility of the population as a premise and linked all the nation's cities by means of a transportation and information-communications network. I further proposed the formation of a network of individual cities. Indeed, the creation of a nationwide network is already in progress in the form of widespread use of automobiles and ownership of television sets, the system of super-fast express trains, and the construction of expressways. Despite the irregular terrain of the Japanese archipelago transport systems and facilities are excellent. They include a national network of air service, ferry and cargo transport service by ship, super-fast express trains, and older regular railways. Their very excellence is indicative of the Japanese fondness for mobility and movement.

Le Corbusier said that cities consist of living spaces, working spaces, and recreational spaces connected by methods of transport. Methods of transport should be re-examined as parts of the space in which we live. Here, the important feature is not the road, which has the sole function of providing a place for vehicles to pass, but the street, which is part of daily-life space and has many functions. In 1960, before working on the design of the Nishijin Labour Centre, I undertook a study of Japanese city streets – particularly those of Kyoto. After the Nishijin centre was completed (in 1962), I published an article, 'Architecture of the Street', in a Japanese architectural magazine (see Chapter 3). In this I discussed how one characteristic of Japanese architecture and cities throughout their long history has been the development of methods for using places designed to facilitate movement – corridors, streets, and so on – as an integral part of daily life. The verandah of the Katsura Imperial Palace (built in the seventeenth century) and the corridors of temples are more than mere passages between one room and another.

A study of the streets of Japan's ancient capital, Kyoto, showed that the organization of the streets provided a useful contrast between Western and Japanese cities. *Urban Design* (1965) argued that the traditional Japanese city lacks squares or plazas as its streets serve the function of plazas. Jane Jacobs, in *The Death and Life of Great American Cities*, makes the same observation.

Seen from a different angle, the fluidity of Japanese society stems from its homogeneity and its history of centralized authority. In contrast to Italy and Germany, both of which became nations through the unification of city states or principalities, Japan has a long history of repeated concentrations of central authority. The Japanese pyramidal power structure has discouraged people from living in fixed places within regions, as was the case in nations composed of city states. Japan has known very little of the kind of warring that set one city state against another elsewhere in the world. Traditional Japanese society is basically egalitarian and is homogeneous in that there is little class distinction in the usual Western sense and all people have an equal chance of rising to power. There has been little need to form groups or communities on the basis of race or class to protect vested interests. These conditions have made it possible for the Japanese to move about with freedom wherever they have wanted to travel.

The third characteristic of Japanese society is the importance of technology and its influence on the body politic. This is illustrated by the Meiji restoration, when the shogun, the master of Japan for centuries, was removed from power and the emperor, previously only a figurehead, took his place. These changes arose from a recognition of the need to alter the political system in order to open Japan to Western influences and in this way to introduce Western technology and promote the industrialization of the nation. Some political change was necessary for Japan to become capable of absorbing a new way of life, but the historical tradition was kept intact; the event is tellingly referred to not as a revolution but as a restoration. The people who wielded power in the new government were eager to adopt Western technology but without Western philosophical rationalism or Western social systems; they intended to incorporate bare technology into the old Japanese tradition. This attitude was extremely optimistic; we can understand it when we remember that these men saw technology, humanity, and nature as forming a single ring. They did not imagine that technology could be in opposition to humanity and nature, or that it could bring harm to mankind.

The scale of the technology that has been introduced into and developed by Japan since 1945 has been immense. Industries like shipbuilding, steel production, automobile manufacture, and electronics have so grown and developed as to acquire autonomous power sufficient to affect the entire nation. At the same time that the rapid economic development of Japan began, in 1960, the Metabolist group advocated the creation of a new relationship between humanity and technology. Thinking that the time would come when technology would develop autonomously to the point where it ruled human life, the group aimed at producing a system whereby man would maintain control over technology.

Rapid economic growth in an industrial nation such as Japan promotes the development of technology of a kind more dynamic than anything previously known. As long as such economic growth continues, even production facilities completed only a few years previously can be made obsolete and replaced by newer facilities. This process can go on reproducing itself. Similarly, spaces that are still capable of serving society adequately are rapidly abandoned in the search for newer, more highly effective and serviceable spaces.

In this no thought is given to the social significance of spaces or to value judgements about providing people with pleasing symbols. The sole consideration is economic efficiency and profit. We advocated the application of metabolic cycle theory as a way of avoiding these conditions. This theory proposed a reorganization which divides architectural and urban spaces into levels extending from the major to the subordinate and which makes it easy for human beings to control their own environments.

By distinguishing between the parts that do not change and the parts that must be preserved, it is possible to ascertain the parts that must periodically be replaced. In our plan for a prefabricated apartment building project (1962) (see pp. 92–4), we devised a way of assembling a number of basic elements so as to create such major spaces as bedrooms and living rooms. Capsule units, attached from the outside, were used for subordinate spaces like those of the kitchen and the service units. This kind of breakdown and recomposition of architecture enables individual expression and the production of character for the individual rooms and their contents; it establishes a kind of identity by means of things that, in the case of buildings in the modern architecture style, were buried within box-like forms.

The intention in using this method is not merely to apply industrialization and prefabrication to mass-produce spaces at low cost. It is intended to make use of prefabrication techniques to restore to architectural spaces something of the characteristics and feelings of the individual human being – characteristics and feelings that are lost when architecture is made anonymous and impersonal. The Nakagin Capsule Tower is not strictly an architectural expression of an apartment house; it is an expression of the 144 people who reside in its 144 units.

Although Metabolism emphasizes the principle of replaceability and changeability of parts, the reasons for doing so derive from a philosophy entirely different from the use-and-discard approach sometimes justified by economics in mass-consumption societies. I know of many instances in which entire buildings have been wastefully destroyed because portions of them were no longer serviceable. If spaces were composed on the basis of the theory of the metabolic cycle, it would be possible to replace only those parts that had lost their usefulness and in this way to contribute to the conservation of resources by using buildings longer.

For still other reasons, technology will continue to be important to Japan in the future. One of these reasons is Japan's dependence on imports for more than 30 per cent of her food and more than 95 per cent of her energy. To guarantee the living standards of her people while relying heavily on other nations for such basics as raw materials and primary energy resources, Japan has no resource but to put the human brain, and technology, to full use and to export the industrial products thereby obtained. Technology is both rice and oil to the Japanese people.

Geography accounts for still another aspect of the Japanese reliance on technology. Japan consists of thirty-seven million hectares of land, of which only one fifth is flat; the remaining area is mountainous. Of the flat area, 6 million hectares are devoted to agricultural use, and 1.1 million hectares are rivers, lakes, or other areas of water or marsh. To increase food production, even by a slight degree, non-agricultural use of land has to be minimized. The area available for human habitation, including that used by railroads, highways, and other social infrastructures, is a mere two million hectares. This is the area on which 105 million people live today and on which, by the year 2000, 135 million will probably make their homes.

Land in Japan is more expensive than in almost any other part of the world. The high density of economic activity partially accounts for this phenomenon, but overall shortage of land is also an important factor. Without considering these special conditions, it is impossible to grasp the full importance to Japan of technological development in high-density building, man-made land, cities over the sea, and high-rise structures.

The fourth characteristic of Japanese society is its reliance on wood in contrast to the reliance on stone that characterizes cultures of the West. The foremost symbol of the Japanese wooden building is the Ise Shrine. The buildings of the shrine are prototypes, and for a thousand years their basic forms have been preserved unaltered. But to preserve the prototype it has been necessary to replace the buildings with exact replicas at twenty-year intervals. The buildings at Ise today, therefore, are not original in the sense of being composed of the materials from which they were first built as the temples and other buildings on the Acropolis are original. Since wood, the traditional building material, rots easily, the Japanese have never felt that the materials themselves have a sense of eternity. On the contrary, they are and always have been conscious of the spirit and philosophy beyond the materials and regard the form as an intermediary conveying that spirit and philosophy to human beings. The faithful reproduction of the Ise buildings may be thought of as a ceremony through which the philosophy and spirit of the old buildings are transmitted into new spaces.

The carpenters of the past, who were the equivalent of architects until Western architecture was introduced, did not draw plans but relied on written instruction sheets called *sashizu*. They were able to build successfully on the basis of nothing more than *sashizu* because of the existence of a system of standardizations (*kiwari*) and detailed specifications (*shikuchi*). Furthermore the workmen could see and feel the rebuilding and replacement process in the finished building. Such is the strength of this tradition that several wooden masterpieces from as early as the seventh century remain standing today.

Japanese cypress (*hinoki*) is used in many Japanese bathrooms to surface walls and for the bathtub, so that the pleasing fragrance of the wood can be enjoyed. To introduce the tactile pleasure of the material into daily life wood used elsewhere in private homes is rarely stripped of its bark and finished to mechanically precise smoothness. In short, wood is regarded less as an architectural material than as a part of the world of nature. The use of bark-edged trunks and limbs of trees in *tokonoma* alcoves and ceilings of rooms and huts for the tea ceremony reflects this same pleasure in material.

The feeling of unity with nature extends to materials other than wood. When the bronze fittings have acquired a patina, and the thatch roofs have become speckled with moss, then they acquire special value because these alterations reveal continuity and unity with nature. The aesthetic fondness for the plain, unadorned, natural, rustic, and slightly sad, expressed by the word *sabi*, is related to this sense of values. My own term for this aesthetic philosophy is 'the aesthetics of metabolism' or 'the aesthetics of time' (*Aesthetics of Metabolism*, 1967) – terms selected to indicate a philosophy which finds value in the preservation of relations between architecture, society, and nature while constantly changing with the passing of time. The relation between society and nature is an open one. Beauty is not created solely by the artist; it is completed by the citizens, the users, and the spectators, who by so doing contribute to its creation. I employ industrialization, prefabrication, and capsulization as ways of evoking this kind of participation. The technology of movement (mobility) can also be used as a participation system.

I believe that what I call media space (or *en*-space, a term using the Japanese word *en*, which means connection or relationship) and in-between space are important in making the relation between architecture, and society and nature an open one (see Chapter 4).

In my Central Lodge (1964), Memorial House for Hans Christian Andersen (1965), and Sagae City Hall (1967), I made very high ceilings for the space they created and natural lighting through skylights to establish a sense of continuity between the buildings and nature. In the Yamagata Resort Centre (1967), the Daido Insurance Building (Sapporo, 1975), the Head Office of the Fukuoka Bank (1975), and Waki-cho City Hall (1975), I employed in-between spaces, which are neither exterior nor interior, to establish continuity between the buildings and their surrounding environment.

This philosophy of continuity, characteristic of wood-based culture, is lacking in stone-based culture. Instead of using the material in such a way as to make full use of its natural characteristics, stone-based culture processes the material, and physically alters it. For instance, stone cut to make sculpture no longer seems to be stone at all. In this respect, Greek and Roman architecture would have been the same if it had used steel and concrete instead of stone. Furthermore, unlike wood-based culture, stone-based culture opposes nature; its architecture uses walls to protect the interior from the exterior. According to this approach, architecture and nature are discontinuous. Human beings do not live with architecture for architecture is only a container for human beings. This aspect of the traditional stone-based culture is directly connected to modern rationalism and to functionalist architecture.

Undeniably, rationalism, functionalism, and individualism have benefited modern society, but they have also produced great losses. It is now time to incorporate in contemporary architecture the kinds of anti-dualism, anti-functionalism, and anti-individualism latent in wood-based culture.

The fifth characteristic of Japanese society is its Buddhist culture. Buddhism was introduced into Japan from the Asian continent in the sixth century. In the twelfth century a number of distinctly Japanese Buddhist schools were established: Jodo-shu, founded by Honen (born in 1133); Jodo-shin-shu, founded by Shinran (1173–1262); Ji-shu, founded by Ippen (1239–89); and Nichiren-shu, founded by Nichiren (1222–82). The power exerted over the people of Japan today by Buddhism is considerably weaker than that historically exercised by the various sects and branches of Christianity over the West. Nonetheless, throughout its long history in Japan, even though it produced many different schools, Buddhism has provided a consistent, profound spiritual and philosophical basis to Japanese culture. It is impossible to discuss the essence of Japanese architecture, music, drama, painting, or literature without referring to Buddhist philosophy. Although Buddhist influences are seen in tangible form in such things as wooden temple architecture, its deepest penetration is into the effects and methods of spatial composition and into ways of establishing relations between nature and architecture and between technology and humanity.

The second of my characteristics of Japanese society – mobility – and fourth – extensive use of wood – are closely related to Buddhist philosophy. The 'Diamond Sutra' (*Kongō-kyō* in Japanese; *Vajra-sūtra* in Sanskrit), a most important Buddhist classic, sets forth the nature of wisdom and of *śūnyatā*. (*Śūnyatā-ku* in Japanese is sometimes translated too simply as 'emptiness' or 'nothingness', but its true meaning is a paradox: that emptiness which is substantial existence.) The Diamond Sutra contains an important teaching: that the truly non-existent (*shinkū* in Japanese) is the basic meaning of existence and that there is no differentiation between life and death as there are no forms or characteristic essences. This theory is called *musō* in Japanese and *lakṣaṇa-alakṣaṇatas* in Sanskrit. It means that life and death are one and that human beings should not become too attached to any one idea or place but should always remain aware of being in eternal time. According to this philosophy, the total greater life surpasses time and space and life and death to become part of the greater flow of transmigration, which is called *rinne* in Japanese and *samsāra* in Sanskrit.

The differences between the seasons are clearly marked in Japan, instilling in everyday activities and experiences a powerful awareness of the changeableness of life. The people of Japan also live with the threat of earthquake, flood, typhoon, and other natural disasters. Since forests cover much of the land, the Japanese have long used wood to make their architecture, their bridges, ships. Wood – and therefore the buildings, bridges, ships, and cities made of it – is liable to be destroyed by natural deterioration. Because of the great extent to which the Japanese have used wood over the centuries, they are accustomed to this kind of gradual destruction. Possibly for this reason it was easy for the Buddhist concept of *samsāra* and the idea that essential emptiness (*śūnyatā*) is true existence to take deep spiritual root in Japanese culture.

In recent years awareness that the limitations to the natural resources of the world and that human society and all of its environment are one in a great life system has become widespread. But to Buddhism this is nothing new: it is a basic tenet of the concept of *samsāra*. The acknowledgement by the Metabolist group in its 1960 book that society is part of a greater circle of life coincides with this doctrine. The principle that architecture should change with time, the principle of replaceability and interchangeability, and the principle of the metabolic cycle, as well as the belief that architecture, cities, and humanity itself are ephemeral, are all in accord with the doctrines of *samsāra* and *lakṣaṇa-alakṣaṇatas*.

The thought of Metabolism is theoretical and philosophical. We do not intend to create forms or styles, because these are only the provisional manifestation of thoughts. Forms and styles occur in consequence of historical, temporal, spatial, material, geographical, social, and sometimes purely personal conditions. Philosophy and thought, however, are transmitted in intangible ways.

The notions of in-between space and *en*-space have persisted in my works from the architecture-of-the-street philosophy of the Nishijin Labour Centre of 1962, through the main office of the Fukuoka Bank (1975) and to the Tokyo building of the Daido Insurance Company (started in 1975).

Individualism, developed as a fundamental condition in the modern society of the West, is not found in the historical spiritual make-up of Buddhist culture. As I have explained, a tendency to concentrate and centralize authority and to produce a society that is basically homogeneous characterizes the history of Japan. This has meant that guarding the responsibilities and duties of the individual and attempting to advance individual interests have not been features of the Japanese national character. This in turn has developed a characteristic tendency for the Japanese to act in groups and to be, to use David Riesman's term, other-directed. He points out that in Western society today the individual in the organization drifts towards self-direction in his or her action pattern. But because the philosophy of *samsāra* does not recognize the existence of the individual as it is held in the West, the non-individualistic conditions prevailing in Japan have led to lack of democracy in the basic, classical sense and to a different concept of communal responsibility. It is for this reason that I feel it is important to discover the *jiga* (the oriental individuum).

The oriental individuum is not the independent individual of the West. Oriental thought does not find the basis of existence of the self in the self but seeks the true existence in a supra-individual that transcends the individual. The contradiction of the individual and the supra-individual remains unresolved, but the two preserve their oneness. In the Sanskrit version (Max Müller, 1881) of the 'Diamond Sutra' this passage occurs:

> *Ya eva subhūte, Prajnapāramitā*
> *Tathāgatiena bhāsitā sā eva apāramitā*
> *Tathāgatena bhāsitā*
> *Tena ucyate Prajnāpāramitā iti.*

When the purely ornamental words are omitted from the text, what remains is *A eva a-A Tena ucyate A iti*. This means *A* is non-*A*; therefore it is called *A*. This simultaneous affirmation and negation is the basis of the theory of identity and differences, or *sokuhi* in Japanese (*soku* means non-diversity and *hi* means non-identity).

The individual does not exist as part of the organization called society, and neither does the total exist for the individual part. The *jiga*, or oriental individuum, consists of a relationship in which the individual and society, while being contradictory, include each other.

The tea room for the tea ceremony shows how the concept of the individuum is applied to building. The space itself is minimal (about two or three metres by four metres). In it the participants perform a ritual of preparing and drinking tea which is a kind of spiritual exercise. But from within the tiny room they can enjoy sensing the vastness of all nature. In the tea ceremony awareness of nature teaches the participants that the scale of their surroundings bears no relation to the scope of spiritual activity. The greater – the world of nature coexists with the lesser – the tea house – neither comprehending nor excluding the other yet each an essential part of the other.

I intended my capsule spaces to be a declaration of war in support of the restoration of the oriental individuum, which has been lost in the process of modernization (*Capsule Declaration*, 1969). It is once again necessary to reject the mystification implanted in such ideas as abstracted universal space. By examining spaces for individuals we must seek new relations between the individual and society. The capsule space, which is a representation of the oriental individuum, is not a part of the piece of architecture to which it is attached. The capsule and the building exist in contradiction yet mutually include each other. Architecture that is a representation of the oriental individuum would not be a part of the city. Such architecture and cities would exist in contradiction but would mutually include one another. The same kind of relation should exist between architecture and nature and between human beings and technology. The philosophy of in-between spaces and *en*-spaces should help make possible a change of direction towards attaining such relations.

A key feature of this is the oriental street, which is different from its Western counterpart in that it does not establish clear boundaries between different city zones, and because it is a multivalent space where many functions are performed, giving it a nature between public and private. A street will be the scene of such intimate family occasions as dining or playing. In festivals the partitions will be removed from the fronts of houses to make the interiors public. The open corridors and covered corridors of traditional Japanese architecture act to establish continuity between nature and architecture and link together different architectural groups.

Its linking character relates *en*-space to the Japanese concept of *ma*. *Ma* has various meanings, amongst them: timing, silence, buffer zone, boundary zone, and void. In addition, it carries the same connotations as *en*-space, or in-between space. In the Noh drama, when an expression of tragedy or grief is suddenly changed to one of joy, there is a moment of immobility in which change is indicated; that moment too is known as *ma*. Or, in the case of oriental music, such as the *Gagaku*, the ancient court music of Japan, silent intervals provide the necessary adjustment which makes it possible to have a series of discordant sounds. And when Japanese people sing traditional popular songs (*enka*), they hum or lengthen a sound by one bar to provide the step from one sound to another. It is *en*-space or *ma* which gives oriental calligraphy and painting much of its character. In the monochrome ink painting technique introduced to Japan from China long ago areas are left untouched on the paper or silk to make an interval of no-statement, and as such stimulate the imagination of the beholder all the more. In all of these instances the silence or space is the Buddhist *kū* (*śūnyatā*). The same thing applies to calligraphy. The written ideograph is itself real, but it is the spatial proportion and balance of single ideographs or groups of them that give the ideograph power.

Thus, the philosophy of *en*-space, nurtured and developed in all of the fine arts, in the performing arts, the tea ceremony, flower arranging, and also architecture and city planning, is well established in Japanese society. It is also thought to be effective as a philosophy for today, as it enables the peaceful coexistence of the individuum and the whole, or of contradictory elements.

The time has come for a re-examination of the role of the individual in society together with the role of architecture in the city. In this connection, the conception of the individual as explained in Buddhist teachings and in the in-between theory as found in the idea of identity and non-identity (*sokuhi*) should be of great value.

I have explained the theory of Metabolism against the background of five characteristics of Japanese society. I hope that the Metabolist theory will give new meaning to the architecture of today, but I do not intend to try to produce an international style. Nor do I hope to establish standards that can be used everywhere. On the contrary, I believe that it is the historical characteristics of each people, nation, and region which through their own uniqueness are of international significance.

1979

Philip Steadman

What Remains of the Analogy? The History and Science of the Artificial

Philip Steadman is Professor of Urban and Built Form Studies at University College London and Director of the Bartlett School's graduate research program. He was trained as an architect, and has taught at the University of Cambridge and the Open University. His research interests are in geometrical description of the forms of buildings and cities, and the relationship between form and function. He has been particularly concerned with the relationship between form and the use of energy at both the architectural and urban scales. His publications include (with Lionel March) *The Geometry of Environment* (1975), *Energy, Environment and Building* (1975), *The Evolution of Designs: Biological analogy in architecture and the applied arts* (1979), *Architectural Morphology* (1983), (with Joe Rooney) *Principles of Computer-Aided Design* (1987), and (with Gerry Lynn Roberts) *American Cities and Technology: Wilderness to Wired City* (1999), and *Vermeer's Camera* (2001).

Like many scholars who have looked seriously at the relationship between technologies and architecture, Steadman was drawn to examine the use of biological analogies in the field. This excerpt forms the final chapter of *The Evolution of Designs*, in which he assessed the many different kinds of analogies that had been used in the design arts – organic, biological, evolutionary. While he determined that most usages were largely metaphorical, or inspirational, he never dismisses the analogy outright, seeking instead the 'Immanent laws' that inspired the similarities. The book was written just as the enthusiasm of cybernetics was being exhausted, but before the explosion of complexity studies was made possible by the availability of ready computational facilities. His appreciation of Christopher Alexander and Herbert Simon suggests precisely the kind of operative biological analogy that contemporary architects continue to pursue.

What remains then, that is useful and true, out of the variety of analogies made between biology and the applied arts? It will help to clarify matters if the answer to this question is divided into two parts, the one concerned with history, the other with science. Indeed the confusion as to whether a theory of the design of artefacts should, or could be a scientific, as opposed to a historical theory is something which has bedevilled biological analogies since they were first formulated – as this account has, I hope, shown. In making this division I propose to differentiate between history and science in terms of the *actual*, as against the *possible*. As W. C. Kneale has said,

> it seems possible to maintain that science should be distinguished from history (in the largest sense of that word), not as the study of universal truths from the study of singular truths, but rather as the study of what is possible or impossible from the study of what has been or actually is the case. Speaking metaphorically, we may say that science is about the frame of nature, while history is about the content.[1]

The biological analogy, despite its association with functionalist and historicist fallacies, leaves us with an overall picture of the history of technology – particularly in its earlier phases – which can, I believe, still be extremely helpful in guiding theory and research.

The starting point from which the 'evolutionary' aspects of the analogy began was the simple fact that in the production of many artefacts, especially in the craft or vernacular traditions, one object is very often *copied* in its design (perhaps with minor differences) from another. The truth of this observation is not altered by any of the criticisms of the last two chapters. The fact of copying gives rise to a *continuity* in form and appearance, when the 'genetic' links are followed between a series of artefacts successively copied, each from the last. The characteristic form of the artefact may undergo a gradual *transformation* as a result of the small alterations introduced at each stage. The fact of many similar artefacts being thus produced with related but not identical forms (and functions) results in the appearance of what may be termed 'populations' of objects, amongst which it may be possible to identify 'types' according either to functional or to morphological criteria.

There may be geographical 'diffusion' of these populations, as a result of the objects (or the knowledge of how to make them) being carried by migrating artisans, or being transported through trade, captured in war and so on. It is perhaps possible for the series formed by repeated copyings to diverge into two or more branches, such that later members along the divided branches are functionally and formally quite distinct. (It may also be possible for branches to converge; this is a point we will come back to.) Thus far, to the extent to which a biological metaphor fits the case, it is not seriously misleading.

A programme something along these lines was proposed for the study of manmade objects by the critic George Kubler, in his *The Shape of Time: Remarks on the History of Things*.[2] Kubler in turn drew inspiration both from the work of the anthropologist Kroeber,[3] and from Henri Focillon's *Vie des Formes*,[4] which Kubler calls 'the boldest and most poetic affirmation of a biological conception of the nature of the history of art'.[5] Focillon's concern was exclusively with the fine arts; with the 'internal' laws by which the forms employed in art are governed and organised, and how they develop in time. But Kubler defines his area of interest to include tools and other useful objects, and the purpose of his book is 'to draw attention to some of the morphological problems of duration in series and sequence'.[6] A further purpose is to offer some corrective or counter-balance to the amount of attention given in twentieth-century art history to iconographical study, to the relative neglect of formal or morphological questions.

This emphasis throughout is on continuity in the history of the forms of artefacts, as they are replicated and their designs transmitted so as to produce sequences which may extend in some instances over extremely long periods. 'Everything made now', Kubler says, 'is either a replica or a variant of something made a little time ago and so on back without break to the first morning of human time.'[7] He introduces the notion of a 'prime' object, which possesses a degree of novelty and original invention in its form (it is a 'mutant'); to be distinguished from the mass of replicas in which the same form is reproduced and perhaps degraded.[8] Certain sets of objects or works may be grouped into 'form-classes'. Complex objects may be made up from assemblies of separate parts or 'traits', each with its own sequential development:

> The closest definition of a formal sequence that we can now venture is to affirm it as a historical network of gradually altered repetitions of the same trait. The sequence might therefore be described as having an armature. In cross section let us say that it shows a network, a mesh, or a cluster of subordinate traits; and in long section that it has a fiber-like structure of temporal stages, all recognisably similar, yet altering in their mesh from beginning to end.[9]

David Clarke in *Analytical Archaeology* has lately offered a theoretical approach to the treatment of archaeological material which has many affinities with Kubler's proposals, but is set within a more precisely quantitative and statistical framework and supported by some applications to real data. The validity of this part of Clarke's work is not in my view affected by the criticisms made of his 'general model' in the last chapter; although this method, as we shall see, is essentially descriptive only, and provides no real explanation of the phenomena involved.

The central image in Clarke's work in this respect is of a population of artefacts (or perhaps of larger aggregations of artefacts) distributed in space (both in physical space and in abstract 'classificatory space'), and undergoing gradual transformation in time, through growth or decline in numbers, through change in possession of different attributes, and thus through a gradual transition from one artefact type to another.[10]

In making such an analysis, the need for independent means of dating, other than the criteria of typological similarity or morphological relationship themselves, is paramount if the danger of circularity in the argument is to be avoided. This point is made strongly by Child[11] – perhaps in reaction to the progressive evolutionism of the nineteenth century – when he emphasises how chronological evidence will be decisive in determining the evolutionary relationship in each instance, whether it represents a transformation towards more efficient or more complex and elaborated forms, perhaps, or whether it is alternatively a degenerating or 'devolutionary' series (as with the Celtic coins studied by Evans, for example).

Taking a single type of artefact, the known occurring examples may be tabulated by their typological characteristics and by their occurrence in time to yield a description of the changes occurring in that type. Clarke describes the kind of pattern of change which might be expected in some highly idealised hypothetical case.[12] At each period of time, or 'phase', there is a population of artefacts distributed normally around a dominant category and for each particular category or characteristic combination of attributes possessed by the artefact, there is a process of gradual increase in numbers over time, from rather few to a point where the category is dominant and most numerous, and then dwindling away again to disappearance.

Thus at each stage there is, in this very simplified and regular idealisation, a dominant category for the artefact type, and there are most examples to be found of this dominant form. But at the same time there are some rather fewer numbers of residual representatives of now 'archaic' and disappearing categories, and there are correspondingly a few representatives of emerging, new and 'prototypical' forms. Clarke has produced detailed empirical evidence, from finds of pottery and flint tools, to show the application of this theoretical model to actual data.[13] And he has other examples of a comparable process, in which the individual attributes of more complex artefacts, such as the decorative motifs on pottery and gravestones, are shown to lie in a similar 'lenticular' distribution about a moving modal form.[14]

What Kubler and Clarke provide us with is a descriptive account of historical sequence in the development of artefact types, and means for describing their morphological change. We have learned from the argument of previous chapters to reject the idea that there is any simple or single *necessary* direction to such changes, or any deterministic character to the process of change in itself. Why then is it that directional changes *are* to be observed in certain sequences of objects – as is undoubtedly the case?

A number of possibilities present themselves. The first is a rather speculative suggestion: that there is some feature of the actual *process* of copying, as applied to particular forms, which results in similar distortions – the same *kind* of miscopying, in effect – being introduced at every stage. This explanation is essentially a psychological one. There is something in the way in which certain forms are perceived and reproduced by copyists which gives rise to (unconscious yet systematic) transformations, conforming to some regular trend. My own copying experiments with drawings (see chapter 7) have demonstrated this effect at a perhaps rather trivial level. In a different field, that of linguistics, it is well established that highly regular types of change occur historically in pronunciation, for example, in the same direction in separate languages and in a way which is unrelated to changes in meaning.

A second possible cause of such systematic trends is to do with the play of fashion. This is an area which has been brilliantly illuminated from a theoretical point of view by Gombrich; as in his essay 'The Logic of Vanity Fair'.[15] The arts as much as economic life may be the scene of competition, in which each artist or craftsman strives to outdo his predecessors in the production of certain results or impressions. One example familiar from the evolutionary histories of architecture might, as Gombrich points out, be very plausibly interpreted in this light:[16] the sequence of French Gothic cathedrals, specifically the progressive increase in the heights of their naves, from 114 feet at Notre Dame, to 119 at Chartres, to 124 feet at Rheims, to 138 feet at Amiens. The sequence culminates in the spectacular attempt to vault the choir at Beauvais at a height of 157 feet; a project ending in disaster. The towers built by rival families in some Tuscan towns such as San Gimignano offer another precisely comparable case.

From the sublime to the (often) ridiculous, a field notoriously liable to competitive trends of this kind is the design of women's clothing – the subject of a celebrated quantitative study made by Kroeber in collaboration with Jane Richardson.[17] It might be expected that sequences here would be extremely fickle; however, by measuring the positions of hemlines, waistlines and necklines over a period of three centuries, Kroeber and Richardson were able to show a regular fluctuation in the dimensions of dresses, moving back and forth between limits set by the constraints of decency at one extreme, and complete coverage of the body at the other. Clarke has a sequence illustrating the elaboration of decoration on English grandfather clocks between the seventeenth and nineteenth centuries which might be interpreted in similar terms; from a simple and austere treatment, to baroque elaboration, and back to simplicity.[18]

Gombrich shows how these competitive trends are subject to a kind of artistic 'inflation', by which the attempts to achieve ever more pronounced and emphatic effects in themselves devalue the currency in which they are bought. 'Competition for attention can lead to the unintended consequence of simply lowering the value of what you have been doing before.'[19] For this reason, as well as owing to technical limitations, the competitive spiral may lead to a crisis, at which the sequence abruptly ends. Alternatively, the excesses provoked in the one-upmanship of fashion may themselves create the circumstance in which a striking impression may be produced by moving in the exact opposite direction. Where a flamboyant luxuriance of decoration is the norm, then an unadorned purity of form will seem all the more dramatic. So the pendulum swings back once more.

There is a third possible cause of directional sequence in changes of artefact design, which for our subject here is by far the most important. It is to do with actual technical improvements, increases in the efficiency or performance of utilitarian objects or machines. We do not have to attribute any automatism to the process, nor do we have to deny the imaginative contributions of individual designers, to allow that repeated attempts to design some specific type of tool or apparatus may be progressively more successful in achieving the desired practical ends. Thus the first steam engines might have been ramshackle, inefficient and unreliable, whereas later models incorporated many improvements, to increase speed, power, strength, or economy.

Such sequences of technical progress might in principle arise in two ways. The end-point of the series might be consciously envisaged by the craftsman or engineer from the beginning, with each successive try reaching nearer and nearer towards that goal. Gombrich offers two examples of this: the aeroplane, where the general ambition to make a powered flying machine long preceded the development of the requisite material and engineering means; and an example drawn from his own special area of interest, the evolution of technical means in painting for the achievement of a realistic illusion of the third dimension.[20] These processes of technical change, as Gombrich says, are thus genuinely 'Lamarckian' in the way in which their direction is the result of deliberate efforts to meet some perceived 'need' or to achieve some practical intention.

Alternatively, a series of unanticipated discoveries might occur along the course of the development of the object, which might be recognised to offer improvements and would thus be incorporated. 'Once bronze was shown to cut better than stones, iron better than bronze and steel better than iron, these alternatives had only to be invented and presented for rational men to use them for their cutting tools.'[21]

Instead of attributing increases in efficiency (or 'fitness') in artefacts to selection exercised by some abstract 'functional environment', this view brings attention back to the designer himself, and the rational choices which he makes amongst available technical means so as to achieve definite practical ends. The designer finds himself in a specific historical situation, facing some particular *problem*. He responds to the logic of that situation with some design solution, and this in itself produces a change in the problem: it creates a new problem. Meanwhile parallel developments in other technologies change the repertoire of possible materials, manufacturing methods, mechanical devices, components, and so on, available to the designer; and social or cultural changes perhaps alter the functional demands which the artefact is designed to meet. Both Popper and Gombrich have argued in favour of an 'analysis of situations' in some such terms as a methodological alternative to historicist theories and an answer to their 'poverty' – Popper for the social sciences, Gombrich for the history of art and by extension for technological and design history.[22]

Henri Focillon has given an account of our favourite evolutionary topic, Gothic architecture and its engineering, according to essentially this method.[23] He treats developments in cathedral construction as a series of *experiments*, the results of each one informing the next. 'By experiment', he says, 'I mean an investigation that is supported by prior knowledge, based upon a hypothesis, conducted with intelligent reason, and carried out in the realm of technique.' Some experiments may have been inconclusive, wasted. Others showed the feasibility of various structural expedients, such as the flying buttress or certain variants of the ribbed vault. It should not be assumed that the logic according to which the results were judged was wholly an engineering logic; it might be the 'logic of the eye', or the 'logic of the intellect', all of which might either coincide, or be in conflict.

> But it is, nevertheless, admissible to suppose that the experiments of Gothic art, bound powerfully one to the other, and in their royal progress discarding all solutions that were either hazardous or unpromising, constitute by their very sequence and concatenation a kind of logic – an irresistible logic that eventually expresses itself in stone with a classic decisiveness.[24]

Another period in the history of architecture of which we might imagine an account being profitably made through an analysis of its 'situational logic' is the development of the skyscraper office building in the Chicago of the 1880s and 1890s. Such an analysis would treat the basic problem set by restricted sites, the constraints of the requirement for daylight, the economic demand for a maximum of floor space;

the mechanical inventions required to make buildings of such a height possible – principally the elevator and new designs of foundation; the limits of masonry construction, and their transcendence with the introduction of the steel frame; the contribution of electric light, fireproofing, improvements in plumbing; the competitive element in the drive towards even greater heights.

In general we can see that an approach to 'artificial history' through the logic of situations can provide an understanding of progressive development in the technical aspects, without resort to any deterministic theory of the necessity of one step following upon another. There is a *logic of priorities*, by which it is necessary for certain inventions or discoveries to be made before others are possible (thus the construction of the high-speed computer, though its principles were worked out a century earlier, had to wait on certain developments in electronics; or, another example, the pneumatic tyre exploited prior progress in the technology of vulcanising rubber). Again the logic of the matter clearly demands, in the kind of sequence represented by the substitution of stone by iron by bronze by steel in cutting instruments, that the historical order follow the relative merits of the materials in question. But this logic only defines the *preconditions* under which opportunities for various new technological 'moves' are created, and it does not determine their nature or future direction.

To sum up: the explanations of artefact sequences made according to a situational logic will be related to the cultural and social circumstances in which the demand for the artefact is created and given meaning; to the constraints imposed on design by technological and material means available at each historical juncture; and to the body of knowledge, scientific or otherwise, by which the designer is informed and on the basis of which his design 'hypotheses' are made and tested. Changes in form, and the emergence of 'types', will be the result of processes which represent responses to *problems*, and which must be referred to *purposes*. The study of typology and morphology *by themselves* (which is in effect what Clarke proposes) provide no such explanation, and, as Kubler says, avoid 'the principal aim of history, which ... has been to identify and reconstruct the particular problems to which any action or thing must correspond as a solution'.[25]

If the craft tradition provides many examples of nicely graduated series in which the changes in the forms of artefacts are small and slow, it is nevertheless easy, and increasingly so with the advance of technology, to point to abrupt transitions, radical innovations, large jumps which serve to break these sequences and which leave the analogy with biological evolution rather hard to sustain. It is quite beyond the scope of this study to try to give any theoretical account of the processes of technical invention or the nature of creativity in design – which are large enough subjects for books in their own right. Perhaps at the most general level, however, without going very far into the psychology of the question, it is possible to attribute radical novelty in the design of artefacts to two kinds of mental operation.

The first is through existing parts or components of a designed object, themselves perhaps produced by slow processes of technical evolution, being put together into new arrangements: a principle of *fusion* and/or *recombination*. Included within this category would be the kinds of inventive process alluded to briefly in chapter 6, by which new types of object are created by the amalgamation of two or more old ones: the convergence of several lines in the family tree of artefact evolution. Where the designer has access to a substantial body of information about artefacts from cultures remote both geographically and historically from his own, then even if he replicates such designs in their entirety (like the facsimile of the Parthenon in Nashville, Tennessee), the chain of copying is by this fact enabled to cross large gaps in time and space; and if he recombines elements or parts of designs drawn from many eclectic sources, the sequences become correspondingly more complex.

The second operation depends in a different way on the accumulation of historical, cultural, and scientific knowledge. Empirical experience of a range of related designs provides a body of knowledge and understanding on the basis of which it is possible to build a generalised *theory* of that class of artefacts, and so use the theory to extrapolate, beyond the tried cases, to hypothetical but related designs as yet not constructed.

An imaginary example drawn from the history of cookery may serve to illustrate these ideas. We might suppose that in primitive, stable, or isolated cultures, culinary recipes are transmitted from one generation to the next with changes occurring only gradually (perhaps occasioned by the changing availability of different foodstuffs, changes in cooking technology, the vagaries of fashion in eating habits, etc.); so that the 'artefact sequence' represented by the succession of many versions of the same meal would show a genuinely evolutionary character. (Notice, incidentally, the very clear illustration which this example offers of the distinction between the inherited 'design', the recipe – in biological terms the 'genotype' – and the particular individual artefacts, the meals – or 'phenotypes' – in which the recipe is realised. Also the description of the dish which the recipe constitutes comprises no representation of what it tastes like, or even necessarily any picture of what it looks like, but only a set of instructions by which to make it.) Each generation of cooks makes the dish 'like mother used to make it', in the sure knowledge that by following the same procedure it will come out just as before.

When cooks become aware of recipes from other countries or other historical periods than their own – perhaps through the circulation of printed cookery books – then they are freed from the limitations of their particular traditional culinary culture, and they can experiment with an eclectic cuisine, perhaps combining separate elements of dishes from different regions and traditions. These combinations may be more or less successful, to the degree to which the cook can achieve a coherence, a 'correlation of the parts'. The creation of wholly new dishes, and not just minor rearrangements of existing ones, will be dependent on the cook having a general understanding of the *principles* of different cooking processes, of the chemical and biological reactions, perhaps, which various ingredients undergo, and of the general kinds of effect and taste which novel combinations and treatments will produce. As the prophet of scientific gastronomy, Brillat-Savarin, put it:

> The sciences are not like Minerva, who sprang fully armed from the brain of Jupiter; they are daughters of time, and take shape imperceptibly, first, by the combination of methods learned from experience, and later by the discovery of principles derived from the combination of these methods.[26]

We might say in general of the transition from craft procedures to 'selfconscious' methods that empirical knowledge, gradually codified perhaps into scientific knowledge, about the performance of actual past designs begins to allow predictions to be made about the engineering performance of new, hypothetical designs which differ substantially from tested precedent – so much so as to make a simple slight extrapolation unreliable. In the craft tradition, since there are no radical departures from the repeated type, it is possible for artefacts to be made which are technically very sophisticated, which exploit physical principles, chemical processes, or the properties of materials in very subtle ways – but without any of their makers having any *theoretical* understanding of how these effects are achieved. The principles have been discovered empirically, and *are embodied in the inherited design*. We might speak, in a sense, of information being conveyed within the forms of the artefacts themselves. The craftsman knows how to make the object, he follows the traditional procedure (the recipe); but in many respects he literally *does not know what he is doing*.

It is rather in the nature of the problem that evidence for these observations is somewhat hard to come across, since we have little recorded documentary evidence from craftsmen of their actual methods of working; and, of course, they will in any case not have set down what they themselves do not know. Here and there it is possible, nevertheless, to pick up scraps of information which are sufficient to demonstrate beyond much doubt that these assertions are broadly true. These instances are mostly cases where the change is actually being made from a craft-based design process to a more consciously theory-based approach; and the actual individuals who have made this transition in their own lifetime are able to articulate what has happened.

Possibly the most striking illustration of this kind is provided by George Sturt's remarkable and fascinating book *The Wheelwright's Shop*, to which new attention has been drawn by the design theorist Christopher Jones, among others.[27] Sturt worked building farm wagons in the nineteenth century, and the book is his collected reminiscences, written at a time (the 1920s) when the old craft was finally disappearing. It provides a detailed testimony to the role of traditional technique – the knowledge of how, but without the knowledge of why – passed on from craftsman to craftsman, in preserving the continuity of tried and tested forms.

From what Sturt says it appears that the detailed information required for the construction of this extremely complex and subtly designed object, the wagon – no one timber of which was straight or square, but all precisely curved and tapered – was not stored in written records or in drawings at all, except for a few templates for particular components. Instead the information was stored in the heads of the wheelwrights, in their almost instinctive skills in cutting and shaping each piece; stored as an accumulated body of lore and tradition, shared between many men; learnt, through apprenticeship, either by verbal rules, or as physical actions and through the sequences of steps required to make each different part of the work; and stored above all in the shapes of existing wagons themselves, which were there to copy and to follow.

With certain features of the designs, it is quite clear from Sturt's account that no one, no one at all, knew their explanation, not even Sturt himself, although he was an educated man in charge of the shop and had many years of learning from the example of master craftsmen. It took him years of reasoning and reflection to appreciate exactly why it was that a wheel must be 'dished' to a certain degree, what it was that fixed the diameter of wheels or the particular curve or taper of each plank. The experienced craftsman knew that these features were necessary, but did not question them or understand their meaning analytically.

> There was nothing for it but practice and experience of every difficulty. Reasoned science for us did not exist. 'Theirs is not to reason why.' What we had to do was to live and work to the measurements, which had been tested and corrected long before our time in every village shop across the country.[28]

An equivalent example to that of Sturt, in the field of ship-building, is provided by J. S. Russell, who acted as naval architect to Brunel on the Great Eastern (an unhappy collaboration), and who was the author of the first systematic treatise on his subject in English, a magnificent book entitled *The Modern System of Naval Architecture*.[29] Russell had seen a revolution in the building of large ships during his own career, the change from timber construction to iron, and from sailing ships to steam propulsion. He had also been instrumental in the creation of the new profession of naval architecture, distinct from the craft of ship-building; and he makes the – not unexpected – analogy with the equivalent professional and craft distinction in the design of buildings.* The new naval architect will work by science, calculation, 'headwork', where the craftsman worked by imitation, by copying, and by inherited, manual skills.[30]

The need for theoretical texts such as his own, Russell says, is occasioned by the enormous changes in scale, techniques and materials of ship construction, to which the traditional methods and craft knowledge have become inapplicable. Only by means of a theoretical, scientific understanding can the performance of new unbuilt designs of ships be predicted with accuracy. In fact, as Russell recounts, a great number of the early experiments during the 1820s with large iron ships had been disasters – ships which overturned on launching, ships calamitously underpowered, ships whose stability could be ensured only by adding extra floats or masses of cement ballast. The design of these vessels had been based on erroneous rules of thumb, and on principles supposedly drawn from traditional boat-building experience but which were as it turned out 'misknown, and misbelieved, and mistaught'.[31] The true but only tacitly known principles of the old craft techniques had been lost or ignored and were only to be rediscovered through scientific experiment and calculation (for example, by Russell's own work in hydrodynamics, and that of such contemporaries as Froude, Griffiths, and others). As Russell says:

> The forms and proportions of ships, prescribed by traditional knowledge, and universally employed in the early parts of this century, have either ceased to exist, or are preserved as relics. Some even of the principles, which prescribed these forms, and were called Science, have lost their hold on the minds of men, and are abandoned.[32]

Our earlier example from cookery provides another case in point. It is quite possible to bake bread, to brew beer or to make an omelette without the slightest chemical knowledge of the (extremely complex) reactions and biological processes which go on in each case. The same is true of those 'recipes' which are used in the building trade. Vitruvius gives a detailed account of the chemical reactions of lime and of pozzolana when they are mixed with water for making mortar and concrete – an account which is, as one might imagine from the general state of Roman chemical theory, completely erroneous.[33] This was unimportant so long as the means of manufacture and the structural characteristics of these building materials were known by empirical experience.

It is an implication of this general point – that in craft production a degree of knowledge relating to effective or well-adapted designs is embodied in the craft products themselves and in traditional methods of manufacture, without that knowledge being appreciated or recorded consciously – that if the craft techniques and forms are abruptly abandoned, then that knowledge, acquired through many generations of trial and error, is altogether and unwittingly lost. (The effect is the same as in Popper's 'thought experiments', where tools and machines were destroyed along with libraries.) It is not necessarily assimilated, in its entirety, into the consciously held analytical, scientific body of knowledge, set down in writing, which informs the self-conscious designer or engineer who replaces the craftsman and who works from principle rather than precedent. Russell's account of nineteenth-century ship-building illustrates this point.

When new designs are made which represent substantial departures from precedent, and where there is no opportunity for the testing of these designs in the real world through constructing prototypes – as in the case of buildings, civil engineering works, large and costly 'one-off' machines and vehicles, and so on – then it becomes necessary to make certain theoretical *predictions* during the course of the design process about their anticipated performance. These predictions may be of a more or less specific or precise nature. They will have to do with the appearance of the artefact; with its physical behaviour in certain respects (perhaps strengths of structural members, weather-resisting properties of materials, the physics of the heating or lighting of buildings, whatever it may be); with the way in which people will use the object or building or behave in relation to it, how they will perceive it, even what their aesthetic judgements about it may be.

It would be reasonable to expect that such predictions would be more difficult to make, and less trustworthy (though no less important, for that) towards the psychological and aesthetic end of this range, and more reliable towards the geometrical, material and physical end. Indeed the predictions in these latter respects may be based in time on scientific knowledge of the properties of materials and structures, and the principles governing the behaviour of classes of related designs, of which the artefact in question represents one instance. This scientific knowledge is of course a formalisation, a generalisation and an extension of the empirical knowledge gained originally through the trial-and-error processes of the craft tradition – developed and tested perhaps in programmes of deliberate controlled experiment. We can see how an increase in generalised or scientific knowledge about the performance of artefacts in these physical and material aspects may be the cause of a departure from craft methods, since it opens up opportunities for radical innovation. At the same time, looked at in another way, such knowledge may be called for precisely because of the breakdown or abandonment of craft procedures (as in the case of hydrodynamics and ship design in the nineteenth century).

I offered the suggestion, in a previous chapter, that the architects and design theorists of the modern movement were guilty, in the view which they took of scientific method, of falling into the 'inductive fallacy'. A second related misunderstanding about science and about its possible applications in design has been that somehow design method could be made scientific, and that there was some possibility of an equation of *design method* with *scientific method.* Whatever parallels might be made between design and scientific procedures in terms of Popper's scheme of alternating hypothesis and test, of 'conjecture and refutation', the fact is that the nature and purpose of the two enterprises are fundamentally distinct. Design is concerned with making unique material objects to answer to specific purposes; while science is concerned with making statements about the characteristic behaviour of general classes of objects or phenomena under given conditions, and defining the limits on these classes and this behaviour. The relationship of the two has been made very clear by Lionel March in his essay 'The Logic of Design'.[34]

If the modern movement theorists had actually gone to the engineers whom they admired and asked them how they went about the business of design, the engineers would have told them that, in all but the simplest and most highly constrained problems, their methods involved essentially the same element of intuition and speculation as did architecture or the applied arts. The key difference between the design processes of engineers and architects was not in the logic of their respective design methods, which was largely the same, but in the body of scientific knowledge which informed and constrained design in either case. The engineers were, and are, the possessors of a body of scientific theory about structures and machines; and it is this body of understanding which provides them with their design hypotheses, with their first preliminary sketch proposals or germinal ideas for designs. Furthermore, once an initial sketch is somewhat developed and begins to be filled out with detail, then the body of scientific understanding is brought to bear again, because it is in the light of this knowledge and the predictions of performance which it allows that the proposed design can be reliably and rigorously criticised and tested and accordingly modified.

The architects were unable to make use of science in architecture, because there was no science *of* architecture – or at least only a rather undeveloped science. That is to say, there did not exist, nor does there exist, a corpus of knowledge of a scientific kind at an abstract level about classes of existing and hypothetical buildings and their behaviour which can be compared with the mechanical and structural theory of the engineers.

This is not to say that *no* body of general or collective knowledge ('World Three' knowledge) exists in architecture. It quite evidently does, in what is transmitted in architectural education, in architectural literature and, not least, embodied in the designs of existing and historical buildings themselves. But such knowledge is not, with certain areas of exception, of an organised, explicit, communally available and, most important, *scientific* nature. The exceptions are provided by the findings of what has traditionally been distinguished as 'building science': studies of building performance in relation to physical and meteorological environment, the properties of building materials, and the engineering behaviour of architectural structures. If we are to talk about an architectural science in more general terms, then certainly such a science must start from, and incorporate, this existing building science.

We are brought back to the questions which I posed in Chapter 1: how far can and should the project of a building or architectural science extend? More generally, what are the features or properties of artefacts of all kinds to which scientific study should be directed, and about which scientific predictions might be made?

My answers will necessarily have to be abbreviated and tentative ones. First, I suggest that, if we are to interpret the history of artefacts through a 'logic of situations', then we must accept as a corollary that certain features of artefacts, of their perception and mode of use by those who experience and employ them, are *in principle* beyond the reach of scientific predictions. The particular meanings which attach to artefacts, the aims which they serve, the exact ways in which they are seen and evaluated aesthetically, hence to some extent people's behaviour in relation to them, are all products of the specific historical situations in which the objects or buildings are made and from which the observers or users come, and they are changed at every step by the new problems which those situations throw up and by the new and individual responses which those problems evoke.

The relation between the observer and the work of art or man-made object is mediated by the cultural structures of World Three; and these structures, and the meanings which they generate, are continually being transformed. We do not have to espouse a complete aesthetic relativism – as Gombrich shows – to accept that, because of our education and our awareness of history, every 'move', the appearance of every new work, alters the context in which we understand and appreciate not only that work itself, but in principle all other works as well.[35] Kubler calls this the 'T. S. Eliot effect', after Eliot's observation of how 'every major work of art forces upon us a reassessment of all previous works'.[36]

For these sorts of reason I am extremely sceptical of a great deal of the work which has been done in the last few years in architectural and environmental psychology and sociology. This seems to hold as a working assumption the belief that regularities of a reproducible and universal, presumably biologically based, kind may be determined in the behavioural or aesthetic responses which people make to certain architectural forms, spatial arrangements, uses of colour, and so on. As Hillier and Leaman suggest, such research is carried on within a biologically conceived 'man–environment paradigm' which actually removes the middle term, removes the structures of World Three, through which the man-made world acquires significance and is understood at all.[37]

The most that could be expected in this direction, in my view, is that certain very general perceptual constancies or behavioural dispositions might be attributable to the human physiological makeup; and that physiology would set the 'outer limits' – to speak very vaguely – on the broad *ways*, or would establish the general *logic*, according to which objects might be seen and given meaning. But this is obviously a long way from making detailed predictions about aesthetic or behavioural responses to particular works on particular historical occasions.

Incidentally, this criticism is not intended to deny the value of certain types of sociological research whose purpose is to canvass the users' assessment of some building *after* it has been put up and occupied, or perhaps to seek consumers' opinions of a given commercial product. It is only to argue that the findings of such studies – a kind of dignified 'market research' if you will – are essentially retrospective; and that if any predictions or response to future objects or buildings are made on their basis, then these are by way of short-term extrapolations only and lack any wide or long-range application.

Meanwhile the laws of physics, the laws of chemistry, the geometrical laws of three-dimensional space – on which rest, ultimately, the applied sciences of the engineering disciplines, including traditional building science – are clearly not altered by the course of technological or cultural history. (Only our *knowledge* of them may change.) Thus the predictions which may be made with their support about future buildings or other artefacts must be accorded a quite different status.

I offer this argument in support of the proposition that 'sciences of the artificial' must confine themselves, in so far as they aim to have long-term or reliable predictive powers, to the physical and material behaviour and attributes of artefacts. Whether the argument be accepted or not, there is a second and much more pragmatic reason for starting from this engineering end of the subject, and that is to do with what Medawar would refer to as the 'agenda of research'.[38] It makes sense to begin in research with problems which there is actually some hope of solving in the short run rather than those which may possibly be of the most pressing political or immediate practical concern (which fact does not guarantee that they are readily soluble, since science works always at the frontiers which it has presently reached and cannot jump far beyond these). What is more there is a 'logic of priorities' in science, by which the investigation of certain sorts of question is absolutely dependent on having answers to other questions which are logically prior.

To take an illustration from architecture: supposing we wish to predict something about the way in which people will perceive or experience an architectural interior which is not yet built, it is obviously essential at the very least to be able to make accurate physical predictions first about that interior itself – about its dimensions; about its colours, which will depend on the materials of the surfaces and on the ways in which light enters the room and is reflected; about the temperature of the air, which will depend on a host of meteorological, material, mechanical, and thermodynamic factors, and so on. Such predictions are by no means trivial, and some rather sophisticated physical and geometrical models are required in order to make them.

What then are the directions in which building science should now move? It is fair to say that most work in the subject in the past has concentrated on the behaviour of isolated building elements and the physical, chemical, and structural properties of different building materials. More recently, and particularly with the development of computer models, efforts have been made to take a more holistic view, and to study the complex behaviour of the various *systems* from which the building is made up: the structural system, the ventilating and heating systems, the lighting system. This work has created needs for the description and classification of the geometric forms of building, since in order to generalise the results of experiments about the relation of physical performance of buildings to their shapes it is clearly necessary to have some way of characterising their designs in geometrical terms.

These results would be applied in the design process in the evaluation of schemes, as for instance when a hypothetical proposal for a building is put forward and the knowledge gained from the scientific activity is applied to making calculations of its particular anticipated performance in the various respects. But at the same time this knowledge informs the making of the 'hypotheses' in the first place, since these are not produced blindly but on the basis of a general understanding of the *sorts* of structures which may be appropriate for the building in question, the *sorts* of lighting which will be achieved by certain kinds of geometric arrangements of windows and walls, or whatever.

Thus the contribution of building science to architectural design is at two levels: it provides the means for predicting the behaviour of *particular* proposed designs in the physical and engineering aspects; but beyond this, and more broadly, it defines (in principle) the limits on *possible* designs which the given constraints impose. The design problem as a whole is only determined, as we have seen, by the purposes which the artefact is to serve, and in relation to some cultural framework which gives the object meaning. This is even true of artefacts as utilitarian-seeming as the beams or columns of an architectural structure. The need for the structural element is referrable to the purposes which the building as a whole is meant to serve, and these purposes are in turn created culturally.

Nevertheless, once a requirement for buildings, and hence beams, is decided upon, then it is the role of structural engineering to determine the possibilities for their design – which will be a function of the materials used, the patterns of loading, their profile in cross-section, and so on. Tests of the strengths of beams can show the limits on their lengths or slenderness – beyond which the members fail or are unsafe – given certain values for these constraining factors.

It is not just limitations on the material possibilities in design which may be susceptible to systematic investigation. Design is concerned above all with the *arrangement* of elements or components – material or spatial – in different two- or three-dimensional configurations. Here the laws of geometry or topology also place restrictions, possibly quite severe ones, on the range and number of spatial arrangements which are possible for certain classes of design. We can thus distinguish, in structural design for example, between the selection of an appropriate configuration (the study of possible configurations from which the designer's choice is made being essentially a matter for combinatorial analysis) and the assignment of appropriate sizes to the elements of that configuration – what Spillers has termed 'parametric design'.[39] (The distinction is exemplified in any handbook of structural steel tables, where the configuration of each available steel member is given as a schematic cross-section and the possible sizes are listed separately in tabular form.)

Some nineteenth-century work on mechanisms, such as that of Reuleaux, had the purpose of enumerating possible arrangements of such mechanical elements as gears and linkages by means of a formal 'algebra of machines'.[40] Developments in combinatorial mathematics over the last few decades, together with the introduction of computer techniques, have provided the tools with which to carry forward Reuleaux's programme. For instance, the representation of mechanisms in mathematical form as graphs separates out the structural relationships between the components (i.e. how the drive or movement is transmitted from one part to another) from the incidental details of their specific sizes, shapes, or materials of construction. As Freudenstein and Woo have argued, this opens up the very interesting possibility of an abstract classification of machines according to their structure, and independent of the particular functional uses – hoists, baby carriages, typewriters – to which they might be put.[41] Furthermore in certain limited areas it offers the prospect of being able to list *all* possible machines of a given class.

Some equivalent exercises in the enumeration of combinatorial possibilities of arrangement in design have been carried out in other areas of engineering, as for instance in cataloguing possible configurations for electrical circuits or possible ways of bracing framed structures for buildings. In the architectural context, I myself have made some studies of the possibilities for the topological arrangement of rectangular rooms in small rectangular plan layouts, these possibilities being regarded as distinct where the relationships of adjacency between the rooms are different, and no account being taken of dimensions.[42] These investigations have been taken further by various colleagues, and we have been able to enumerate all such plan arrangements with up to nine rooms and to classify these possibilities according to different properties of architectural interest. Meanwhile it has been shown, by March and Earl, how in principle a similar enumeration may be made of all topologically distinct sub-divisions of the plane into regions – i.e. all 'plans' in a very general sense – without restriction to any particular geometric discipline, rectangular or otherwise.[43]

The implications of this sort of work are that, if for example it is decided to design a house layout in which there are to be a given number of rectangular rooms adjacent to each other in certain specified ways and arranged all within a surrounding rectangular boundary, then the number of topologically distinct possibilities for that layout is finite, and they may be exhaustively tabulated. Limitations on the areas, proportions or dimensions of the rooms will set further bounds on the 'solution space' within which all admissible arrangements are contained. Of course the designer may not wish to restrict himself in this way to a rectangular geometry, or he may change his mind about the adjacencies or sizes of rooms – in which case the number and character of the possible solutions, and the boundaries of the solution space, will alter correspondingly.

The fact that houses in Western industrial societies consist very frequently of sub-divisions of an enclosed volume into rectangular spaces is a cultural peculiarity. Indeed the way in which the artefact 'house' may be identified at all in any society is by reference only to some cultural definition. Houses are obviously not limited in any obligatory or absolute way to consisting of rectangular rooms arranged together in various ways; but *where they are so made up* the limitations set by the geometry of the situation, as revealed by an investigation of the sort described, must necessarily apply. Topological and dimensional constraints set bounds, one might say, on what is allowable or feasible in the design of such plans, but within these bounds they do not by any means define what is desirable, let alone what particular choices the architect might actually make in any given scheme.

Whether such intrinsic material or spatial constraints are of real practical consequence for design will depend very much on how severe or restrictive they turn out to be. At this stage the question is one of open debate. In a rather different, anthropological context (and in the course of an anti-functionalist polemic) Sahlins expresses doubt on whether the 'negative determinations' set by physics, chemistry, and biology on culture are of very great interest, since, as he argues, they are generally so loose and permissive.[44] But he is talking of institutions, forms of behaviour and language, not of material artefacts, which we might well imagine would be much more narrowly constrained by the laws of physics or geometry. Certainly, the limitations on architectural arrangement of the kind outlined above are much more constraining than most architects would intuitively – and without seeing the mathematical demonstration – allow.

Where might all this involve any biological analogy? There is a shared body of mechanical and structural theory which would apply equally to the study of artefacts and to the anatomy or 'engineering' of animals and plants. Indeed in biology there is currently something of a revival of interest in the kind of engineering analysis of organic structures and mechanisms which D'Arcy Thompson pioneered, and through which, as we saw in the account of the 'principle of similitude', it is possible to determine limitations on the possible forms and structures of organisms – their sizes, weights, strengths, speeds of locomotion, and the like. Some attention has also been given, for instance by Rosen, to the question of whether the designs of organic systems approach 'optimality' in an engineering sense.[45] (The difficulty in a mathematical treatment of such problems, as Rosen says, is in the definition of appropriate cost functions according to which the success of the structures or organic processes in question can be measured.)

It is perhaps not quite right to speak of an *analogy* here anymore, rather of two separate fields of study in which the same theoretical and analytical tools might be brought to bear. Still, there are certain broad similarities between artefacts and organisms to do with the coordination and purposiveness of their designs, the integration of their functioning parts and systems, which would possibly require the same kinds of analytic approach for their understanding in mechanical or engineering terms. And certainly there would seem to be potential application of some ideas from the theory of systems, coming from biology, to the sciences of the artificial.[46]

Where the two subjects have the most in common, in my view, as the foregoing has perhaps indicated, is in the study of *morphology*; not morphology in the purely descriptive sense in which Goethe originally conceived it, but morphology in the sense of the study of possible forms, of which the actual historical forms of organisms and of artefacts represent particular cases. After an explanation of homology – similarity of form due to common descent – was provided by *The Origin of Species*, the question of analogy in biology, as the zoologist C. F. A. Pantin has observed, was rather brushed aside. But, as Pantin says, 'within the animal kingdom functionally analogous organs may achieve a remarkable similarity in quite unrelated creatures. The analogy is closest where the imposed functional specification is the most detailed.'[47]

It is Pantin's argument that, for example, the functional specification for an eye must be very precise and cannot vary much between animals (assuming an equivalent standard of vision is to be provided), since the nature of the incoming light stimuli and the optics of the situation remain exactly similar, and the different ways the stimuli can be satisfactorily received are rather few. On the other hand there are very many different ways in which the problem of locomotion can be solved, and the specification here is much looser. It is the motor organs, though, which give any creature its characteristic general appearance, and this accounts for a great deal of the variety in animal form. Meanwhile the brains and eyes are always much the same. 'An octopus is obviously staring at you – it is its arms that make it so inhuman and uncouth.' Analogies are 'far from trivial', Pantin says, and the physiologist does not hesitate to argue 'by analogy' from the details of the octopus nerve or brain to the same organs in man, despite the great evolutionary gulf between the two species.

In a paper on 'Organic Design' Pantin comes to some very interesting conclusions provoked by the subject of analogy.[48] He suggests that some nineteenth-century views of evolutionary change were of a process through which the forms of organisms could be almost indefinitely and continuously deformed in any direction. As he puts it, 'The older conceptions of evolutionary morphology stressed the graded adaptation of which the organism is capable, just as putty can be moulded to any desired shape'[49] (Darwin used the term 'plastic').

Pantin's reflections on the matter suggest, as we have seen, that on the contrary there are only certain ways of meeting given functional specifications, that the materials available are of a restricted variety, and that there are strict constructional limitations set by 'engineering' considerations. His chosen metaphor for the morphological possibilities of organic structure would not be modelling clay, but rather a child's constructional toy such as 'Meccano': 'a set consisting of standard parts with unique properties, of strips, plates and wheels, which can be utilised for various objectives such as cranes and locomotives'.[50]

We see that Pantin puts a new kind of interpretation on the 'conditions of existence' of Aristotle and Cuvier. For them the conditions of existence were a theoretical teleological device, an appeal to final causes, whose only explanation could be metaphysical. What Pantin suggests is that there are 'conditions of existence' embodied in the material basis of life and in the physical laws which govern organic structure and process (indeed inorganic structure too). These conditions account for that distinctness of animal (or vegetable) species which Cuvier had insisted on, and for similar reasons. While Cuvier had argued that certain combinations of parts or organs were impossible functionally, Pantin is widening this argument to assert that in fact only certain structures or parts in themselves are possible.

What emerges is some sort of synthesis of, or compromise between, two views: at the one extreme the complete unalterable functional integrity of each separate species and the impossibility of any transformation of one into another, which had been the Cuvierian position; and at the other extreme, a complete evolutionary plasticity allowing transformation in any direction and with any result. Transformation clearly is possible, and has taken place through what appears to be a slow moulding, pushing and pulling, pinching and squeezing of organic forms into new shapes. But this process is channelled, Pantin says, along certain given routes, whose direction is constrained by the permutational possibilities of the 'component parts' and by the limited engineering possibilities available for the solution of any given functional problem.

Transferring all this back to the discussion of the design of man-made artefacts, we see that an evolutionary view of their history, in the craft tradition, would have to take account of the material, geometrical, and mechanical limitations within which this evolutionary process must be constrained – the 'conditions of existence' of each artefact type. The same would apply where artefacts of novel form are created by recombination, amalgamation, or on the basis of generalised engineering principle. This fact would provide a logical explanation of 'analogies' in the designs of man-made objects – similarities of form not attributable to any connection through common cultural influence.

As Hermann Weyl has said, evolution is a historical process, and an account of its evolution alone does not offer scientific explanation of any phenomenon. 'Explanation . . . is to be sought not in its origin but in its immanent law. Knowledge of the laws and of the inner constitution of things must be far advanced before one may hope to understand or hypothetically to reconstruct their genesis.'[51] In architecture this needs what W. R. Lethaby was calling for over sixty years ago, 'a systematic research into the possibilities of walls and vaults, and of the relations between the walls and the cell, and between one cell and another'.[52] 'Some day we shall get a morphology of the art by some architectural Linnaeus or Darwin, who will start from the simple cell and relate it to the most complex structure.'[53]

*   The car designer Raymond Dietrich has described how he and his colleague T. L. Hibbard, who together set up LeBaron Carrossiers in the early 1920s, wanted to make the break from the carriage-building tradition which had continued to dominate car body design technique up to that date; and how they turned for their model to the design process in architecture. 'We wanted to be to cars what architects are to buildings.' 'The Dietrich Story – Part 1', *Veteran and Vintage* (February, 1974), 156–62.

1   W. C. Kneale, 'The Demarcation of Science', *The Philosophy of Karl Popper*, ed. P. A. Schilpp, vol. 1, pp. 205–17. See p. 208.

2   G. Kubler, *The Shape of Time: Remarks on the History of Things* (New Haven and London, 1962).

3   Especially A. L. Kroeber, *Style and Civilizations*. See G. Kubler, *The Shape of Time*, p. 2, n. 1.

4   H. Focillon, *Vie des Formes* (Paris, 1934), trans. C. B. Hogan and G. Kubler as *The Life of Forms in Art* (New Haven, 1942; references are to revised edn, New York, 1948).

5   G. Kubler, *The Shape of Time*, p. 32.

6   Ibid., p. viii.

7   Ibid., p. 2.

8   Ibid., pp. 39–40.

9   Ibid., pp. 37–8.

10 D. L. Clarke, *Analytical Archaeology*. See chapter 4, 'Material culture systems – attribute and artefact', and chapter 5, 'Artefact and type'.

11 V. Gordon Childe, *Piecing Together the Past: The Interpretation of Archaeological Data*, p. 164.

12 D. L. Clarke, *Analytical Archaeology*, p. 172 and Figure 34.

13 Ibid., Figures 45–8, pp. 202–9.

14   Ibid., Figure 30, p. 168; Figure 37, p. 176; Figure 38, p. 177.

15   E. H. Gombrich, 'The Logic of Vanity Fair'.

16   Ibid., p. 929. The data are quoted from J. Gimpel, *The Cathedral Builders* (New York, 1961), p. 44.

17   A. L. Kroeber and J. Richardson, 'Three Centuries of Women's Dress Fashions, a Quantitative Analysis', *Anthropological Records*, 5: 2 (1940), 111–53.

18   D. L. Clarke, *Analytical Archaeology*, Figure 29, p. 164.

19   E. H. Gombrich, 'The Logic of Vanity Fair', p. 929.

20   E. H. Gombrich, 'Evolution in the Arts', p. 267.

21   E. H. Gombrich, 'The Logic of Vanity Fair', p. 941.

22   Ibid., and E. H. Gombrich, entry under 'Style' in *International Encyclopaedia of the Social Sciences*. K. R. Popper, *The Poverty of Historicism*, in particular p. 149.

23   H. Focillon, *The Life of Forms in Art*, p. 9.

24   Ibid., p. 9.

25   G. Kubler, *The Shape of Time*, p. 8.

26   J.-A. Brillat-Savarin, *La Physiologie du Goat* (2 vols, Paris, 1826), trans. A. Drayton as *The Philosopher in the Kitchen* (Harmondsworth, 1970). See p. 50.

27   G. Sturt, *The Wheelwright's Shop* (Cambridge, 1923). For J. C. Jones's comments, see his *Design Methods: Seeds of Human Futures*, pp. 17–19. Sturt wrote another book, under the pseudonym of George Bourne, on the subject of evolution in art, entitled *The Ascending Effort* (London, 1910); but in it, curiously, he makes little reference to craft evolution – even though he makes use of biological metaphor in *The Wheelwright's Shop*, referring to the wagon as an 'organism', to its 'adaptation' and so on.

28   G. Sturt, *The Wheelwright's Shop*, p. 19.

29   J. S. Russell, *The Modern System of Naval Architecture* (3 vols, London, 1865).

30   Ibid., p. 301.

31   Ibid., p. xxiv.

32   Ibid., p. xxix.

33   Vitruvius, *The Ten Books on Architecture*, book 2, chapters 5 and 6.

34   L. March, 'The Logic of Design and the Question of Value', *The Architecture of Form*, ed. L. March (Cambridge, 1976), pp. 1–40.

35   E. H. Gombrich, 'The Logic of Vanity Fair', pp. 945–7. See also Popper's comments in the same volume, pp. 1174–80.

36   G. Kubler, *The Shape of Time*, p. 35 and n. 4.

37   B. Hillier and A. Leaman, 'The Man–Environment Paradigm and its Paradoxes', *Architectural Design*, 8 (1973), 507–11.

38   P. B. Medawar, *The Art of the Soluble* (London, 1967). See pp. 7, 87.

39   W. R. Spillers, 'Some Problems of Structural Design', *Basic Questions of Design Theory*, ed. W. R. Spillers (Amsterdam and New York, 1974), pp. 103–17.

40   F. Reuleaux, *The Kinematics of Machinery: Outlines of a Theory of Machines* (London, 1876).

41   F. Freudenstein and L. S. Woo, 'Kinematic Structure of Mechanisms', *Basic Questions of Design Theory*, ed. W. R. Spillers, pp. 241–64.

42   P. Steadman, 'Graph-Theoretic Representation of Architectural Arrangement', *Architectural Research and Teaching,* 2/3 (1973), 161–72. Also W. Mitchell, P. Steadman, and R. S. Liggett, 'Synthesis and Optimisation of Small Rectangular Floor Plans', *Environment and Planning B*, 3 (1976), 37–70.

43   L. March and C. F. Earl, 'On Counting Architectural Plans', *Environment and Planning B*, 4 (1977), 57–80.

44   M. Sahlins, *The Use and Abuse of Biology* (London, 1977), pp. 63–6.

45   R. Rosen, *Optimality Principles in Biology* (London, 1967), pp. 6–7.

46   For a discussion see H. A. Simon, *The Sciences of the Artificial*, 'The Architecture of Complexity', pp. 84–118.

47   C. F. A. Pantin, *The Relations Between the Sciences* (Cambridge, 1968), p. 93.

48   C. F. A. Pantin, 'Organic Design', *The Advancement of Science*, 8 (1951), 138–50.

49   C. F. A. Pantin, *The Relations Between the Sciences*, p. *93*.

50   Ibid., pp. 93–4.

51   H. Weyl, *Philosophy of Mathematics and Natural Science* (Princeton, 1949), p. 286.

52   W. R. Lethaby, *Form in Civilization* (Oxford, 1922), p. 90.

53   W. R. Lethaby, *Architecture* (London, 1911; 3rd edn, Oxford, 1955), p. 2.

Alan Colquhoun (born 1921) studied architecture at the Edinburgh College of Art, 1939–42, and the Architectural Association, London, 1947–49. He practised in the 1950s with London County Council before setting up in partnership in 1961 with John Miller as Colquhoun & Miller. Perhaps their best known completed project is the refurbishment of the Whitechapel Art Gallery in London (1984). Colquhoun retired from practice in 1990. He has also taught and lectured widely in Europe and North America, including periods at the Architectural Association (1957–64), Polytechnic of Central London (now University of Westminster) (1976–78) and Princeton University, New Jersey, from 1978 onwards.

Since the early 1950s, Colquhoun's published critical writings have gradually accumulated to form a comprehensive "theory of modernism" in architecture. Renowned for his intellectual rigor and conceptual clarity Colquhoun has concentrated on themes that appeared central to the modernist attitude in architecture – language, typology, and the structure of form. More recently his approach has been to try to relate these issues to current practice and to analyze the nature of architectural expression in relation to wider social and cultural forces.

The essay reproduced here originally appeared in the journal *Architectural Design* in 1962 and was reprinted in the book *Essays in Architectural Criticism: Modern Architecture and Historical Change* (1981), part of the influential "Oppositions" book series produced by MIT Press and edited by Peter Eisenman. Colquhoun questions the common distinction between utilitarian and aesthetic criteria in architecture, highlighting the fundamentally metaphorical level on which architectural form must communicate. This piece could also be seen as an antidote to the less critical celebration of techniques represented by another significant strand of mid-century modernist thinking – inspired in large part by the writings of Le Corbusier (1923, 1929), Siegfried Giedion (1928, 1941, 1948) and Richard Buckminster Fuller (1929, 1969). See also Reyner Banham, 1960, 1965) and Peter Cook (1970).

1981

Alan Colquhoun

Symbolic and Literal Aspects of Technology

One of the remarkable facts about the architecture of the mid-twentieth century is that so many of its buildings exploit heavy and traditional methods of construction. From the point of view of building technique this would seem to be a regression from the ideals of the early period of the Modern Movement, which aimed at an expression of the lightness inherent in tensile structure and synthetic materials.

It is true that architects for the majority of buildings put up today make use of the simple principle of a concrete or steel frame sheathed in some form of curtain wall and in doing so appear to be putting into practice the theories formulated by Le Corbusier in the 1920s. Yet the architectural qualities of most of these buildings are so meager that one is forced to ask whether, in the mere application of an apparently logical system, the essential features of good architecture are not being overlooked. And indeed, there is a tendency among architects, whenever the program allows, to break away from the simple frame structure with panel infill to some form of structure that allows the building a greater plastic flexibility and gives to its forms a greater density.

Both the Caius College hostel and the Royal College of Art building, by Sir Leslie Martin and H. T. Cadbury-Brown respectively, aim at and achieve an effect of mass which is not a necessary product of the program and its structural interpretation. In the case of the Royal College of Art building the reinforced concrete frame is partly covered by brick panels which, together with the studios at roof level, emphasize the vertical axis and create an impression of ambiguity as to whether the structure consists of a frame or of solid load-bearing walls. In the case of Caius College a brick structure is used in such a way as to exaggerate the massiveness of this kind of construction and to create a feeling of enclosure and protection reminiscent of a walled town or a Roman amphitheater.

A layman might conclude that such buildings were a reaction against the "glass box" architecture rising in our cities, and that there must be a split in the architectural profession reflecting the sort of chasm that seems to him to exist between what is proper to "office" architecture on the one hand and private houses on the other. But if there is such a split, it is probably in the mind of each architect. Every architect today is torn between two concepts of architecture. On the one hand, architecture is seen to consist of unique works of art, the creation of individual sensibility. On the other, it is seen as belonging to the public sphere, where private sensibility is under the control of "techniques" in the broadest sense of that word.

In spite of its theory, the Modern Movement failed to establish a substantial relationship between these two concepts. To see why this is so, it is necessary to look more closely at the real conditions that sustained it. It has become a truism to say that the buildings of the Rationalist Movement were, whatever their mystique, built largely of traditional materials. The only real innovation brought into use in the early twentieth century was tensile structure. The other changes that were brought about in the organization and appearance of buildings were the result of a priori theories about the nature of the Machine Age and the social purpose of architecture. Together they formed a "functionalist" architecture of enormous power, whose image was created largely out of Expressionist, Cubist, or Neoclassical aesthetic theory. This architecture formed the active wing of the avant-garde movement as a whole and was concerned as much with the salvation of society through art as through technique. The "functionalist" building was, in fact, a pure work of art, freed from the arbitrary rules of craft and of individual fantasy and raised to the level of Platonic form by means of the machine – a work of pure exactitude.

We cannot grasp the meaning of the Modern Movement unless we understand that the role which symbolic expression played in it was fundamentally the same as it had been in previous architecture. There is a tendency in criticism to distinguish between utilitarian and moral criteria, on the one hand, and aesthetic criteria, on the other. According to this conception, aesthetics is concerned with "form," while the logical, technical, and sociological problems of building belong to the world of empirical action. This distinction is false, because it ignores the fact that architecture belongs to a world of symbolic forms in which every aspect of building is presented metaphorically, not literally. There is a logic of forms, but it is not identical with the logic which comes into play in the solution of the empirical problems of construction. The two systems of thought are not consecutive but parallel.

This was as true of the Modern Movement as it was of any other period of architecture. In it the new technology was an idea rather than a fact. It became part of its content as a work of art and not merely or principally a means to its construction. Our admiration of the buildings it created is due more to their success as symbolic representations than to the extent to which they solved technical problems. However much the materials they used were conceived to be the products of machine techniques, these architects never regarded them as "ready-mades" but adapted them to a preconceived plastic form, even though this form itself was triggered by a notion of machine technology. One might quote as an example Le Corbusier's use of the curtain wall, in which the glazing bars are so profiled and proportioned as to preserve the integrity of the plane and to create the feeling of a tight skin stretched over the entire surface of the building.

The fact that the technical and social revolution assumed by the Modern Movement did not take place brought to the surface the extent to which its work was the result of private will to form – and the extent to which *all* architecture must be so based. Once this was admitted, the ontological link between art and technique was broken, and the architect was free to enlarge the theoretical context within which he designed. This link had been forged from a utopian and eschatological view of society, art, and technique which was no longer tenable. What inevitably followed was a more empirical attitude toward construction and researches into form which was fundamentally unconcerned with the problem of advanced techniques.

To some extent this has been forced on designers by economic necessity, but there is no doubt that the feeling of mass and permanence which traditional or semi-traditional materials give has been sought after for its own sake, and that the technical and public aspects of building have fallen progressively outside the field of private symbolic expression. It is as if the urge to create the world anew by means of structures which had the lightness and tenuousness of pure thought had given way to the desire to create solid hideouts of the human spirit in a world of uncertainty and change, each one in itself a microcosm of an ideal world.

And here we come to the dilemma which was at the root of the Modern

Movement and which is still present today. If buildings are to retain their quality of uniqueness as symbols, how can they also be the end products of an industrial system whose purpose is to find general solutions? In the 1920s a series of unique solutions stood as symbols for a universal idea which could not be put into practice. Today we are faced with an imminent revolution in building technique, the very existence of which may make the unique solution impossible.

All forms of symbolic expression emerge from and feed on the world of fact. Architecture can exist only in the context of its sociological, technical, and economic conditions, and as soon as it ceases to do this it dies. But up till now the means of construction at the disposal of society, out of which it has created its symbolic forms, has always been malleable to the will of the designer. This condition can exist whether the method of manufacture is manual or mechanical, but to make this possible the designer must participate at the beginning of the process.

When we are discussing architecture, we are discussing whole buildings or building complexes. Therefore any element which is being designed must be thought of in the context of the whole of which it is to be a part. A simple system of components based on an additive module, which are interchangeable to suit any situation, does not give this essential condition, since the character of the form of a building alters according to its size, situation, and program, and a building is not the mere sum of its parts.

Yet it is just such a system that would recommend itself to an organization concerned with the economic carrying out of a large and complex building program. A number of such systems have, in fact, been operated, and while they satisfy more or less the needs of flexibility of assembly, they do not satisfy the need of flexibility in design. They are only capable of solving the simplest arrangement of all the possible arrangements to which they apply. Thus, a structural grid with infill panels becomes inexpressive to the extent that these panels are simply additive. It is true that a building which is an agglomeration of units can achieve great intensity and unity, but this can only be achieved if the design of each unit anticipates the complex as a whole. This will require modifications which are neither economical nor logical from the point of view of the simple operation of joining one unit to another in an additive series. We have here a confusion between technology as a means to construction and technology as the content of the building form itself. Such systems render a building incapable of symbolizing plastically the utopian ideals which undoubtedly inspire them.

Le Corbusier, in discussing the design of a series of metal houses at Lagny says, "The problem here is utterly commonplace." Yet these houses have a charm which derives from their uniqueness. They are designed for repetition, but the components have been subject to a control which has always kept a certain plastic and expressive end in view and which does not allow for extension or diminution. Similarly, in the design of cars a particular model is unique however many times it is repeated. Whether or not such examples are relevant to the problem, say, of mass-produced housing or schools, there is no doubt that serious thought will have to be given to the question of component design in relation to a particular architectural intention if architecture is not to lose all possibility of symbolic expression.

In a fluid situation where the decisions on fundamental questions seem to be outside the control of the architect, there is a tendency for him to flee into escapist backwaters of irrelevant symbolism. But it is not the urge to symbolism itself which is wrong, for without it architecture would cease to exist. However much society needs an architecture which expresses its ideals and which provides for the human spirit, there is a danger that its economic mechanisms may make such an architecture impossible. This is particularly so because many architects considering themselves to be the heirs of the Modern Movement fundamentally misinterpret its aims and its virtues. The science of building, the rationalization of construction and assembly, however vital in themselves, remain in the world of literal action. It is only when the architect, seizing this world, organizes it according to the logic of symbolic forms that architecture results.

1982

Luis Fernández-Galiano

Organisms and Mechanisms, Metaphors of Architecture

Mechanical, Thermal, and Cybernetic Machines versus the Living and the Built

Luis Fernández-Galiano (born 1950) is a Spanish architect, teacher, and editor. He is Professor of Architecture at the Universidad Politécnica de Madrid (ETSAM), editor of the journal *Arquitectura Viva*, and the architecture critic for Spain's leading newspaper, *El País*. He has been Cullinan Professor at Rice University, a visiting scholar at the Getty Center of Los Angeles, and a visiting critic at Princeton, Harvard, and the Berlage Institute. As an editor and professor, he has written widely on architecture and the city.

This excerpt is drawn from the English translation of his book *Fire and Memory: On Architecture and Energy* (2000), originally published in Spanish in 1991, and (by his own account) written in 1982. In general, the work offers a profoundly architectural inquiry into the issues raised by the energy supply crises of the 1970s, looking at the ways in which buildings have always been characterized by their mode of energy consumption. However, the importance of the book extends well beyond the narrow terms of consumption and efficiency with which energy is normally considered in architecture. The chapter explores changes in the imaginative paradigms, the metaphors, with which the activities and operations of buildings are collectively understood. From mechanical to thermodynamic to cybernetic, Fernández-Galiano charts the emergence of the age of systems in the parallel explanations of mechanisms and organisms, showing their interrelationship and their importance to design.

› On the fraternity between buildings, living beings, and machines

Biological and mechanical quotations are omnipresent in architecture, occurring, moreover, with singular simultaneity. Organisms and mechanisms frequently appear in plans or sketches of buildings, punctuating, emphasizing, offering metaphors, or suggesting comments. After all, the building is an artifact meant to shelter living beings, and there is nothing strange about the mechanical or natural universe serving as a model, a contrast, or a stimulus in the design process. Nevertheless, the extent to which they overlap and coincide is astonishing.

In what are probably the two most famous notebooks of architectural history[1] separated by more than six centuries, living beings and machines are juxtaposed and entangled among construction sketches. The oldest known clock with an escapement device[2] and the first frame saw appear in Villard de Honnecourt's *Album* of the late thirteenth century, but so do drawings of a lion and a bear, a lobster and a swan, a dragonfly and a fly, parrots and dogs, cats and horses. . . . Organic and mechanical metaphors notoriously abound in Le Corbusier's notebooks, while skeletons and automobiles, fish and airships proliferate among his designs for buildings. Of course the two had very different approaches. Whereas the medieval builder contemplated architectural solutions and mechanical devices with the same degree of interest and drew decorative details and exotic animals with equally avid curiosity, the contemporary architect established conscious, explicit parallelisms and formulated pedagogical or polemical analogies between buildings and the mechanical or natural world. Both, notwithstanding, pursued a conception that makes architecture have a share in a world inhabited by living beings and mechanical contrivances. This said, what links are there between organisms and machines that explain their frequent and simultaneous presence in the mind and pencil of builders?

Before proceeding further, note that the idea here is not so much to explore organic and mechanical references in architecture as to reflect on the parallelisms and reciprocal relationship between the very conceptions of organisms and machines, and this from two perspectives: the fluctuations in their dialogue through history, and the opinions and works of two architects of this century who exemplify these opposed approaches.

One can rightly engage in a historical examination of the dialogue between organisms and machines because the relationship between them has suffered major modifications, as a result of the contrast between the extraordinarily slow evolution of organisms and the accelerated rate of change that the world of machines has been subjected to in the last few centuries. That is, in a reduced span of time the mechanical universe has undergone radical transformation, whereas the organic universe, in the sense used here, has remained practically unchanged. It is thus the machine, and its successive versions, that have determined the different conceptions of the relationship between organism and mechanism: mechanical, thermal, and cybernetic machines[3] have generated the views of the organism as mechanism, motor, and automaton, respectively.[4]

In each of these historically successive metaphors, energy plays a different role:

whereas in the world of mechanisms energy is above all work, mechanical motion, in the world of thermal machines it is basically heat, and in that of cybernetic machines it is information. Similarly, the old analogy between artifact/building and organism/body takes on different lines, with the building considered a body composed of parts, a body that nourishes itself, and an intelligent body, respectively.

From an architectural perspective, the importance of these considerations lies in the fact that organic references are almost always influenced by the way the organism is viewed through the machines of the age. If the organism is contemplated through the perspective of the machine, the distance between organic and mechanical analogies of architecture can be understood to be more symbolic than functional, as will be shown in the parallel analyses of the paradigmatic cases of Frank Lloyd Wright and Le Corbusier, the theme of the second half of this chapter.

› Mechanical organisms

From the bête machine to the automaton

The dialogue between machine and life is first manifest in the conception of the organism as a mechanical artifact, and there are few better examples of this dialogue than Leonardo's. His drawings not only juxtapose heads and machines, hygrometers and figures, Madonnas and hydraulic wheels, but also formulate detailed parallelisms between the organic and mechanical worlds, such as the famous one about fish and ships, or those expressed in flying machine designs. These parallelisms are not accidental. In fact, Leonardo's unexecuted book titled the *Elements of Machines* was to have presented the elemental parts of mechanical devices and served as a prelude to his treatise *On the Human Body*. He described the relationship between the two thus:

> Do not forget that the book on the elements of machines with its beneficial functions should precede proofs relating to the motion and power of man and other animals; then on their bases, you will be able to verify your propositions.[5]

Indeed in Leonardo's opinion, as Benevolo points out

> machines were not a world of independent objects, with laws and development to be studied, but artificial extensions of man's capacities for movement and work, similar to the limbs of the body and reducible to the same vital principles, as the limbs, in their turn, are reducible to mechanisms which are moved directly by the "soul." The real objective of [his] research lay in comparing and giving a single interpretation of the biological universe and the mechanical universe.[6]

In any case, from the Leonardo who proclaims his conviction that "the bird is an instrument operating through mathematical laws"[7] or the Gómez Pereira whose *Antoniana Margarita* of 1554 defends the thesis that all creatures except man are automata without a soul, there is a long history of contemplating living nature in mechanical terms.[8] Unquestionable milestones in this history are *la bête machine* (the animal as a machine) of the *Discourse on Method* and the detailed elaboration that Julien de La Mettrie made almost a century later, in *L'homme machine* of 1748.[9]

During this long period the proliferation of rudimentary automata served as a symbolic bridge between machines and organisms;[10] the clockwork or hydraulic mechanisms of Juanelo, De Calls,[11] Kircher, or Vaucanson fascinated their contemporaries and continue to amaze us today.

It is astonishing to see how tenaciously these makers of machines seek to create replicas of living beings. Juanelo Turriano, for example, is known for the water lifter he built for the city of Toledo[12] but probably spent more time contriving automata: flying birds, shepherdesses playing the lute, and swordsmen for the entertainment of Charles V. Jacques de Vaucanson, to mention another case, invented a new type of lathe and revolutionized textile machinery, but these technological advances cannot be separated from his work as a builder of automata, which gave him popularity and fortune. Through them, moreover, he was able to offer an admirable material illustration of Descartes's philosophical theses,[13] accurately interpreted by his *anatomie mouvante*: machines that ape the organism make it possible to think of the organism as a machine. Mumford is perhaps not altogether fair, therefore, when he says that "technology remembers Vaucanson for his loom, more than for his mechanical duck that seemed alive and could not only eat but also digest and excrete."[14] Despite their apparent frivolity, automata are technical ideological grounds; even more importantly, they are eloquent ideological manifestos – more accessible than philosophical treatises – through which mechanical and clockwork views of the organism are diffused and generalized.

› From the clock to the steam machine

thermodynamic Freud

The mechanical *Weltbild* underwent a deep transformation with the advent of the steam machine. While maintaining a significant continuity with the mechanical paradigm and thus confirming Lewis Mumford's opinion that "the clock, not the steam-machine, is the key-machine of the modern industrial age,"[15] the invention of this machine brought about a major shift in the functional and symbolic realm.

Prigogine and Stengers have described such shifting of emphasis thus:

> Developing from an automaton nature, which is as alien to man as a clock is to a clockmaker, in the course of the nineteenth century we witness the transformation of that mechanical nature into a motor nature, with the new, distressing question regarding the exhaustion of resources and the descent into conflict with the rival perspective of progress – precisely what has allowed the transit from the clock to the igneous machine.[16]

In this way, the transit from the mechanism to the engine introduces the second expression of the dialogue between the machine and life, which consists of viewing the organism as a thermal machine.

The diffusion of the steam engine, and even more so of the science of energy built on the heat of thermal machines, gave rise to a vigorous cultural shake-up as much as to a far-reaching technical and economic mutation. Thermodynamics transformed our conception of the world: man, society, nature, from then on, would be reflected in the mirror of energy. If the scientific importance of thermo-dynamics was great,

> its cultural resonance was also immense: a new conception of man as an energy machine [Jacques Lacan, for example, has shown to what extent Freudian theory rested on this view]; a new conception of society as an engine . . . a new conception of nature itself as *energy*, that is, the creative and productive capacity of qualitative differences.[17]

The mention of Freud in this context is not casual. His anthropology conceives the subject as a tangle of fluxes and energy drives; the role of the libido, or the rela-tionship between the principle of pleasure and the reduction of tension, as Rudolf Arnheim has shown, establishes a direct parallel with the second thermodynamic principle.[18] Long before Lacan, this parallelism was noted by disciples of Freud like Siegfried Bernfeld, who as early as 1934 wrote that "physical systems for which the entropy principle holds behave as if they had an impulse to reduce their internal quantities of tension within the system as a whole."

In the same way, Freud's theory about irrational and unpredictable com-ponents in the mind and in human conduct creates a kinship between his work and the concept of thermodynamic causality, which substitutes chance and probability for the necessary relationships of Newtonian mechanicism. Norbert Wiener indi-cated the points of contact between Freud's view and Gibbs's statistical approaches, stressing that "in recognizing chance as a basic element incorporated into the very fabric of the universe, these men come close to one another, and close as well to the tradition of St. Augustine."[19]

Sigmund Freud, in any case, has only been cited as an example, especially relevant, perhaps, but by no means the only one, since the thermodynamic con-ception of organisms penetrates the entire cultural fabric of the nineteenth century and survives to our days. Suffice it to quote the description of a living being offered by a contemporary philosopher, Edgar Morin: "The living being is a thermo-hydraulic machine in slow combustion operating between zero and sixty degrees Celsius, eighty percent of which consists of circulating and soaking water, incessantly consuming itself and being consumed." He adds: "It is defi-nitely a well-tempered, multi-regulated machine with a formidable informational device."[20] This last phrase already implies what would be the third expression of the dialogue between the machine and life, the contemplation of the organism as a cybernetic machine.

› From the engine to the servomechanism

A cybernetic anthropology
The example of psychoanalytic theory also serves to illustrate the informational view of the organism. It was Wiener himself who induced Gregory Bateson to con-sider psychoanalytic practice in cybernetic terms. According to Heims, "Wiener put forward the idea that in communication systems the crucial concept is information rather than energy, and that therefore Freud's emphasis on libido was inappropri-

ate."[21] Along these lines, Bateson elaborated a set of theories including that of the double link in schizophrenia, the treatment of alcoholism, and the application of family therapy, all extraordinarily influential and based on the conception of the human being as a cybernetic machine.

Indeed, this cybernetic view of organisms exceeds the anthropological limits of our example and extends to any vital phenomenon. It tends to be interpreted in terms of feedback, servomechanisms, circular processes, etc. The very popularity of these terms testifies to the diffusion of the set of theories formulated in the heat of the development of computer technology during World War II, among which we must mention – besides Wiener's cybernetics, systematically presented in 1948 – the game theory of von Neumann and Morgenstern (1947) and the information theory of Shannon and Weaver (1949). Although these theoretical constructs have numerous antecedents,[22] the most relevant probably being the concept of homeostasis that was elaborated by Cannon in the late 1920s, only with them, from 1950 onward, was the conception of the organism as a servo-regulated automaton generalized.

As Morin has written,

> the idea of the cybernetic machine moved through the track of molecular biology to become the basis of the new conception of life. . . . The incorporation of cybernetics into biology constituted an incorporation of biology into cybernetics. The living being from then on could be conceived, and was conceived, as the most complete cybernetic machine and even as the most complete automaton [von Neumann, 1966], exceeding the most modern of automatic fabrications [Rosnay, 1966] in complexity, perfection, and efficiency, even in the least of bacteria.[23]

In mentioning the transit from the clock to the steam machine, we noted that this transformation did not contradict a certain continuity of the mechanical paradigm. The same thing applies now as we consider the passage from the steam machine to the computer, from the engine to the cybernetic automaton; in this case, too, the mechanical paradigm survives, hidden but omnipresent, as the true thread of an entire age. Far from denying it, the cybernetic view confirms the persistence of the mechanism. As Ludwig von Bertalanffy has indicated, there is an evident relationship between the model of the "organism as servomechanism" and the zeitgeist of a mechanized society: "the domination of the machine, the theoretical view of living beings as machines and the mechanization of man himself" are closely related to the "mechanistic world picture."[24]

This mechanistic conception has bequeathed us a submissive, predictable, manipulable automaton nature: "a dull affair, soundless, scentless, colourless, merely the hurrying of matter, endlessly, meaninglessly," in the words of Whitehead.[25] In the final analysis, this is the very world view that underlies the analogies between the organism and mechanical, thermal, and cybernetic machines which we have described.

Note, however, that all these analogies have been put forward with the organism as the subject, or at least the image of the organism viewed through a mechanical magnifying glass, through the smoked glass of thermodynamics, or through the frosted and analytical glass of information. It is equally possible and even necessary to run the process in reverse, to scan the inside of each analogy

and describe the reflections of the different categories of machines in the revealing looking glass of organic life. In this way we can understand that if there are
mechanical organisms, so are there organic mechanisms; that if there is an
automaton nature, so is there a natural automaton, and that the two are interrelated.

The mechanical face therefore has an organic back; the organism is perceived
through the machine, but the machine is likewise perceived through the organism.
The fact that both belong to the same functional realm must be understood in the
context of this mutual reflection, this inextricable interweaving, this interminable
dialogue of misted-up or shattered concave mirrors that distort, diffuse, and fragment – in the kaleidoscope of history – the inseparable and confronted images of
mechanisms and organisms.

› Organic mechanisms

mechanical machines and mechanizing machines

What emerges from the foregoing is that the machine can be contemplated *from*
the organism and *as* an organism. Having spoken of the *bête machine*, we shall
now describe the mechanism as a prolongation of the organic body and as a materialization of the organization of the social body. If we have mentioned the conception of the organism as a thermal machine, so is it possible to speak of thermal
machines as organisms and extensions of the organic; finally, in the same way that
we have dwelt on the contemplation of living beings as servomechanisms, so shall
we on that of cybernetic automata as living machines.

Even at its very origins, the machine was indebted to the organism for at least
two reasons. On one hand, as Ernst Kapp suggested more than a century ago in his
Grundlinien einer Philosophie der Technik, it can be said that machines come
about as projections of organs, so that the hammer, for instance, is an extension of
the fist, and the assemblage comprising the hammer and the hand that clenches it
is the equivalent of an elemental machine.[26] On the other hand, as repeatedly
stressed by Lewis Mumford, the mechanism appeared as an element of social life
long before the peoples of the Western world turned to the machine.[27] What he
calls the "social megamachine" – the mechanical organization of the slaves who
built the pyramids, the soldiers of the Macedonian phalanx, or the oarsmen of
Roman galleys – entailed the creation of organic machines to precede and prepare
the way for mechanical devices.

Subsequently, the artificial creation of a moving agent with the first steam
engine meant a qualitative leap in the mechanical evolution and an opportunity
to renew and reinforce the organic conception of the machine, as enthusiastically
expressed by Bernard Forest de Belidor,[28] who contemplated one of the first Newcomen machines in France and shared his experience in *L'architecture
hydraulique*:

> So here is the most marvelous of all machines; its mechanism resembles that of animals. Heat is
> the principle behind its movement; the circulation produced in its conduits is like that of blood

> in veins, with valves that open and close according to need; it nourishes itself and excretes on
> its own at an established rate, and extracts from its work everything it needs in order to subsist.

A century later in 1853, describing the opening of a factory in the industrial community of Saltaire, a British clergyman adopts the same fervent tone: "Finally the large steam machines began to move, transmitting energy to all parts of the vast organism which, as if touched by a mysterious hand, woke up to life. . . . What a marvelous scene!"[29]

In both cases, in contrast to the conception of the Cartesian *bête machine* (and despite their being inscribed in the same mechanical universe), it was no longer the organism that was interpreted as a machine, but the machine that was explained in organic terms. In this context, as Mumford noted, it was perhaps more than anecdotal that Giovanni Branca's engraving representing one of the steam machine's antecedents should depict an anthropomorphic cauldron. Thermal machines constitute a spectacular second approximation to the organism, and it comes as no surprise that they are understood in terms of it.

Samuel Butler was keenly aware of this approximation when he put the following words in the mouth of the author of the book of machines:

> The vapour-engine must be fed with food and consume it by fire even as man consumes it; it
> supports its combustion by air as man supports it; it has a pulse and circulation as man has. It
> may be granted that man's body is as yet the more versatile of the two, but then man's body is
> an older thing; give the steam-engine but half the time that man has had, give it also a
> continuance of our present infatuation, and what may it not ere long attain to?[30]

As we see, the phantom of the rebellion of machines already weighs over industrial culture: the shadow of the automaton hovers over a world where the line between organisms and mechanisms is progressively blurred.

Finally, the third great approximation of the machine to the organism can be associated to cybernetics and what Morin has called "the Wienerian revolution: contemplating the machine as a living being," with a use of terms that endeavors to be more rigorous than metaphorical. This concept of Wiener and its extrapolation in the works of Maturana, Varela, and Morin himself affirms that "today we must conceive the machine not as mechanism but as praxis, production, and *poiesis*," since "in the machine there exists not only the *mechanical* (repetitive) but also the *mechanizing* (inventive)."[31]

This organic, creative, *mechanizing* view of the cybernetic automaton tries to break what Mumford has called "the ominous bond" between the "automaton" and the "other," an irremediable consequence of the gestation of the automaton, which "is the last step in a process that began with the use of one part or another of the human body as a tool,"[32] that is, in a process of increasing alienation from the organism of man. Nevertheless, this "mechanizing machine" has no tranquilizing effect; far from making the mechanical automaton more attractive, it gives it the outlines of a nightmare.

› Living machines

between the golem and the cultural fact

The growing mechanization of the organic[33] and the parallel biologization of the mechanical have preoccupied not only the apostles of vitalism – who criticize the metaphor of the automaton because, unlike the living being, it has an end that lies outside of itself – but also the creators of the latest generation of automata, the cybernetic machines, and particularly the greatest of them all, the mathematician Norbert Wiener.

Wiener indeed was aware of the risks involved in a biological interpretation of cybernetic machines, which he associated with the golem, the disturbing animated robot of Jewish legends.[34] "The machine," he said, "is the modern equivalent of the *golem* of the rabbi of Prague." And his own work "with mechanical analogies between organisms, or the nervous systems of organisms, and automata or formal or mechanical models," as Heims has observed, made him resemble "the maker of a *golem*."[35]

In this preoccupation with the supplanting of life by the mechanism, Wiener's attitude contrasts with that of another great mathematician, the creator of the theory of automata John von Neumann, whose career ran parallel to Wiener's in many ways but who accepted the protagonism of the machine. Perhaps better than anyone else, von Neumann represents the survival of the mechanical paradigm in this third cybernetic phase, as proven by his own epistemological position, since unlike Wiener, who "considered random processes and chaos fundamental, von Neumann saw the mechanism and the logic underlying it in all scientific phenomena."[36]

Live cybernetic machines are the ultimate expression of the kidnapping of life by mechanism, but they simultaneously and paradoxically present the possibility of defeating the mechanical automaton and the view of nature "as a stupid and passive mechanism, essentially alien to freedom and the aim of the human spirit,"[37] and of replacing, as Prigogine has proposed, the classical description of the world as an automaton with the Greek paradigm of the world as a work of art.

Note that the world is referred to as a work of art, a product of culture, and not as a biological organism. Up to this point we have shown that if organisms can be contemplated as machines, so machines can be interpreted as living beings. Here, following Mumford's advice, we have avoided "the false notion that the mechanism has nothing to learn from life" and "the equally false notion that life has nothing to learn from the mechanism," and tried to stress the close bond that renders the organism and the machine inseparable. In the light of this bond, it would be inconsistent to substitute a totally organic for a totally mechanical conception of the world, one being practically equivalent to the other. We ought to pay heed to Ludwig von Bertalanffy's warning: "After having overthrown the mechanistic view, we are careful not to slide into 'biologism,' that is, into considering mental, sociological and cultural phenomena from a merely biological standpoint."[38]

In architecture and urbanism it is mechanical analogies that have been worn out by overuse, but we must keep in mind that buildings, like cities, in the words of Kevin Lynch when referring to the latter, are not machines and neither are they organisms, and perhaps resemble them even less. . . . Rather than communities of non-thinking organisms undergoing inevitable phases until they reach a certain iron limit . . . cities are the product of beings capable of learning. Culture can stabilize or alter the habitat system, and it is not clear whether we wish it to be otherwise.[39]

Such a capacity to learn and such a cultural dimension of the transformation of the environment, both of which require the protagonism of human freedom, can be said to be incorporated in that version of thermodynamic architecture which made time and memory its axis, and which we previously called rehabilitative architecture to distinguish it from heliotechnical and bioclimatic architectures, expressions of the mechanical and organic paradigms.

> Mechanical Wright, organic Le Corbusier

the biotechnical unanimity

The foregoing has offered many examples of the links between organisms and mechanisms that enable us to situate the corresponding analogies in a common functional space. We can now verify the hypothesis through a parallel reading of the two architects of this century who best represent these opposed approaches: Frank Lloyd Wright and Le Corbusier.

In Chapter 1 a comparison between Wright and Le Corbusier was drawn to present the characteristics of igneous and solar architecture, using terms that referred us to the cosmological opposition that had taken shape between the world of combustion and that of trajectories. A similar opposition between identical poles is present in our discussion of the bond that links the conceptual pair organism/mechanism to the architectural pair bioclimatic/heliotechnical.

We say "a similar opposition" because the organic, an inevitable reference in bioclimatic architecture, is an evident expression of the aleatory and unstable world of combustion, while the mechanical, besides being a characteristic feature of heliotechnical architecture, is a necessary component of the obligatory, clockwork world of celestial orbits. And we say "identical poles" because both architectures admirably reflect the fire/sun and the organism/mechanism dialectic. These intertwined analogies are what enable us to consider Frank Lloyd Wright a representative example of the bioclimatic school and Le Corbusier a perfect paradigm of heliotechnical architecture.

The association of these names to the organic and mechanical views of architecture is of course a commonplace in architectural criticism, so we shall refrain from dwelling on it further. Suffice it to remember, in the words of Peter Collins, that "in the present century the biological analogy has been associated primarily with Frank Lloyd Wright," whereas "we are mainly familiar with the mechanical analogy as expressed by Le Corbusier in *Towards a New Architecture*,"[40] although both analogies, as Collins notes, have their roots in the last century.[41]

Nonetheless, the idea here is not so much to dwell on what is specifically organic in Wright or mechanical in Le Corbusier as to examine the points of contact between both perspectives which allow us to encapsulate them within a common

field. Chapter 1 ended with a mention of the double dimension, functional and symbolic, that characterizes the relationship between architecture and energy. The
examination to be undertaken must necessarily begin with a parallel verification
that the lines separating organism from mechanism are as vague and blurred in the
functional field as they are clear and sharp in the symbolic field.

In fact we could say that if the mechanism appears as a mediator between
architecture and biology, so does the organic serve as a bridge between the building and the machine. To Wright, the mechanical imitates the organic;[42] to Le Corbusier, it is the organic that must be contemplated in mechanical terms:[43] organism
and artifact intertwine and intersect, quoting each other, reflecting and explaining
one another. Wright wrote that "a chair *is* a machine to sit in ... a tree *is* a machine
to bear fruit ... they are that before they are anything else. And to violate that
mechanical requirement ... is to finish before anything of higher purpose can
happen."[44] Le Corbusier, in turn, did not hesitate to define the city as "biologie
cimentée."[45]

Words of one could easily be taken for words of the other. The fervent organicist
of Broadacre spoke of the city as a "great machine" that has been formed in "blind
obedience" to the cosmic laws of a universe that in a sense is also an "obedient
machine";[46] the propagandist of the *machine à habiter* described his Ville Radieuse
as an "organized body" supporting a "biological organization" in an eighty-three-
page text where, according to Françoise Choay, the words *vie* and *vivre* appear sixty-
five times (not counting verbal conjugations and derived adjectives!).[47]

"Any house is a far too complicated, clumsy, fussy, mechanical counterfeit of
the human body. Electric wiring for nervous system, plumbing for bowels, heating
system and fireplaces for arteries and heart, and windows for eyes, nose, and lungs
generally. The structure of the house, too, is a kind of cellular tissue stack full of
bones."[48] When Frank Lloyd Wright writes this, we hear echoes of Le Corbusier's
biological comparisons "of the physiology of breathing with the ventilation of
buildings; of the nervous system with the networks of electricity supply, communication and telephone services in a building or city, of the bowels with sewer pipes
and refuse systems; and, favourite analogy of all, the circulation of the blood with
the circulation of people or traffic."[49]

In fact such parallelisms obey the deeper connections that bring biological
and mechanical analogies together, as we have already said, in the common functionalist stream of the modern movement,[50] always underlying which, as Alan
Colquhoun and Philip Steadman have written, is "an implied belief in biotechnical
determinism."[51] A biotechnical determinism, incidentally, that is present as much
in heliotechnical architecture, whether the equinoxes and solstices that govern Le
Corbusier's *brise-soleil* or the solar charts that define the design of solar collectors
in the latest generation of autonomous houses, as it is in bioclimatic architecture,
whether the influence of site and region on Wright's desert houses[52] or the microcli-
matic detail of the passive architecture of recent years.

› Mechanical cathedrals

the functional machine and the symbolic machine

If organism and mechanism, as we have seen, interpenetrate and merge in the functional field, in the symbolic realm their respective images move away from one another and polarize into a state of permanent confrontation. It is this expressive, aesthetic, symbolic conflict that makes Wright criticize the "childish attempt to make buildings resemble steamships, flying machines, or locomotives," in what is a clear allusion to the proliferation of mechanical images in *Vers une Architecture.* He writes:

> Nor should we outrage the machine by trying to make dwelling places too complementary to machinery. . . . The machine . . . should build the building, if the building is such that the machine may build it naturally and therefore build it supremely well. But it is not necessary for that reason to build as though the building, too, were a machine – because, except in a very low sense, indeed, it is not a machine, nor at all like one. . . . Let us not forget that the simplicity of the universe is very different from the simplicity of a machine.[53]

Thus the polemic takes shape above all in the plane of images. Underlying either set of stylistic codes is a deep-rooted acceptance of industrial production processes and adherence to Taylorist methods.[54]

In the case of Le Corbusier, the matter is so clear that it will suffice to recall the hymn to Taylorism he intoned in some famous paragraphs after visiting the Ford assembly lines in Detroit:

> When the cathedrals were white, collaboration was complete. . . . In the Ford factory, everything is collaboration, unity of views, unity of purpose, a perfect convergence of the totality of gestures and ideas. With us, in building, there is nothing but contradictions, hostilities, dispersion, divergence of views, affirmation of opposed purposes, pawing the ground. . . . Let the hitherto contradictory currents line up in a single procession. . . . Let the ghosts stop blocking the road![55]

To the architect, overwhelmed as he is by the great American dream, the dilemma can be expressed clearly: "On one side barbarism, on the other – here at the Ford plant – modern times!"[56]

Wright's case, however, is more complex. His ferocious diatribes against the machine did not preclude his occasional use of the Model T to explain what his "assembled house" would be;[57] and the same architect who warned against the machine becoming "a way of life instead of being used by life as a tool" built what is surely the most eloquent monument to the mechanical way of life, the Larkin company headquarters.

Completed in 1904, two years before the publication of Frederick Winslow Taylor's chief work, *Principles of Scientific Management*, the Larkin building is indeed the physical materialization of mechanical space. The machine is present in this huge administrative container ("a cathedral of work"), not so much as artifact but as mechanized social organization. The rigid Fordian regimentation of office employees operating in a single space, the strict arrangement of work stations, and even the furniture contribute to what Mumford called a social megamachine, the mechanical organization of human labor that historically preceded the emergence of the machine as a mechanical artifact.

In this sense it could be said that the Larkin building has more merits as a product of the universe of machines than much of what we have called engineering architecture. On entering the building, the employees find themselves in a mechanical universe where their individual freedom is reduced to a minimum (despite the emphatic inscription engraved over the main entrance: "Honest labour needs no master"). They have no control over its artificially homogeneous thermal and lighting conditions; the height of the windows prevents them from having any visual contact with the outside; they may not modify the furniture arrangement, with the desks rigorously lined up in rows and the filing cabinets stuffed under the sills; there is no privacy whatsoever in the vast supervised spaces; not even the seats can be moved around as they are attached to the desks on one arm.[58]

Mechanization of work and organizational Taylorism are therefore the protagonists of this architecture: an architecture that contains a formidable social machine, but which, paradoxically or perhaps consequently, *does not express it.*

Le Corbusier deemed the industrial assembly line to be the contemporary equivalent of the building of medieval cathedrals; Wright built his work cathedral – conceived with the same reverential attitude as the contemporary Unity Temple – as a tribute to the scientific organization of administrative labor. These mechanical cathedrals are the dream shared by the two architects: the industrial factory that processes matter at the Ford plant and the administrative factory that processes information at the Larkin building[59] belong, in the final analysis, to the same material and philosophical paradigm.

Chosen here as representatives of mechanical and organic poetics, Le Corbusier and Wright are clearly one in accepting the *functional machine*, though their attitudes continue to vary when it comes to the *symbolic machine*. Similarly, heliotechnical architecture and bioclimatic architecture – which we have associated with these two architects from the very beginning – are situated in what could be called a broad functionalist position; readily assuming functionalism, both locate the machine-life polemic in a decidedly symbolic realm. Otherwise, organism and mechanism are by all means equivalent, and the architectural analogies made about them interchangeable.

› Environment and form

between tabula rasa and the memory of place
All this emphasis on the equivalence of the mechanical and organic approaches might be judged to be rather excessive. We should thus probably qualify it by recognizing, in Peter Collins's words, that "one great advantage of the biological [over the mechanical] analogy was that it laid particular emphasis on the importance of environment, since clearly all living organisms depend on environments for their existence, and constitute in themselves environments which influence other organisms nearby."[60] In our case, the notorious advantage of bioclimatic architecture over heliotechnical architecture rests on a similar reasoning.

But Collins himself qualifies this observation at another point of his text: "So far as [Darwin's] biological theory of the relationship of form to environment is concerned, the relevance of Darwinism to architecture has tended to decrease. Improvements in air-conditioning equipment[61] are making architectural form increasingly independent of climatic considerations."[62]

In any case, the environment that biological metaphors give importance to tends to be exclusively the natural environment, and only rarely that involving the built domain. In fact, the two architects we have used as paradigms significantly express themselves in very similar terms when calling for a "fresh start" (Le Corbusier) or a "radical elimination" (Wright) of all existing construction.[63]

Such a tabula rasa stance appears in both heliotechnical and bioclimatic architecture, which deal with an exclusive dialogue between the building and the outside world (whether that of trajectories or that of climate), ignoring its relation with other buildings. A permanent dialogue with the built domain only appears in that variant of thermodynamic architecture we have conveniently called the architecture of rehabilitation, with the priority it gives to existing things and its attention to memory, and where the term "environment" acquires all the rich historical, cultural, and collective connotations that are absent in mechanical and organic reductionisms.

The mention of the collective dimension here is not at all casual. On the contrary, one of the most important aspects of rehabilitative architecture is precisely the shift of emphasis from individual buildings to communities of buildings. Here we are following up on what Morin calls the superposition of a collective "macroorder" and an individual "micro-disorder,"[64] in order to approach the existing (and remembered) environment with its varied buildings, which Alberti rightly said was produced not by the diversity of uses or desires but by the diversity of people.[65]

If people are diverse and buildings heterogeneous, the reconciliation of the latter's micro-disorder with the geographical and historical macro-order in which they are inserted becomes the main task of an architecture that endeavors to rehabilitate places and memories, in quest of a climatic and technical but also social and cultural genius loci.

Given their irreducible uniqueness, an energy-oriented examination of *individual buildings* would require analytical tools of a symbolic and perceptive nature far transcending the intentions and possibilities of this text. The relationship between energy and style in the context of the search for a possible thermal aesthetic; the importance of perceiving energy and embracing temperature versus the contemporary dominion of the visual that constitutes a true "dictatorship of the eye"; the influence of energy on the shape of space, from the protagonism of climate to that of fossil fuels; the shift from thermal variety to thermal homogeneity, from the space hierarchized by the central hearth and articulated by the positioning of rooms to the space/time uniformity generated by artificial light and peripheral heating; the symbolic dialectic between the transparent architecture of glass and the opaque architecture of the fireplace, between the greenhouse and the cave, lightness and thermal inertia; energy understood as a repairing *pharmakon* in the Albertian framework of a rehabilitation theory: all these themes require extensive elaboration not to be undertaken here. Chapter 7, which serves as an epilogue, simply sketches some of the issues in the context of a quick history of thermal space in architecture.

As for the energy-oriented analysis of *communities of buildings*, it introduces questions of an economic and sociological nature that are difficult to avoid. Rehabilitative architecture that endeavors to value the existing while proposing technical and symbolical alternatives involves the conservation of whatever energetic capital – physical or informational – has accumulated through time in the built domain.

Among the questions raised by such an analysis, none is as important as that concerning energy accounting. This has played a major role since the 1970s as the key to a possible technical and social alternative by which, in the context of an ecological economy, "arbitrary" monetary calculation would give way to "objective" energy computation. In the field of construction, energy accounting gave rise to hopes for the discovery of a scientific standard that would make it possible to quantitatively compare different technical options, thereby clearing the road toward an environment-conserving architecture: one that is an enemy of waste, jealous in the preservation of inherited knowledge, careful in the use of material and energy resources; an architecture reconciled with both nature and culture.

The progress that energy accounting has made in this context cannot be overstated. The next chapter will therefore tackle the historical origins of the concept.

1   Leonardo da Vinci's manuscripts do not count here because they transcend the specific field of architectural history (although they do fall under it; we have no examples of built work by Leonardo – apart from a few models and works of uncertain authorship – but Vasari called him an "excellent architect" and his manuscripts abound with projects, sketches, and theoretical studies).

2   At least in the West, for in China the clock escapement was invented early in the eighth century by I-Hsing, a Tantric Buddhist monk who was the greatest mathematician and astronomer of his time, and Liang Ling-Tsan. Joseph Needham, *Science in Traditional China: A Comparative Perspective* (Hong Kong: Chinese University Press, 1981), p. 15.

3   Ludwig von Bertalanffy adds "molecular machines," meaning the mechanical structures that operate at a molecular level (*General System Theory: Foundations, Development, Applications* [New York: Braziller, 1968], p. 140).

4   Prigogine and Stengers point out that each of these images, taken from the technology of its time, contradicts the idea of an immanent organizing intelligence. Ilya Prigogine and Isabelle Stengers, *La nouvelle alliance: Métamorphose de la science* (Paris: Gallimard, 1979), p. 171.

5   Manuscript A, 10r, quoted in Martin Kemp, *Leonardo do Vinci: The Marvellous Works of Nature and Man* (London: Dent, 1981), p. 119.

6   Leonardo Benevolo, *The Architecture of the Renaissance* (Boulder, CO: Westview Press, 1978), p. 242.

7   Codex Atlanticus, 161ra, quoted in Kemp, *Leonardo do Vinci*, p. 122. The flight of birds is described *in extenso* in the Turin manuscript *Codice sul volo degli uccelli e varie altre materie*.

8   Interspersed, to be sure, with bits and pieces of premechanistic thought. For example, William Harvey's *De motu cordis* of 1628 mixes the evident mechanism of the description of the heart as a pump with an amalgam of cosmological ideas derived from hermeticism, Neoplatonism, and natural magic. And Gómez Pereira bases his mechanical study of animal behavior on the natural philosophy of *calculatores* and medieval medicine.

9   There seems to be a parallel tradition, from Galileo to D'Arcy Thompson, that also uses physical and mechanical knowledge for the study of living things, but does not apply to them the analogy of the machine.

10   Automata were being constructed long before the Renaissance, to be sure. Suffice it to remember the important Alexandrine tradition and its Islamic continuation, or medieval carillons with animal figures – Villard de Honnecourt himself drew a mechanical eagle whose movements were to accompany the reading of the Scriptures in churches. Nevertheless it was in the sixteenth century that automata became popular, and too few samples dated before that have come down to our days. Moreover, only in the Renaissance did they begin to be describable as "metaphors of the organic." Previously, when not mere entertainment objects, automata had a magical or religious dimension, and sometimes, like many of the mechanisms described by Hero in his *Pneumatica*, they were made for the sole purpose of serving the "scientific production of miracles" that Farrington so vehemently denounced. Even in mannerism, as Paolo Portoghesi states in *Infanza delle macchine* (Bari: Laterza, 1981) and Marcello Fagiolo develops in *Natura e artificio* (Rome: Officina Edizioni, 1979), automata or those "blasphemous variants of the human" represented the wonder and enigma of movement, the mystery of artificial life. D'Alembert and Diderot's enthusiastic response to the androids of Vaucanson was still a long way off.

11   De Caus, incidentally, was also a pioneer of environmental technology. Early in the seventeenth century, this French engineer built one of the oldest *orangeries* we know of for the Elector Palatine in Heidelberg (see John Hix, *The Glass House* [Cambridge, MA: MIT Press, 1974], p. 10), in the same gardens where he had installed his famous and much-copied grottoes with moving figures, true automatic theaters along the lines of those Hero of Alexandria describes in his *Automatopoietike*.

12   Which, to be sure, had certain organic echoes, proof of which is the maker's own description of it as "a machine that dances." Seventy-five years after its construction, in 1645, a new show was presented in Madrid by the name of *El Mago* (The Magician) in which the dancers imitated the movements of Juanelo's contrivance.

13   The philosopher himself was rumored to have constructed a mechanical woman, which legend called Francine. So named was an illegitimate daughter of Descartes, and the myth has as much truth to it as the golden servants of Hephaestus, the mechanical cow of Daedalus, or the androids that Roger Bacon and Albertus Magnus are said to have fabricated (not to mention Juanelo's "stick man").

14   Giedion seems to share this opinion ("it is Vaucanson's practical activities that are historically the most interesting") when he puts emphasis on the transition "from the miraculous to the utilitarian." Nevertheless he goes on to stress the admiration Vaucanson's automata drew among the likes of Condorcet, Diderot, or D'Alembert, who described the famous duck in the *Encyclopedie*. See Siegfried Giedion, *Mechanization Takes Command: A Contribution to Anonymous History* (New York: Oxford University Press, 1948), pp. 34–6.

15   Lewis Mumford, *Technics and Civilization* (1934; New York: Harcourt, Brace and World, 1963), p. 14.

16   Prigogine and Stengers, *La nouvelle alliance*, p. 28.

17   Ibid., pp. 126–9.

18   Rudolf Arnheim, *Entropy and Art: An Essay on Disorder and Order* (Berkeley, CA: University of California Press, 1971), p. 44. Arnheim also quotes David Riesman: "It seems clear that Freud, when he looked at love or work, understood man's physical and psychic behavior in the light of the physics of entropy and the economics of scarcity" (Riesman, *Individualism Reconsidered* [Glencoe, IL: Free Press, 1954], p. 325).

19   Quoted in Steve J. Heims, *John von Neumann and Norbert Wiener: From Mathematics to the Technologies of Life and Death* (Cambridge, MA: MIT Press, 1980), p. 155.

20   Edgar Morin, *La methode*, vol: 1: *La nature de la nature* (Paris: Seuil, 1977), pp. 229–30.

21   Heims, *John von Neumann and Norbert Wiener*, p. 304

22   We must mention the engineer Leonardo Torres Quevedo, who deserves to be called the precursor of cybernetics, as much for his theoretical works – including *Ensayos sobre*

automática (1914) – as for his practical constructions, such as the Telekino, the Electro-mechanic Arithmometer, and the famous Automatic Chess Players, all of which were put together during the first two decades of the century. The most complete descriptions are provided by Jose García Santesmases in *Obra e inventos de Torres Quevedo* (Madrid: Instituto de Espana, 1980).

23   Morin, *La nature de la nature*, pp. 165–6.

24   Ludwig von Bertalanffy, *General System Theory: Foundations, Development, Applications* (New York: Braziller, 1968), pp. 161, 259.

25   Alfred North Whitehead, *Science and the Modern World* (New York: Free Press, 1967), p. 54.

26   Quoted in Sigvard Strandh, *Machines* (London: Mitchell Beazley, 1979), pp. 3, 54. It is this same idea that Samuel Butler develops in his satire *Erewhon* (London, 1872). Jorge Luis Borges and Adolfo Bioy Casares summed it up cheerfully in a dense paragraph of their *Crónicas de Bustos Domecq* that located in Butler the roots of "functionalism" (a term now rather discredited, they nevertheless warn us, in the small world of architects). Such a conception of machines presents striking similarities to Lotka's "exosomatic instruments," as Philip Steadman shows. Steadman, too, believes that *Erewhon* contains antecedents of some key notions of functionalism, including the Lamarckian evolution of the artifacts that underlay the *objet-type* of Ozenfant and Le Corbusier – whose purist magazine *L'Esprit Nouveau*, by the way, mentioned Butler in a favorable light. See Steadman, *The Evolution of Designs: Biological Analogy in Architecture and the Applied Arts* (Cambridge: Cambridge University Press, 1979), chapter 8, pp. 124–36.

27   Mumford, *Technics and Civilization*, p. 41. The search for the military origins of machines also led Mumford to reckon the cannon as the first steam engine.

28   Born in Catalonia in 1698, Bernard Forest de Belidor was a typical product of the military schools founded in France during the early decades of the eighteenth century. In 1729 he published *La science des ingenieurs*, a widely disseminated treatise that was considered an exemplary work for more than a hundred years. Both this book and *L'architecture hydraulique* (1737–53) were republished by Navier, with updated notes, as late as 1813 and 1819, respectively – a testimony to his popularity and continued validity. See Edoardo Benvenuto, *La scienza delle costruzioni* (Florence: Sansoni, 1981), pp. 274, 418.

29   Rev. R. Balgarnie, *Sir Titus Salt, Baronett: His Life and Its Lessons* (London: Hodder and Stoughton, 1878), quoted by Ornella Selvafolta in "Lo spazio del lavoro 1750–1910," in *La macchina arrugginita* (Milan: Feltrinelli, 1982), p. 54. Ludwig Boltzmann himself writes in 1900:

> We cannot shake off the idea that nature is something animate. Don't today's machines work like conscious beings? They puff, pant, howl, groan, they emit sounds of complaint, fear, warning, and they whistle shrilly when the force applied on them increases. To maintain their strength they take from their surroundings the necessary materials, and eliminate what is not necessary, all the while going by the same laws that our own bodies do.

Boltzmann, *Escritos de mecánica y termodinámica* (Madrid: Alianza, 1986), p. 192.

30   Samuel Butler, *Erewhon* (1872; Newark, DE: University of Delaware Press, 1981), p. 190.

31   Morin, *La nature de la nature*, pp. 160, 165, 161.

32   Mumford, *Technics and Civilization*, pp. 4, 10.

33   This includes not only the mechanization of man or the conception of organisms as machines, but also the very penetration of mechanisms into the organic world, as attested in part III ("Mechanization Meets the Organic") of Giedion's book *Mechanization Takes Command*.

34   Legends that have to be linked to old traditions of building androids, from classical mythology to the most popular contemporary version of the theme of *Frankenstein*, the famous Gothic novel of Mary Shelley (where the leading character, by the way, the doctor who

creates the humanoid monster, admits to having searched for the secret of life in Albertus Magnus and Paracelsus, two figures also said to have built artificial men).

35   Heims, *John von Neumann and Norbert Wiener*, pp. 374–5. An interesting interpretation of the golem myth can be found in André Robinet, *Le défi cybernétique* (Paris: Gallimard, 1973). This work also explores the automaton theme in Pascal, Descartes, Malebranche, and Leibniz.

36   Heims, *John von Neumann and Norbert Wiener*, p. 154.

37   Prigogine and Stengers, *La nouvelle alliance*, p. 57.

38   Von Bertalanffy, *General System Theory*, p. 88.

39   Kevin Lynch, *A Theory of Good City Form* (Cambridge, MA: MIT Press, 1981), pp. 95, 97. The importance of the cultural dimension in artifacts – including cities and buildings – was expressed very clearly by Baudrillard in *The System of Objects:* "Our practical objects . . . are continuously fleeing from technical structurality toward secondary meanings, from the technological system to a cultural system."

40   Peter Collins, *Changing Ideals of Modern Architecture*, 2nd ed. (Montreal: McGill-Queen's University Press, 1998), pp. 156, 159.

41   In fact, in Collins's opinion they were the most significant analogies of that period: "Of the various analogies used in the last century to clarify the principles of a new architecture, probably the only one to equal in importance the biological analogy has been the analogy between buildings and machines" (ibid., p. 159). We know that as early as 1914 Geoffrey Scott devoted two chapters of *The Architecture of Humanism* to criticizing the "mechanical fallacy" and the "biological fallacy," thereby testifying to the popularity of both analogies.

42   "This thing we call the machine, contrary to the principle of organic growth, *but imitating it.*" Frank Lloyd Wright, *The Future of Architecture* (1953; New York: Mentor, 1963), p. 90; my italics. The quote is taken from the first of the Princeton lectures of 1930.

43   As when he speaks, for instance, of "la ville vivante, totale, fonctionnante avec ses organs *qui son ceux de la société machiniste.*" Le Corbusier, *La ville radieuse* (Paris: Vincent Fréal, 1933), p. 140; my italics.

44   Wright, *The Future of Architecture*, p. 159. The quote is from the fourth of the Princeton lectures.

45   Le Corbusier, *La ville radieuse*, p. 111.

46   Wright, *The Future of Architecture*, p. 92. From the first Princeton lecture.

47   Le Corbusier, *La ville radieuse*, pp. 134, 139; Françoise Choay, *La règle et le modèle* (Paris: Seuil, 1980), p. 295.

48   Wright, *The Future of Architecture*, p. 143. From the fourth Princeton lecture.

49   Le Corbusier's biological analogies are expressed here in the words of Philip Steadman, *The Evolution of Designs: Biological Analogy in Architecture and the Applied Arts* (Cambridge: Cambridge University Press, 1979), p. 48.

50   Collins in fact puts them together under the heading "functionalism," along with analogies he names "gastronomic" and "linguistic." Also see Steadman, *The Evolution of Designs*, p. 16.

51   Alan Colquhoun, "Typology and Design Method," *Perspecta* 12 (1969), p. 72, quoted in Steadman, *The Evolution of Designs*, p. 1.

52   "The site determined the features and character of Taliesin. . . . Taliesin is now a stone house and it is a house of the North – really built for the North. . . . Taliesin was built *to belong to the region*" (Wright's emphasis). Then, moving from Wisconsin to Arizona, "the terrain now changed absolutely. Here we came to the absolute desert. . . . Taliesin West had to be absolutely according to the desert. So Taliesin there is according to its site again, according to its environment." Wright, *The Future of Architecture*, pp. 19, 21. Indeed, on numerous occasions Wright insisted that "climate means something to man"; see Frank Lloyd Wright, *The Natural House* (New York: Horizon Press, 1954), p. 178.

53   Wright, *The Future of Architecture*, pp. 144–5, 160. From the fourth Princeton lecture.

54   An adherence that is often more rhetorical than pragmatic. The "balloon frame" or the Levitt
     homes, for example, are better adapted to industrial production than Le Corbusier's Dom-ino
     House. As for Wright, it will suffice to remember the construction fiasco in the dwellings of
     concrete blocks that he called Usonian Automatic, built between 1921 and 1924. Seeking
     "the elimination of specialized work" (Wright in his *Natural House*), he delegated all execu-
     tion of building services to the factory. This proved so inefficient that the project has been
     compared with Wright's "loathsome furniture," where "efforts reap better results as a *plastic
     idea* than as a solution to a *pragmatic fact*" (James Tice in *Architectural Design*, 8–9/1981,
     p. 62).

55   Le Corbusier, *When the Cathedrals Were White*, trans. Francis E. Hyslop, Jr. (New York: Reynal
     and Hitchcock, 1947), pp. 167–70.

56   Ibid., p. 167. With those same two words for a title (*Modern Times*), a year after Le Corbusier
     visited America in 1935, Charlie Chaplin launched the most scathing criticism of Taylorism on
     the big screen, culminating in a cinematographic reflection on the mechanization of life that
     was already present in the caustic Keaton of *The Electric House* (1922), but which had its ear-
     liest philosophical manifestation in Paul Wegener's *The Golem* (1920), antecedent of numer-
     ous Frankensteins, and its crowning expression in Fritz Lang's classic *Metropolis* (1927).
     Chaplin's irony in *Modern Times* and William Cameron Menzies's disturbing images in *The
     Future Life* (also previewed in 1936) were ignored by the enlightened, optimistic European
     architect then traveling through the United States, whose prophetic redemptorism would be
     better represented by the phrase "The machine is saving us, long live the machine!" of Sergei
     Eisenstein's *The General Line* (1929) than by the cultural mood of mechanism-and-Ford
     America.

57   See John Sergeant, *Frank Lloyd Wright, Usonian Houses: The Case for Organic Architecture*
     (New York: Watson-Guptill, 1976), p. 146.

58   Francis Duffy has drawn attention to some of these conditions in "Office Buildings and
     Organizational Change," in Anthony D. King, ed., *Buildings and Society* (London: Routledge
     and Kegan Paul, 1980), pp. 266–9, whereas Kenneth Frampton, a critic normally concerned
     with the relation between architecture, work, and production, overlooks them completely
     when describing the Larkin building in *Modern Architecture: A Critical History* (London:
     Thames and Hudson, 1980), pp. 61–2.

59   Wright deplored the clients' ordering changes in the building that made it just "another of
     their factories," but this only brings to light the rhetorical component of his acceptance of
     industrial production processes (see note 54).

60   Collins, *Changing Ideals in Modern Architecture*, p. 166. Philip Steadman has explored in
     detail what he calls "ecological analogy," in reference to the way the environment has a
     bearing on the shape of artifacts and organisms through their functions. This is crystallized in
     Louis Sullivan's famous "form follows function." In Steadman's opinion, there is a thread
     that connects Cuvier's comparative anatomy and Darwin's theory of evolution to Greenough,
     Viollet-le-Duc, and Semper, and on to Sullivan and Wright (*The Evolution of Designs*, pp.
     57ff.). Joseph Rykwert goes farther back in time to suggest that the idea reaches Greenough,
     via Milizia, from the Lodoli of Algarotti, who preached the need to unite *rappresentazione*
     and *funzione* ("Lodoli on Function and Representation," in Rykwert, *The Necessity of Artifice*
     [London: Academy Editions, 1982], pp. 114–21). I myself would say that in the notorious func-
     tionalist slogan one even perceives echoes of the theory of signatures that Paracelsus took
     from Pliny, which invested plants with the curative properties that their very shapes sug-
     gested, so that one resembling a heart would be a cardiac tonic, one that suggested sexual
     organs an antidote to sterility, etc. There is still much research to be done on the mythical
     and archaic origins of a good portion of contemporary architectural thought, the fruits of
     which may prove far more important than we currently believe.

61   A milestone among these advances is, significantly and paradoxically, the Larkin building.
     Wright was never a staunch supporter of the system:

> To me air conditioning is a dangerous circumstance. . . . [It] has to be done with a good
> deal of intelligent care. . . . I think it far better to go *with* the natural climate than try to fix a
> special artificial climate of your own. . . . I doubt that you can ignore climate completely, by
> reversal make a climate of your own and get away with it without harm to yourself.

Wright, *The Natural House*, pp. 175–8.

62   Collins, *Changing Ideals in Modern Architecture*, p. 154.

63   On this point they are in agreement with one of the most illustrious fathers of functionalism,
the architectural theorist and reformer Carlo Lodoli, a Franciscan friar who lived in eighteenth-century Venice and is said to have coined the terms "organic" and "functional." The
portrait Alessandro Longhi did of him was accompanied by two panels with inscriptions that
summed up the philosopher's thought. A quotation from the book of Jeremiah – Ut *eruas et
destruas* . . . – expressed that a building is preceded by the destruction of everything preex-
isting (Rykwert, *The Necessity of Artifice*, p. 116).

64   Morin, *La nature de la nature*, p. 74.

65   Leon Battista Alberti, *De re aedificatoria*, Book IV; Chapter I.

1985

Steve Ternoey

The Patterns of
Innovation and
Change

Steve Ternoey (born 1949) is an American architect and daylighting consultant, who was working for the Solar Energy Research Institute (SERI) in the 1980s when architects and government agencies were responding to the energy supply crises of the late 1970s. This excerpt was drawn from a manual on The Design of Energy Responsive Commercial Buildings, which explained that large office buildings were "internal-load dominated" rather than "climate-dominated" and so required a different logic of environmental design.

In itself the manual is an unexceptional, if useful, technical guide for professional designers, except for the brief section on innovation and change authored by Ternoey that cites the work on the diffusion of technology by Shoemaker and Rogers. This is a now well-established field largely initiated by Everett Rogers with his 1962 book *Diffusion of Innovations*, now in its fifth edition (2003), which began with his effort to understand why some farmers adopted new farming technologies and others did not. He standardized the descriptions used by other researchers, producing the now well-known terms used to name the rate at which different individuals begin using a specific technique or device: innovators (2.5 percent), early adopters (13.5 percent), early majority (34 percent), late majority (34 percent) and laggards (16 percent). Each adopter's willingness and ability to adopt an innovation depends on their awareness, interest, evaluation, trial, and adoption. And different people can fall into different categories for different innovations – a farmer might be an early adopter of hybrid corn, but a late majority adopter of the iPod. Rogers demonstrated how these innovations would spread through society in the equally well-known "S curve" of adoption. The language and concepts quickly crossed from analytical description to techniques for developing and marketing products, which is the spirit in which Ternoey and SERI used it to assess the state of energy-responsive construction.

Understanding the mechanisms by which technological innovations spread clearly marks the transition from technology as a tool, to technology as system of users and techniques.

Understanding how major new design challenges typically proceed from initial ideas and concepts to resolved solutions can provide important insights into the current state of the art in the design of energy-responsive commercial buildings. In addition, this process leads to a clearer picture of the challenge that remains. In this section, a model is presented that traces new ideas or products from their early emergence through resolution.

The model presented here is based on diffusion research, a field of study that explores how social systems are changed through the diffusion of new ideas. Diffusion research traces the flow and change of new ideas from their originator to potential users.[1]

Innovations (i.e., some new idea, process, or technology) follow an evolutionary sequence of events between the origin of a general concept to the adoption of sets of users or behaviors by the majority. Diffusion research indicates that innovations are seldom directly conveyed to the majority by the original Innovator. Rather, innovations are invented or initiated by one group or type of people, the Innovators, and modified, reinvented, or resolved by another group of people, the Adopters. Innovations follow a process toward wide acceptance and use and are not a single act performed by one person. For instance, Sullivan developed a paradigm for the skyscraper, but did not invent the elevator or the steel frame.

One way to define the start and end of the diffusion process is by assessing the changing identity of the innovation as it migrates through and beyond specific groups of people. At the start of the process, innovations have a distinct, separate identity. An Innovator typically focuses much energy on a single new or unique element that is often an incremental improvement to an existing problem or need. Later, Adopters of innovations are typically concerned with broader issues than is the Innovator. In one form or another, innovations are modified or changed to make them appropriate for a larger and often different set of concerns. Final resolution of a new innovation is reached when it loses its separate identity and is absorbed into much larger everyday concerns and procedures.

Reinvention is the term used to define the act of changing an innovation by an Adopter in the process of its use and implementation. Reinvention is necessary to make an innovation more appropriate to a larger set of concerns or objectives. Reinvention is both general acceptance of an innovation and a rejection of some of its elements. Reinvention is associated with implementation, a point in time after the origination of a specific innovation.

Innovators and Adopters can be distinguished by their personal motives and the degree of risk they are willing to take. The changing level of risk associated with an innovation is the cue that signals or invites the participation of people beyond the originator of the idea.

Innovations always begin as high-risk ventures. Without a proven track record, early innovations are developed by people who like or are motivated to take that risk. Innovators are motivated by a need to be first and intentionally seek uncertainty and change. Innovators are the first to adopt new ideas in their communities and professions. However, they tend to be innovative in only one focused facet of their lives. Innovators' focus on change often isolates their interactions to those of national or international peer groups, since the local community may not reflect an equal desire for change.

Since innovations are a speculative venture, many experimental mistakes are made compared to the final number of successes that emerge. Once a limited number of successes is achieved, the level of risk is reduced and a new sequence of activities begins. A new set of people, who desire constructive change but are less venturesome than the high-risk-taking Innovators, begin to adopt and change innovations to make them appropriate for their needs and uses. Diffusion research calls the first set of such people Early Adopters.

Early Adopters are prestigious, respectable leaders of their business and community. To maintain this position of respectability, Early Adopters are willing to take some risk to explore the new and useful; yet, accountability and success are important too, and the degree of risk that is acceptable is much less than that of the Innovators. Early Adopters usually track the activities of Innovators but pick and choose only a limited number of innovations that appear useful to the wider concerns of the Early Adopter and offer an appropriate degree of risk.

The value of the Early Adopters is threefold. First, they filter and approve the work of Innovators for the rest of the community and/or profession. Second, they experiment with the new ideas or products and arrive at generalized principles that increase the probability of success. Third, they integrate the new information with a larger set of concerns that may not be important to Innovators. All three of these activities result in the reinvention of the original idea, product, or concept.

Diffusion theory offers a good model to appraise the present state of the art in the design of energy-responsive commercial buildings. At present, a rapidly expanding information base is being formulated that reflects the abilities and benefits of individual energy-related components, systems, and concepts. This research has produced both successes and failures and has been generated at a very high level of risk. Component-specific patterns of success are emerging, and this large collective effort at the component level is reducing the risk associated with energy-related options. At present, our research base is topical or technology-specific, documented under titles such as energy analysis, active solar, passive solar, daylighting, and improved conventional systems. Results that focus on the pieces rather than on the overall final product are characteristic of the work of Innovators. Yet, to reach resolution we need to know more than how the pieces work. What remains to be done is to define the appropriate role and impact of energy-related design issues in the context of the overall intended result at the whole-building level. In the language of diffusion research, the design of energy-responsive commercial buildings is at the Innovator/Early Adopter transition point. The state of the art is reinvention. What we need now are models of success at the whole-building level, an area of concern better understood by Early Adopters.

This book is specifically written to inform the Early Adopters of the benefits and liabilities of existing energy-related alternatives on a whole-building level. The goal is to translate existing energy-related design information into a format that supports the Early Adopters' role of integrating and reinventing innovations to respond to wider concerns and needs. To achieve this goal, two levels of information are presented. In Part One, Reviewing and Interpreting Our Collective Learning Experience, many case studies and examples are presented to summarize the abilities and impact on design of existing alternative means of environmental control and the methods that are being used to assess this problem. In this part of the book, the intent is to present the principles, advantages, and disadvantages of alternative solution types and design approaches as well as to comparatively assess them. In effect, Part One represents a source book or seeds for reinvention.

Part Two, A Framework for Design, presents a detailed, non-technology-specific approach to the design of energy-responsive commercial buildings. Based on the lessons learned from the most successful examples of Part One, this framework provides a logical basis for considering energy issues in the design process and is formulated to encourage innovation and reinvention by the reader. A major premise of this book and the framework presented in Part Two is that through reinvention major energy-related innovations beyond our present collective knowledge and experience are not only possible but probable. These new innovations will be the product of the work of Early Adopters.

1   The material presented here has been adopted from personal conversations with Dr. Floyd Shoemaker, and from *Communication of Innovations/A Cross-Cultural Approach* by Rogers and Shoemaker, *Reinvention in the Innovation Process* by Rice and Rogers, and *The Diffusion of Innovations: An Assessment* by Radnor, Feller, and Rogers.

1987

Martin Pawley

Technology
Transfer

Martin Pawley (born 1938) is an English architectural writer, critic, and broadcaster. He studied architecture at the Oxford School of Architecture (now Oxford Brookes University), at the Ecole Normale Supérieure des Beaux Arts in Paris and at the Architectural Association, London. After spending several years in practice he taught architecture at the AA, at Cornell University, Rensselaer Polytechnic Institute, Florida A&M University and UCLA. He served as architecture critic of The *Guardian* newspaper from 1984 to 1991, and The *Observer* newspaper from 1992 to 1995. In the early 1990s he was a regular contributor to the BBC television arts program *The Late Show*. He is currently a consultant editor to *World Architecture* magazine, which he edited from 1992 to 1996.

Pawley is perhaps best known for his outspoken views on the allegedly stifling influence of the "art-historical" agenda in late twentieth-century British architecture. He advocates the abandonment of apparently obsolete historical precedents in favour of an enthusiastic and unselfconscious celebration of the possibilities of new materials and techniques. In the text reproduced here he offers a survey of recent attempts to deploy cutting-edge technologies from fields such as space exploration, aeronautics and shipbuilding as a means to finally make good on the early modernists' promise of creating a truly "machine age" architecture. As he has written in a more recent work: "The act of building can be better understood, and valued, as the provision of 'terminals' for the systems and networks that sustain modern life, rather than as the creation of cultural monuments."[1] His central thesis contains strong echoes of Le Corbusier's "engineer's aesthetic" (see Le Corbusier, 1923), as well as the more direct influence of Archigram and Reyner Banham (see Cook, 1970 and Banham, 1960).

1   Martin Pawley, *Terminal Architecture* (London: Reaktion Books, 1998), p. 7.

Whenever the principles of architecture become unclear, the rudder of history is moved until they can be understood again. In periods of uncertainty these movements often describe a circle, as they are doing today, until the past once again falls into place. That is what happened when the theorists of the ancient world convinced themselves that Classical architecture was the progressive refinement of prehistoric construction; when the theorists of the Renaissance claimed the rediscovery of the Classical past; and when the Gothic revivalists of the nineteenth century claimed the patrimony of the Dark Ages. More recently, in a gyration apparently so drastic as to have no remembered precedent, it happened when the Modernists of the twentieth century claimed – like mutineers – that science, technology, and socialism had entirely changed the cosmos so that the whole of architectural history could be compressed into a single category called the past, and cast adrift in an open boat.[1]

› The great mutiny and its consequences

The architects of the generation of 1914, the monocled mutineers who lived through the invention of the automobile and aeroplane, were the first to embrace science and technology as a substitute for their accumulated cultural legacy, bringing these matters into the mainstream of architectural thought for the first time. They took this step as artists, licensed to find inspiration where they chose, but they soon found that immersion in science and technology threatened their old identity. Before they died, the mutineers came to realise that their art had been summoned by the machine, and not the other way around. As the twentieth century progressed advances in materials engineering, environmental controls and information technology meant that buildings served up as homage to the industry of 1914 were soon as hopelessly obsolete as their Victorian predecessors. Alison and Peter Smithson might naively write in the report that accompanied their 1951 Coventry Cathedral competition design: 'Modern Architecture has at its disposal means of expression which would have sent Brunelleschi wild with joy' – but more insightfully Maxwell Fry had written seven years earlier: 'If the developments that had led to our present technical skill were to continue at the same pace into this century, at a pace that is exceeding our capacity as artists to assimilate them, then our hopes of establishing a workable architecture would be slight.'[2]

Seizing upon the means of expression that would have sent Brunelleschi wild, and yet at the same time assimilating them as artists, proved to be impossible. The logic of their position urged the mutineer architects to make another quantum forward leap, and then another, and another, until a breathless race to keep up with the materials and methods of science and industry became the identity of architecture itself. But chiefly because they chose to remain a collection of individual artists instead of becoming an industry, the architects of the generation of 1914 never did initiate an architecture of continuous technological revolution.[3] Instead the mutineers fell out, and in what can now be seen as something like the restoration of a monarchy, a large part of their number reverted to the concept of a building technology chained to the limitations of artistic assimilation.

As a result we live today in an age of Restoration architecture, a period popu-
lated by frightened practitioners who, in Charles Jencks' felicitous phrase, know
just how far too far they can go; and theorists who believe that their task is to heal
the gigantic breach caused by the Modern mutiny. Where once the break with tradition was seen as thrilling and final, now creeping tendrils of sentiment are encouraged to grow over it, concealing it from view like a crack in a wall. Long-lived
practitioners, veterans of the exciting days of the mutiny, now face career prospects
like those of French army officers after the defeat of Napoleon. Venerable surviving
Modernists are urged, as by priests at their deathbed, to give their blessing to the
Restoration – the triumph of the voyage in the open boat.[4] Who can blame them
when they consent? 'No memory of having starred atones for later disregard' wrote
Robert Frost. In return for denying their golden dawn they receive a moment of brief
media attention and the fickle adulation of young architects. And if they refuse?
Edmund Burke truly wrote of those who find themselves at odds with the fashion of
the times in which they live: 'They seem deserted by mankind, overpowered by a
conspiracy of their whole species.'

› Problems of restoration life and times

The great weakness of Restoration architecture is its lack of ideology. It has no unifying theory – 'a supposition explaining something, based on principles independent of the phenomenon to be explained' – as the *Concise Oxford Dictionary* put it.
This is despite a veritable explosion of writing about architecture that has taken
place since the collapse of consensus support for modern design some fifteen
years ago, much of it glorying in the present state of wild opportunism.

A few short years of creeping incorporation and stylistic anarchy has been
enough to sink the once clinically lucid language of modern architecture to the
level of banality of the fashion page.[5] Fuelled by the unromantic threat of insurance
claims; incorporation with shareholders control; the growth of circumscribed
'design consultancy' work; the consumerisation of minor works, and the migration
of so much architectural terminology that the word architecture may be found
under 'computer' in the dictionaries of the twenty-first century, a terminal demystification of the profession seems entirely possible. Perhaps the darkest portent of
all is the fact that it is now widely believed that there is no longer any need for
expert judgement where the design of buildings is concerned. 'I know that what I
feel in spirit about a building is just as valid a criticism as any professional or technical point of view', the Prince of Wales wrote to Peter Palumbo at the height of the
battle for Mansion House Square. And in this as in so many other matters there is
no reason to suppose that his opinion differs greatly from those of his future subjects.

Restoration architecture combines a superficial glorification of variety and
ornament with a concealed convergence of identity between buildings that can be
compared to the process of homogenisation that began in the motor industry
twenty years ago. With or without regard to the pace at which 'artists' can assimilate it, global product distribution is overwhelming the construction industry, and

with it the architectural profession. Today, just as the removal of the badge from the nose or tail of a car can reveal its shared parentage with a different make, so can the peeling away of a decorated facade reveal the homogenisation of serviced floorspace beneath the skin.

Restoration architects have conceded creative hegemony everywhere except in this 'badge engineering' of buildings, the so-called 'signature building' of American architecture. Carbon-copy engineering – in terms of the names of the consultants responsible as well as the structural and environmental control systems used – is now accepted as the norm. From penetrating deep into the genesis of the building, as it did during the Modern mutiny, the power of the architect over construction has shrunk to the literally superficial: a thin skin on the front of a new building, like the badge on the nose of a car; a small feature on the outside of a refurbished building; a bureaucratic role in the filing of applications and the authorisation of payments. An architect's 'capacity as an artist', still offers him this role, but today only inertia saves him from the modified cry of the small boy: 'The Emperor is as expendable as a light bulb.'

Compared to the great days of the mutiny, when heads of state appealed to architects to re-plan capital cities, design satellite towns and solve the global housing problem, the role of the architect is tragically diminished. In engineering terms he is hardly a designer at all, his work oscillating uneasily between envelopment by a burgeoning design profession and surrender to the reactionary forces of conservation and historicism. For him there is no future apart from buttondown slavery as a corporate executive, or the thankless task of acting unpaid adviser to community enterprise.

No future, that is, unless something that research scientists call a reordering of the data takes place. For in architecture as well as politics the quickest and most effective way of overcoming humiliation and loss of power is a revised perception of the events that brought it about.

› Towards an architecture of technology transfer

Technology transfer is a term that is used in different ways, but a generally agreed definition might be any process whereby the techniques and materials developed in one field or industry are applied to other fields and industries. A process with a vast unwritten history, technology transfer either results from serendipitous curiosity on the part of individuals, or from a serious marketing effort by corporations intent on developing new outlets for materials or techniques. Modern examples of the second category in building include the use of insulation material as roofing, various spin-offs from aerospace research – like the Teflon coatings and flat wiring now used in a vast range of product applications – and the use of motor industry-developed cold rolled steel structural members for lightweight construction. Perhaps the neatest illustration of the first category comes from one tiny but crucial component in the elaborate NASA unmanned Mars landing programme, where the problem of designing a simple lightweight soil-sampling scoop was brilliantly solved by the adaptation of a coiled steel carpenter's rule, whose dished, semi-rigid extending arm provided the model for the light, retractable scoop that was eventually used.[6]

Few architectural historians have concerned themselves with the role of technology transfer in architectural design, even though its implications can be of the first importance. In fact the only critical assessment of the phenomenon in recent years occurs in Reyner Banham's *Theory and Design in the First Machine Age*,[7] which was first published in the heyday of the Modern mutiny in 1961. While Banham himself takes the view that architecture and technology may have different evolutionary patterns, so that he stands aside from the suggestion that the collapse of Modernism resulted from its failure to keep pace with technology, he alone among historians writing at the time foresaw that collapse. In *Theory and Design* he drew attention to the already worrying obsolescence of the 'new technologies' annexed by Modern architecture from nineteenth-century engineering, and identified this area as the one in which its greatest weakness lay. In the final chapter of *Theory and Design*, he quotes from Richard Buckminster Fuller's 1938 book *Nine Chains to the Moon*, to show that the failure of Modern architects to grasp the *endlessness* of technological evolution had sowed the seeds of their decline as early as 1927, when Fuller's revolutionary light metal, air-deliverable Dymaxion House adumbrated the frame-hung component structures that were to dominate most other fields of engineering design within twenty years. 'The International Style brought to America by the Bauhaus', wrote Fuller in 1938, 'demonstrated fashion-innoculation without the necessary knowledge of the scientific fundamentals of structural mechanics and chemistry.'[8] Or as Banham interpretatively puts it, Modern architecture 'produced machine-age architecture only in the sense that its monuments were built in a machine age, and expressed an attitude to machinery – in the sense that one might stand on French soil and discuss French politics, and still be speaking English'.

As we know from developments in related fields, the next step in advanced construction technology after glass, steel, and concrete should have been light-frame and monocoque enclosures using laminated wood, aluminium alloys, and plastics developed during the Second World War ('Enter alloy – exit rust' as Fuller put it in 1944[9]). But whether a handful of *avant-garde* architects could have dragged the construction industry into a pattern of continuous technological evolution at that time, even with the help of the massive development of light engineering that the War brought about, must remain an open question. Light-frame and monocoque enclosures flowered briefly in the post-war emergency schools and housing programmes, but in the fifty years from Fuller's Dymaxion House to the end of the collapse of the Modern Movement, only a small number of architects published or carried out work based on this method.

Mindful of Fry's wartime dictum, it is tempting to say that failure to keep up with science was the price Modern architecture paid for artistic integrity. Banham is more cautious in suggesting that 'What we have hitherto understood as architecture, and what we are beginning to understand of technology may be incompatible disciplines.' But either way the fact remains that one generation – however much it may have misunderstood what it was doing – seized the initiative in technology transfer, and the next let it slip away. For the generation of Le Corbusier, Walter

Gropius, Mies van der Rohe, and Richard Neutra, steel, glass, and concrete were revolutionary new materials that cried out to be used in buildings as different from their brick, stone, and timber predecessors as a motor car was different from a horse-drawn wagon. With varying degrees of single-mindedness they spent their lives developing new ways to build using these same materials. But when it transpired that steel, glass, and concrete were merely the forerunners of high-strength alloys and composites grown from a science and technology leaping daily further ahead, the ingenuity of their followers was overwhelmed. Tragically it was assumed by the politicians who elevated Modern architecture to global supremacy in the thirty years after 1945 that architects held technological mastery in their hands like an Olympic torch that could be passed on from generation to generation. Seldom can faith in expertise have been more naively placed. Not only did the generation of 1914 misunderstand the process of technology transfer, as Banham suggests, but the majority of them did not even think it was a matter of much importance. Taking the permanent architecture of antiquity as their model the Modern masters expected, rightly, that it might take a century to learn to build properly with concrete and steel. They did not expect, within their lifetimes, to be called upon to explore construction using synthetic materials like nylon, carbon fibre, kevlar, mylar, nomex, or Teflon; or to have to contend with a massive explosion of information technology within buildings, let alone electronic intelligence itself. Only a very few, like Maxwell Fry, even understood how difficult such a task might be.

For a complex of reasons Modern architecture tried to ignore the demands of technological assimilation in an age of science. Like surgeons operating without anaesthetic in a modern teaching hospital, the architects of the great mutiny became dangerously obsolete in their own environment. Towards the end of their lives this became evident in their work, just as Fuller had predicted. Despite the spectacular output of synthetic materials and new structural technologies that marked the post-war period, their palette remained limited, as did that of their immediate successors. In spite of the spirited defence of their design studio methodology that is still occasionally advanced, notably by Schon, who still speaks of architects 'knowing how to act correctly in conditions of information overflow',[10] it was precisely because the sons of the pioneers concentrated on formal inventiveness rather than exploring the process of technology transfer that had given them their new ways to build, that Modern architecture died of ignorance while new information was exploding all around it.

› What Howard Roark really did

The idea that the collapse of Modern architecture was an information failure throws new light on the nature of the great mutiny. Seen as the result of a temporary coincidence of science and building, the equally temporary success of the Moderns assumes less mythological proportions. What Howard Roark, the composite Modern architect hero of the first half of the century, really did was not so much triumph over critics and philistines to bring a new enlightenment, as specify new products and enlarge the market for new materials. Indeed the financial and political support without which he could never have displaced the entrenched forces of traditional construction, came precisely from these materials producers. Two World Wars created massive production capacity in the cement and concrete industry – likewise steel, light metals, plywood, plastics, and synthetic fibres; Modern architects created an outlet for them in the civilian economy by rendering their use culturally acceptable in building. That at any rate was the irreversible effect of their work, however far removed it may be from Ayn Rand's conception of their existential struggle.[11] With the hindsight of forty years it is possible to re-order the data of the Modern era so as to see the careers of its great individualists simply as the dramatic, populist elements in an essentially undramatic process – the adaptation of industrial and engineering materials and methods to the design of commercial, cultural, and domestic buildings.

What we know about the techniques employed by the most successful of the Modern pioneers is entirely consistent with this view. We know that they literally copied the design and construction of grain silos; stripped the masonry cladding from structural steelwork and put in glass instead; and borrowed from the 'look' of the design of ships and aeroplanes[12] to create 'a new aesthetic'. All these processes involved artistic controversy and public debate but their cultural significance was far less than their economic consequence. In essence they were a resource-shift in building technology, part of the historic process of technology transfer whose aesthetic effects have always been better documented than its substance. While sudden and traumatic, the Modern episode can still be shown to take its place in a long line of technology transfers in building whose very antiquity throws doubt on the idea that architecture and technology are incompatible. For if they are, is it not strange that their encounters against the vast backdrop of history have been so frequent and so one-sided in their results?

› Head-smashing winds of change

Architectural culture is a vast shock absorber against change; like the boom on a gybing yacht it comes over last and it comes over hard, but the driving force, the sail itself, has already taken up its new position by the time the swing occurs.

Perhaps the conversion of timber-frame construction into stone decoration in the ancient world was attended by dramas to match the frustrations endured by Modern architects in the 1920s and 1930s, when their work was as fanatically opposed as is the demolition of historic buildings today. Perhaps the use of light-weight earthenware pots in the construction of the dome of S Vitale of Ravenna over 1,500 years ago had to be fought through the medieval equivalent of a series of public inquiries. More plausibly the outrage caused by the generation of 1914 came from the pent-up surge of innovation that it directed into building. After the mainsail of industrial production had already swung over onto a new tack, the boom of *avant-garde* architecture finally smashed the head of academic revivalism – making it possible (as it were) to turn the entire technological legacy of the nineteenth century into architecture in an afternoon. The whole process was an architectural transplant of the great nineteenth-century engineering boom in which iron shipbuilding took to the land. Camouflaged as an artistic revolution, the Modern Movement in architecture did no more than break free from the technical suppression of nineteenth-century revivalism and restore building construction to its correct relationship with the new production industries. In this sense the 'mutiny' was a sudden change in the 'genetic frequency' of technology transfer in building.

Seen in this way, as a largely unrecognised logistical process, the history of technology transfer in architecture assumes a new importance. But so too do the difficulties that must be overcome in any attempt to bring it, undisguised, to the forefront of design. For not only must the trappings of 'artistic assimilation' be abandoned, but even the idea that the process of building design is 'creative' in the fine art, as opposed to the engineering sense.

› In praise of uncreative methods

As the failure of the Archigram Group proves, problems of credibility dog all attempts to separate architecture from permanence. Between 1961 and 1967 this loose alliance of five principal partners produced a dazzling array of projects based on contemporary technology transfer, freely drawing on the materials and methods of the Apollo programme. Ultimately none of it came to fruition except in the context of the market for architectural drawings, where the original designs were subsequently sold. Comparable in their predictive authority with the 1914 drawings of Sant' Elia and Chiattone, these projects for an indeterminate, intermediate architecture of lightweight mobile enclosure connected with the briefly flowering youth movement of the late 1960s, but failed to enlist the kind of industrial marketing support that once underpinned concrete construction or the idea of prefabrication. With the collapse of the youth movement and the growth of a reactionary investment market in housing after the energy crisis of 1973 the group abandoned its search for real clients to concentrate exclusively on the art market.

The lesson of Archigram's failure to attract investment was that technology transfer, even when based on a considerable knowledge of the products of advanced technology, cannot succeed without the support of an industrial base. In the 1960s the nascent aerospace industry itself survived on public funding and lacked anything that might be described as surplus production capacity. What Archigram tried to do was to swing the cultural boom over, against the wind of construction investment. In doing so it found itself opposed by the full force of the heavyweight permanent construction industry and its attendant architectural value system. The contrast between Archigram's lightweight, transitional architecture and the heavyweight, 'High-Tech', late Modern architecture of, say, Norman Foster or Richard Rogers is instructive. Conceived ten years later than the best-known Archigram projects, Richard Rogers' Lloyd's Building, for example, was designed as a permanent, flexibly serviced enclosure which promised a fifty-year capability to withstand developments in information technology – an absurd claim, as events since its opening in 1986 have already shown.[13] But a truer diagnosis, that only radical flexibility could cope with the space needs of the mushrooming financial services industry, would have produced no £150 million masterpiece. Without a driving mainsail – like the resources of the cement and concrete industry – the case was hopeless. Archigram offered temporary, flexible enclosure and failed: Rogers offered flexible servicing for a heavy concrete-frame structure squarely in a tradition of permanence, and succeeded.

› Surplus production can supplant permanence

The obstacle presented by permanence is as great ten years after the design of the Lloyd's Building as it was ten years before it, but the means to overcome it remain the same. Buckminster Fuller was the first to grasp that weight was not irrelevant to building, but ultimately controlled its cost. He saw that true flexibility or continuous replacement could supplant the concept of permanence, but only with the support of industries with surplus production capacity. Thus it had been with the evolution of machine production under the impact of continual technological innovation, and thus it would be mutatis mutandis with architecture. What was needed to establish an architecture of technology transfer was neither more nor less than a real time engineering value base. Unlike the 'historic' contribution of permanent architecture, the architecture of the future must be in continual transition. To make itself financially viable it must draw its value from its performance, which in turn must be as exactly measurable as that of a car or an aeroplane, and be calculated like any other engineering system.

Architects who successfully use technology transfer against the background of a Restoration culture do so by compromise with the fine art tradition of permanence. Norman Foster is well known for his ingenious use of components and materials that have their origin in industries far removed from construction: solvent-welded pvc roofing derived originally from swimming-pool liners; flexible neoprene gaskets using a material developed originally for cable-jacketing; adhesive-fixed glazing transferred from the automobile industry; superplastic aluminium panels and metallised fabric fireproofing from aerospace; tensioning devices from trailer sidescreens; raised-floor systems from jetliners; photochromic glazing from jet bombers. All these and more, including techniques of presentation and colour schemes drawn from aviation magazines, are to be found in his projects and his buildings. But Norman Foster will not agree that his work is a more or less organised search for technology that can be transferred. In his view there is a conflict between this 'redneck' definition of design and the prior claims of the fine art tradition and the role of engineering. As Peter Rice has commented: 'High-Tech architects have concluded that the discipline provided by the engineer is the best framework in which to conduct architecture.' Or, as Michael Hopkins puts it: 'Our architecture comes out of our engineering and our engineering comes out of our engineers.' Perhaps underlying this faith in engineering is a doubt that technology transfer can stand on its own as a creative process; a reciprocal of the doubt expressed by some critics that the construction of 108 concept models for a major commission (Foster's abandoned BBC Radio Centre at Langham Place) is either intellectual or creative in the traditional fine art sense.[14]

› Down with the heritage value system

To find total acceptance of the priority of technology transfer in architecture today it is necessary to study the work of a former Foster associate Richard Horden, the designer of the purest technology-transfer building yet constructed in Britain. Horden's 1984 'Yacht House' in Hampshire embodies the principles of technology transfer that have been sporadically applied by Norman Foster, but concentrated into the generating structural frame of a small domestic building. Horden finds his materials and methods in the high-performance components produced by the yacht spar and standing rigging industry. His unique structure, intended to form the basis of an omni-functional enclosure system, shows not only that architectural design developed from the central principle of multi-sourced industrial component combination is feasible, but that its results can still be culturally acceptable within a fine art design tradition. With it Horden has gone further than any living architect to show that a true architecture of technology transfer need neither be impoverished nor primitive.

Like Horden the London and Los Angeles practice of Future Systems Inc, with its two partners Jan Kaplicky and David Nixon, has striven for nearly ten years to develop an architecture of technology transfer. Future Systems has as yet no completed building to mark the achievement of commercial viability, but it does have the distinction of being the only British firm of architects involved in the design of the 1992 NASA manned space station. Future Systems projects, like the seminal projects of Archigram, lean clearly on technology transplanted from aerospace design, but they reach further into the emulation of organic structures and the inclusion of flexibility in the form of articulated movement.

Recently the deliberate presentation of their advanced structural system projects in the context of conventional Restoration architectural competitions – such as the 1985 Grand Buildings contest for Trafalgar Square, which Horden also entered – has begun to enable them to quantify the benefits of monocoque construction in commercial terms. Exoskeletal construction enabled their Grand Buildings entry to achieve a far higher net-to-gross ratio of serviced floorspace than any other competitor, as well as providing a capacity for rapid internal reconfiguration to deal with information technology changes that would put both Lloyd's and the Hongkong & Shanghai Bank to shame.[15]

Today it is only by such acts of stealth as Foster's carefully metered inclusions of alien technology within a fine art dominated culture, Future Systems' competition entries, and Horden's unique house design, that the architecture of technology transfer remains visible under the obsolete heritage value system that has ruled architecture since the Restoration. In reality, because it is a theory of architecture as economic, multi-sourced element combination, it belongs to a different and more appropriate value system alongside production engineering, automobile, marine, and aerospace design. Eventually, Horden believes, the entire spectrum of manufactured components, from the smallest rigging screw to the largest offshore oil-rig assembly, will become a hunting ground for transferable technology. He tends to draw elements for his designs from the smaller end of the component size continuum, but sees the vast – as yet uncompiled – data base of all products as the proper area of search for the architect of the future.

None of this can be done without the construction of a bridge from the rotting hulk of contemporary Restoration architecture to this new conception of building as the product of cross-industry component and material combination. At present such a bridge can only be built upon the ability of architects like Foster, Horden, Kaplicky, and Nixon to make its results culturally acceptable. But by itself this ability is not enough. It needs the support of expanding industries and, most important of all, an ideological certainty equal to that which enabled the Modern Movement to temporarily overcome obsolete heritage values.

› The need for a work of history

To begin the process of developing an ideology for this new architecture the best starting point would be a substantial study of its history. Such a document could become the first reference work of the architecture of the information age; a technological and methodological – rather than an art historical – study of technology transfer in architecture. A partial model is to be found in Marian Bowley's 1960 *Innovations in Building Materials*,[16] the last authoritative study of technology transfer in construction. But this volume has, characteristically, little reference to the actual or possible role of architects. Unlike the historians of construction, architectural historians (with the exception of Banham) have only recognised technology transfer as a peripheral matter, remarked in such ancient events as the conversion of the form of decorated tree-trunks into stone columns, or the transfer of plant-derived decoration into carving. No one, Banham included, has ever treated it as a unitary phenomenon, a continuous process whose evolution can be traced through centuries of craft-construction until, with the coming of the Industrial Revolution, it begins to accelerate out of control.

It is one of the many serious consequences of the crucial cultural gap that has separated historians and theorists of architecture from the reality of practice, that no such architectural history on the model of Bowley's has ever been written. Even though a pattern of well-documented examples shows this quickening wave-motion with the clarity of an evolutionary diagram. The adaptation of wooden boat building into roof construction in the Middle Ages, for example, took place over hundreds of years: the development of reinforced-concrete boat construction into reinforced-concrete building took fifty years;[17] but the adaptation of offshore oil-rig technology to building types in the present century was achieved in less than a decade. The process is clearly identical and clearly important; only the wave frequency of the transfer has speeded up.

There is a clear relationship between the absence of this crucial field of study and the present predicament of technology transfer in the age of Restoration architecture. Without it the delusions of significance that still append to the obsolete categorisation of architecture by style instead of content cannot be swept away, and the progressive marginalisation of architecture will continue.

Compared to the trivial works of style-history that presently crowd out genuine theory in the body of architectural knowledge, a serious analysis of technology transfer in buildings would have the immediate authority of a stock-market analysis

coupled with the direct applicability of a consumer report. It would unravel myster-
ies and explode myths with the clarity and force of the early writings of Adolf Loos
or Le Corbusier. From the outset it would provide a quantifiable base from which to
compare the evolutionary and economic significance of pre-Modern, Modern, and
Post-Modern architecture. Placed in a material historical context some Post-Modern
buildings, for example, might show themselves to be more fertile in technology
transfers than their High-Tech counterparts – consider Terry Farrell's temporary
Clifton Nurseries building in Covent Garden, with its Teflon-coated glass-fibre roof
membrane and its Proctor mast roof beams for example. Classical Revival
envelopes executed in profiled composite panels might be more impressive still,
representing an ingenious way of 'culturalising' the architectural use of such
advanced boat building composites as kevlar, epoxy, and carbon fibre.

By setting aside the obfuscating camouflage of style, a deep study of the
architecture of technology transfer would expose the massive material similarity
that characterises contemporary architecture, and show more clearly than ever
before what are the deep structures and what are the surface structures in the
design of buildings.

› The architect and the bee

By opening such a revolutionary field of study the Byzantine world of Restoration
architecture would suddenly become accessible to the quantitative analytical techniques that rule the late twentieth-century world of engineering design and manufacture. Architecture, which is now an occult world of ignorance and obsolete
mystery, shot through with individual acts of achievement, could become an open-
access field of competition. The mighty ocean of product information that presently
relies on fragmented, peripheral awareness could be given accessibility with the
simplicity and directness of a video game. Architects freed of the tyranny of history
for the second time in a century could concentrate on design by assembly, identify-
ing the availability of new materials and techniques, and 'specifying them into
culture' with a squeeze on the joystick button. Like bees, architects would be seen
to have been carrying out an evolutionary as well as a productive task. Their genetic
role: the cross-pollination of materials and methods from a one-world product
economy to a new architecture.

Those who doubt that the emergence of this new field of study in architecture
could create its own ideology should consider the power of history, which is not
only the story of the past but the ultimate proof of the present. When such a record
is absent, our actions become as cyclical and unchanging as those of plants and
animals, whose history endlessly repeats itself, and our adaptability is forfeit. Nor
would such a change in the story of architecture make it untrue – it would make it
true again and again for successive generations – just as the movements of the
rudder of a ship, in response to changing winds, changing seas or changing orders,
enable it to keep a true course.

1   Barbara Miller Lane (*Architecture and Politics in Germany: 1918 to 1945*. Harvard University
     Press, 1968) quotes Walter Gropius in 1919: 'The old forms are in ruins, the benumbed world
     is shaken up, the old human spirit is invalidated and in a flux towards a new form.' Conrads
     and Sperlich (*Fantastic Architecture*, Architectural Press, 1963) quote Bruno Taut in the first
     issue of the magazine *Dawn* in 1920: 'Space. Homeland. Style. To hell with them, odious con-
     cepts! Destroy them, break them up! Nothing shall remain! Break up your academies, spew
     out the old fogeys . . . Let our North wind blow through this musty, threadbare tattered world.'
     Anatole Kopp (*Constructivist Architecture in the USSR*, Academy Editions, 1985) provides
     similar quotations from the Russian Constructivists.

2   E. Maxwell Fry, *Fine Building*, Faber & Faber, 1944.

3   Industrialisation of the process of construction is a question of new materials . . . Our
     technologists must and will succeed in inventing materials that can be industrially
     manufactured and processed and that will be weatherproof, soundproof and insulating. I
     am convinced that traditional methods of building will disappear.

                                                                     Mies van der Rohe, *G*, No 3, 1924

     Or in a later version: 'It will soon be possible to break altogether with the tradition of putting
     stone on stone or brick on brick and move in the direction of rational fabrication.' J. D. Bernal,
     The Social Function of Science, 1939 (quoted in Andrew Saint, *Towards a Social Architecture*,
     Yale, 1987). As late as 1962 Herbert Ohl, the German expert on industrialised building, wrote:

     The artistic and formal interests of the last hundred years have taken the task of the
     architect away from productivity, in spite of all attempts to rescue him . . . The architect
     must realise that the machines, processes and appropriate materials of industry are the
     effective means for the production of buildings.

                                                                Architectural Design, April 1962, p. 162

4   I have twice witnessed this process at work when Berthold Lubetkin addressed groups of
     younger architects in 1985 and 1986. Many of the questions put to him take the form: 'But
     surely if you were in practice now, you would behave as we do and not be as intransigent as
     you were then?' To his credit, Lubetkin never concedes this point.

5   The outside of a house should be dictated by the inside, as the form of the animal body is
     dictated by the skeleton, the disposition of the organs and the functioning of the various
     systems – blood circulation, nervous and muscular systems and so on . . .
     communications, drainage, services and so on.

     (Anthony Bertram, *The House: a machine for living in*, A. & C. Black, London, 1935.) There are
     more famous examples of such clarity of thought in the *oeuvre* of Le Corbusier, and earlier
     ones in Loos, but Bertram is particularly robust, mocking the occupants of 'Tudorbethan'
     dwellings by demanding to know why they do not wear doublet and hose *etc.*

6   I am indebted to Richard Horden for this example.

7   Reyner Benham, *Theory and Design in the First Machine Age*, Architectural Press, London,
     1961.

8   Quoted in *Theory and Design*, pp. 325–6. A similar thought can be discerned in a quote from
     Edwin Lutyens dating from seven years earlier still. 'The modern architecture of so-called
     functionalism does not seem to me to . . . show yet a genuine sense of style – a style rooted
     in feeling for the right use of materials.' (*Country Life*, 20 June 1931.)

9   Quoted in *The Buckminster Fuller Reader*, ed. James Meller, Jonathan Cape, London, 1970.

10   Donald Schon, *The Reflective Practitioner: how professionals think in action*, Temple Smith,
     1983.

11   Ayn Rand, *The Fountainhead*, New York, 1943. An exhaustive study of the relationship
     between Rand's hero and stylistic rationality is to be found in Andrew Saint's *The* Image *of
     the Architect*, Yale University Press, 1983.

12   The repeated appearance of the Farman 'Goliath' in *Vers une Architecture* is a case in point.

Le Corbusier made no effort to employ the materials and methods of contemporary aircraft construction, but he did emulate the appearance of wing struts seen obliquely – using them as columns – and the visual relationship of planes to solids – as with wings and fuselage. The ability to net complex structures in this formal, unanalytical way may be uniquely architectural. The engineer Peter Rice has described it as 'A fine visual appreciation of the way the engineer's design is perceived. [The architect] refines the form in relation to an image so that ultimately it is explainable at a simpler level.' One of the very few direct technical influences of the aircraft industry on construction in Britain during the Modern period came from the Great War airship programme, when the task of solving the large number of simultaneous equations generated by segmented circular space frames led to the development of new methods of calculation for lattice girders. Richard Southwell, *Methods of Calculating Tension Coefficients*, London, 1920. The direct copying of American industrial building by the European pioneers is discussed in detail by Reyner Benham in *A Concrete Atlantis*, MIT Press, 1986.

13   For a discussion of the obsolescence of the Lloyd's Building see Martin Pawley, 'Into the Unknown', AR October 1986, p. 88.

14   The Peter Rice quote is from a profile of the engineer published in *The Architects' Journal*, 21 and 28 December 1983. The Michael Hopkins quote from *Building*, 8 November 1985. The concept of technology transfer as a limited, non-intellectual, non-creative approach to design emerged in conversation with Norman Foster during 1985. The term 'redneck' to describe it was contributed by AR editor Peter Davey.

15   The 1985 Grand Buildings competition brief called for a minimum gross floor area of 18,000 m² within the framework of plot ratio, daylight angles and fire regulations governing the site. Future Systems' design provided 23,000 m² gross with a remarkable 89 per cent, 20,500 m² net lettable. In addition the repositioning of the suspended floors within the envelope offered unprecedented flexibility – of a type crucially relevant to designs like Lloyd's.

16   Marian Bewley, *Innovations in Building Materials*, Duckworth, London, 1960.

17   The exhibition of a rowing boat made of concrete reinforced with a rectangular mesh of iron rods at the Paris *Exposition Universelle* of 1855 is recorded by S. B. Hamilton (*A Note on the History of Reinforced Concrete in Buildings*, HMSO, 1956) as antedating the first reinforced-concrete building by ten years, and the first large-scale use of reinforced concrete for building by forty years.

1988

Bruno Latour

Mixing Humans
and Nonhumans
Together: The
Sociology of a
Door-Closer

Bruno Latour (born 1947) is a French philosopher-sociologist-anthropologist, who has developed a variety of innovative approaches to the study of science and technology. Since 1982, he has been Professor at the Centre de Sociologie de l'Innovation at the Ecole Nationale Supérieure des Mines in Paris and has been a visiting Professor at UCSD, at the London School of Economics and in the History of Science department of Harvard University.

Beginning with the established traditions of the sociology of science in the 1970s, Latour produced a pioneering anthropological study of the Salk Institute called "Laboratory Life," but in the 1980s moved to studies that probed the underlying structures of technical and scientific knowledge. This shift developed into a body of work called Actor-Network Theory (ANT), of which this essay is an example. While he has since disavowed that formulation, he continues to find novel ways to investigate *"these strange situations that the intellectual culture in which we live does not know how to categorize."*[1]

This essay is one of the few we have included from recognized philosophers of technology and have done so because it so directly addresses an element of building construction. In the terms of ANT, it develops a number of useful concepts for describing the behaviors embedded into building mechanisms, the "behavior imposed back onto the human by nonhuman delegates," and ultimately explores "the moral and ethical dimension of mechanisms."

1   Bruno Latour, *We Have Never Been Modern* (Harlow: Pearson Education, 1993), p. 3.

Is sociology the study of social questions, or is it the study of associations? In this paper the author takes the second position and extends the study of our associations to nonhumans. To make the argument clearer, the author chooses one very humble nonhuman, a door-closer, and analyzes how this "Purely" technical artifact is a highly moral, highly social actor that deserves careful consideration. Then the author proposes a vocabulary to follow human and nonhuman relations without stopping at artificial divides between what is purely technical and what is social. The author builds "its" or "his" own text in such a way that the text itself is a machine that exemplifies several of the points made by the author. In particular, the author is constructed and deconstructed several times to show how many social actors are inscribed or prescribed by machines and automatisms.

The most liberal sociologist often discriminates against nonhumans. Ready to study the most bizarre, exotic, or convoluted social behavior, he or she balks at studying nuclear plants, robots, or pills. Although sociology is expert at dealing with human groupings, when it comes to nonhumans, it is less sure of itself. The temptation is to leave the nonhuman to the care of technologists or to study the impact of black-boxed techniques upon the evolution of social groups. In spite of the works of Marx or Lewis Mumford and the more recent development of a sociology of techniques (MacKenzie and Wacjman, 1985; Bijker, Hughes, and Pinch, 1986; Winner, 1986; Latour, 1987), sociologists still feel estranged when they fall upon the bizarre associations of humans with nonhumans. Part of their uneasiness has to do with the technicalities of complex objects and with the absence of a convenient vocabulary allowing them to move freely from studying associations of humans to associations of nonhumans. In this paper I want to contribute to the reinsertion of nonhumans into the mainstream of American sociology by examining an extremely simple technique and offering a coherent vocabulary that could be applied to more complex imbroglios of humans and nonhumans.

› Reinventing the door

On a freezing day in February, posted on the door of the Sociology Department at Walla Walla University, Washington, could be seen a small hand-written notice: "The door-closer is on strike, for God's sake, keep the door closed." This fusion of labor relations, religion, advertisement, semiotics, and technique in one single insignificant fact is exactly the sort of thing I want to help describe. As a technologist teaching in an engineering school in Columbus, Ohio, I want to challenge some of the assumptions sociologists often hold about the "social context" of machines.

Walls are a nice invention, but if there were no holes in them, there would be no way to get in or out; they would be mausoleums or tombs. The problem is that, if you make holes in the walls, anything and anyone can get in and out (bears, visitors, dust, rats, noise). So architects invented this hybrid: a hole-wall, often called a door, which, although common enough, has always struck me as a miracle of technology. The cleverness of the invention hinges upon the hinge-pin: instead of driving a hole through walls with a sledge hammer or a pick, you simply gently push the door (I am supposing here that the lock has not been invented; this would over-complicate the already highly complex story of this door). Furthermore, and here is the real trick, once you have passed through the door, you do not have to find trowel and cement to rebuild the wall you have just destroyed; you simply push the door gently back (I ignore for now the added complication of the "pull" and "push" signs).

So, to size up the work done by hinges, you simply have to imagine that every time you want to get in or out of the building you have to do the same work as a prisoner trying to escape or a gangster trying to rob a bank, plus the work of those who rebuild either the prison's or the bank's walls.

If you do not want to imagine people destroying walls and rebuilding them every time they wish to leave or enter a building, then imagine the work that would have to be done in order to keep inside or to keep outside all the things and people that, left to themselves, would go the wrong way. As Maxwell could have said, imagine his demon working *without* a door. Anything could escape from or pene-trate into the department, and there would soon be complete equilibrium between the depressing and noisy surrounding area and the inside of the building. Tech-niques are always involved when asymmetry or irreversibility is the goal; it might appear that doors are a striking counter example since they maintain the hole-wall in a reversible state, but the allusion to Maxwell's demon clearly shows that such is not the case. The reversible door is the only way to irreversibly trap inside a differ-ential accumulation of warm sociologists, knowledge, papers, and also, alas, paperwork; the hinged door allows a selection of what gets in and what gets out so as to locally increase order or information. If you let the drafts get inside, the drafts will never get outside to the publishers.

Now, draw two columns (if I am not allowed to give orders to the reader of *Social Problems* then take it as a piece of strongly worded advice). In the right column, list the work people would have to do if they had no door; in the left column write down the gentle pushing (or pulling) they have to do in order to fulfill the same tasks. Compare the two columns; the enormous effort on the right is bal-anced by the little one on the left, and this thanks to hinges. I will define this trans-formation of a major effort into a minor one by the word *translation* or *delegation*; I will say that we have delegated (or translated or displaced or shifted out) to the hinge the work of reversibly solving the hole-wall dilemma. Calling on a sociologist friend, I do not have to do this work nor even to think about it; it was delegated by the carpenter to a character, the hinge, that I will call a nonhuman (notice that I did not say "inhuman"). I simply enter the department of sociology. As a more general descriptive rule, every time you want to know what a nonhuman does, simply imagine what other humans or other nonhumans would have to do were this char-acter not present. This imaginary substitution exactly sizes up the role, or function, of this little figure.

Before going on, let me cash out one of the side benefits of this table: in effect, we have drawn a scale balance where tiny efforts balance out mighty weights. The scale we drew (at least the one that you drew if you have obeyed my orders – I mean, followed my advice) reproduces the very leverage allowed by hinges. That the small be made stronger than the large is a very moral story indeed (think of David and Goliath). By the same token, this is also, since at least Archimedes' days, a very good definition of a lever and of power: the minimum you need to hold and deploy astutely in order to produce the maximum effect. Am I alluding to machines or to Syracuse's King? I don't know, and it does not matter since the King and Archimedes fused the two "minimaxes" into one single story told by Plutarch: the defense of Syracuse. I contend that this reversal of forces is what sociologists should look at in order to understand the "social construction" of techniques and not at a hypothetical social context they are not equipped to grasp. This little point having been made, let me go on with the story (we will understand later why I do not really need your permission to go on and why, nevertheless, you are free not to go on, although only *relatively* so).

› Delegating to humans

There is a problem with doors. Visitors push them to get in or pull on them to get out (or vice versa), but then the door remains open. That is, instead of the door you have a gaping hole in the wall through which, for instance, cold rushes in and heat rushes out. Of course, you could imagine that people living in the building or visiting the department of sociology would be a well disciplined lot (after all, sociologists are meticulous people). They will learn to close the door behind them and retransform the momentary hole into a well-sealed wall. The problem is that discipline is not the main characteristic of people. Are they going to be so well-behaved? Closing a door would appear to be a simple enough piece of know-how once hinges have been invented; but, considering the amount of work, innovations, sign-posts, recriminations that go on endlessly everywhere to keep them closed (at least in Northern regions), it seems to be rather poorly disseminated.

  This is where the age-old choice, so well analyzed by Mumford (1966), is offered to you: either to discipline the people or to *substitute* for the unreliable people another *delegated human character* whose only function is to open and close the door. This is called a groom or a porter (from the French word for door) or a gatekeeper, or a janitor, or a concierge, or a turnkey, or a gaoler. The advantage is that you now have to discipline only one human and may safely leave the others to their erratic behavior. No matter who these others are and where they come from, the groom will always take care of the door. A nonhuman (the hinges) plus a human (the groom) have solved the hole-wall dilemma.

  Solved? Not quite. First of all, if the department pays for a porter, they will have no money left to buy coffee or books or to invite eminent foreigners to give lectures. If they give the poor little boy other duties besides that of porter, then he will not be present most of the; time, and the damned door will stay open. Even if they had money to keep him there, we are now faced with a problem that two hundred years of capitalism has not completely solved: how to discipline a youngster to reliably fulfill a boring and underpaid duty. Although there is now only one human to be disciplined instead of hundreds (in practice only dozens because Walla Walla is rather difficult to locate), the weak point of the tactic is now revealed: if this one lad is unreliable then the whole chain breaks down. If he falls asleep on the job or goes walkabout, there will be no appeal; the damned door will stay open (remember that locking is no solution since this would turn it into a wall, and then providing every visitor with the right key is an impossible task). Of course, the little rat may be punished or even flogged. But imagine the headlines: "Sociologists of science flog porter from poor working class back-ground." And what if he is black, which might very well be the case, given the low pay? No, disciplining a groom is an enormous and costly task that only Hilton Hotels can tackle, and that for other reasons that have nothing to do with keeping the door properly closed.

If we compare the work of disciplining the groom with the work he substitutes for, according to the list defined above, we see that this delegated character has the opposite effect to that of the hinge. A simple task, forcing people to close the door, is now performed at an incredible cost; the minimum effect is obtained with maximum spending and spanking. We also notice, when drawing the two lists, an interesting difference. In the first relationship (hinges vis-à-vis work of many people), you not only had a reversal of forces (the lever allows gentle manipulations to heavy weights) but also a reversal of *time*. Once the hinges are in place, nothing more has to be done apart from maintenance (oiling them from time to time). In the second set of relations (groom's work versus many people's work), not only do you fail to reverse the forces, but you also fail to modify the time schedule. Nothing can be done to prevent the groom who has been reliable for two months from failing on the sixty-second day; at this point it is not maintenance work that has to be done, but the same work as on the first day – apart from the few habits that you might have been able to *incorporate* into his body. Although they appear to be two similar delegations, the first one is concentrated in time, whereas the other is continuous; more exactly, the first one creates a clear-cut distinction between production and maintenance, whereas in the other the distinction between training and keeping in operation is either fuzzy or nil. The first one evokes the past perfect ("once hinges had been installed"); the second the present tense ("when the groom is at his post"). There is a built-in inertia in the first that is largely lacking in the second. A profound temporal shift takes place when non-humans are appealed to: time is folded.

› Disciplining the door-closer

It is at this point that you have this relatively new choice: either to discipline the people or to substitute for the unreliable humans a delegated nonhuman character whose only function is to open and close the door. This is called a door-closer or a "groom." The advantage is that you now have to discipline only one nonhuman and may safely leave the others (bell-boys included) to their erratic behavior. No matter who they are and where they come from – polite or rude, quick or slow, friends or foes – the nonhuman groom will always take care of the door in any weather and at any time of the day. A nonhuman (hinges) plus another nonhuman (groom) have solved the hole-wall dilemma.

Solved? Well, not quite. Here comes the deskilling question so dear to social historians of technology: thousands of human grooms have been put on the dole by their nonhuman brethren. Have they been replaced? This depends on the kind of action that has been translated or delegated to them. In other words, when humans are displaced and deskilled, nonhumans have to be upgraded and reskilled. This is not an easy task, as we shall now see.

We have all experienced having a door with a powerful spring mechanism slam in our face. For sure, springs do the job of replacing grooms, but they play the role of a very rude, uneducated porter who obviously prefers the wall version of the door to its hole version. They simply slam the door shut. The interesting thing with such impolite doors is this: if they slam shut so violently, it means that you, the visitor, *have* to be very quick in passing through and that you *should* not be at someone else's heels; otherwise your nose will get shorter and bloody. An unskilled nonhuman groom thus presupposes a skilled human user. It is always a trade-off. I will call, after Madeleine Akrich, the behavior imposed back onto the human by nonhuman delegates *prescription* (Akrich, 1987). How can these pre- scriptions be brought out? By replacing them by strings of sentences (usually in the imperative) that are uttered (silently and continuously) by the mechanisms for the benefit of those who are mechanized: do this, do that, behave this way, don't go that way. Such sentences look very much like a programming language. This substi- tution of words for silence can be made in the analyst's thought experiments, but also by instruction booklets or explicitly in any training session through the voice of a demonstrator or instructor or teacher. The military are especially good at shouting them out through the mouthpiece of human instructors who delegate back to them- selves the task of explaining, in the rifle's name, the characteristics of the rifle's ideal user. As Akrich notes, prescription is the moral and ethical dimension of mechanisms. In spite of the constant weeping of moralists, no human is as relent- lessly moral as a machine, especially if it is (she is, he is, they are) as "user friendly" as my computer.

The results of such distributions of skills between humans and nonhumans is well known: members of the department of sociology will safely pass through the slamming door at a good distance from one another; visitors, unaware of the *local cultural condition*, will crowd through the door and will get bloody noses. This story is of the same form as that about the buses loaded with poor blacks that could not pass under driveways leading to Manhattan parks (Winner, 1980). So, inventors get back to their drawing board and try to imagine a nonhuman character that will not prescribe the same rare local cultural skills to its human users. A weak spring might appear to be a good solution. Such is not the case because it would substitute for another type of very unskilled and undecided groom who is never sure about the door's (or his own) status: is it a hole or a wall? Am I a closer or an opener? If it is both at once, you can forget about the heat. In computer parlance, a door is an OR, not an AND *gate*.

I am a great fan of hinges, but I must confess that I admire hydraulic door- closers much more, especially the old copper plated heavy one that slowly closed the main door of our house in Columbus, Ohio. I am enchanted by the addition to the spring of an hydraulic piston which easily draws up the energy of those who open the door and retains it, then gives it back slowly with a subtle variety of implacable firmness that one could expect from a well trained butler. Especially clever is Its way of extracting energy from each and every unwilling, unwitting passer-by. My military friends at the academy call such a clever extraction an "obligatory passage point," which is a very fitting name for a door; no matter what you feel, think, or do, you have to leave a bit of your energy, literally, at the door. This is as clever as a toll booth.

This does not quite solve all the problems, though. To be sure the hydraulic door-closer does not bang the noses of those who are not aware of local conditions, so its prescriptions may be said to be less restrictive. But it still leaves aside segments of human populations. Neither my little nephews nor my grandmother could get in unaided because our groom needed the force of an able-bodied person to accumulate enough energy to close the door. To use the classic Langdon Winner motto (1980), because of their prescriptions these doors *discriminate* against very little and very old persons. Also, if there is no way to keep them open for good, they discriminate against furniture removers and in general everyone with packages, which usually means, in our late capitalist society, working or lower-middle class employees (who, even coming from a higher stratum, has not been cornered by an automated butler when he or she had their hands full of packages?). There are solutions though: the groom's delegation may be written off (usually by blocking its arm) or, more prosaically, its delegated action may be opposed by a foot (salesman are said to be expert at this). The foot may in turn be delegated to a carpet or anything that keeps the butler in check (although I am always amazed by the number of objects that fail this trial of force, and I have very often seen the door I just wedged open politely closing when I turned my back to it).

As a technologist, I could claim that, provided you put aside maintenance and the few sectors of population that are discriminated against, the groom does its job well, closing the door behind you constantly, firmly, and slowly. It shows in its humble way how three rows of delegated nonhuman actants (hinges, springs, and hydraulic pistons) replace, 90 percent of the time, either an undisciplined bell-boy who is never there when needed or, for the general public, the program instructions, that have to do with remembering-to-close-the-door-when-it-is-cold. The hinge plus the groom is the technologist's dream of efficient action, at least it was until the sad day when I saw the note posted on Walla Walla Sociology Department's door with which I started this article: "the groom is on strike." So not only have we been able to delegate the act of closing the door from the human to the nonhuman, we have also been able to delegate the little rat's lack of discipline (and maybe the union that goes with it). On strike? Fancy that! Nonhumans stopping work and claiming what? Pension payments? Time off? Landscaped offices? Yet it is no use being indignant because it is very true that nonhumans are not so reliable that the irreversibility we would like to grant them is complete. We did not want ever to have to think about this door again – apart from regularly scheduled routine maintenance (which is another way of saying that we did not have to bother about it) – and here we are, worrying again about how to keep the door closed and drafts outside.

What is interesting in the note on the door is the humor of attributing a human character to a failure that is usually considered as "purely technical." This humor, however, is more profound than the synonymous notice they could have posted "the groom is not working." I constantly talk with my computer, who answers back; I am sure you swear at your old car; we are constantly granting mysterious faculties to gremlins inside every conceivable home appliance, not to mention cracks in the concrete belt of our nuclear plants. Yet, this behavior is considered by moralists, I mean sociologists, as a scandalous breach of natural barriers. When you write that a groom is "on strike," this is only seen as a "projection," as they say, of a human behavior onto a nonhuman cold technical object, one by nature impervious to any feeling. They call such a projection anthropomorphism, which for them is a sin akin to zoophily but much worse.

It is this sort of moralizing that is so irritating for technologists because the automatic groom is already anthropomorphic through and through. "Anthropos" and "morphos" together mean either what has human shape or what gives shape to humans. Well the groom is indeed anthropomorphic, and in three senses: first, it has been made by men, it is a construction; second it substitutes for the actions of people, and is a delegate that permanently occupies the position of a human; and third, it shapes human action by prescribing back what sort of people should pass through the door. And yet some would forbid us to ascribe feelings to this thoroughly anthropomorphic creature, to delegate labor relations, to "project" – that is to say, to translate – *other* human properties to the groom. What of those many other innovations that have endowed much more sophisticated doors with the ability to see you arrive in advance (electronic eyes), or to ask for your identity (electronic passes), or to slam shut – or open – in case of danger? But anyway, who are you, you the sociologists, to decide forever the real and final shape of humans, to trace with confidence the boundary between what is a "real" delegation and what is a "mere" projection, to sort out forever and without due inquiry the three different kinds of anthropomorphism I listed above? Are we not shaped by nonhuman grooms, although, I admit, only a very little bit? Are they not our brethren? Do they not deserve consideration? With your self-serving and self-righteous social problems, you always plead against machines and for deskilled workers; are you aware of *your* discriminatory biases? You discriminate between the human and the inhuman. I do not hold this bias but see only actors – some human, some nonhuman, some skilled, some unskilled – that exchange their properties.

So the note posted on the door is an accurate one. It gives a humorous but exact rendering of the groom's behavior: it is not working; it is on strike (notice, that the word "strike" is also an anthropomorphism carried from the nonhuman repertoire to the human one, which proves again that the divide is untenable). What happens is that sociologists confuse the dichotomy human/inhuman with another one: *figurative/non-figurative*. If I say that Hamlet is the figuration of "depression among the aristocratic class," I move from a personal figure to a less personal one (class). If I say that Hamlet stands for doom and gloom, I use less figurative entities; and if I claim that he represents western civilization, I use non-figurative abstractions. Still, they all are equally actants, that is to say entities that *do* things, either in Shakespeare's artful plays or in the commentators' more tedious tomes. The choice of granting actants figurativity or not is left entirely to the authors. It is exactly the same for techniques. We engineers are the authors of these subtle plots or *scenariis*, as Madeleine Akrich (1987) calls them, of dozens of delegated and interlocking characters so few people know how to appreciate. The label "inhuman" applied to techniques simply overlooks translation mechanisms and the many choices that exist for figuring or de-figuring, personifying or abstracting, embodying or disembodying actors.

For instance, on the freeway the other day, I slowed down because there was a guy in a yellow suit and a red helmet waving a red flag. Well, the guy's moves were so regular and he was located so dangerously and had such a pale although smiling face that, when I passed by, I recognized it to be a machine (it failed the Turing test, a cognitivist would say). Not only was the red flag delegated, not only was the arm waving the flag also delegated, but the body appearance was also added to the machine. We engineers could move much further in the direction of figuration, although at a cost; we could have given him/her (careful here, no sexual discrimination of robots) electronic eyes to wave only when there is a car approaching or regulated the movement so that it is faster when cars do not obey. Also we could have added – why not? – a furious stare or a recognizable face like a mask of President Reagan, which would have certainly slowed drivers down very efficiently. But we could also have moved the other way, to a *less* figurative delegation; the flag by itself could have done the job. And why a flag? Why not simply a sign: "work in progress"? And why a sign at all? Drivers, if they are circumspect, disciplined, and watchful will see for themselves that there is work in progress and will slow down.

The *enunciator* (a general word for the author of a text or for the mechanics who devised the machine) is free to place or not a representation of himself or herself in the script (texts or machines). The engineer may delegate or not in the flag-mover a shape that is similar to him/herself. This is exactly the same operation as the one I did in pretending that the author of this article was a hardcore technologist from Columbus, Ohio. If I say "we, the technologists," I propose a picture of the author-of-the-text which has only a vague relation with the author-in-the-flesh, in the same way as the engineer delegates in his flag-mover a picture of him that bears little resemblance to him/her.[3] But it would have been perfectly possible for me and for the mechanics to position no figurated character at all as the author *in* the scripts *of* our scripts (in semiotic parlance there would be no narrator). I would just have had to say things like "recent developments in sociology of science have shown that" instead of "I," and the mechanics would simply have had to take out the dummy worker and replace it by cranks and pullies.

› Appealing to gods

Here comes the most interesting and saddest lesson of the note posted on the door: people are not circumspect, disciplined, and watchful, especially not Walla Walla drivers after the happy-hour on Friday night. Well, that's exactly the point that the note made: "The groom is on strike, *for God's sake*, keep the door closed." In our societies, there are two systems of appeal: nonhuman and super-human, that is machines and gods. This note indicates how desperate its frozen and anonymous authors were (I have never been able to trace them back and to honor them as they deserved). They first relied on the inner morality and common sense; of humans. This failed; the door was always left open. Then they appealed to what we technologists consider the supreme court of appeal, that is, to a nonhuman who regularly and conveniently does the job in place of unfaithful humans. To our shame, we must confess that it also failed after a while. The door was again always left open. How poignant their line of thought is! They moved up and backward to the oldest and firmest court of appeal there is, there was, and ever will be. If human and nonhuman have failed, certainly God will not deceive them. I am ashamed to say that, when I crossed the hallway this fatal February day, the door *was* open. Do not accuse God, though, because the note did not appeal directly to Him (I know I should have added "Her" for affirmative action reasons, but I wonder how theologians would react). God is not accessible without mediators. The anonymous authors knew their catechisms well, so instead of asking for a direct miracle (God Him/Herself holding the door firmly closed or doing so through the mediation of an angel, as has happened on several occasions, for instance when Paul was delivered from his prison), they appeal to the respect for God in human hearts. This was their mistake. In our secular times, this is no longer enough.

Nowadays nothing seems to do the job of disciplining men and women and forcing them simply to close doors in cold weather. It is a similar despair that pushed the road engineer to add a Golem to the red flag to force drivers to beware – although the only way to slow drivers is still a good traffic jam. You seem to always need more and more of these figurated delegates aligned in rows. It is the same with delegates as with drugs; you start with soft ones and end by shooting up. There is an inflation for delegated characters too. After a while they weaken. In the old days it might have been enough just to have a door for people to know how to close it. But then, the embodied skills somehow disappeared; people had to be reminded of their training. Still, the simple inscription "keep the door closed" might have been sufficient in the good old days. But you know people; they no longer pay attention to such notices and need to be reminded by stronger devices. It is then that you install automatic grooms, since electric shocks are not as acceptable for men as for cows. In the old times, when quality was still good, it might have been enough just to oil it from time to time, but nowadays even automatisms go on strike.

It is not, however, that the movement is always from softer to harder devices, that is, from an autonomous body of knowledge to force through the intermediary situation of worded injunctions, as the Walla Walla door would suggest. It also goes the other way. Although the deskilling thesis appears to be the general case (always go from intra-somatic to extra-somatic skills; never rely on undisciplined men, but always on safe delegated nonhumans), this is far from true. For instance, red lights are usually respected, at least when they are sophisticated enough to integrate traffic flows through sensors. The delegated policeman standing there day and night is respected even though it has no whistles, gloved hands, and body to enforce this respect. Imagined collisions with the other cars or with the absent policeman are enough to keep drivers and cars in check. The thought experiment "what would happen if the delegated character was not there," is the same as the one I recommended above to size up its function. The same incorporation from written injunction to body skills is at work with car user manuals. No one, I guess, will cast more than a cursory glance at the manual before igniting the engine. There is a large body of skills that we have now so well embodied or incorporated that the mediations of the written instructions are useless. From extra-somatic they have become intra-somatic. Incorporation in human or in nonhuman bodies is also left to the authors/engineers.

› Offering a coherent vocabulary

It is because humans, nonhumans, and even angels are never sufficient in themselves and because there is no one direction going from one type of delegation to the other, that it is so useless to impose a priori divisions between which skills are human and which ones are not human, which characters are personified and which remain abstract, which delegation is forbidden and which is permissible, which type of delegation is stronger or more durable than the other. In place of these many cumbersome distinctions why not take up a few simple descriptive tools?

Following Madeleine Akrich's lead, we will speak only in terms of *scripts* or scenes or scenarios played by human or nonhuman actors, which may be either figurative or non-figurative. Humans are not necessarily figurative; for instance you are not allowed to take the highway policeman as an individual chum. He/she is the representative of authority, and if he/she is really dumb, he/she will reject any individualizing efforts from you, like smiles, jokes, bribes, or fits of anger. He/she will fully play the administrative *machinery*.

Following Akrich, I will call the retrieval of the script from the situation *description*. These descriptions are always in words and appear very much like semiotic commentaries on a text or like a programming language. They define actors, endow them with competences and make them do things, and evaluate the sanction of these actions very much like the narrative program of semioticians.

Although most of the scripts are in practice silent either because they are intra- or extra-somatic, the written descriptions are not an artifact of the analyst (technologist, sociologist, or semiotician) because there exist many states of affairs in which they are explicitly uttered. The gradient going from intra-somatic to extra-somatic skills through discourse is never fully stabilized and allows many entries revealing the process of translation. I have already listed several entries: user manuals, instruction, demonstration or drilling situations (in this case a human or a speech-synthesizer speaks out the user manual), practical thought experiments ("what would happen if instead of the red light a policeman were there"). To this should be added the innovator's workshop where most of the objects to be devised are still at the stage of projects committed to paper ("if we had a device doing this and that, we could then do this and that"); market analysis in which consumers are confronted with the new device; and, naturally, the training situation studied by anthropologists where people faced with a foreign device talk to themselves while trying out various combinations ("what will happen if I attach this lead here to the mains?"). The analyst has to capture these situations in order to write down the scripts. The analyst makes a thought experiment by comparing presence/absence tables and collating all the actions done by actants: if I take this one away, this and that other action will be modified.

I will call the translation of any script from one repertoire to a more durable one *transcription* or *inscription* or encoding. Translation does not have here only its linguistic meaning but also the religious one, "translation of the remains of St Christel," and the artistic one, "translating the feelings of Calder into bronze." This definition does not imply that the direction always goes from soft bodies to hard machines, but simply that it goes from a provisional, less reliable one to a longer-lasting, more faithful one. For instance, the embodiment in cultural tradition of the user manual of a car is a transcription, but so is the replacement of a policeman by a traffic-light. One goes from machines to bodies, whereas the other goes the other way. Specialists of robotics have very much abandoned the pipe dream of total automation; they learned the hard way that many skills are better delegated to humans than to nonhumans, whereas others may be moved away from incompetent humans.

I will call *prescription* whatever a scene presupposes from its transcribed actors and authors (this is very much like "role expectation" in sociology, except that it may be inscribed or encoded in the machine). For instance, a Renaissance Italian painting is designed to be viewed from a specific angle of view prescribed by the vanishing lines, exactly like a traffic light expects that its users will watch it from the street and not sideways. In the same way as they presuppose a user, traffic lights presuppose that there is someone who has regulated the lights so that they have a regular rhythm. When the mechanism is stuck it is very amusing to see how long it takes drivers before deciding that the traffic light is no longer mastered by a reliable author. "User input" in programming language is another very telling example of this inscription in the automatism of a living character whose behavior is both free and predetermined.

This inscription of author and users in the scene is very much the same as that of a text that already showed how the author of this article was ascribed (wrongly) to be a technologist from Ohio. It is the same for the reader. I have many times used "you" and even "you sociologists." If you remember well, I even ordered you to draw up a table (or advised you to do so). I also asked your permission to go on with the story. In doing so, I built up an *inscribed reader* to whom I prescribed qualities and behavior as surely as the traffic light or the painting prepared a position for those looking at them. Did you *subscribe* to this definition of yourself? Or worse, is there any one at all to read this text and occupy the position prepared for the reader? This question is a source of constant difficulties for those who do not grasp the basics of semiotics. Nothing in a given scene can prevent the inscribed user or reader from behaving differently from what was expected (nothing, that is, until the next paragraph). The reader-in-the-flesh may totally ignore my definition of him or her. The user of the traffic light may well cross on the red. Even visitors to the department of sociology may never show up because Walla Walla is too far away, *in spite of* the fact that their behavior and trajectory have been perfectly anticipated by the groom. As for the computer user input, the cursor might flash for ever without the user being there or knowing what to do. There might be an enormous gap between the prescribed user and the user-in-the-flesh, a difference as big as the one between the "I" of a novel and the novelist. It is exactly this difference that so much upset the authors of the anonymous appeal posted on the door. It is because they could not discipline people with words, notes, and grooms, that they had to appeal to God. On another occasion, however, the gap between the two may be nil: the prescribed user is so well anticipated, so carefully nested inside the scenes, so exactly dovetailed, that it does what is expected. To stay within the same etymological root, I would be tempted to call the way actors (human or nonhuman) tend to extirpate themselves from the prescribed behavior *des-inscription* and the way they accept or happily acquiesce to their lot *subscription*.

The problem with scenes is that they are usually well prepared for anticipating users or readers who are at close quarter. For instance, the groom is quite good in its anticipation that people will push the door open and give it the energy to reclose it. It is very bad at doing anything to help people arrive there. After fifty centimeters, it is helpless and cannot act, for example, to bring people to Washington state. Still, no scene is prepared without a preconceived idea of what sort of actors will come to occupy the prescribed positions. This is why I said that, although *you* were free not to go on with this paper, *you* were only "relatively" so. Why? Because I know you are hard-working, serious American sociologists, reading a serious issue of *Social Problems* on the sociology of science and technology. So, I can safely bet that I have a good chance of having you read the paper thoroughly! So my injunction "read the paper up to the end, you sociologist" is not very risky. I will call *pre-inscription* all the work that has to be done upstream of the scene and all the things assimilated by an actor (human or nonhuman) before coming to the scene as a user or as an author. For instance, how to drive a car is basically pre-inscribed in any (western) youth years before he or she comes to passing the driving license test; hydraulic pistons were also pre-inscribed for slowly giving back the energy gathered years before innovators brought them to bear on automated grooms. Engineers can bet on this pre-determination when they draw up their prescriptions. This is what Gerson and his colleagues call "articulation work" (Fujimura, 1987). A lovely example of efforts at pre-inscription is provided by Orson Welles in *Citizen Kane*, where the hero not only bought a theater for his singing wife to be applauded in, but also bought the journals that were to do the reviews, bought off the art critics themselves, and paid the audience to show up − all to no avail, since the wife eventually quit. Humans and nonhumans are very, very undisciplined no matter what you do and how many predeterminations you are able to control upstream of the action.

Drawing a side-conclusion in passing, we can call *sociologism* the claim that, given the competence and pre-inscription of human users and authors, you can read out the scripts nonhuman actors have to play; and *technologism* the symmetric claim that, given the competence and pre-inscription of the nonhuman actors, you can easily read out and deduce the behavior prescribed to authors and users. From now on, these two absurdities will, I hope, disappear from the scene, since the actors at any point may be human or nonhuman and since the displacement (or translation, or transcription) makes the easy reading-out of one repertoire into the next impossible. The bizarre idea that society might be made up of human relations is a mirror image of the other no less bizarre idea that techniques might be made up of nonhuman relations. We deal with characters, delegates, representatives, or, more nicely, lieutenants (from the French "lieu" "tenant," i.e., holding the place of, for, someone else); some figurative, others nonfigurative; some human, others nonhuman; some competent, others incompetent. You want to cut through this rich diversity of delegates and artificially create two heaps of refuse: "society" on one side and "technology" on the other? That's your privilege, but I have a less messy task in mind.

A scene, a text, an automatism can do a lot of things to their prescribed users at close range, but most of the effect finally ascribed to them depends on a range of other set-ups being aligned. For instance, the groom closes the door only if there are people reaching the Sociology Department of Walla Walla. These people arrive in front of the door only if they have found maps and only if there are roads leading to it; and, of course, people will start bothering about reading the maps, getting to Washington state and pushing the door open only if they are convinced that the department is worth visiting. I will call this *gradient* of aligned set-ups that endow actors with the pre-inscribed competences to find its users a *chreod* (a "necessary path" in the biologist Waddington's Greek): people effortlessly flow through the door, and the groom, hundreds of times a day, recloses the door – when it is not stuck. The result of such an alignment of set-ups is to decrease the number of occasions in which words are used; most of the actions become silent, familiar, incorporated (in human or in nonhuman bodies) – making the analyst's job so much harder. Even the classic debates about freedom, determination, predetermination, brute force, or efficient will – debates which are the twentieth-century version of seventeenth-century discussions on grace – will be slowly eroded away. (Since *you* have reached this point, it means I was right in saying earlier that you were not at all free to stop reading the paper. Positioning thyself cleverly along a chreod, and adding a few other tricks of my own, I led you *here* . . . or did I? Maybe you skipped most of it; maybe you did not understand a word of it, oh you undisciplined American sociologist readers!)

There is one loose end in my story: why did the little (automatic) rat go on strike? The answer to this is the same as for the question earlier of why few people show up in Walla Walla. It is not because a piece of behavior is prescribed by an inscription that the predetermined characters will show up on time and do the job expected of them. This is true of humans, but it is truer of nonhumans. In this case the hydraulic piston did its job, but not the spring that collaborated with it. Any of the words above may be used to describe a set-up at any level and not only at the simple one I chose for the sake of clarity. It does not have to be limited to the case where a human deals with a series of nonhuman delegates; it can also be true of relations among nonhumans. In other words, when we get into a more complicated lash-up than the groom, we do not have to stop doing sociology; we go on studying "role expectation," behavior, social relations. The non-figurative character of the actors should not intimidate us.

› The lieutenants of our societies

I used the story of the door-closer to make a nonhuman delegate familiar to the ears and eyes of sociologists. I also used reflexively the semiotic of a story to explain the relation between inscription, prescription, pre-inscription, and chreods. There is, however, a crucial difference between texts and machines that I have to point out. Machines are lieutenants; they hold the places and the roles delegated to them, but this way of shifting is very different from other types (Latour, 1988b).

In story-telling, one calls *shifting out* any displacement of a character either to another space or to another time or to another character. If I tell you "Millikan entered the aula," I translate the present setting – you and me – and shift it to another space, another time, and to other characters (Millikan and his audience). "I," the enunciator, may decide to appear or to disappear or to be represented by a narrator who tells the story ("that day, I was sitting on the upper row of the aula"); "I" may also decide to position you and any reader inside the story ("had you been there, you would have been convinced by Millikan's experiments"). There is no limit to the number of shiftings out a story may be built with. For instance, "I" may well stage a dialogue inside the aula between two characters who are telling a story about what happened at the Academy of Science in Washington, D.C. In that case, the aula is the place *from which* narrators shift out to tell a story about the Academy, and they may or may not *shift back in* the aula to resume the first story about Millikan. "I" may also *shift in* the entire series of nested stories to close mine and come back to the situation I started from: you and me. All these displacements are well known in literature departments and make up the craft of talented writers.

No matter how clever and crafty are our novelists, they are no match for engineers. Engineers constantly shift out characters in other spaces and other times, devise positions for human and nonhuman users, break down competences that they then redistribute to many different actants, build complicated narrative programs and sub-programs that are evaluated and judged. Unfortunately, there are many more literary critiques than there are technologists and the subtle beauties of techno-social imbroglios escape the attention of the literate public. One of the reasons for this lack of concern may be the peculiar nature of the shifting out that generates machines and devices. Instead of sending the listener of a story into another world, the technical shifting out inscribes the words into another matter. Instead of allowing the reader of the story to be at the same time away (in the story's frame of reference) and here (in his armchair), the technical shifting out forces him to choose between frames of reference. Instead of allowing enunciators and enunciatees a sort of simultaneous presence and communion with other actors, technics allow both of them to ignore the delegated actors and to walk away without even feeling their presence.[4]

To understand this difference in the two directions of shifting out, let us venture out once more onto a Columbus freeway. For the umpteenth time I have screamed to Robin, "don't sit on the middle of the rear seat; if I brake too hard, you're dead." In an auto shop further along the freeway I come across a device *made for* tired-and-angry-parents-driving-cars-with-kids-between-two-and-five (that is too old for a baby seat and not old enough for a seat belt) and-from-small-families (that is without other persons to hold them safely) and-having-cars-with-two-separated-front-seats-and-head-rests. It is a small market but nicely analyzed by these Japanese fellows and, given the price, it surely pays off handsomely. This description of myself and the small category into which I am happy to *subscribe* is *transcribed* in the device – a steel bar with strong attachments to the head rests – and in the advertisement on the outside of the box. It is also pre-inscribed in about the only place where I could have realized that I needed it, the freeway. Making a short story already too long, I no longer scream at Robin and I no longer try to foolishly stop him with my extended right arm: he firmly holds the bar that protects him – or so I believe – against my braking. I have delegated the continuous injunction of my voice and extension of my right arm (with diminishing results as we know from Feschner's law) to a reinforced, padded, steel bar. Of course, I had to make two detours: one to my wallet, the second to my tool box. Thirty bucks and five minutes later I had fixed the device (after making sense of the instructions encoded with Japanese ideograms). The detour plus the translation of words and extended arm to steel is a shifting out to be sure, but not of the same type as that of a story. The steel bar has now taken over my competence a far as keeping my son at arm's length is concerned.

If, in our societies, there are thousands of such lieutenants to which we have delegated competences, it means that what defines our social relations is, for the most part, prescribed back to us by nonhumans. Knowledge, morality, craft, force, sociability are not properties of humans but of humans *accompanied* by their retinue of delegated characters. Since each of those delegates ties together part of our social world, it means that studying social relation without the nonhumans is impossible (Latour, 1988a) or adapted only to complex primate societies like those of baboons (Strum and Latour, 1987). One of the tasks of sociology is to do for the masses of nonhumans that make up our modern societies what it did so well for the masses of ordinary and despised humans that make up our society. To the people and ordinary folks should now be added the lively, fascinating, and honorable ordinary mechanism. If the concepts, habits, and preferred fields of sociologists have to be modified a bit to accommodate these new masses, it is a small price to pay.

1   The paper was initially published under the pseudonym Jim Johnson, Columbus Ohio School of Mines. See note below for Latour's "social deconstruction" of the authors.

2   A version of this paper was delivered at Twente, Holland, in September, 1987. It owes a lot to Madeleine Akrich's work.

3   The author-in-the text is Jim Johnson, technologist in Columbus, Ohio, who went to Walla Walla University, whereas the author-in-the-flesh is Bruno Latour, sociologist, from Paris, France, who never went to Columbus nor Walla Walla University. The distance between the two is great but similar to that between Steven Jobs, the inventor of Macintosh, and the figurative nonhuman character who/which says "welcome to Macintosh" when you switch on your computer. The reason for this use of pseudonym was the opinion of the editors that no American sociologist is willing to read things that refer to specific places and times which are not American. Thus I inscribed in my text American scenes so as to decrease the gap between the prescribed reader and the pre-inscribed one. (*Editors' Note:* Since we believed these locations to be unimportant to Bruno Latour's argument, we urged him to remove specific

place references that might have been unfamiliar to US readers and thus possibly distracting. His solution seems to have proven our point. Correspondents to the author-in-the-flesh should go to Centre de Sociologie de l'Innovation, Ecole Nationale Supérieure des Mines, 62 boulevard Saint-Michel, 75006 Paris, France.)

4   To the shame of our trade, it is an art historian, Michael Baxandall (1985), who offers the most precise description of a technical artifact (a Scottish Iron Bridge) and who shows in most detail the basic distinctions between delegated actors which remain silent (black-boxed) and the rich series of mediators who remain *present* in a work of art.

Akrich, Madeleine. 1987. "Comment decrite les objets techniques." *Technique et Culture* 5:49–63.

Baxandall, Michael. 1985. *Patterns of Invention. On the Historical Explanation of Pictures.* New Haven, CT: Yale University Press.

Bijker, Wiebe, Thomas Hughes, and Trevor Pinch, eds. 1986. *New Developments in the Social Studies of Technology.* Cambridge, MA: MIT Press.

Fujimura, Joan. 1987. "Constructing 'do-able' problems in cancer research: articulating alignment." *Social Studies of Science* 17:157–93.

Latour, Bruno. 1987. *Science in Action.* Cambridge, MA: Harvard University Press.

—— 1988a. "How to write *The Prince* for machines as well as for machinations." Pp. 20–63 in Brian Elliot (ed.), *Technology and Social Change.* Edinburgh: Edinburgh University Press.

—— 1988b. "A relativistic account of Einstein's relativity." *Social Studies of Science* 18:3–44.

MacKenzie, Donald and Judy Wacjman, eds. 1985. *The Social Shaping of Technology. A Reader.* Philadelphia and Milton Keynes: Open University Press.

Mumford, Lewis. 1966. *The Myth of the Machine.* New York: Harcourt.

Strum, Shirley and Bruno Latour. 1987. "Redefining the social link: from baboons to humans." *Social Science Information* 26:783–802.

Winner, Langdon. 1986. *The Whale and the Reactor: A Search for the Limits in an Age of High Technology.* Chicago, IL: University of Chicago Press.

—— 1980. "Do artefacts have politics?" *Daedalus* 109:121–36.

Peter McCleary (born 1938), a native of Scotland, holds degrees in engineering from the Universities of Glasgow, Strathclyde, and Imperial College, London. He spent five years in architectural practice, mainly with Frank Newby, Ove Arup, and Arup Associates in London. In 1965, McCleary joined the faculty at the University of Pennsylvania where his studies in structures were enlarged by working with the visionary engineer Robert Le Ricolais, and his interests in phenomenology were cultivated by teaching alongside Louis Kahn. From 1974 to 1982, he was chairman of the Department of Architecture and in that role founded its Paris Program in 1980, and the Program in Historic Preservation, serving as its first chairman in 1981. From 1982 to 1988, he served as chairman of Penn's PhD Program in Architecture. He has held research grants from the NEA, the National Trust for Historic Preservation, and the French Ministry of Culture, as well as fellowships from the Fulbright and Graham Foundations. He was founder of the ACSA Annual Technology Conference and was also awarded the ACSA Distinguished Professor Medal.

In this essay McCleary takes up the phenomenological analysis of the built environment as a mediating interface between the human body and the world. With reference to Maurice Merleau-Ponty's description of technology as a kind of prosthetic extension of the body, he also develops Heidegger's notion of *techne* as a mode of "revealing." The idea that knowledge is created in the production of the built environment presents technology as more than just a means to a practical end – once we realize this potential for revelation it becomes worthy of philosophical questioning. In the course of the essay McCleary develops a series of dialectical models to describe the impact of new technological developments on the fundamental experience of space.

1988

Peter McCleary

Some Characteristics of a New Concept of Technology

A partial answer or 'informing response' is sought to the question, 'what are the characteristics of knowledge derived during the production of the built environment?'

The dialectical relationship between builders and their environments leads to a special kind of knowledge which comes from their use of technical equipment, processes and theories.

Understanding of production ranges from experience of the environment through the equipment, etc., to experience of the equipment, etc., i.e. from transparent to opaque mediation.

The concepts of amplification and reduction explain the extensions and losses of experience that result from the particular functional characteristics of equipment, processes and theories.

Architects as builders respond to the scientific, ethical and aesthetic agenda of societies; and their production exhibits an appropriateness – to and/or appropriation – of aspects of those agenda.

Understanding the characteristics of the new concept of technology leads to new questions and identifies new dangers. Is there an inexorable shift from transparency to opacity? Can it be reversed? Should we return to a pre-technical condition? And among the dangers that mediation presents are: a fragmentation of perception and experience; the abstract seems more real than the mundane; the persistence of the codifications of architectural languages from archaic perceptions; and the separation of the professional from the layman.

The new concept of technology will generate new perceptions that might lead to new concepts of architectural space and time which will demand a new language of architecture.

› Introduction

The significance of technology has returned to the discourse on the purpose and meaning of architecture. In the interim a new concept of technology has arisen, one that does not limit itself to building materials and processes, but defines technology more broadly as the understanding of skills and knowledge of the dialectical relationship between humans and their environments (natural and built) in the production of a new superimposed built environment.

Neither the pre-modern architect as master-builder, not the Modernist coordinator of production, nor the fragmented perception of the Post-Modernist, have yielded a concept of technology useful to both designing and building. A useful, and new, concept demands a new way of thinking about the productive relationship between humans and their environments. In his 'analysis of environmentality', Martin Heidegger suggests that our productive encounters with the environment are 'the kind of dealing which is closest to us (and it) is not a bare perceptual cognition, but rather that kind of concern which manipulates things and puts them to use; and this had its own kind of "knowledge" '.[2]

This implies such a knowledge derives from the activities of designing and building, that is, both reflection and action. It further recognizes that the architect,

in a productive interchange with the environment, experiences it through the mediation of *technics* (i.e. technical equipment) which are contextually arranged as
techniques (i.e. technical processes); and that experience is conceptualized from
the architect's reflection-in-action and then formalized as *technology* (i.e. technical
theories).

If questioning is the primary tool of thinking, then new ways of thinking which
lead to new concepts perhaps need to ask new questions.

This paper asks a few new 'questions concerning technology' without 'implying . . . the guarantee of an answer, but at least that of an informing response'.[3]

The first and most general question we ask is, 'what are the characteristics of
the particular kind of knowledge or experience derived from the dialectic between
builders and their environments and acquired through the use of technical equipment, processes, and theories?'

Just as Maurice Merleau-Ponty's blind man experiences the world at the tip of
his cane, the architect-builder acquires knowledge of the environment through the
mediation of equipment, processes, and theories. We will examine the levels of
mediation ranging from *transparency* (where the environment is experienced
through the equipment etc.) to *opacity* (where the experience is of the equipment
and not of the environment).

Examination of Martin Heidegger's claim that man lives in the space opened
up by equipment shows that each mediator explores a different aspect of the
environment. These aspects then stand out from the totality of the world, i.e., there
is an *amplification*. At the same time there is a reduction in experience of the unex-
plored aspects. We will examine the types of mediation or 'directionality' of equipment, processes, and theories and the tendency to fragmentation of perception
with a concomitant reduction in holistic experience.

Jose Ortega y Gasset's observation that 'man finds that the world surrounds
him as an intricate new woven of both facilities and difficulties' suggests that questions about the classification of the human realm and its environments might yield
'informing responses' to additional characteristics of technical experience.[4]

What is 'architecture without architects'? 'Building' is probably the answer.
Architecture that is anonymous, indigenous, vernacular, folk, and so forth, makes
use of 'appropriate technology'. This 'technology' responds to knowledge (from
empirical to the natural sciences) of the physicality of the natural environment; to
understanding (from craft to the applied sciences) of the built environment; and to
the ethical codifications of the civil and cultural agenda of societies.

Architecture 'with' architects is the product of a more autonomous act. The
intentionality of the builder as architect includes aesthetic concerns along with
ethical and scientific matters.

When the production limits its response to the natural and built environments,
and to the civil and cultural context, the product has an *appropriateness-to* the
environments.

When the architect's intentions dominate the production, the product is an
appropriation-of aspects from the environments.

To explore this new concept of technology, we will examine the characteristics
of transparency–opacity, amplification–reduction, and appropriateness–appropriation.

› The first characteristic: transparency and opacity

Merleau-Ponty's blind man probes aspects of the world (in this case, the ground under-foot) at the tip of his cane. Is the ground high or low, wet or dry, hard or soft, hot or cold; is the opening wide or narrow, and so forth? It is critical to note that the characteristics the blind man seeks are those of the ground, and not of the cane; the cane, in fact, should go unnoticed. He wishes it to 'withdraw', that is, to become a *transparent* technic.

If, in using a hammer, 'that with which we concern ourselves primarily is the work', and it is the materiality of the 'work' that we experience, then the hammer 'with-draws', i.e., becomes 'transparent'.[5] Perfect transparency exists only where there is no mediation, as in the 'face-to-face' meeting or situation. We can experience either of two extremes; perfect transparency (that is, no mediation); or its opposite, the totally mediated situation in which one experiences not the world, but the machine-itself, which in turn encounters the world. In the latter case, the world is opaque to man, and it is not the technic that 'withdraws' but rather the world itself 'recedes' from man.

A vast range of experiences exist between these two extremes; from that of encountering the world *through* technics (i.e., transparency) or, as a result of a loss in transparency, to that of an experience of technics (i.e., opacity, where man experiences the characteristic of the technic and the technic encounters the world).

I claim that the historical development of the builders' mode of production led to a loss of transparency and a concomitant gain in opacity. A personal example: my grandfather cut grass with the short-handled sickle; my father reduced the stress in his back by using the long-handled scythe; I experienced the cutting of grass with a hand-driven lawnmower; my daughter encounters the characteristics of grass with a fuel-powered (or self-propelled) hand-guided mower; her child will use a lawn-tractor where the experience is of driving and not of cutting; my great-grandchildren will, in all likelihood use, if anything, an automaton or mechanical goat.

In making and using transparent technics (e.g., the worker's hammer and sickle), the builder designs, controls, and provides the power. The earliest machines replaced human energy with a source of power other than the human body. The introduction of servo-mechanisms (e.g., the governor, the thermostat, etc.) made human control redundant. The builder was reduced to a designer, who in turn will be replaced by self-designing automata.

We find similar losses in transparency in our move from the 'primitive hut' through the log cabin to the heavy timber frame, and finally to the balloon and plat-form frame house. The shift occurs within the mode of production in cutting and preparing the lumber: the technics change from the maul (hammer) and wedge, to the adze, the axe, the hand-saw, the band-saw, and finally to the saw-mill.

Technique (or technical processes) too, has shifted towards opacity. As the skilled worker, or artisan, in the steel industry has been replaced by the chemist, metallurgist, and systems engineer, the manufacture of steel products has become applied scientific. In the fabrication phase, the template shop has been replaced by computer graphics, and the chalk marking and hand cutting of steel has been replaced by robots and laser-cutting and yet fabrication remains partially empirical.

The assembly of steel buildings, CPM and PERT notwithstanding, is still choreographed by the craft and experience of the contractor.

Similarly, *technology* (or technical theories), takes part in the 'enframing', with the reciprocal loss in transparency. Over time, our ways of understanding have shifted from knowing through doing (i.e., *craft*) to borrowing intellectual frameworks from other disciplines or phenomena (i.e., *empiricism*) to constructing theories based on systematic, methodological observations of the material-at-hand itself (i.e., *applied science*, or engineering). Where an applied scientific theory exists, the reality of the built design must conform to that theory (albeit with a factor of safety to compensate for the lack of correlation between theory and practice). when no 'true' engineering theory is available, the builder is controlled by regulations, such as standard, codes, etc., all based on the collective experience. In some cases where 'good' practice was established in 'pre-scientific' times, the common sense 'rules-of-thumb' of craft were, and often still are, sufficient justification to build.

Our knowledge of structural steel is applied scientific, our knowledge of reinforced concrete is empirical, and our understanding of brick masonry is still based on craft. Craft knowledge of the performance characteristics of materials becomes less acceptable each day as we move inexorably towards the precision, but limited concerns, of the applied sciences. Thus, technical theories, too, have become opaque, and increasingly we can perceive the realm of building only through the filters of the theories of applied science.

Among those involved in building, the architects, many of whom favor the primacy of perception, prefer, I believe, the transparency of arts and crafts to the opacity of machine reproduction.

It would seem that many architects continue to favor: transparent passive solar heating and cooling over opaque mechanical heating, ventilation, and air-conditioning systems; adze fluted stone columns to rolled and fabricated steel sections; charcoal, ink, and water color drawings to computer graphics; the phenomenological investigations of Kahn, Barragan, and Scarpa to Revivalism, Modernism, Post-Modernism, and 'low and high-tech'; sailing with the power of the wind and the feel of the water to the steering of a powered motor boat; the Porsche and the country road to the Mercedes and the turnpike.

Is, then, the architect's position one that says that a 'good' technology is one that withdraws? Is a preference for transparency and low technology a desire for *no* technology? This preference for transparency is not unexpected, since architects derived their theories, for the most part, from buildings of the past. And since most of those buildings were built using transparent equipment, processes, and theories, the architect's perception not illogically has been, and continues to be, receptive of and encouraging to transparent modes of production.

› The second characteristic: amplification and reduction

In *Being and Time*, Heidegger notes that humans encounter the world through equipment and since they use this equipment 'in-order-to do something', it has the character of 'closeness', and the 'closeness of the equipment has been given directionality'.[6]

If the concepts of transparency and opacity explain, in part, the characteristics of 'closeness', what are the characteristics of this 'directionality' between equipment (technics) and the place or space that humans encounter in using equipment?

Let us take an example from visual perception. At the scale of the body, the human eye is the mediator, or technic, that receives the visual information regarding the world. Whether the world as perceived is chaos or cosmos is decided by another mediator, the schema of the brain. To encounter, visually, the microcosm (Gk. kosmos = world), the microscope (Gk. scopes = target, skopeo = to look at) is used as a technic. To encounter, visually, the macrocosm, one uses the macro- or tele-scope.

The prefix 'tele' indicates 'far, or at a distance', and the scope acts in 'bringing far things nearer' in the sense, here, of the visual. Similarly, in tele-communications, one encounters the telephone (sound), telegraph (drawn message), television (pictures), telephoto; additionally, there is telemechanics (mechanical movement by radio), telepathy (emotional influence), telekinesis (no material connection), and so forth.

Essentially, some aspect of the thing 'at a distance' enters the space of the observer – and there is an extension, an emphasis, or *amplification* of a sense through the use of the technic.

In using the telescope, one encounters only the visual characteristics of the distant object. The activity (or object) cannot be heard (the missing technic is the telephone); it can't be touched (tele-'tact'?); smelled (tele-'fume'?), tasted (tele-'gust', or 'tang'?). Phenomenologically, the greatest loss is that the 'sense of place' cannot be experienced. Thus, any technic (in the above case, the telescope) in amplifying an isolated sense, or senses, brings with it the concomitant loss or *reduction* in total experience.

As with transparency and opacity, we might ask here whether techniques and technology mediate in the same manner as technics, that is, through the directionality of amplification and reduction.

In the case of the *technical processes* of production of so-called industrialized building, the means most often dominate the ends and even become the ends when the precise logic of building supersedes the imprecise concerns of dwelling. This leads to the amplification of production efficiency, tolerances, fit, modular coordination, fast-tracking, and so on; and the reduction or diminution of the ends of 'spaces good to be in'.

Considering *technical theories*: There was a time when all theories of architecture were based on geometry. It was the time when 'art' meant the craft of building and 'science' meant the theory of architecture. Art as craft, and science as theory, were united through geometry which represented the science of the art, the theory of craft. Geometry could represent the dialectic between theory and practice. Also, in that time, construction was not one thing and structure another; they were two aspects of building with concern for strength, stability, and durability. The development of theories of strictures and the strength and performance of materials and their separation from the act of construction is in part responsible for the separation of load-bearing and space defining elements in nineteenth- and twentieth-century architecture.

Further, there was no separation between a Euclidean geometry that described the location of elements in space and the geometry used by the carpenter in laying out the scaffolding, centering, formwork, and the stereotomy used by the mason. When the notion of measurable connected geometric spaces entered our awareness, there arose a new geometry, i.e., topology, and we shifted from an intuition of the phenomenology of spaces to a conscious exploration and articulation of those connectivity relationships.

Each particular geometry represents a conscious exploration of an aspect of the spatial relationships of elements. Technical equipment, processes, and theories all lead to an amplification of some aspect of, or relationship to, the world and the concomitant reduction of emphasis on the other aspects or relationships.

› The third characteristic: *appropriateness-to* and *appropriation-of* context

Beyond the question of the relationship of technics, techniques, and technology to perception, it is also 'worth asking', what are the *environments* or *contexts* in which humans and their equipment are embedded; contexts, aspects of which are amplified and reduced, and which are experienced either through transparent media or which recede behind opaque media.

Since technical equipment, processes, and theories mediate between individuals, and societies, and their natural worlds, the builder must respond to at least three contexts: (i) the physicality of that natural world; (ii) the civil and cultural agenda of those societies; (iii) and the intentionality of the individual builder.

In their encounters with the *physicality of the natural world*, builders find themselves surrounded by difficulties which threaten survival and with facilities available for overcoming such difficulties. When these difficulties are severe, as in extreme climates of either heat or cold, the technic seeks mainly to overcome the characteristic of the difficulty. In extreme dry heat, the builder constructs a thick wall and a court-yard (wet-heat demands a different response); in extreme cold he builds an igloo (or a fire). When the difficulty is not climatic, but concerns effort, as in spanning a great distance with a bridge or reaching a great height with a tower, the characteristic of the mediator is once more towards amplification of the special characteristic (in this case, a structural solution is the major determinant of form). If the world surrounds us with facilities, then there is need to intervene. As already noted Ortega y Gasset said 'what in reality prevails . . . man finds that the world surrounds him as an intricate net woven of both facilities and difficulties'. In such a case, specific conditions do not insist on recognition, and hence there are no obvious or determinate characteristics to the technics, techniques, and technology. In such a case, what is the source of the characteristics of mediators such as a wall, a roof, or any other element of building?[7]

Consider the case of *the roof*. It is an element which mediates between several realms. Between humans, it serves as a territorial or defensive technic. Between the human and the built world there are issues of ethics and aesthetics (i.e., proportion and composition), and it must relate to both the scale of the built world and to the scale of the human (as in the external and internal domes of St. Paul's cathedral, London).

Between the human realm and the natural world, there could be this 'intricate net of' many difficulties and facilities.

When we consider this roof, what experience do we seek of the path of the sun, the moon, and the stars; the patterns of clouds and light; the path of the birds; the rain, hail, and snow, and so forth? Since the difficulties (often, rain and snow) insist on our recognition, the response of the mediating technic to such difficulties is to 'stand-out' and become the salient characteristic of the roof; while the other aspects of the world are diminished, or more often, completely dismissed.

The *civil* and *cultural* agenda of the society offers the second set of contexts. There are many known studies of the socio-economic-political-religious context, e.g., the similarities and differences in the productions of labor-intensive, capital-saving societies with those which are labor-saving, capital-intensive. Equally well documented are the effects of the structure of a society on its modes of production and vice-versa, e.g., the discourse on historical, dialectical, and mechanical materialism. Less well studied is the dialectical relationship between the culture and its modes of production and its products.

Further examination of the roof, as mediator, shows that Northern Americans, Northern Europeans, and Japanese all have chosen to make rain the major determinant of the shape in their design. All, too, seek to control the pathway of the 'run-off'; in order to isolate it from the human realm. The Northern Americans and Europeans take the 'run-off', in as short a distance as possible, from the three-dimensional volume of the rainfall to the two-dimensional plane of the roof, to the one-dimensional line of gutter channels to the point of a closed-channel pipe and finally into the zero-dimensional 'hiddenness' of a storm-drainage system. It is as if to isolate, not only spatially but also visually, the 'offending' material from their realms of senses. The Japanese traditionally have isolated the run-off spatially but amplified it visually as it cascades over the eaves of the roof. On the occasions when they have channeled rain into a linear path, it has been done with an open link chain as its guideline. And when they reduce the rain to a point it is amplified as a drop in a stone bowl. The Japanese desire a more transparent relationship to the natural elements than do Northern Americans or Europeans. Thus it would seem that cultures have a range of phenomenological relationships to the world and their degree of transparency or opacity influences their attitude to amplification and reduction.

The third context of mediation, leading to amplification of one aspect from the totality of possible experiences, derives from the *intentionality* of the designer. While the 'Northerner's' encounter with rain results in a fixed set of architectural responses, there has been a more varied response to *light* as the 'difficulty'.

As with all aspects of our experience with the world, the encounter with light has been partitioned into the concerns of science, ethics, aesthetics and, in a previous age, metaphysics. *Science* has proposed solutions that answer to the needs of the physiology of human comfort and the physics of light. *Ethics* and societal values have struggled to free themselves from the limited concerns of human productivity and task lighting. Science and Ethics have generated a measurable and finite set of solutions to the mediation of the window, or 'wind-eye'. *Aesthetic* concerns have ranged from Le Corbusier's 'masterly, correct and magnificent play of forms in light' to Kahn's more ordered, or perceptible, interplay between light and shadow, opening and structure, space and place, and so forth.

The *metaphysical* dimension of light would seem no longer to be part of a language of architecture. In deleting the tribune layer and attempting to integrate the triforium into the clerestorey window at Chartres, Thierry of Chartres was responding to the metaphysics of Grosseteste who said, 'light is the mediator between bodiless and bodily substances' and that, 'the objective value of a thing is determined by the degree to which it partakes of light'.[8]

The architect who followed the metaphysics of Dionysius the Pseudo-Areopagite sought no separation in science, ethics, aesthetics, and metaphysics. For the architect of the twentieth century there has been a fragmentation of perception, so that the transcendental has been eliminated. Of the remaining dimensions, the physical-measurable and the ethical-codifiable aspects have primacy over the aesthetic.

It is, of course, in this realm of the aesthetic that architects declare their autonomy or individual *intentionality*.

Louis Kahn's hierarchical interpenetration of the geometries of the structure, pathways of people and equipment, finishes, and so forth was composed to lock in the experience or *repose*. This 'silence to light, light to silence' could be achieved only with homogeneously modulated windows.

In contrast, Le Corbusier achieved *drama* in his interiors by using articulated light scoops or heterogeneous fenestration.

Neither Kahn nor Le Corbusier limited their architectural expressions as responses to the anonymous contexts of the physical, social, and cultural. Favoring their own, i.e., autonomous, aesthetic system, they willed or appropriated a form which became a salient characteristic of their mediating roof.

In general, when the context of the physicality of the natural world and/or the socio-cultural agenda specifies what needs to be amplified and what reduced, the technics, techniques, and technology are considered *appropriate* (i.e., L. propriare, near; suitable, proper, belonging). In the absence of clues as to what is appropriate, the builder *appropriates* (i.e., take possession of, L. proprius, own, self) a *text* from the context; that is, the builder's intentionality abstracts through amplification and reduction, and presents finally, *interpretations* of aspects of the world.

› Summary and conclusions

The 'informing response' to the question on the characteristics of knowledge derived from our experience of production says, in part, that our perception is a function of transparency and opacity, amplification and reduction, and appropriateness and appropriation. Some other characteristics to be explored in the future are: *exten-sions* of experience (where new aspects of the environment are presented); *trans-formations* of experience (where the environments are presented in a new way); *deconstruction* and reformulation of *space* and *time*; the *speed* of production and its relationship to *style*; and *spatializing* as a result of the *division of labor*.

It would be useful (for the designer who needs prescriptive rather than descriptive explanations) to address the questions to empirical cases of architec-tural production which are also grounded physically, socially, culturally, and histor-ically.

The characteristics, *transparency* and *opacity*, as levels of mediation, are part of the new concept of technology. As such they are responses to the original ques-tion, but our analysis makes them the source of a set of new questions. First, what could result from the seemingly inexorable shift from transparency to opacity? Newer, more opaque, technics, techniques, and technologies will need new techni-cians. Previous shifts gave birth to the structural, mechanical, electrical, and acoustical engineers; and construction managers presently seek to legitimize their knowledge.

Second, can the loss in transparency and the concomitant gain in opacity be reversed? The revival of neo-classicism and the return to the pre-technical of ver-nacular are among the efforts to recapture a condition of 'closeness'. To achieve a true transparency and not a pastiche, the building materials, equipment, processes, and theories will need to revert to a prior state. A more complex and profound belief is that the architect's task is to explore the fundamental ontology of humans which can be revealed only through the act of building. Here one seeks experience both of the world and of the technical realm, i.e. transparency and opacity at the same time, but of different realms.

Third, is it necessary to return to a pre-technical condition of transparency? Other disciplines have confronted a similar dilemma. Herbie Hancock, the jazz pianist, refutes any notion that older musical instruments are somehow more natural than their electronic counterparts. He says, 'the pianist is probably farthest away from the thing that actually produces the sound . . . there's a whole series of mechanical things between the player and the sound'. According to music critic John Rockwell, for Hancock and others what makes a 'piano' is the mechanism by which one addresses the instrument; theoretically the hammers and strings and the resonating wooden chamber are irrelevant. Of course new compositional possibilities 'open-up' where the performer is not limited by the dexterity of the body and where the logic of the mediation is 'theoretically irrelevant'.[9]

Can there be a space, like sound, not limited by the experience of the human body and where the logic of the building materiality and its processes of construc-tion are theoretically irrelevant? Such space would yield a new language of archi-tecture not influenced by building construction; an architecture where composition is everything and construction is almost nothing. Are Russian constructivism and the new deconstructivism such an architecture?

Walter Benjamin imagined a similar space when, with respect to film, he wrote: 'evidently a different nature opens itself to the camera than opens to the naked eye – if only because an unconsciously penetrated space is substituted for a space consciously explored by man'. If architects were to deconstruct and reconstruct their perceptions similar to 'filmic' space and time, then a 'different nature' would open for architecture.[10]

Amplification and *reduction* as further characteristics of the new concept of technology reveal some dangerous effects of mediation.

Through amplification, whether it be in response to the physicality of the natural world, the social and cultural agenda, or the intentionality of the builder, certain aspects of the world stand-out (ek-sist), that is to say that technics, techniques, and technology offer up a *drama*; and our perception converts the world into a *spectacle*, made dramatic through our mediation.

The first danger is that 'isolations' of a sense are 'amputations' which lead to the 'blocking of perception' and a condition where humans are 'fragmented by their technologies'.

The second danger from forgetting the reductive effort of mediation lies not only in the 'fragmentation of living' but also in the emphasis given to the amplifications. These abstracted texts, delivered through mediation, seem more precise, even perfect, and 'the temptation is to take the new features ... as "more real" than those features which are more mundane' or worldly (that is, those features revealed by ordinary perception).[11]

A third danger is that these texts or amplifications or abstractions can be organized and even codified into a language; a language which has 'silenced' the remainder of perception and a language which was codified for the perception of its time. The Classical language of architecture is such a codification. To validate its reuse is to accept its perception of the dialectic with the world, its balance of transparency and opacity, and the amplification and reduction given by *its* technical equipment, processes, and theories.

A fourth danger lies in the separation of the layman from the literati. The layman experiences only the mundane; the architect, who knows the codified language, transforms reality and re-presents a text, the logic and validity of which is accessible only to those who speak or read the unembodied language.

As each culture abstracts and codifies its perception, its architects construct a language of architecture from its deconstructions. If it is a revived language, Classical or Modern, then looking at the past becomes thinking like the past and even living in the past. Such architects may be out of phase with the perception of their time.

The new concept of technology says that all technical experience is *appropriated* from contexts which are physical, social, cultural, or intentional.

Both 'low-tech' and 'high-tech' are 'appropriate technologies'; their difference lies in the contexts from which they appropriate their products and processes. Steel was produced originally from iron ore mined from nature and today it is increasingly produced from scrap steel taken from the built environment. Similarly, while 'low-technology' appropriates from natural materials and craft skills and knowledge, 'high-technology' appropriates from newer human-made materials, and engineering skills and knowledge.

In general, what seems to be the case is that cultures, whether their relationship to the world is 'unconsciously penetrated' phenomenological or 'consciously explored' scientific, express a desire for a level of transparency and opacity which influences their choice of amplified and reduced technics, techniques, and technology. Conversely, as the types of mediation become more universal, the range of cultural levels diminish and an 'international' expression results. Internationalism will be modified only by the intentionality of the architect.

The concept of technology discussed in this paper is new perhaps only to the building professions. Its main schema was proposed, more than fifty years ago, by Walter Benjamin, Martin Heidegger, Ernst Mach, Michael Polanyi, and others. Benjamin began his seminal 1936 article on 'The Work of Art in the Age of Mechanical Reproduction' with a quote from Paul Valery's 'Aesthetics'. Valery's words are worthy of repetition.

> Our fine arts were developed, their types and uses were established, in times very different from the present by men whose power of action upon things was insignificant in comparison with ours. But the amazing growth of our techniques . . . make it a certainty that profound changes are impending in the ancient craft of the Beautiful . . . For the last twenty years neither matter nor space nor time has been what it was from time immemorial. We must expect great innovations to transform the entire technique of the arts, thereby affecting artistic invention itself and perhaps even bringing about an amazing change in our very notion of art.[12]

According to the new concept of technology, perception and production are not the vanguard and the rear-guard of the dialectic between humans and their environments but they are mutually dependent experiences.

1   This article is derived from a paper 'Metamorphosis of Perception Through Technics, Techniques, and Technology' which was written in Fall 1984 and presented to the University of Pennsylvania Faculty Mellon Seminar on Technology and Culture on 22 April, 1985.

2   Heidegger, Martin, *Being and Time* (Translated by John Macquarrie and Edward Robinson). Harper and Row (New York) 1962, p. 95.

3   Steiner, George, *Martin Heidegger*. Penguin Books (London) 1980, p. 24.

4   Ortega y Gasset, Jose, 'Man The Technician'. In *History as a System and other Essays toward a Philosophy of History*. W. W. Norton (New York) 1962, p. 110.

5   Heidegger, op. cit., p. 99.

6   Ibid., p. 135.

7   Ortega y Gasset, op. cit., p. 110.

8   von Simson, Otto, *The Gothic Cathedral*. Harper and Row (New York) 1964, pp. 51–2.

9   Rockwell, John, 'Electronics is Challenging Traditions of Music', *New York Times* (November 1986).

10 Benjamin, Walter, 'The Work Of Art In The Aged Mechanical Reproduction', *Illuminations* (Edited by Hannah Arendt). Fontana (Great Britain) 1982, p. 238.

11 Ihde, Don, *Technics and Praxis*. D. Reidel (Dordrecht, Holland) 1979, p. 22.

12 Valery, Paul, 'The Conquest of Ubiquity'. In *Aesthetics*. (Translated by Ralph Manheim); Pantheon (New York) 1964, p. 225.

Joseph Rykwert (born 1926) is an English architect, historian, and theorist. He received his architectural education at the Bartlett, University College London and the Architectural Association and went on to earn a doctorate at the Royal College of Art in London in 1970. After a brief period in practice as an architect Rykwert taught at the Royal College of Art before becoming Chairman and Professor of Art History at the University of Essex in 1967. In 1979 he was appointed Slade Professor of Fine Arts at Cambridge University, where he was also a Reader in Architecture until 1987.

Rykwert is currently Paul Philippe Cret Professor of Architecture and Professor of Art History, Emeritus at the University of Pennsylvania. He has also held visiting professorships and fellowships at numerous European and American institutions, including the Bollingen Foundation (1966), the National Gallery of Art in Washington, D.C. (1981), the Graham Foundation (1983–86), and the Getty Center for the History of Art and the Humanities in Santa Monica, California (1990). Other honors include the Chevalier de l'Ordre des Artes et Lettres (1984). He has written and published extensively on the history of architectural theory, most notably a series of influential books for MIT Press, including *The Idea of a Town* (1963), *On Adam's House in Paradise* (1972), *The First Moderns* (1980), and *The Dancing Column: On Order in Architecture* (1996), all of which have been published in several languages.

The text included here on the "Organic and Mechanical" provides a concise historical analysis of the use of these notoriously slippery terms. With a strong focus on the eighteenth and nineteenth centuries – combined with references back to classical antiquity and the Renaissance – the argument highlights the persistent recourse to both natural and mechanical metaphors as a source of architectural authority. In this broad cultural perspective the more recent debates within modernism can be seen as part of an ongoing historical process. This discussion can also be seen as a prelude to some of today's digitally driven concerns (see also Collins, 1959; Fernández-Galiano, 1982; De Landa, 2002).

1992

Joseph Rykwert

Organic and Mechanical

For R. M. (who got it wrong, but did not apologize) and in memoriam M. F.

In a lecture he gave at UCLA in 1948, nearly thirty years after the event, Erich Mendelsohn recalled the great scientist's one-word approval at the opening of the Einstein tower: "Einstein in person pronounces his scientific judgement: 'organic!' ... I understand what he means: that you can't change or take away a part without destroying the whole." And a few sentences later he adds: "The principle of elasticity is dictated by nature. Upon it nature works in all her organisms – in her material, vegetable and animal kingdoms: in man and plant. This is the *structural* meaning of 'organic' architecture."[1]

Einstein's one-word compliment to Mendelsohn was very much in vogue, but it also had a long and involved intellectual history: Organic is from *organon*, the Greek word meaning "instrument" or "tool." Aristotle's group of writings, which is his main contribution to logic, came to be called the *Organon*, because it was thought to be a "collection of treatises constituting an instrument for the accurate verbal enunciation of all mental conceptions whatsoever," as a nineteenth-century commentator put it.[2]

The Scholastics and many later philosophers took over this notion of instrumentality. It was refined by Francis Bacon into a new logic, his *Novum Organum*, which was first printed in 1620. This was to be the second installment of his *Instauratio Magna*, in which deduction was replaced by induction as the primary intellectual procedure.

Much later, Immanuel Kant did not think that an *organon* of Pure Reason, a "compendium of those principles according to which all pure cognitions *a priori* can be obtained,"[3] could yet (or perhaps ever!) be attempted; his Critique – being concerned only with synthetic, and therefore excluding analytic a priori knowledge – could not be a complete system of pure reason. In Kant's conception, *Organon* had therefore shifted from being considered an instrumental compendium of the pieces that regulated the procedure of verbal formulation to a much more articulated notion: that of a systematic and complete exposition of mental operation.

Kant had a distinct and crucial conception of the *organism*, moreover – as of a thing that has an interior binding purpose – a purpose that secured the different parts of the whole to one another in an intimate interdependence, and yet has a design unto itself. This is in opposition to an instrument, whose purpose is relative and accidental. In the *Critique of Judgement*,[4] Kant was restating a traditional (by then) use of the term. Indeed, his observation can be read as a brilliant gloss on a passage of Aristotle's on the body as a function of the soul.[5]

The generation after Kant's produced another and quite different mutation of the word. Johann Gottfried Herder was the first to posit an organic principle of political life: the collective (particularly the nation) is its plantlike and basic – its organic – unit, not the individual, who has been the true subject of natural history but can only ever act as a member of that organism. Mankind is the "vast and multifarious organ" of the Godhead,[6] but each member-group of humanity has grown according to its particular inner and plantlike necessity. Understanding between individuals belonging to different communities is possible only through a deliberate act of empathy, of *Einfuhlung* – a term that Herder coined to label the effort of historical and cross-cultural understanding. With Herder the notion of nationhood that made the unity of language, climate, and soil into an organic whole, which had already been implicit in the writings of Vico and Montesquieu, was made into the one concept essential for understanding the diversity of all cultural phenomena.

In a book that he produced jointly with Herder, *Von Deutscher Art and Kunst*,[7] Goethe (perhaps inevitably therefore) attempted to eradicate the concept of *mimesis* in architecture. The essay "Ueber Deutsche Architektur" is now chiefly remembered as a paean to Master Erwin von Steinbach, the quasi-mythical master designer of Strasbourg cathedral. What now seems most important about it, however, is its emphatic rejection of the theory, which had been fashionably advanced a generation earlier by the Abbe Laugier, that the primitive hut, made of upright tree trunks and crowned by a double-pitched roof, lay at the origin of all architecture, and any renewal of the art of building must ultimately appeal to it.

The time-honored notion that human need is typically – and in some sense definitively – answered by the hut, a primal artifact but also the product of rational human reflection, Goethe eagerly contradicts: the true work of art springs almost spontaneously out of the spirit of a people, as a plant does from the soil. It is not columns that grow out of the ground to support the roof, but walls which a genius such as Erwin can transform, so that they "rise against the sky like a sublime, overspreading tree of God, which, with its thousands of branches, millions of twigs, and leaves as numerous as the sands of the sea declares the beauty of the Lord its *Master*."[8]

Goethe's argument echoes another one that had been advanced nearly a century earlier by the younger Felibien: that the thin columns of Gothic churches retain the character of the leafy shelters to which the peoples of temperate climates so often resorted. Curiously enough, Felibien seems to think that northern peoples (such as the Germans) commemorated ancestral caves rather than the leafy woods by their buildings. Subsequently, it was not Felibien's but Goethe's notion (of which he later repented) that passed into the romantic commonplace: genius is directly inspired by the beauty of nature to wholly original creation. This idea particularly suited the Schlegel Brothers and their followers, although this highly developed analogy of plant life was not yet seen as suggesting the word *organic.*

There was a quite different but even more violent change in the use of the word among Kant's immediate disciples. Both Fichte and Schelling concerned themselves with its implications. Schelling particularly thought of the whole world in terms of a special application of the one general concept: organism. "Organism," he says, "is the principle of things. It is not the property of any single object ... [because there] are separate modes of apprehending universal organism – and universal organism is the precondition of the mechanical working of the whole physical world. Furthermore, since this life is the precondition of all things, even those things in nature which seem dead are in fact only extinct of life."[9] Schelling's organism is a process in which the essential polarities, the conflicts of nature, are reconciled; it is the highest natural power, below which exist those of mechanism and of matter.

In Hegel's somewhat later systematic account, Organic Physics makes up the third part of his description of the outside world: the first is Mechanics, which deals with general notions of space and time, material and movement; the second, Physics, speaks of individuality and differentiation. Organic Physics in its turn is also tripartite – it concerns itself with Idea as unmediated existence. Geological nature describes form, or the general representation of life, whereas vegetable nature is particular and formal subjectivity. Individual concrete subjectivity is animal organism.[10] Moreover, the primal organism is not a living thing, Hegel further maintains, in that it is quite unlike Goethe's seed of all being, that Urpflanze that he had reconstructed as the stock from which all life had sprung, and which – to his mind – already had the characteristics of a formed vegetable. The separate members of Hegel's primal organism do not contain the life process. The dead, or at least the "unliving" organism, earth, is a crystal of life. It is, however, subject to the meteorological process that fructifies it into vitality.[11] To understand the true nature of his use of the word, it is important to appreciate how Hegel determined another force, disease, which he considered an inorganic, or perhaps more accurately, an antiorganic, principle.[12] Disease results from one individual part setting itself against the totality of the manifold, and insistently isolating its activity against the unity. In his very conception of disease, Hegel shows that he is concerned with a force that has no exact physical (or, at any rate, tangible and observable) equivalent. The nature of the Hegelian conception of what is, and what is not, organic was closely dependent on the way in which natural phenomena had been studied in the preceding generations.

The determining factor in those earlier studies is the absence, until the end of the eighteenth century at least, of a specific scientific discipline dealing with living beings, or at any rate with what was particularly "living" about them. Natural historians had, of course, been around since antiquity. Yet natural history could not (or would not) deal with what the eye could not see: it was very much *la nomination du visible*, so that the natural historians' main enterprise was taxonomic.[13] The distinction between animal, vegetable, and mineral which is the basis of Hegel's Organic Physics was first established by Nicolas Lemery in his *Cours de Chimie* of 1675, and became the fundamental classification of naturalia – as opposed to heavenly or elemental phenomena – which required instrumental help for the eye, or even inference from observation.

Although two of the greatest natural historians of the seventeenth century, Claude Perrault and Christopher Wren, were also very influential and prolific architects, all this affected architecture – and even architectural theory – only indirectly. In antiquity the word *organic* had entered architectural discussion as a rather lowly by-product of the Aristotelian notions: organic referred to *organon* in the sense of instrument only, so that Vitruvius found it necessary to distinguish between *machinae* and *organa*.[14] The first were moved, mostly cyclically, by a great force, the second could be moved by one man alone. Both words were of Greek origin: *machina* and *mechanicus* (from the Greek mechos – a means, an expedient, or a remedy) referred to any kind of contrivance; *organon* came from an archaic term, *ergon*, work. It followed that the Latin *organicus* did not mean anything very different from *mechanicus:* something done by means of instruments, indirectly. Organic music, therefore, was any kind of music played on instruments; and an *organon* or *organus*, any kind of musical instrument. In this Vitruvius was following the normal Latin usage of his day, whereas the Greek word had simply followed Gresham's law of language – the coarser word will always drive out the more complex: Vitruvius's contemporary Lucretius, in discussing the body as a sensorium, uses the word *organicos*[15] to mean only "musicians," and makes no reference to the organs of the body there.

In late antiquity people who played on *organa* were called *organici*; but people who made them were *mechanici*. This late antique usage of both words continued throughout the Middle Ages; it reached a kind of paroxysm in a musical instrument invented in the mid-seventeenth century by Michele Todini (or Todino, a famous virtuoso on the bagpipe); he called it the *mechanical organ*. It occupied a whole apartment housing numerous freestanding string as well as wind instruments. None of the instruments were connected physically.[16] Only one of the instruments was played, the others sounded sympathetically: in its quasi-magical working this mechanical organ combined the attributes of a natural and an artificial object. *Organa*, like Todini's one, often astonished even the initiated: the astonishment at the quasi-magical feats of *mechanici* is evident in Hellenistic books of instruction on the subject, such as those of Hero of Alexandria or Philo of Byzantium. The feeling of surprise at such man-made miracles is echoed by sixteenth- and seventeenth-century writers.[17]

It is paradoxical that the two words that are now taken to be diametrically opposed were almost synonymous for such a long time. In Byzantium, in the Islamic world, as well as in the West, the parallel was maintained. The singing golden birds and the roaring lions round the Imperial throne of Constantinople; the robot servant of Albertus Magnus; the talking head of Roger Bacon; the fluttering (and airborne) iron fly of Regimontanus – all these mythical and semimythical automata were fairly rare. The increasing diffusion of precise, refined mechanical skills at the outset of the industrial revolution – which produced new and cheaper clockwork, for instance – also culminated in an explosion of interest in, and of skill in making, *androids*. Automatic "writers," flautists, trumpeters, and even a swimming, quacking, digesting duck coincided with the intellectual elaboration of Descartes's view, that corporal man is just one special case of *res extensa*, into the doctrinally materialist *homme-machine*[18] and *homme-plante* of Julien de la Mettrie.

But La Mettrie's use of the machine as an interpretative analogue of the living body was in a sense much the same as Aristotle's. The miracle-machines of the Hellenistic and medieval engineers imitated the effects rather than the movements or the structure of animate beings. The idea of constructing machines that replicate or emulate (or even improve on) the movement of bodies begins with the anatomist-engineers of the Renaissance.

Aristotle had used *organon* as the exalted title of his great logical *summa*; but he had also used the term (in a sense that he took over from Anaxagoras) to signify every bodily member as an instrument, and especially the hand, because it is the more particularly human faculty, in "that [it] was not one instrument but many, an instrument that represents many instruments."[19] The commonest unqualified use of the word *organ* for any part of the body was to signify the mouth and tongue, the instruments of the human voice, from which by metonymy the term was transferred to several other parts, until it was used for most of the harder-working organs, both human and animal, in silver Latin. But because it was also used as a synonym for "agent," it came to be used as a metaphor of the whole body (and of the human body in particular) in the eighteenth century. And that is how it was assimilated to the much later notion that the building is in some ways organic.

Yet the passages in Vitruvius[20] that set out this idea (and dependent ones in many later writers) have been used very selectively in recent controversy to assimilate this ancient notion of the body image to an "organic" theory of architecture. I hope to show that this assimilation is misleading because the body image in antique and in "humanist" theory was used as an abstract model – mathematical and functional – for imitation in building, with no plastic, formal implications. The old dictionary of the French Academy still says for *organique* "*terme de physique ... qui se dit du corps de l'animal, en tant qu'il agit par le moyen d'organes,*"[21] and it is in that straightforward sense that the term enters architectural theory proper. The Venetian Carlo Lodoli, a Franciscan friar who attempted to revolutionize thinking about architecture, and succeeded (if not entirely in the way he expected), is the first to have spoken of an "organic architecture" (even if he applied the term only to furniture) because he was using the word more in the Latin than in the Greek sense. Furniture, he considered, should take the concave form of those parts of the body that come into contact with it. And he had indeed made (or at least had made) a chair with a curved shoulder-rest like the antique ones that were known from sculptures, and what is more, that chair (which was in advance of the fashions of the day) also had the seat hollowed out "as the English are now beginning to do." In the same passage, his follower Andrea Memmo, who is our main authority for Lodoli's ideas, also discusses other kinds of architecture the master considered beside the organic: topiary, or garden architecture, curule architecture, or coach-building, and so on.[22] Memmo's influence on the next two or three generations of theorists was enormous, if almost entirely indirect, and it helped to transmit his reading of Lodoli and the way it related the body to the members of a building, and suggests that a building must be, among other things, also a visible working out of its mechanical forces, its functions.

However, a quite different development outside the direct control of philosophers was giving the transformation of the word yet another unexpected turn: that same Nicolas Lemery, who had formulated that tripartition of *naturalia* into animal, vegetable, and mineral, had also first seen that the distinction between acid and alkali implied the universality of chemical action. More than a century later, Antoine Lavoisier and Johann Jakob Berzelius showed the fundamental unity of vegetable and animal matter, as well as the general validity of chemical laws, which applied to living matter in the same way as they did to inanimate. But it was Friedrich Wohler's synthesis of urea in 1828 that demonstrated (even if the notion was not immediately accepted) that organic material did not require the presence of the imponderable "vital force" for its existence, although its action, animating inert matter, had been inferred by many chemists.

It was generally agreed by the middle of the nineteenth century that as a result of the chemists' onslaught the word *organic* could no longer be used in the venerable sense it had retained until the eighteenth century, but would be needed for matters that were either directly concerned with, or were being regarded as an analogy of, things in the vegetable and animal world. And, some maintained, the term could be extended as a representation: "To say that physiology is the physics of animals is to give a highly inaccurate idea of it; I might as well say that astronomy is the physiology of stars," wrote the surgeon-anatomist Marie-Francois Bichat in 1800.[23]

Physics and engineering had dominated the scientific life of the seventeenth and eighteenth centuries, and chemistry had still been suspect (as being akin to alchemy and sorcery, as well as tainted by immediate commercial applications of various kinds), yet quite a new interest in it brought about the separation of natural history into the twin disciplines of organic and inorganic chemistry. This extended the realm of *naturalia* to the stars and to microorganisms, but also separated animal and vegetable decisively from mineral study. And in fact, organic chemistry was also subject to a new discipline: biology. That word was coined by Gottfried Reinhold Treviranus for the title of a book, *Biologie; oder die Philosophie der Lebenden Natur*, which appeared in parts beginning in 1802; it was immediately taken up by the great French naturalist Lamarck for his own use, and he gave it universal currency.

The arrival of this new organic chemistry finally transformed the word out of recognition. It was first circumscribed in the 1830s and 1840s by Friedrich Wohler and then more definitively by Justus von Liebig, whose *Handbuch der Organischen Chemie* of 1839 (although one of a series of academic textbooks) became the classic statement. Now the "vital force" of eighteenth- and nineteenth-century chemists also had its Aristotelian and Scholastic antecedents and was understood to be not only the life-giving agency, inherent in matter, but also the presumed cause of the temporal articulation of life into generation, growth, and decay. When this *vis vitalis* was no longer required as part of the chemists' conceptual baggage, it was taken over by philosophers: Schopenhauer identified it with the will.[24] The biologist-philosopher Hans Driesch created a summa of the doctrine in his *Science and Philosophy of the Organism*,[25] although a few years later Driesch saw entelechy "accomplished" when an organic society, a nation, located its vital part in a leader; his accession to the *Fuhrerprinzip* did not help to validate his theories more generally. Although by the 1920s it was scorned by many scientific biologists, it was seen as a valuable hypothesis more recently, particularly by some Marxist biologists (notably Trofim Lysenko), who regarded it as the agent that made its transactions with the environment in spite, as it were, of chemically transmitted heredity.

For all the divisive action of the new sciences, there is another field of study that develops at the end of the eighteenth century, and which suggests that the *Naturphilosoph*'s unified conception of nature had not exhausted its scientific, even empiric, usefulness: morphology, or the science of forms. The word was newly minted by Goethe already in the 1790s, although it was first printed in 1800 by another scientist.[26] This morphology was the study of organic form in the old sense, and various attempts were made in the nineteenth century to suggest that in fact organic and inorganic matter were organized on closely analogous mechanical principles. The term was also taken up by Lamarck and his followers in a study of the development of species through time, which came to be called evolution, and which was to absorb such vast intellectual energy in the nineteenth century. Lamarck first formulated the notion that species were modified by their own activity under the influence of environment. The two conflicting developments from Lamarck, that of Georges Cuvier and of Etienne Geoffroy de Saint-Hilaire, were discussed by the aged Goethe in 1830.[27]

Such transformations of late eighteenth-century taxonomy lead up to the general possibility of a *history of nature*, as against the old notion of *natural history*, in which evolution referred only to the transformations of the individual in the course of his or her development. Georges Cuvier, who in fact belonged to the generation immediately after Lamarck, suggested a completely new system of classifying living beings. It was his notion that all plant and animal organs should be classified not by surface similarity but by their relation to the elemental functions of the individual. Breathing, digestion, movement, circulation, and nervous excitation were the essential functions. The most important organs, therefore, were not the visible (and on the whole symmetrical) ones on which the whole classificatory scheme of the old natural history had been based but the more complex asymmetrical ones that (in the larger animals, at any rate) are only visible on dissection. Cuvier supposes a classification by the topology of functions: the way in which they relate to each other, and the way in which the primary internal and secondary external organs are conditioned by these interrelationships is what makes the unity of a natural class.[28] Within the organism, therefore, it is function that determines the form. Cuvier's systematic account of organisms presupposed that they were not a continuous chain of resemblances but a discontinuity of groupings organized around a functional nexus. Whatever the fate of his "catastrophic" theory of evolution (which was overshadowed by Darwinian accounts of a gradual natural selection), still his primary account of the relation between organ and function remained of the greatest interest; and of course his influence on architecture, wholly unintended, was capital. Unfortunately, the one architect who was known to be a friend of Cuvier's, Theodore Brogniart, although he was an enormously prolific designer, left no writings, and his only scientific contact with Cuvier was said to be in discussion about fossils in building stones. However, Cuvier's doctrine was to have a powerful effect on both Gottfried Semper and Eugene Viollet-le-Duc in the next generation, so that virtually no architect in the second half of the nineteenth century escaped its influence in one way or another. Moreover, a vast popular and semi-popular literature followed in the wake of Cuvier's theorizing, of which even architects could not remain ignorant.

Yet the most influential carrier of such ideas turned out to be not an architect but a sculptor, "the Yankee Stonecutter," Horatio Greenough. The author of very many not very distinguished (and sometimes very large) sculptures, which might quite accurately be called neoclassical (although he certainly would not have liked that), Greenough is now best remembered for having coined the dictum or precept that "form follows function." Although this was destined to become an all-purpose slogan, it was very clear to Greenough that he was formulating a strictly *organic* notion. Indeed, he continued to think of artifacts in terms of the relation between function and organ – which explains his definition of beauty as the promise of function, action as its presence.[29] And the very promise of function is agreeable to the senses because all forms of organic life require an envelope, a protection greater than they can in fact support – and therefore the envelope is always stretched to the limit of its economic possibility – as is the sail billowing in the wind. Organic life was, to Greenough, much the same as organized life, and meant in the first place human life, which is why it is articulated into three phases: of beauty, of action, and of character. This division reflects the much later introduction of the action of time into the static taxonomies of natural history.

Greenough's influence is well chronicled; one of the few intellectual debts that Louis Sullivan was prepared to acknowledge was to him, and his ideas were generally received through Transcendentalist writers, particularly through his friend Emerson. Those very striking insights of Greenough's owed more to his eclectic reading than to the Florentine milieu in which he worked – the circle of Luigi Bartolini at the Florentine Academy – although his sculpture owes everything to his Florentine contemporaries.

Another carrier of such ideas, much less known but perhaps even more influential, was Leopold Eidlitz, who was born in Prague and, after being trained as an architect in Vienna, arrived in New York as a young man in the 1840s. He was a convinced practitioner of the Romanesque–Early Christian manner that had been developing in Germany; the best-known episode in his career is his partnership with H. H. Richardson in the building of the New York State Capitol at Albany. But Eidlitz was also something of a scientific thinker and a theorist; his "Nature and Function of Art, More Especially of Architecture"[30] restated themes that had been bandied about by idealist philosophers and romantic writers. The work of art must be a *realized idea*, like a natural phenomenon. The artist in his godlike way attempts to create a new organism, but just "because it is new it cannot be an imitation of any work of nature." On the other hand, being an organism, it must be developed according to the methods of nature. Ornament must be an integral part of the structure as "a flower appears amid leaves of a plant."[31] That thought could have been Eidlitz's, but the metaphor is definitely Sullivan's. And the metaphor suggests what had already been implicit in Eidlitz's thinking. For the ornament to be seen as functioning like a plant, it would also have to look like one. How else would the visual metaphor be presented to the innocent spectator? The scaling down from the grand philosophical and scientific questions of principle to their architectural application was slow but definite. Greenough would talk about form following function, and even extol the beauty of sailing ships, yet he made sculptures for buildings with great Corinthian porticoes.

Still, his express preference was for buildings that would have the beauty of a Yankee clipper, "the beauty of her bows, the symmetry and rich tracery of her masts and rigging – and those grand wind muscles, her sails!"[32] Elsewhere Greenough asserts, "The men who have reduced locomotion to its simplest elements ... are nearer to Athens ... than they who would bend the Greek temple to every use. I contend for Greek principles, not Greek things."[33] The model of beauty is doubled: it is the animal, the organism, but also its rival, the machine, the clipper, the *architectura currule*, of the new light coach builders that Lodoli had also admired. The machine is in some sense present as the "Greek," the perfectly spare and economic answer to need – much as the organism is shaped when its function is adapted to environment.

"Machine" is, therefore, again presented as an analogue of "organism," although in quite a different sense from that of Vitruvius. Nor is Greenough interested in the quarrel between Gothic and Greek. When he died in 1852, he was still a young man, and the Gothic revival had barely started in the United States. Before his death, practically all the major public buildings on the East Coast were more or less "classical." Of course, Ruskin's *Stones of Venice* had only been published the year before. Ruskin was to become – if anything – more popular in the United States later in the century than he was in England, and he proposed, in the wake of Pugin (it was a debt that he was not prepared to acknowledge), that all ornament derived directly from nature, and that the superiority of Gothic architecture over classical is guaranteed by its closer imitation of natural forms. In the course of the decade, another highly influential book, *The Grammar of Ornament* by Owen Jones, one of the designers associated with the Crystal Palace and the government-sponsored teaching of art in England (all of which was abhorrent to Ruskin), offered ten plates of natural forms, of leaves and flowers, as a token of the sort of ornament that future architects, weary of the imitation of the past, would want to devise. Jones had known and (briefly, unhappily) collaborated with Gottfried Semper, whom I mentioned earlier, and whose great work *Der Stil* first appeared in 1863. In the United States it was not read or translated until much later, when the Chicago architect John Wellborn Root (a friend of Louis Sullivan's, and familiar with the ideas of Eidlitz and Greenough) first published passages from Semper's texts in English. Semper's views, issuing from the new biology and the linguistic ideas of the Romantics, and neutral on the problems of industry in architecture, were in fact much more acceptable to the midwestern architects than Ruskin's – involved as they were with his insistence on the value of manual work.

Ideas about a new way of imitating nature, of relating the organism to the built form, were therefore cultivated in the third quarter of the nineteenth century in a fairly close circle with which Louis Sullivan had constant contact. When young, Sullivan had found employment in the office of Frank Furness, to whom he was drawn by admiration for the buildings that he saw while walking around Philadelphia. Of his generation, Furness was perhaps the most idiosyncratic Gothic revivalist: in his work and thinking, many of the notions I have been discussing were gathered up. He was a self-confessed admirer of Viollet-le-Duc, but also an emulator of Owen Jones. His father, who was a Bostonian Transcendentalist and a Unitarian minister, was a friend of Emerson as well as an acquaintance of Eidlitz. The influence of the Anglo-German theorists and of Viollet-le-Duc, which presented both the invention of ornament and the creative digestion of new material as the substantial problems of a new architecture, was a most important counterweight to the ornamental conventionality and structural indifference of the French and Frenchified academies. Against their cosmopolitan gloss, it came to appear as a native and even home-spun philosophy, and Frank Furness was seen as one of its earliest and most inspiring representatives.

Organic architecture, then, in the 1880s and 1890s had its focus in the inventions of Sullivan and Root, their Chicago (and later their West Coast) contemporaries, as well as in the burgeoning Art Nouveau movement in Europe – although that would be food enough for another article. In the work of the Belgian and the French designers particularly, the obsessional interest in the devising of an ornament that would flow, or "grow" like real plant forms without depending on any obvious model, became a compulsion. Art Nouveau was a movement that was over very quickly: in the United States and in Britain it was succeeded by a philistine classicism; in Germany by a more learned and refined revival of Prussian post-Napoleonic sobriety; in France by the return to academic "normality" that was leavened by Auguste Perret's particular understanding of archaic post-and-beam construction.

Although it had been born out of the anticipation of the grandeurs of the forthcoming twentieth century, by the time the century had taken measure of itself, Art Nouveau was finished and its leaders converted to other ways, dispersed, or dead. The First World War made the enlightened sobriety of that first decade unacceptable, particularly in Germany. Erich Mendelsohn, when visiting the United States in the early 1920s, had appreciated the work of Sullivan and Wright – he had particular praise for the Larkin Building, the Unity Temple, and the Heurtley House. But Mendelsohn also saw himself as a product, or perhaps even as the heir, of Art Nouveau. Already in 1914, at the time of the great Werkbund exhibition in Cologne, he had turned against Peter Behrens, whose contribution, the Festival Hall, he thought a "total failure," whereas the theater by Henri van de Velde seemed to him the only really worthwhile thing in the exhibition. And of course this accomplishment of the Art Nouveau masters, the introduction of plant forms, of curvilinear and irregular, yes *organic*, elements, into the horizontal plane – even the structuring of the whole plan on such forms – was their specific accomplishment, without which there would have been no Einstein tower. The Einstein tower, which still owes much to the plan convolutions and the linear flow of van de Velde, had been praised as "organic!" by Einstein. That word Mendelsohn had understood to mean "that you can't change or take away a part without destroying the whole," a definition that echoes almost literally Leon Battista Alberti's slightly more elastic definition of beauty five centuries earlier as "that reasoned harmony of all the parts . . . so that nothing may be added, taken away, or altered but for the worse."[34] This definition and its many derivatives that have the human body as their model are often quoted in the literature of modernism as being typically "classical" and, therefore, also paradoxically "inorganic" and antigrowth.

Both Mendelsohn and Wright (in their different ways) elide the Art Nouveau appeal to the plant form as a visible and structuring model: Wright's organic architecture was in fact to be a total work of art – a notion that he introduced into architectural thinking from the theatrical writing of Edward Gordon Craig and Adolph Appia, long before the much-abused Gropius; in Wright's organic theory, the formal inclusion of heating and ventilation (as well as of lighting), which had in fact been very imaginatively treated in the early buildings of Horta (in the Hotel Solvay and in his own house), was to become part of a "complete work of art" together with structure *and* furniture.[35] It is a doctrine of which he did not repent, and which he drew from his immediate Art Nouveau predecessors. He could not accept with it the strange version of empathy that Sullivan had developed out of the earlier ideas of

Eidlitz and Greenough, and in fact was not really interested in the body image, which was so obsessively investigated by some of his contemporaries, such as the mystical Claude Bragdon.

The development that was brought about in Germany by the First World War and its aftermath came to Scandinavia a decade later when a number of architects, most notably the mature Gunnar Asplund and the young Alvar Aalto, shed their highly accomplished, and sometimes very lyrical, version of Schinkelian classicism for the charms of "free form," which owed much to the Germans, and something to the Americans: their eager use of "natural" as against man-made materials something that they did not share with Wright, for instance, who always used concrete enthusiastically – transferred the Art Nouveau dependence on plant form into a tactile naturalism.

There is, therefore, no identifiable organic theory of architecture (based on a direct appeal to nature, at any rate to the nature that biology and chemistry study) that can be usefully summarized. Yet the constant appeal to the notion of the organism, particularly as it relates to the body image in architecture, seems to be an important recurring theme in speculation about building. Mendelsohn's forms may have seemed new and disturbing in the 1920s, yet his justifying appeal to the conception of an organism was almost pedestrianly old-fashioned because it was glimpsed only through the flowery veil drawn by the masters of Art Nouveau. To them it seemed that a kind of Goetheian *Urpflanze* was the seed of all formal thinking, as it had once been the germ of vegetable, as well as animal, being. And yet perhaps the notion might again be isolated from vegetative implications. The wider importance of a conception of organism will perhaps then be seen as central to architectural thinking.

1   E. Mendelsohn, *Letters of an Architect*, ed. O. Beyer, trans. G. Strachan (New York, 1967), p. 166.

2   Octavius Freire Owen, in his introduction to the *Organon* for Bohn's (London, 1853).

3   *Kritik der reinen Vernunft*, Introduction, 2d ed. (Riga, 1787), sec. vii, p. 25: "Ein Inbegriff der-jenigen Prinzipien . . . nach denen alle reinen Erkenntnisse a priori konen erworben and wirklich zu Stande gebracht werden."

4   Immanuel Kant, *Kritik der Urtheilskraft*, 2d ed. (Berlin, 1793), sec. 80, pp. 368ff.

5   Aristotle, *De Partibus Animalium*, I, 5; 645b, 14ff.

6   Johann Gottfried Herder, *Briefe zu Befdrderung der Humanitat*, I (Riga, 1793), p. 17: "Grosses and vielfaches Organ."

7   1773; this very important publication pioneered the revaluation of medieval German art and architecture.

8   Johann Wolfgang von Goethe, *Werke*, ed. E. Beutler, vol. XIII (Zurich, 1948–54), pp. 16ff.

9   F. W. Schelling, *Werke* II (Stuttgart/Augsburg, 1856–61), pp. 350ff.

10 Hegel, in *Encyclopadie der Philosophischen Wissenschaften im Grundrisse*, ed. G. Lasson (Leipzig, 1911), sec. 337ff.

11 Ibid., p. 341.

12 Ibid., pp. 371ff.

13 M. Foucault, *Les Mots et les Choses* (Paris, 1966), p. 144.

14 Vitruvius, *De Architectura*, X, 1, iii.

15 Lucretius, *De Rerum Natura*, III, 132.

16 Michele Todini, *Dichiarazione della Galleria Armonica* (Rome, 1676); Athanasius Kircher, S. J.,

Phonurgia Nova sive Conjugium Mechanico-Physicum Artis et Naturae (Campidona [Kempten], 1673), pp. 167ff.

17   "We marvel at something," says Thomas Aquinas, "when we see an effect, but do not know its cause" (*Summa contra gentiles*, III, 101). On medieval marvelous machines, see most recently Michel Camille, *The Gothic Idol, Ideology and Image-Making in Medieval* Art (Cambridge, 1989), pp. 244ff.

    Kircher writes of Todini's *celebra et pené prodigiose machina* that he *exhibet is . . . artis suae specimina summo omnium stupore & admiratione eorum, qui magno . . . numero confluunt.*

18   La Mettrie's concept of the *homme-machine* is found in a book of that title that was first published in Leiden in 1747; for the best modern edition, see Aram Vartanian, *L'Homme Machine: A Study in the Origins of an Idea* (Princeton, 1960).

19   Aristotle, *De Partibus Animalium*, 687a.

20   Vitruvius, *De Architectura*, III, 1; IV, 1ff.

21   Dictionnaire de l'Academie (Paris, 1778).

22   A. Memmo, *Elementi di Architettura Lodoliana*, vol. 1 (Zara [Zadar], 1833), pp. 84ff.

23   "Dire que la physiologie est la physique des animaux c'est en donner une idee extremement inexacte; j'aimerais autant dire que l'astronomie est la physiologie des astres," quoted in G. Cangouilhem, *La Connaissance de la Vie* (Paris, 1965), p. 95.

24   Arthur Schopenhauer, "Parerga and paralipomena," in *Samtliche Werke*, vol. VI (Leipzig, n.d.), pp. 94ff, 186ff.

25   Hans Driesch, *Science and Philosophy of the Organism*, 2d English ed. (London, 1929), esp. pp. 269ff. on "Entelechy and Mechanics."

26   Goethe, *Werke*, op. cit., vol. XVII, pp. 114ff. The anatomist Karl-Friedrich Burdach is usually credited with having first printed the word in 1800.

27   Ibid., pp. 380ff.

28   Georges Cuvier, *The Animal Kingdom* (London, 1849), pp. 16ff; the first edition was published in Paris in 1816.

29   Horatio Greenough, *Form and Function*, ed. Harold A. Small (Berkeley, 1947), pp. 71ff.

30   Leopold Eidlitz, *Nature and Function of Art, More Especially of Architecture* (New York, 1881), but based on earlier articles.

31   Louis Sullivan, "Ornament in Architecture," in *Kindergarten Chats* (New York, 1947), p. 189.

32   Horatio Greenough, *Aesthetics in Washington* (1851), quoted by Nathalia Wright, *Horatio Greenough: The First American Sculptor* (Philadelphia, 1963), pp. 267ff.

33   Horatio Greenough, *Aesthetics in Washington* (1851), p. 22.

34   Leon Battista Alberti, *De Re Aedificatoria*, VI, 2; cf. Prologue and IV, 2. For a modern translation, see Leon Battista Alberti, "On the art of building," in *On the Art of Building in Ten Books*, trans. J. Rykwert, R. Tavernor, and N. Leach (Cambridge, MA, 1988).

35   Frank Lloyd Wright, *Ausgefuhrte Bauten* (Berlin, 1911). C. R. Ashbee quoting Wright's words (p. 10) in the introductory essay.

1994

Stewart Brand

Shearing

Stewart Brand (born 1938 in Rockford, Illinois) is perhaps best known as the founder and editor of the Whole Earth Catalog (1968–85), which began his long career of adventurous and topical publications. The immediate activities of Whole Earth spawned the *Co-Evolution Quarterly* (1974) and *Two Cybernetic Frontiers* (1974), and also led to a very early involvement with computer networks in activities such as the WELL (Whole Earth 'Lectronic Link) (1984) and *The Media Lab: Inventing the Future at MIT* (1987), all of which drew on his broad interest in biology, ecology, technology, computers, and systems theory. In 1988 he co-founded the *Global Business Network* and ran a series of private conferences on "Learning in Complex Systems," sponsored by strategic planners at Royal Dutch/Shell, AT&T, and Volvo. In 1988 he joined the Board of Trustees of the Santa Fe Institute, an organization dedicated to multi-disciplinary research in the sciences of complexity. Brand's enthusiastic support of geodesic domes in the Whole Earth Catalog, and later recognition of their failures, led him to a long study of buildings and architecture that culminated in the book from which this excerpt is drawn, *How Buildings Learn: What Happens After They're Built* (1994).

Brand approached the subject of architecture like a biologist or ecologist, documenting building changes through time, and drew heavily on two bodies of work in the field that paid close attention to time effects in building: office planning and historic preservation. Since the 1950s, office planners had been increasingly influenced by management theories of organizational communication and rapid response, and the architect Frank Duffy (see Duffy, 1997) had carefully studied and described techniques for accommodating the many kinds and rates of change in offices. Preservationists, on the other hand, were dealing with buildings that had been changed over decades and generations, and so had to develop concepts for understanding and describing those changes.

This brief excerpt is an elaboration of Duffy's four layers of "longevity of built components": "Shell, Services, Scenery, and Set," which Brand extended to six: "site, structure, skin, services, space plan, stuff." SLA has subsequently extended them to seven, but however many there are, the velocity of different layers becomes a key description of the building as system of adaptation to change.

HERE'S A PUZZLE. On most American magazine racks you'll find a slick monthly called *Architectural Digest.* Inside are furniture and decor ads and articles with titles like "Unstudied Spaces in Malibu" and "Paris, New York (20th-Century French Pieces Transform an East Side Apartment)." Almost no architecture. The magazine's subtitle reads: "The International Magazine of Fine Interior Design."

Architects and interior designers revile and battle each other. Interior design as a profession is not even taught in architecture departments. At the enormous University of California, Berkeley, with its prestigious Environmental Design departments and programs, architecture students can find no course on interior design anywhere. They could take a bus several miles to the California College of Arts and Crafts, which does teach interior design, but no one takes that bus.

How did *Architectural Digest* manage to jump the chasm? Advertisers, the market, and a profound peculiarity of buildings did it. Originally, back in 1920, it *was* an architecture magazine, though for a public rather than a strictly professional audience. Gradually the magazine noticed that its affluent readers rebuilt interiors much more often than they built houses. After 1960, the advertisers, followed dutifully by the editors, migrated away from exterior vision toward interior revision – toward decorous remodeling – where the action and the money were. The peculiarity of buildings that turned *Architectural Digest* into a contradiction of itself is that different parts of buildings change at different rates.

The leading theorist – practically the only theorist – of change rate in buildings is Frank Duffy, cofounder of a British design firm called DEGW (he's the "D"), and president of the Royal Institute of British Architects from 1993 to 1995. "Our basic argument is that there isn't such a thing as a building," says Duffy. "A building properly conceived is several layers of longevity of built components." He distinguishes four layers, which he calls Shell, Services, Scenery, and Set. Shell is the structure, which lasts the lifetime of the building (fifty years in Britain, closer to thirty-five in North America). Services are the cabling, plumbing, air conditioning, and elevators ("lifts"), which have to be replaced every fifteen years or so. Scenery is the layout of partitions, dropped ceilings, etc., which changes every five to seven years. Set is the shifting of furniture by the occupants, often a matter of months or weeks.

Like the advertisers of *Architectural Digest*, Duffy and his architectural partners steered their firm toward the action and the money. DEGW helps rethink and reshape work environments for corporate offices, these days with a global clientele. "We try to have long-term relationships with clients," Duffy says. "The unit of analysis for us isn't the building, it's the use of the building through time. Time is the essence of the real design problem."

I've taken the liberty of expanding Duffy's "four S's" – which are oriented toward interior work in commercial buildings – into a slightly revised, general-purpose "six S's":

- SITE – This is the geographical setting, the urban location, and the legally defined lot, whose boundaries and context outlast generations of ephemeral buildings. "Site is eternal," Duffy agrees.
- STRUCTURE – The foundation and load-bearing elements are perilous and

expensive to change, so people don't. These are the building. Structural life ranges from thirty to 300 years (but few buildings make it past sixty, for other reasons).

- SKIN – Exterior surfaces now change every twenty years or so, to keep up with fashion or technology, or for wholesale repair. Recent focus on energy costs has led to re-engineered Skins that are air-tight and better-insulated.
- SERVICES – These are the working guts of a building: communications wiring, electrical wiring, plumbing, sprinkler system, HVAC (heating, ventilating, and air conditioning), and moving parts like elevators and escalators. They wear out or obsolesce every seven to fifteen years. Many buildings are demolished early if their outdated systems are too deeply embedded to replace easily.
- SPACE PLAN – The interior layout – where walls, ceilings, floors, and doors go. Turbulent commercial space can change every three years or so; exceptionally quiet homes might wait thirty years.
- STUFF – Chairs, desks, phones, pictures; kitchen appliances, lamps, hair brushes; all the things that switch around daily to monthly. Furniture is called mobilia in Italian for good reason.

Duffy's time-layered perspective is fundamental to understanding how buildings actually behave. The 6-S sequence is precisely followed in both design and construction. As the architect proceeds from drawing to drawing – layer after layer of tracing paper – "What stays fixed in the drawings will stay fixed in the building over time," says architect Peter Calthorpe. "The column grid will be in the bottom layer." Likewise the construction sequence is strictly in order: Site preparation, then foundation and framing the Structure, followed by Skin to keep out the weather, installation of Services, and finally Space plan. Then the tenants truck in their Stuff.

Frank Duffy:

> Thinking about buildings in this time-laden way is very practical. As a designer you avoid such classic mistakes as solving a five-minute problem with a fifty-year solution, or vice versa. It legitimizes the existence of different design skills – architects, service engineers, space planners, interior designers – all with their different agendas defined by this time scale. It means you invent building forms which are very adaptive.

The layering also defines how a building relates to people. Organizational levels of responsibility match the pace levels. The building interacts with individuals at the level of Stuff; with the tenant organization (or family) at the Space plan level; with the landlord via the Services (and slower levels) which must be maintained; with the public via the Skin and entry; and with the whole community through city or county decisions about the footprint and volume of the Structure and restrictions on the Site. The community does not tell you where to put your desk or your bed: you do not tell the community where the building will go on the Site (unless you're way out in the country).

Buildings rule us via their time layering at least as much as we rule them, and in a surprising way. This idea comes from Robert V. O'Neill's *A Hierarchical Concept of Ecosystems*. O'Neill and his co-authors noted that ecosystems could be better understood by observing the rates of change of different components. Hummingbirds and flowers are quick, redwood trees slow, and whole redwood forests even slower. Most interaction is within the same pace level – hummingbirds and flowers pay attention to each other, oblivious to redwoods, who are oblivious to them. Meanwhile the forest is attentive to climate change but not to the hasty fate of individual trees. The insight is this: "*The dynamics of the system will be dominated by the slow components, with the rapid components simply following along.*"[1] Slow constrains quick; slow controls quick.

The same goes with buildings: the lethargic slow parts are in charge, not the dazzling rapid ones. Site dominates the Structure, which dominates the Skin, which dominates the Services, which dominate the Space plan, which dominates the Stuff. How a room is heated depends on how it relates to the heating and cooling Services, which depend on the energy efficiency of the Skin, which depends on the constraints of the Structure. You could add a seventh "S" – human Souls at the very end of the hierarchy, servants to our Stuff.

Still, influence does percolate the other direction. The slower processes of a building gradually integrate trends of rapid change within them. The speedy components propose, and the slow dispose. If an office keeps replacing its electronic Stuff often enough, finally management will insist that the Space plan acquire a raised floor to make the constant recabling easier, and that's when the air conditioning and electrical Services will be revamped to handle the higher load. Ecologist Buzz Holling points out that it is at the times of major changes in a system that the quick processes can most influence the slow.

1   R. V. O'Neill, D. L. DeAngelis, J. B. Wade, T. F. H. Allen, *A Hierarchical Concept of Ecosystems* (Princeton: Princeton University Press, 1986), p. 98.

354

1995

Rem Koolhaas

Speculations on
Structures and
Services

Rem Koolhaas was born in Rotterdam in the Netherlands in 1944 and spent the early part of his childhood in Indonesia where his father was employed by the government as Cultural Director. After working as a journalist in The Hague – and writing several film screenplays – he moved to London in 1968 to study at the Architectural Association, graduating in 1972. He founded the Office for Metropolitan Architecture (OMA) in 1975 with his wife, the painter Madelon Vriesendorp, and Elia and Zoe Zenghelis. Since 1995 he has taught at Harvard University's Graduate School of Design and he has also been a Visiting Professor at Tsinghua University in Beijing.

Koolhaas and OMA have completed a number of critically acclaimed buildings including: the National Dance Theater in The Hague (1987); the Kunsthal, Rotterdam (1992); the master plan and the Grand Palais for Lille (1994); the Educatorium, a "factory for learning" at the University of Utrecht in the Netherlands (1997); a residence in Bordeaux (1998); the Prada store in New York (2002) and the Campus Center at the Illinois Institute of Technology in Chicago (2003). He was awarded the Pritzker Prize in 2000 and the RIBA Gold Medal in 2004.

His parallel career in architectural research and writing began in 1972 when he received a Harkness Fellowship to study in the United States. After working with Colin Rowe and O. M. Ungers at Cornell University he became a Visiting Fellow at the Institute for Architecture and Urban Studies (directed by Peter Eisenman) in New York. While there he researched and wrote the book *Delirious New York: a Retroactive Manifesto for Manhattan*, which was first published in 1978. More recently he has collaborated with the graphic artist Bruce Mau on the book *S, M, L, XL* (1995) from which the following extract is taken. The title describes the grouping of projects within the book according to their size and also reflects a preoccupation with issues of urban scale and density as opposed to the traditional architectural concerns of composition, materiality, and detail. Koolhaas celebrates the link between technology, economics, and progress, and – while reveling in the often unexpected products of rapid global development – questions architects' frequent lack of engagement with the real forces of change in contemporary society.

Architects will be the last for whom the apples fall . . .

Since gravity works as a sum, the theoretical shape of a column is a cone; to deal with accumulating forces, it is thin at the top and fat at the bottom.

The *taller* the building, the more the structural inheritance from the upper regions dictates decisions below. Each high-rise represents the systematic reduction of freedom toward where it matters most: on the ground.

The *deeper*[1] the building, the more it depends on artifice for its servicing. Air is injected into its interior, used (i.e., turned into poison), and extracted; the inside core, inaccessible to daylight, is lit by fluorescent tubes (gasses in a permanent state of explosion). In the conventional solution – combining the claims of structure and services – the ducts that carry air to and from the center are hung from the floor, then hidden behind a false ceiling. This zone of darkness is further stuffed with equipment for lighting, electricity, smoke detectors, sprinklers, computers, and other building "controls."

The section is no longer simply divided by the discrete demarcations of individual floors; it has become a *sandwich*, a kind of conceptual zebra; free zones for human occupancy alternate with inaccessible bands of concrete, wiring, and ducts.

To avoid interference from the columns and their unwelcome inheritance, the structural grid widens, increasing the depth of the floor slabs. Ducts inflate to deliver greater perfection to ever more distant destinations. Wiring proliferates, claiming more space.

The more *sophisticated* the building, the greater the expansion of the inaccessible zones, expropriating ever larger parts of the section. The expertise and autonomy of the advisers (quaint title) parallels this expansion. Suddenly, the architect has to fight on two fronts: on the first, he faces the client, who is already nervous at having started this enterprise – a Big Building; on the second, he confronts the sabotage of engineers, his supposed "teammates," with their tantalizingly vague (if not outright poetic) indications from what is supposedly the domain of pure science. Floors suddenly "have to be . . . millimeters," ducts "probably not less than . . . in diameter," beams "would be a lot safer at . . . meters," stability "*could* be achieved by . . ." Additional "disciplines" claim major reservations in section and plan (nobody knows exactly what for) in a metaphysics of pragmatic precaution against "things" that "might" or "always" happen.

Idealism vs. philistinism: the section becomes battlefield; white and black compete for outright domination. (In some hospitals the dark bands of the section exceed 50 percent of the total and block 75 percent of the budget.) The dark zone is not only strictly "useless" for the future inhabitants of the building; it also becomes conceptually inaccessible to the architect, who has become an intruder in his own project, boxed in, his domain a mere residue of the others' demands. The architect's arguments are always opinions; they cannot compete with the aura of objectivity that shields building technologies from critical probing. (In this reading, "high tech" is not only ridiculous in its decorative posturing, but worse, celebrates the final masochistic surrender of the architect: the substitution by technical impediment of architectural possibility.)

The presence of technique in *Delirious New York* was selective: the book identified the elevator, steel, air-conditioning as a "technology of the fantastic." By surrendering their "objective" status, these inventions could enlist in the experimental enterprise of a new architecture and, in fact, become indissociable from it.

This bonding represented an almost Darwinian adaptation to the demands of the metropolitan ecology: a mutated architecture no longer obsessively committed to form making but to the creation of *conditions*, the fabrication of content – scriptwriting by tectonic means.

In retrospect, Manhattan architects seem to have had an impossibly direct relationship with their profession: a pure alignment with collective forces that they could translate without any distancing tactics, with an apparent absence of second thoughts. While each European building is also comment, reflection, philosophy, theory, hesitation – with a corresponding depth, tension, subtlety – the suspense of American building is the shamelessness of its utilitarian efficiency. Like children playing with matches, New York's architects had invented a way to live innocently with Promethean obligations.

Maybe *that* was the (almost 100-year-old) immanent otherness of the twentieth-century architect: the revelation that instead of adopting the megalomaniac caricatures of constructivist social engineer, Wrightian *Gesamtkunstwerk* maestro, Miesian stoic, Corbusian artist-organizer, he might simply abandon the stage of conventional expectation and reappear in a completely different arena, in fact, change professions.

After *Delirious New York*, it was convenient to treat the book – the transformation of architecture it implied – as an isolated incident. OMA's European beginnings in the early eighties offered no pretexts for its relevance. We were involved in our own on-the-job training, staring the beast of architecture in the face for the first time. The additional weight of proving the book's combined revisions would have been a theoretical millstone. As in cryogenics, this body of work was frozen.

In 1985 we began to collaborate with Cecil Balmond, a Ceylonese engineer, and his structure and services unit at Ove Arup. He was patient with our unreasonable demands, and sometimes took our amateurism seriously. Our growing intimacy with each other's disciplines – in fact, a mutual invasion of territory – and the corresponding blurring of specific professional identities (not always painless) allowed us, at the end of the eighties – when, to our own consternation, Bigness emerged like a sudden iceberg from the mist of deconstructivist discourse and imposed itself as a political, economic, artistic necessity – to defrost earlier ambitions and to explore the redesign and demystification of architecture, this time experimenting on ourselves.

With the cluster of the Very Big Library (250,000 m²), ZKM (two laboratories, a theater, two museums), and the Zeebrugge terminal, it seemed that the impossible constellation of need, means, and naïveté that had triggered New York's miracles had returned.

The simultaneous work in the summer of 1989 on these three competitions forced us to explore the potential of building Big in Europe, with repercussions equally architectural and technical. They were treated, in the newly bonded OMA-Arup team, as aggressive confrontations with the survival of earlier regimes. While other disciplines were gloating over their new freedoms – the hybrid, the local, the informal, chance, the singular, the irregular, the unique – architecture was stuck in the consistent, the repetitive, the regular, the gridded, the general, the overall, the formal, the predetermined. The work became a joint campaign to explore these freedoms for architecture and engineering, to reconquer the section, to address our shared discomfort with services as the sprawling coils of a proliferating unconscious, to abolish the single grandiose solution integrating structure and services. It was also, more secretly, a search for ways to make buildings that would *look* completely different: for genuine newness. This exploration allowed us to explode other unquestioned assumptions, for instance, that the so-called facade is of particular importance in architecture just because it is the interface between the building and the "natural" world (which explains the humiliating fact that across a seventy-year gap in a century marked by incredible change, the *look* of architecture has barely changed).

In these projects – some of them more than 100 meters deep – the facades merely represent four out of an endless series of possible cuts, most of them vastly more important for the building and its performance as a collective object.

As we concentrated on the "settlement" of the program on these unusual territories, their very unnaturalness opened up more new possibilities: we were forced, for the first time, to explore new potentials for the formation of space.

When we realized that we identified 100 percent with these programmatic enterprises that intervene drastically in the cultural and political landscape of Europe, we wondered whether – paradoxically by playing with the real fire of Bigness, even in Europe – it could be again possible to become innocent about architecture, to use architecture to articulate the new, to imagine – no longer paralyzed by knowledge, experience, correctness – the end of the Potemkin world.

---

1   A "deep" plan suggests a condition where the distance between core and facade is considerable, "depth" is also used to indicate the vertical dimension – the height – of structures such as beams and trusses.

1995

Félix Guattari

Machinic
Heterogenesis

(Pierre-) Félix Guattari, psychiatrist and philosopher, was born in 1930 in Villeneuve-les-Sablons in France. Originally trained as a psychoanalyst, Guattari worked during the 1950s at a clinic near Paris called La Borde, noted for its innovative therapeutic practices and its teaching of philosophy, psychology, and ethnology students, alongside social workers. Guattari also studied with the celebrated French psychoanalyst Jacques Lacan, whose reevaluation of the centrality of the "unconscious" in psychoanalytic theory had attracted many disciples. In the mid-1960s Guattari broke with Lacan and founded his own clinics, the Society for Institutional Psychotherapy (1965) and the Centre for Institutional Studies and Research (1970). He also maintained his clinical base at La Borde until his death in 1992.

Guattari established the Federation of Groups for Institutional Study and Research, supporting numerous radical political causes, and in 1967 was one of the founders of OSARLA (Organization of Solidarity and Aid to the Latin-American Revolution). Inspired by the student uprising in Paris in May 1968, Guattari collaborated with the French philosopher Gilles Deleuze (1925–95) to produce a two-volume work of "antipsychoanalytic" philosophy entitled *Capitalism and Schizophrenia*. In volume 1, *Anti-Oedipus* (1972), they suggested that schizophrenia provided one of the few authentic forms of rebellion against the capitalist system. Volume 2, *A Thousand Plateaus* (1980), is characterized by a deliberately disjointed style of inquiry, reflecting the authors' belief that the linear organization of conventional philosophical writing represented a powerful form of conceptual control. The work is presented as a study in what Deleuze and Guattari called "deterritorialization" – an attempt to destabilize the traditionally repressive, unitary, conceptions of identity, meaning, and truth.

In his last book, *Chaosmosis* (originally published in French as *Chaosmose* in 1992), from which the current extract is taken, Guattari again takes up one of his favourite topics – the nature of subjectivity: "How to produce it, collect it, enrich it, reinvent it permanently in order to make it compatible with mutant universes of value?" In this essay he insists that technology must be defined more broadly, that we must abandon the simplistic opposition between the technical and the natural – the distinction between the tool and its human operator. Instead we must try to grasp the "machinic" as a continuum of related elements, containing particular technical devices inseparably embedded within the vast networks of materials, processes, systems and infrastructures (both technical and socio-political) within which they must inevitably operate. True to his cultural roots Guattari devotes a substantial part of his argument to an attack on the binary logic inherent in French structuralist philosophy – its reliance on the oppositional pairings of signifier and signified, natural and artificial, organic and mechanical, etc. In those fields less dominated by the legacy of structuralist thinking, some of this intellectual energy may seem misplaced. But likewise with his contemporary and collaborator, Gilles Deleuze, this process yields many new possibilities of thought across traditional conceptual boundaries.

Common usage suggests that we speak of the machine as a sub-set of technology. We should, however, consider the problematic of technology as dependent on machines, and not the inverse. The machine would become the prerequisite for technology rather than its expression. Machinism is an object of fascination, sometimes of délire, about which there's a whole historical "bestiary." Since the origin of philosophy, the relationship between man and machine has been the object of interrogation. Aristotle thought that the goal of techne was to create what nature found impossible to accomplish. Being of the order of "knowledge" and not of "doing," techne interposes a kind of creative mediation between nature and humanity whose status of intercession is a source of perpetual ambiguity. "Mechanist" conceptions of the machine empty it of everything that would enable it to avoid a simple construction *partes extra partes.* "*Vitalist*" conceptions assimilate the machine to living beings; unless it is living beings that are assimilated to machines. The "cybernetic" perspective developed by Norbert Wiener[1] envisages living systems as particular types of machines equipped with the principle of feedback. More recent "systemic" conceptions (Humberto Maturana and Francisco Varela) develop the concept of autopoiesis (auto-production), reserving it for living machines. Following Heidegger, a philosophical fashion entrusts techne – in its opposition to modern technology – with the mission of "unmasking the truth" that "seeks the true in the exact." Thus it nails techne to an ontological plinth – to *a grund* – and compromises its character of processual opening.

Through these positions, we will attempt to discern various levels of ontological intensity and envisage machinism in its totality, in its technological, social, semiotic, and axiological avatars. And this will involve a reconstruction of the concept of machine that goes far beyond the technical machine. For each type of machine, we will pose a question, not about its vital autonomy – it's not an animal – but about its singular power of enunciation: what I call its specific enunciative consistency. The first type of machine we are going to consider is the material apparatus. They are made by the hand of man – itself taken over by other machines – according to conceptions and plans which respond to the goals of production. These different stages I will call finalised, diagrammatic schemas. But already this montage and these finalisations impose the necessity of expanding the limits of the machine, *stricto sensu*, to the functional ensemble which associates it with man. We will see that this implies taking into account multiple components:

- material and energy components;
- semiotic, diagrammatic, and algorithmic components (plans, formulae, equations, and calculations which lead to the fabrication of the machine);
- components of organs, influx and humours of the human body – individual and collective mental representations and information;
- investments of desiring machines producing a subjectivity adjacent to these components;
- abstract machines installing themselves transversally to the machinic levels previously considered (material, cognitive, affective, and social).

When we speak of abstract machines, by "abstract" we can also understand "extract" in the sense of extracting. They are montages capable of relating all the heterogeneous levels that they traverse and that we have just enumerated. The abstract machine is transversal to them, and it is this abstract machine that will or will not give these levels an existence, an efficiency, a power of ontological auto-affirmation. The different components are swept up and reshaped by a sort of dynamism. Such a functional ensemble will hereafter be described as a machinic assemblage. The term assemblage does not imply any notion of bond, passage, or anastomosis between its components. It is an assemblage of possible fields, of virtual as much as constituted elements, without any notion of generic or species' relation. In this context, utensils, instruments, the most basic tools and the least structured pieces of a machine acquire the status of a proto-machine.

Let us take an example. If we take a hammer apart by removing its handle, it is still a hammer but in a "mutilated" state. The "head" of the hammer – another zoomorphic metaphor – can be reduced by fusion. It will then cross a threshold of formal consistency where it will lose its form: this machinic gestalt works moreover as much on a technological plane as on an imaginary level, to evoke the dated memory of the hammer and sickle. We are simply in the presence of metallic mass returned to smoothness, to the deterritorialisation which precedes its appearance in a machinic form. To go beyond this type of experiment – comparable to the piece of Cartesian wax – let us attempt the inverse, to associate the hammer with the arm, the nail with the anvil. Between them they maintain relations of syntagmatic linkage. And their "collective dance" can bring to life the defunct guild of black-smiths, the sinister epoch of ancient iron mines, the ancestral use of metal-rimmed wheels ... Leroi-Gourhan emphasised that the technical object was nothing outside of the technical ensemble to which it belonged. It is the same for sophistic-ated machines such as robots, which will soon be engendered by other robots. Human action remains adjacent to their gestation, waiting for the breakdown which will require its intervention: this residue of a direct act. But doesn't all this suggest a partial view, a certain taste for a dated period of science fiction? Curiously, in acquiring more and more life, machines demand in return more and more abstract human vitality: and this has occurred throughout their evolutionary development. Computers, expert systems, and artificial intelligence add as much to thought as they subtract from thinking. They relieve thought of inert schemas. The forms of thought assisted by computer are mutant, relating to other musics, other Universes of reference.[2]

It is, then, impossible to deny the participation of human thought in the essence of machinism. But up to what point can this thought still be described as human? Doesn't technico-scientific thought fall within the province of a certain type of mental and semiotic machinism? What we need here is a distinction between on the one hand semiologies that produce significations, the common currency of social groups – like the "human" enunciation of people who work with machines – and on the other, a-signifying semiotics which, regardless of the quantity of signifi-cations they convey, handle figures of expression that might be qualified as "non-human" (such as equations and plans which enunciate the machine and make it act in a diagrammatic capacity on technical and experimental apparatuses). The semiologies of signification play in keys with distinctive oppositions of a phone-matic or scriptural order which transcribe enunciations into materials of signifying expression. Structuralists have been content to erect the Signifier as a category unifying all expressive economies: language, the icon, gesture, urbanism, or the cinema, etc. They have postulated a general signifying translatability for all forms of discursivity. But in so doing, have they not misunderstood the essential dimension of machinic autopoiesis? This continual emergence of sense and effects does not concern the redundancy of mimesis but rather the production of an effect of singular sense, even though indefinitely reproducible.

This autopoietic node in the machine is what separates and differentiates it from structure and gives it value. Structure implies feedback loops, it puts into play a concept of totalisation that it itself masters. It is occupied by inputs and outputs whose purpose is to make the structure function according to a principle of eternal return. It is haunted by a desire for eternity. The machine, on the contrary, is shaped by a desire for abolition. Its emergence is doubled with breakdown, catastrophe – the menace of death. It possesses a supplement: a dimension of alterity which it develops in different forms. This alterity differentiates it from structure, which is based on a principle of homeomorphism. The difference supplied by machinic autopoiesis is based on disequilibrium, the prospection of virtual Universes far from equilibrium. And this doesn't simply involve a rupture of formal equilibrium, but a radical ontological reconversion. The machine always depends on exterior elements in order to be able to exist as such. It implies a complementarity, not just with the man who fabricates it, makes it function, or destroys it, but it is itself in a relation of alterity with other virtual or actual machines – a "non-human" enunciation, a proto-subjective diagram.

This ontological reconversion dismisses the totalising scope of the concept of the Signifier. Because the signifying entities which operate the diverse mutations of the ontological referent – that makes us move from the Universe of molecular chemistry to the Universe of biological chemistry, or from the acoustic world to the world of polyphonic and harmonic music – are not the same. Of course, lines of signifying decoding, composed of discrete figures – binarisable, syntagmatisable and paradigmatisable – sometimes appear in one Universe or another. And we can have the illusion that the same signifying network occupies all these domains. It is, however, totally different when we consider the actual texture of these Universes of reference. They are always stamped with the mark of singularity. From acoustics to polyphonic music, there is a divergence of constellations of expressive intensity. They involve a certain pathic relationship, and convey irreducibly heterogeneous ontological consistencies. We thus discover as many types of deterritorialisation as traits of expressive materials. The signifying articulation hanging over them – in its indifferent neutrality – is incapable of imposing itself as a relation of immanence to machinic intensities, to this non-discursive, auto-enunciating, auto-valorising, autopoietic node. It does not submit to any general syntax of the procedures of deterritorialisation. No couplet – Being–being, Being–Nothingness, being–other – can claim the status of an ontological binary digit. Machinic propositions elude the ordinary games of discursivity and the structural coordinates of energy, time, and space.

Yet an ontological transversality does nonetheless exist in them. What happens at a level of the particulate-cosmic is not without relation to the human soul or events in the socius. But not according to harmonic universals of the Platonic type (Sophist). The composition of deterritorialising intensities is incarnated in abstract machines. We should bear in mind that there is a machinic essence which will incarnate itself in a technical machine, and equally in the social and cognitive environment connected to this machine – social groups are also machines, the body is a machine, there are scientific, theoretical, and information machines. The abstract machine passes through all these heterogeneous components but above all it heterogenises them, beyond any unifying trait and according to a principle of irreversibility, singularity, and necessity. In this respect the Lacanian signifier is struck with a double lack: it is too abstract in that it makes heterogeneous, expressive materials translatable. It lacks ontological heterogenesis, it gratuitously uniformises and syntaxises diverse regions of being, and, at the same time, it is not abstract enough because it is incapable of taking into account the specificity of these machinic autopoietic nodes, to which we must now return.

Francisco Varela characterises a machine by the set of inter-relations of its components independent of the components themselves."[3] The organisation of a machine thus has no connection with its materiality. He distinguishes two types of machines: "allopoietic" machines which produce something other than themselves, and "autopoietic" machines which engender and specify their own organisation and limits. Autopoietic machines undertake an incessant process of the replacement of their components as they must continually compensate for the external perturbations to which they are exposed. In fact, the qualification of autopoietic is reserved by Varela for the biological domain: social systems, technical machines, crystalline systems, etc., are excluded. This is the sense of his distinction between allopoiesis and autopoiesis. But autopoiesis, which uniquely defines autonomous entities – unitary, individuated, and closed to input/output relationships – lacks characteristics essential to living organisms, like the fact that they are born, die, and survive through genetic phylums. Autopoiesis deserves to be rethought in terms of evolutionary, collective entities, which maintain diverse types of relations of alterity, rather than being implacably closed in on themselves. In such a case, institutions and technical machines appear to be allopoietic, but when one considers them in the context of the machinic assemblages they constitute with human beings, they become ipso facto autopoietic. Thus we will view autopoiesis from the perspective of the ontogenesis and phylogenesis proper to a mecanosphere superposed on the biosphere.

The phylogenetic evolution of machinism is expressed, at a primary level, by the fact that machines appear across "generations," one suppressing the other as it becomes obsolete. The filiation of previous generations is prolonged into the future by lines of virtuality and their arborent implications. But this is not a question of a univocal historical causality. Evolutionary lines appear in rhizomes; datings are not synchronic but heterochronic. Example: the industrial "take off" of steam engines happened centuries after the Chinese Empire had used them as children's toys. In fact, these evolutionary rhizomes move in blocks across technical civilisations. A technological innovation may know long periods of stagnation or regression, but there are few cases in which it does not "restart" at a later date.

This is particularly clear with military technological innovations: they frequently punctuate long historical periods that they stamp with the seal of irreversibility, wiping out empires for the benefit of new geopolitical configurations. But, and I repeat it, this was already true of the most humble instruments, utensils, and tools which don't escape this phylogenesis. One could, for example, dedicate an exhibition to the evolution of the hammer since the Iron Age and conjecture about what it will become in the context of new materials and technologies. The hammer that one buys today at the supermarket is, in a way, "drawn out" on a phylogenetic line of infinite, virtual extension.

It is at the intersection of heterogeneous machinic Universes, of different dimensions and with unfamiliar ontological textures, radical innovations and once forgotten, then reactivated, ancestral machinic lines, that the movement of history singularises itself. Among other components, the Neolithic machine associates the machine of spoken language, machines of hewn stone, agrarian machines based on the selection of grains and a village proto-economy. The writing machine will only emerge with the birth of urban mega-machines (Lewis Mumford) correlative to the spread of archaic empires. Parallel to this, the great nomadic machines constituted themselves out of the collusion between the metallurgic machine and new war machines. As for the great capitalistic machines, their foundational machinisms were prolific: urban State machines, then royal machines, commercial and banking machines, navigation machines, monotheist religious machines, deterritorialised musical and plastic machines, scientific and technical machines, etc.

The question of the reproducibility of the machine on an ontogenetic level is more complex. Maintaining a machine's operationality – its functional identity – is never absolutely guaranteed: wear and tear, fine balance, breakdowns, and entropy demand a renewal of its material components, its energy and information components, the latter able to be lost in "noise." Equally, the maintenance of a machinic assemblage's consistency demands that the element of human action and intelligence involved in its composition must also be renewed. The man–machine alterity is thus inextricably linked to a machine–machine alterity which operates in relations of complementarity or agonistic relations (between war machines) or again in the relations of parts or apparatuses. In fact, the wear and tear, accident, death, and resurrection of a machine in a new copy or model are part of its destiny and can become central to its essence in certain aesthetic machines (the "compressions" of César, the "metamechanics," the happening machines, the delirious machines of Jean Tinguely). The reproducibility of the machine is not a pure programmed repetition. The scansions of rupture and indifferentiation, which uncouple a model from any support, introduce their own share of both ontogenetic and phylogenetic difference. It is in this phase of passage to a diagrammatic state, a disincarnate abstract machine, that the "supplements of the soul" of the machinic node are distinguished from simple material agglomerates. A heap of stones is not a machine, whereas a wall is already a static protomachine, manifesting virtual polarities, an inside and outside, an above and below, a right and left ... These diagrammatic virtualities take us beyond Varela's characterisation of machinic autopoiesis as unitary individuation, with neither input nor output; they direct us towards a more collective machinism without delimited unity, whose autonomy accommodates diverse mediums of alterity. The reproducibility of the technical machine differs from that of living beings, in that it is not based on sequential codes perfectly circumscribed in a territorialised genome. Obviously every technological machine has its own plans for conception and assembly. But while these plans keep their distance from the machine, they also move from one machine to another so as to constitute a diagrammatic rhizome which tends to cover the mecanosphere globally. The relations of technological machines between themselves, and the way their respective parts fit together, presuppose a formal serialisation and a certain perdition of their singularity – stronger than that of living machines – correlative to a distance between the machine manifested in energetico-spatio-temporal coordinates and the diagrammatic machine which develops in more deterritorialised coordinates.

This deterritorialising distance and loss of singularity needs to be related to a reciprocal smoothing of the materials constitutive of the technical machine. Of course, singular rough patches belonging to these materials can never be completely abolished but they must only interfere with the machine's "play" if they are required to do so by its diagrammatic functioning. Let us examine these two aspects of machinic separation and smoothing, taking an apparently simple machinic apparatus – the couple formed by a lock and its key. Two types of form, with ontologically heterogeneous textures are at work here: 1) materialised, contingent, concrete and discrete forms, whose singularity is closed in on itself, embodied respectively in the profile Fl of the lock and by the profile Fk of the key. Fl and Fk never quite coincide. They evolve through time, due to wear and oxidation, but both forms must stay within the framework of a separation-type limit beyond which the key would cease to be operational; 2) "formal," diagrammatic forms, subsumed within this separation-type, which appear as a continuum including the whole range of profiles Fl, Fk, compatible with the effective operation of the lock.

One quickly notices that the machinic effect, the passage to the possible act, is entirely concerned with the second type of form. Although ranged across the most restrained separation-type limit possible, these diagrammatic forms appear infinite in number. In fact, it is a matter of an integral of forms Fk, Fl.

This infinite integral form doubles and smooths the contingent forms Fl and Fk which only have value machinically inasmuch as they belong to it. A bridge is thus established "above" the concrete, authorised forms. I call this operation deterritorialised smoothing and it applies as much to the normalisation of the machine's constitutive materials as it does to their "digital" and functional description. Ferric ore which has been insufficiently worked, or deterritorialised, retains irregularities from the milling of the original material, which would distort the ideal profiles of the lock and key. The smoothing of the material has to remove excessive aspects of contingence from it, and make it behave in a way that accurately moulds the formal imprints extrinsic to it. We should add that this moulding – in a way comparable to photography – should not be too evanescent and should conserve a properly sufficient consistency. Here again we find a separation-type phenomenon, putting into play a theoretical diagrammatic consistency. A lead or golden key risks bending in a steel lock. A key that is changed into a liquid or gaseous state immediately loses its pragmatic efficiency and departs from the field of the technical machine.

This phenomenon of a formal threshold can be found at all levels of intra- or inter-machine relations, and in particular with the existence of spare parts. The components of the technical machine are thus like the units of a currency, and this has become more evident since computers started to be used in their conception and design. These machinic forms, these smoothings of material, of a separation-type limit between parts and their functional adjustments, would suggest that form takes precedence over consistency and over material singularities – the technological machine's reproducibility appearing to dictate that each of its elements fit into a pre-established definition of a diagrammatic order. Charles Sanders Peirce, who described the diagram as an "icon of relation" and assimilated it to the function of algorithms, proposed a broader vision that is worth developing further in the present perspective. Here, the diagram is conceived as an autopoietic machine which not only gives it a functional and material consistency, but requires it to deploy its diverse registers of alterity, freeing it from an identity locked into simple structural relations. The machine's proto-subjectivity installs itself in Universes of virtuality which extend far beyond its existential territoriality. Thus we refuse to postulate a formal subjectivity intrinsic to diagrammatic semiotisation, for example, a subjectivity "lodged" in signifying chains according to the well-known Lacanian principle: "a signifier represents the subject for another signifier." For the machine's diverse registers, there is no univocal subjectivity based on cut, lack, or suture, but there are ontologically heterogeneous modes of subjectivity, constellations of incorporeal Universes of reference which take the position of partial enunciators in multiple domains of alterity, or more precisely, domains of alterification.

We have already encountered a certain number of these registers of machinic alterity:

- the alterity of proximity between different machines and between different parts of the same machine;
- the alterity of an internal, material consistency;
- the alterity of formal, diagrammatic consistency;
- the alterity of the evolutionary phylum;
- the agonistic alterity between machines of war, whose prolongation we could associate with the "auto-agonistic" alterity of desiring machines which tend towards their own collapsus and abolition.

Another form of alterity which has only been approached very indirectly, is the alterity of scale, or fractal alterity, which establishes a play of systematic correspondences between machines at different levels.[4] We are not, however, in the process of drawing up a universal table of forms of machinic alterity because, in truth, their ontological modalities are infinite. They organise themselves into constellations of incorporeal Universes of reference with unlimited combinatories and creativity.

Archaic societies are better equipped than White, male, capitalistic subjectivities to produce a cartography of this multivalence of alterity. With regard to this, we could refer to Marc Augé's account of the heterogeneous registers relating to the fetish object Legba in African societies of the Fon. The Legba comes to being transversally in:

- a dimension of destiny;
- a universe of vital principle;
- an ancestral filiation;
- a materialised god;
- a sign of appropriation;
- an entity of individuation;
- a fetish at the entrance to the village, another at the portal of the house and, after initiation, at the entrance to the bedroom . . .

The Legba is a handful of sand, a receptacle, but it's also the expression of a relation to others. One finds it at the door, at the market, in the village square, at crossroads. It can transmit messages, questions, answers. It is also a way of relating to the dead and to ancestors. It is both an individual and a class of individuals; a name and a noun. "Its existence corresponds to the obvious fact that the social is not simply of a relational order but of the order of being."[5] Marc Augé stresses the impossible transparency and translatability of symbolic systems. "The Legba apparatus . . . is constructed on two axes. One is viewed from the exterior to the interior, the other from identity to alterity. Thus being, identity and the relation to the other are constructed, through fetishistic practice, not only on a symbolic basis but also in an openly ontological way."[6]

Contemporary machinic assemblages have even less standard univocal referent than the subjectivity of archaic societies. But we are far less accustomed to the irreducible heterogeneity, or even the heterogenetic character, of their referential components. Capital, Energy, Information, the Signifier are so many categories which would have us believe in the ontological homogeneity of referents (biological, ethological, economic, phonological, scriptural, musical, etc.)

In the context of a reductionist modernity, it is up to us to rediscover that for every promotion of a machinic intersection there corresponds a specific constellation of Universes of value from the moment a partial non-human enunciation has been instituted. Biological machines promote living Universes which differentiate themselves into vegetable becomings, animal becomings. Musical machines establish themselves against a background of sonorous Universes which have been constantly modified since the great polyphonic mutation. Technical machines install themselves at the intersection of the most complex and heterogeneous enunciative components. Heidegger, who turned the world of technology into a kind of malefic destiny resulting from a movement of distancing from being, used the example of a commercial plane on a runway: the visible object conceals "what and how it is." It unveils itself "only as standing-reserve inasmuch as it is ordered to insure the possibility of transportation" and to this end, "it must be in its whole structure and in every one of its constituent parts on call for duty, i.e., ready for take-off."[7] This interpellation, this "ordering" which reveals the real as "standing-reserve" is essentially operated by man and understood in terms of a universal operation, travelling, flying . . . But does this "standing-reserve" of the machine really reside in an already-there, in terms of eternal truths, revealed to the being of man? In fact the machine speaks to the machine before speaking to man and the ontological domains that it reveals and secretes are, on each occasion, singular and precarious.

Let us reconsider the example of a commercial plane, this time not generically but using the technologically dated model baptised as the Concorde. The ontological consistency of this object is essentially composite: it is at the intersection, at the point of constellation and pathic agglomeration of Universes each of which have their own ontological consistency, traits of intensity, their ordinates and coordinates, their specific machinisms. Concorde simultaneously involves:

- a diagrammatic Universe with plans of theoretical "feasibility" – technological Universes transposing this "feasibility" into material terms;
- industrial Universes capable of effectively producing it;
- collective imaginary Universes corresponding to a desire sufficient to make it see the light of day;
- political and economic Universes leading, amongst other things, to the release of credit for its construction . . .

But the bottom line is that the ensemble of these final, material, formal, and efficient causes will not do the job! The Concorde object moves effectively between Paris and New York but remains nailed to the economic ground. This lack of consistency of one of its components has decisively fragilised its global ontological consistency. Concorde only exists within the limited reproducibility of twelve examples and at the root of a possibilist phylum of future supersonics. And this is hardly negligible!

Why are we so insistent about the impossibility of establishing the general translatability of diverse referential and partial enunciative components of assemblage? Why this lack of reverence towards the Lacanian conception of the signifier? Precisely because this theorising which stems from structural linguistics forbids us from entering the real world of the machine. The structuralist signifier is always synonymous with linear discursivity. From one symbol to another, the subjective effect happens without any other ontological guarantee. As opposed to this, heterogeneous machines, as envisaged from our schizoanalytical perspective, do not produce a standard being at the mercy of a universal temporalisation. To clarify this point we should establish some distinctions between the different forms of semiological, semiotic, and coded linearity:

- the codings of the "natural" world, which operate on several spatial dimensions (for example those of crystallography) and which do not imply the extraction of autonomised operators of coding;
- the relative linearity of biological codings, for example, the double helix of DNA which, starting from four basic chemical radicals, develops equally in three dimensions;
- the linearity of pre-signifying semiologies, which develop on relatively autonomous, parallel lines, even if the phonological chains of spoken language appear to always overcode all the others;
- the semiological linearity of the structural signifier which imposes itself despotically over all the other modes of semiotisation, expropriates them and even tends to make them disappear within the framework of a communicational economy dominated by informatics (please note: informatics in its current state, since this state of things is in no way definitive);

- the superlinearity of a-signifying substances of expression, where the signifier loses its despotism. The informational lines of hypertexts can recover a certain dynamic polymorphism and work in direct contact with referent Universes which are in no way linear and, what is more, tend to escape a logic of spatialised sets.

The indicative matter of a-signifying semiotic machines is constituted by "point-signs": these on one hand belong to a semiotic order and on the other intervene directly in a series of material machinic processes. Example: a credit card number which triggers the operation of a bank auto-teller. The a-signifying semiotic figures don't simply secrete significations. They give out stop and start orders but above all activate the "bringing into being" of ontological Universes. Consider for a moment the example of the pentatonic musical refrain which, with only a few notes, catalyses the Debussyst constellation of multiple Universes:

- the Wagnerian Universe surrounding Parsifal, which attaches itself to the existential Territory constituted by Bayreuth;
- the Universe of Gregorian chant;
- that of French music, with the return to favour of Rameau and Couperin;
- that of Chopin, due to a nationalist transposition (Ravel, for his part, appropriating Liszt);
- the Javanese music Debussy discovered at the Universal Exposition of 1889;
- the world of Manet and Mallarmé, which is associated with Debussy's stay at the Villa Medicis.

It would be appropriate to add to these past and present influences the prospective resonances which constituted the reinvention of polyphony from the time of the Ars Nova, its repercussions on the French musical phylum of Ravel, Duparc, Messiaen, etc., and on the sonorous mutation triggered by Stravinsky, his presence in the work of Proust . . .

We can clearly see that there is no bi-univocal correspondence between linear signifying links or archi-writing, depending on the author, and this multireferential, multidimensional machinic catalysis. The symmetry of scale, the transversality, the pathic non-discursive character of their expansion: all these dimensions remove us from the logic of the excluded middle and reinforce us in our dismissal of the ontological binarism we criticised previously. A machinic assemblage, through its diverse components, extracts its consistency by crossing ontological thresholds, non-linear thresholds of irreversibility, ontological and phylogenetic thresholds, creative thresholds of heterogenesis and autopoiesis. The notion of scale needs to be expanded to consider fractal symmetries in ontological terms. What fractal machines traverse are substantial scales. They traverse them in engendering them. But, and this should be noted, the existential ordinates that they "invent" were always already there. How can this paradox be sustained? It's because everything becomes possible (including the recessive smoothing of time, evoked by René Thom) the moment one allows the assemblage to escape from energetico-spatio-temporal coordinates. And, here again, we need to rediscover a manner of being of Being – before, after, here, and everywhere else – without being, however, identical to itself; a processual, polyphonic Being singularisable by infinitely complexifiable textures, according to the infinite speeds which animate its virtual compositions.

The ontological relativity advocated here is inseparable from an enunciative relativity. Knowledge of a Universe (in an astrophysical or axiological sense) is only possible through the mediation of autopoietic machines. A zone of self-belonging needs to exist somewhere for the coming into cognitive existence of any being or any modality of being. Outside of this machine/Universe coupling, beings only have the pure status of a virtual entity. And it is the same for their enunciative coordinates. The biosphere and mecanosphere, coupled on this planet, focus a point of view of space, time, and energy. They trace an angle of the constitution of our galaxy. Outside of this particularised point of view, the rest of the Universe exists (in the sense that we understand existence here-below) only through the virtual existence of other autopoietic machines at the heart of other bio-mecanospheres scattered throughout the cosmos. The relativity of points of view of space, time and energy do not, for all that, absorb the real into the dream. The category of Time dissolves in cosmological reflections on the Big Bang even as the category of irreversibility is affirmed. Residual objectivity is what resists scanning by the infinite variation of points of view constitutable upon it. Imagine an autopoietic entity whose particles are constructed from galaxies. Or, conversely, a cognitivity constituted on the scale of quarks. A different panorama, another ontological consistency. The mecanosphere draws out and actualises configurations which exist amongst an infinity of others in fields of virtuality. Existential machines are at the same level as being in its intrinsic multiplicity. They are not mediated by transcendent signifiers and subsumed by a univocal ontological foundation. They are to themselves their own material of semiotic expression. Existence, as a process of deterritorialisation, is a specific intermachinic operation which superimposes itself on the promotion of singularised existential intensities. And, I repeat, there is no generalised syntax for these deterritorialisations. Existence is not dialectical, not representable. It is hardly livable!

Desiring machines which break with the great interpersonal and social organic equilibria, which invert orders, play the role of the other as against a politics of auto-centering on the self. For example, the partial drives and perverse polymorphic investments of psychoanalysis don't constitute an exceptional and deviant race of machines. All machinic assemblages harbour – even if in an embryonic state – enunciative zones which are so many desiring proto-machines. To clarify this point we need to extend our transmachinic bridge and understand the smoothing of the ontological texture of machinic material and diagrammatic feedbacks as so many dimensions of intensification that take us beyond the linear causalities of the capitalistic apprehension of machinic Universes. We also need to abandon logics based on the principles of the excluded middle and sufficient reason. Through this smoothing there appears a being beyond, a being-for-the-other which gives consistency to an existent beyond its strict delimitation, here and now. The machine is always synonymous with a nucleus constitutive of an existential Territory against a background of a constellation of incorporeal Universes of reference (or value). The "mechanism" of this turning around of being consists in the fact that some of the machine's discursive segments do not only play a functional or signifying role, but assume the existentialising function of pure intensive repetition that I have called the refrain function. The smoothing is like an ontological refrain, and thus, far from apprehending a univocal truth of being through techne – as Heideggerian ontology would have it – it is a plurality of beings as machines which give themselves to us the moment we acquire the pathic and cartographic means of reaching them. The manifestations – not of Being, but of multitudes of ontological components – are of the order of the machine. And this, without semiological mediation, without transcendent coding, directly as "being's giving of itself," as giving. Acceding to such a "giving" is already to participate ontologically in it as a full right. The term right does not occur here by chance, since at this proto-ontological level it is already necessary to affirm a proto-ethical dimension. The play of intensity of the ontological constellation is, in a way, a choice of being not only for self, but for the whole alterity of the cosmos and for the infinity of times.

If there's choice and freedom at certain "superior" anthropological stages, it's because we will also find them at the most elementary strata of machinic concatenations. But the notions of elements and complexity are susceptible here to being brutally inverted. Those that are most differentiated and undifferentiated coexist within the same chaos which, at infinite speed, plays its virtual registers – one against the other and one with the other. The machinic-technical world, at the "terminal" of which present-day humanity structures itself, is barricaded by horizons of constants and the limitation of the infinite velocities of chaos (the speed of light, the cosmological horizon of the Big Bang, Planck's constant and the elementary quantum of action in quantum physics, the impossibility of going below absolute zero . . .). But, this very same world of semiotic constraints is doubled, tripled, and infinitised by other worlds which under certain conditions seek only to bifurcate out of their Universes of virtuality and engender new fields of the possible.

Just as scientific machines constantly modify our cosmic frontiers, so do the machines of desire and aesthetic creation. As such, they hold an eminent place within assemblages of subjectivation, themselves called to relieve our old social machines which are incapable of keeping up with the efflorescence of machinic revolutions that shatter our epoch.

Rather than adopting a reticent attitude with respect to the immense machinic revolution sweeping the planet (at the risk of destroying it) or of clinging onto traditional systems of value – with the pretence of re-establishing transcendence – the movement of progress, or if one prefers, the movement of process, will endeavour to reconcile values and machines. Values are immanent to machines. The life of machinic Fluxes is not only manifested through cybernetic feedback: it is also correlative to a promotion of incorporeal Universes stemming from an enunciative Territorial incarnation, from a valorising consciousness of being. Machinic autopoiesis asserts itself as a non-human *for-itself* through zones of partial proto-subjectivation and it deploys a *for-others* under the double modality of a "horizontal" eco-systemic alterity (the machinic systems position themselves in a rhizome of reciprocal dependence) and phylogenetic alterity (situating each actual machinic stasis at the conjunction of a passéist filiation and a Phylum of future mutations). All systems of value – religious, aesthetic, scientific, ecosophic – install themselves at this machinic interface between the necessary actual and the possibilist virtual. Thus Universes of value constitute incorporeal enunciators of abstract machinic complexions compossible with discursive realities. The consistency of these zones of proto-subjectivation is then only assured inasmuch as they are embodied, with more or less intensity, in nodes of finitude, Territories of chaosmic grasping, which guarantee, moreover, their possible recharging with processual complexity. Thus a double enunciation: finite, territorialised and incorporeal, infinite.

Nevertheless, these constellations of Universes of value do not constitute Universals. The fact that they are tied into singular existential Territories effectively confers upon them a power of heterogenesis, that is, of opening onto singularising, irreversible processes of necessary differentiation. How does this machinic heterogenesis, which differentiates each colour of being – which makes, for example, from the plane of consistency of a philosophical concept a world quite different from the plane of reference of the scientific function or the plane of aesthetic composition – end up being reduced to the capitalistic homogenesis of generalised equivalence, which leads to all values being valued by the same thing, all appropriative territories being related to the same economic instrument of power, and all existential riches succumbing to clutches of exchange value? The sterile opposition between use value and exchange value will here be relinquished in favour of an axiological complexion including all the machinic modalities of valorisation: the values of desire, aesthetic values, ecological, economic values … Capitalistic value which generally subsumes the ensemble of these machinic surplus values, proceeds with a reterritorialising attack, based on the primacy of economic and monetary semiotics, and corresponds to a sort of general implosion of all existential Territories. In fact, capitalistic value is neither separate nor tangential to systems of valorisation: it constitutes their deathly heart, corresponding to the crossing of the ineffable limit between a controlled, chaosmic deterritorialisation – under the aegis of social, aesthetic, and analytical practices – and a vertiginous collapse into the black hole of the aleatory, understood as a paroxysmically binary reference, implacably dissolving the whole consistency of Universes of value which would claim to escape capitalistic law. It is thus only abusively that one could put economic determinations in a primary position with respect to social relations and productions of subjectivity. Economic law, like juridical law, must be deducted from the ensemble of Universes of value, for whose collapse it continually strives. Its reconstruction, from the scattered debris of planned economies and neo-liberalism and according to new ethico-political finalities (ecosophy) calls for, in contradistinction, an untiring renewal of the consistency of machinic assemblages of valorisation.

1   Norbert Wiener, *Cybernetics, or Control and communication in the animal and the machine.* Technology Press. Cambridge, MA, 1948.

2   Cf. Pierre Levy, *Les Technologies de l'intelligence.* La Découverte, Paris, 1990, *Plisse fractal. Ideographie dynamique* (memoire d'habilitation a diriger des recherches en sciences de l'information et de la communication) et *L'ideographie dynamique.* La Découverte. Paris, 1991.

3   F. Varela op. cit.

4   Leibniz, in his concern to render homogeneous the infinitely large and the infinitely small, thought that the living machine, which he assimilated to a divine machine, continued to be a

machine in its smallest parts until infinity (which would not be the case with a machine made by the art of men), in *Monadologie*, pp. 178–9. Delagrave. Paris, 1962.

5   M. Augé, "Le fétiche et son objet" in *L'Objet en psychanalyse*, presented by Maud Mannoni. Denoël, "L'espace analytique," Paris, 1986.

6   Ibid.

7   Martin Heidegger, *Basic Writings*, edited by David Farrell Krell. Harper. San Francisco, 1977, p. 298.

Francis Duffy (born 1940) is an English architect who has specialized in the planning and design of commercial offices. He studied at the Architectural Association, London and the University of California, Berkeley before completing a PhD at Princeton, where he wrote a dissertation on the radical office planning theories of the 1960s. He founded an architectural practice, DEGW, in 1971, and from 1993 to 1995 was President of the Royal Institute of British Architects. Duffy has spent his career helping business organizations use space more effectively over time, focusing on the effect of changes in organizational situations and information technology on office design. He has published widely on the subject and in recent years has been a visiting Professor at MIT.

This brief excerpt from *The New Office*, one of his many books on the changing nature of offices, describes the different life-spans or rates of change among the different elements in commercial buildings. The accommodation of different rates of change has become quite commonplace in commercial construction, with the base-building or shell produced by one group and the fit-out or infill done subsequently and frequently by tenants. The open building movement has also tried to adopt these techniques for residential construction, allowing residents to more completely customize their dwellings, and to take those customizations with them when they move. The pragmatic efficiencies of these techniques have also made them interesting to environmentalists seeking to minimize the use of material.

It seemed revealing to include this excerpt along with that by Stewart Brand (1994) and SLA (2003), because they make clear just how completely buildings are viewed as complex, changeable systems and then how their parts and sub-systems have been specialized to reflect that systematic interdependence.

1997

Francis Duffy

Time in Office Design

Just as the differences in scale must be fully recognized, so the different lifespans of various office components must be understood if business is going to achieve both the long-term and the shorter-term flexibility it would like office buildings to deliver. While big office buildings may look from the outside as if they are meant to last for ever, many aspects of the interior, and, even these days of the exterior, are being changed all the time. The contrast between the apparently timeless exterior of the Seagram Building and the seething transience of the interior with its constant turnover of diverse tenants, for example, has already been highlighted in Chapter 1. It is essential that the time cycles are allowed to coexist more or less independently. To assume that everything should last for the same length of time is absurd; to attempt to use only short-term elements to solve long-term problems is inherently wasteful; to have to dismantle long-term structures to solve short-term problems is ridiculously expensive.

The appearance of a building – its construction and fit-out, for example – has a different lifespan from its operation – its environmental services, for instance . . . Stewart Brand, the inventor of *The Whole Earth Catalog*, in his recent book *How Buildings Learn* has elaborated this idea further and distinguishes between 'site, structure, skin, services, space plan, and stuff'. The last is close to what we mean by 'settings' – the highly mobile assemblages of 'chairs, desks, phones, pictures, lamps . . . all the things that twitch around daily to monthly' that make up so much of the ordinary office worker's experience of the working environment. Brand also draws attention to the opposite end of the permanence scale – the site – which is the only real fixed element in all this temporal complexity because, whatever may happen on or above it, the site itself can never be replaced. Hence, perhaps, the disproportionate influence of real estate brokers on office design, at least in the English-speaking world.

Those design elements that have the shortest lifespan should be as amenable to control and change by ordinary office workers as possible – there is no excuse for rigidly prescriptive internal layouts which suppress individuality and creativity. This applies both to the arrangement of the furniture (settings) and to the furniture itself (scenery). It should be remembered that the life cycle of personal computers and other electronic office devices is even shorter than that of furniture. Such equipment is always obsolescent and generally replaced within three years.

Paul Virilio – architectural theorist, philosopher and self-proclaimed "critic of the art of technology" – was born in Paris in 1932 to a Breton mother and an Italian Communist father. In 1950 he became a Christian and after training at the Ecole des Metiers d'Art, specialized in stained-glass and worked alongside the sculptor Henri Matisse in various Parisian churches. He was conscripted into the army during the Algerian war of independence (1954–62), and later studied phenomenology with the philosopher Maurice Merleau-Ponty at the Sorbonne.

1997
Paul Virilio
The Third Interval

In 1958 Virilio began a period of research into the nature of military space and the organization of territory, focusing on the "Atlantic Wall" – the 15,000 Nazi bunkers built along the coastline of France during the Second World War. In 1975 this work was presented as part of the *Bunker Archeologie* exhibition at the Decorative Arts Museum in Paris, and published in a book of the same title. He also collaborated with the architect Claude Parent, forming the group "Architecture Principe" in 1963 and together they developed a theory of urban space based on the use of the oblique axis and the inclined plane. After participating in the May 1968 uprising in Paris, Virilio was nominated as Professor by the students at the Ecole Spéciale d'Architecture, where in 1973 he became Director of Studies. In the same year, he also became director of the magazine *L'Espace Critique*. In 1989 he took over the program of studies at the Collège International de Philosophie de Paris, under the direction of Jacques Derrida. In 1998, Virilio officially retired from teaching, although he continued to work on projects with homeless groups in Paris, as well as developing a building to house the first "Museum of the Accident."

Like Jean Baudrillard – his sympathetic adversary in the realm of contemporary cultural theory – Virilio has written numerous short and polemical texts on a wide variety of topics many of which have been translated into English. These include: *Speed & Politics: An Essay on Dromology* (1986), *The Aesthetics of Disappearance* (1991), *War and Cinema: The Logistics of Perception* (1989), *Politics of the Very Worst* (1999), *Polar Inertia* (1999), and *The Information Bomb* (2000).

The present essay is taken from the book *Open Sky*, originally published in 1997 and translated into English in 1999. Here Virilio claims that the new technologies of "telepresence" have created a new category of experience, one that transcends the limitations of the classical concepts of space and time. This new "interval" – subject to the limits of speed and light – emerges from the illusion of simultaneity created by the latest digital communication technologies. The almost-instantaneous availability of "real-time" information challenges our conventional understanding of the experience of the here-and-now.

'Without even leaving, we are already no longer there.'

Nikolai Gogol

Critical *mass*, critical *moment*, critical *temperature*. You don't hear much about critical *space*, though. Why is this if not because we have not yet digested relativity, the very notion of space-time?

And yet critical space, and critical expanse, are now everywhere, due to the acceleration of communications tools that *obliterate the Atlantic* (Concorde), *reduce France to a square one and a half hours across* (Airbus) or *gain time over time* with the TGV, the various advertising slogans signalling perfectly the shrinking of geophysical space of which we are the beneficiaries but also, sometimes, the unwitting victims.

As for telecommunications tools, not content to limit extension, they are also eradicating all duration, any extension of time in the transmission of messages, images.

Mass transportation revolution of the nineteenth century, broadcasting revolution of the twentieth – a mutation and a commutation that affect both public and domestic space at the same time, to the point where we are left in some uncertainty as to their very reality, since the urbanization of *real space* is currently giving way to a preliminary urbanization of *real time*, with teleaction technologies coming on top of the technology of mere conventional television.

This abrupt transfer of technology, from the building of real-space infrastructures (ports, railway stations, airports) to the control of the real-time environment thanks to interactive teletechnologies (teleports), gives new life today to the critical dimension.

Indeed, the question of the real instant of instantaneous teleaction raises once again the philosophical and political problems traditionally associated with the notions of **atopia** and **utopia**, and promotes what is already being referred to as **teletopia**, with all the numerous paradoxes attendant on this, such as:

Meeting at a distance, in other words, **being telepresent**, here and elsewhere, *at the same time*, in this so-called 'real time' which is, however, nothing but a kind of real space-time, since the different events do indeed take place, even if that place is in the end the no-place of teletopical techniques (the man–machine interface, the nodes or packet-switching exchanges of teletransmission).

Immediate teleaction, instantaneous telepresence. Thanks to the new practices of television broadcasting or remote transmission, *acting*, the famous teleacting of remote control, is here facilitated by the maximum performance of electromagnetism and by the radioelectric views of what is now called **optoelectronics**, the perceptual faculties of the individual's body being transferred one by one to machines – but also, most recently, to captors, sensors, and other microprocessor detectors, capable of making up for the lack of tactility at a distance, *widespread remote control* preparing to take up where *permanent telesurveillance* left off.

What then becomes critical is not so much the three dimensions of space, but the fourth dimension of time – more precisely, the dimension of the **present** since, as we will see below, 'real time' is not the opposite of 'delayed time', as electronics engineers claim, but only of the 'present'.

Paul Klee hit the nail on the head: 'To define the present in isolation is to kill it.' This is what the teletechnologies of real time are doing: they are killing 'present' time by isolating it from its here and now, in favour of a commutative elsewhere that no longer has anything to do with our 'concrete presence' in the world, but is the elsewhere of a 'discreet telepresence' that remains a complete mystery.

How can we fail to see how much such radiotechnologies (digital signal, video signal, radio signal) will shortly turn on their heads not only the nature of the human environment, our *territorial body*, but most importantly, the nature of the individual and their *animal body*? For the staking out of the territory with heavy material infrastructure (roads, railroads) is now giving way to control of the immaterial, or practically immaterial, environment (satellites, fibre-optic cables), ending in the *body terminal* of man, of that interactive being who is both transmitter and receiver.

The urbanization of real time is in fact first the urbanization of *one's own body* plugged into various interfaces (keyboard, cathode screen, DataGlove or DataSuit), prostheses that make the super-equipped able-bodied person almost the exact equivalent of the motorized and wired disabled person.

If last century's revolution in transportation saw the emergence and gradual popularization of the dynamic motor vehicle (train, motorbike, car, plane), the current revolution in transmission leads in turn to the innovation of the ultimate vehicle: the static audiovisual vehicle, marking the advent of a behavioural inertia in the sender/receiver that moves us along from the celebrated *retinal persistence* which permits the optical illusion of cinematic projection to the *bodily persistence* of this 'terminal-man'; a prerequisite for the sudden mobilization of the illusion of the world, of a *whole* world, telepresent at each moment, the witness's own body becoming the last urban frontier. Social organization and a kind of conditioning once limited to the space of the city and to the space of the family home finally closing in on the animal body.

This makes it easier to understand the decline in that unit of population, the family, initially extended then nuclearized, that is today becoming a single-parent family, individualism having little to do with the fact of a liberation of values and being more an effect of technological evolution in the development of public and private space, since the more the city expands and spreads its tentacles, the more the family unit dwindles and becomes a minority.

Recent **megalopolitan** hyperconcentration (Mexico City, Tokyo) being itself the result of the increased speed of exchanges, it looks as though we need to reconsider the importance of the notions of **acceleration** and **deceleration** (vector quantities with positive or negative velocities according to the physicists). But we also need to reconsider the less obvious notions of **true velocity** and **virtual velocity** – the speed of that which occurs unexpectedly: a crisis, for instance, an accident – properly to understand the importance of the 'critical transition' of which we are today helpless witnesses.

As we know, speed is not a phenomenon but a relationship between phenomena: in other words, relativity itself. Which is why the constant of the speed of light is so important, not only in physics or astrophysics, but in our daily lives, from the moment we step beyond the transport age into the organization and *electromagnetic conditioning of the territory.*

This is the 'transmission revolution' itself, this control of the environment in real time that has now put paid to traditional development of a real territory.

Speed not only allows us to get around more easily; it enables us above all to see, to hear, to perceive, and thus to conceive the present world more intensely. Tomorrow, it will enable us to act at a distance, beyond the human body's sphere of influence and that of its behavioural ergonomics.

How can we fully take in such a situation without enlisting the aid of a new type of interval, **the interval of the light kind** (neutral sign)? The relativistic innovation of this third interval is actually in itself a sort of unremarked cultural revelation.

If the interval of **time** (positive sign) and the interval of **space** (negative sign) have laid out the geography and history of the world through the geometric design of agrarian areas (fragmentation into plots of land) and urban areas (the cadastral system), the organization of calendars and the measurement of time (clocks) have also presided over a vast chronopolitical regulation of human societies. The very recent emergence of an interval of the third type thus signals a sudden qualitative leap, a profound mutation in the relationship between man and his surroundings.

Time (duration) and **space** (extension) are now inconceivable without **light** (limit-speed), the cosmological constant of the **speed of light**, an absolute philosophical contingency that supersedes, in Einstein's wake, the absolute character till then accorded to space and to time by Newton and many others before him.

Since the turn of the century, the absolute limit of the speed of light has *lit up*, so to speak, both space and time. So it is not so much **light** that illuminates things (the object, the subject, the path); it is the constant nature of light's **limit speed** that conditions the perception of duration and of the world's expanse as phenomena.

Listen to a physicist talking about the logic of elementary particles: 'A display is defined by a complete set of observables that commutate.'[1] It would be hard to find a better description of the macroscopic logic of **real-time** technologies than this 'teletopical commutation' – or 'switch-over' – that completes and perfects the till now fundamentally 'topical' nature of the City of Men.

So, politicians, just as much as urbanists, find themselves torn between the permanent requirements of organizing and constructing real space – with its land problems, the geometric and geographic constraints of the centre and the periphery – and the new requirements of managing the real time of immediacy and ubiquity, with its access protocols, its 'data packet transmissions' and its viruses, as well as the chronogeographic constraints of nodes and network interconnection. Long term for the topical and architectonic interval (the building); short, ultra-short – if not indeed non-existent – term for the **teletopical** interval (the network).

How do we resolve this dilemma? How do we frame these basically spatio-temporal and relativistic problems?

When we look at all the difficulties faced by world money markets and the disasters of electronic share quotation systems, with the 'Program Trading' being responsible for the acceleration of economic chaos – indeed, for the computer crash of October 1987 and the one narrowly averted in October 1989 – it is pretty clear how fraught the present situation is.

Critical transition then is not an empty term: it masks a true crisis in the temporal dimension of *immediate action*. Following the crisis in 'whole' spatial dimensions and the resultant rise of 'fractional' dimensions, we will soon see a crisis, in short, in the temporal dimension of the present moment.

Since **time-light** (the time of the speed of light) is now used as an absolute standard for immediate action, for instantaneous teleaction, the intensive duration of the 'real moment' now dominates duration, the extensive and relatively controllable time of history – in other words, of that long term that used to encompass past, present, and future. This is in the end what we could call a **temporal commutation**, a commutation also related to a sort of **commotion** in present duration, an accident of a so-called 'real' moment that suddenly detaches itself from the place where it happens, from its here and now, and opts for an electronic dazzlement (at once optoelectronic, electroacoustic, and electrotactile) in which remote control, this so-called 'tactile telepresence', will complete the task of the old telesurveillance of whatever stays at a distance, out of our reach.

According to Epicurus, *time is the accident to end all accidents*. If this is so, then with the teletechnologies of general interactivity we are entering the age of the **accident of the present**, this overhyped remote telepresence being only ever the sudden catastrophe of the reality of the present moment that is our sole entry into duration – but also, as everyone knows since Einstein, our entry into the expanse of the real world.

After this, the real time of telecommunications would no longer refer only to delayed time, but also to an *ultra-chronology*. Hence my repeatedly reiterated proposal to round off the chronological (before, during, after) with the **dromological** or, if you like, the **chronoscopic** (underexposed, exposed, overexposed). Indeed, the interval of the light kind (the interface) taking over in future from those of space and time, the notion of exposure in turn takes over from the notion of succession in the measurement of present duration as well as from the notion of extension in the immediate physical expanse.

The exposure speed of time-light might therefore enable us to reinterpret the 'present', this 'real instant' which is, let's not forget, the space-time of a perfectly real action helped along by the feats of electronics and shortly of photonics – that is, by the limit capabilities of electromagnetic radiation and of the light quantum, that frontier-post of access to the reality of the perceptible world (note here the light cone – or illuminating pencil – used by astrophysicists).

The question today posed by teletopical technologies is thus a major one for the planner, since the urbanization of real time permitted by the recent transmission revolution leads to a radical reversal in the order of the movement of displacement and of physical transportation. In fact, if operating remotely allows gradual elimination of the material infrastructures rigging out the territory in favour of the fundamentally immaterial wave trains of telesurveillance and instantaneous remote control, this is because the **journey** and its components are undergoing a veritable mutation-commutation. Where physical displacement from one point to another once supposed departure, a journey and arrival, the transport revolution of last century had already quietly begun to eliminate delay and change the nature of travel itself, arrival at one's destination remaining, however, a 'limited arrival' due to the very time it took to get there.

Currently, with the instantaneous broadcasting revolution, we are seeing the beginnings of a '*generalized arrival*' whereby everything arrives without having to leave, the nineteenth century's elimination of the journey (that is, of the space interval and of time) combining with the abolition of *departure* at the end of the twentieth, the journey thereby losing its successive components and being overtaken by *arrival* alone.

A **general arrival** that explains the unheard-of innovation today of the static vehicle, a vehicle not only audiovisual but also tactile and interactive (radioactive, optoactive, interactive).

One such static vehicle is the 'DataSuit', invented by the American Scott Fisher while he was working for NASA on the development of a human body device that would be capable of transferring actions and sensations by means of an array of sensor-effectors. In other words, capable of producing *presence at a distance*, and this no matter what the distance, since the NASA project was supposed to allow total telemanipulation of a *robotic double* on the surface of planet Mars, thus achieving the individual's effective telepresence in two places at the same time, a split in the personality of the manipulator, whose 'vehicle' was to be this instantaneous interactive vector.

To cite another of Paul Klee's premonitory sayings: 'The viewer's main activity is temporal.'

What can we say about the interactivity of the *teleoperator* other than that, for such a person (at the human–robot interface), as for the now time-honoured *tele-viewer*, activity is not so much spatial as temporal?

Doomed to inertia, the interactive being transfers his natural capacities for movement and displacement to probes and scanners which instantaneously inform him about a remote reality, to the detriment of his own faculties of apprehension of the real, after the example of the para- or quadriplegic who can guide by remote control – *teleguide* – his environment, his abode, which is a model of that home automation, of those 'Smart Houses' that respond to our every whim. Having been first *mobile*, then *motorized*, man will thus become *motile*, deliberately limiting his body's area of influence to a few gestures, a few impulses, like channel-surfing.

This critical situation is no different from that experienced by any number of spastics who thus become by force of circumstance – the critical force of the circumstance of technology – models of the new man, of that inhabitant of the future teletopical city, the **metacity** of a social deregulation the transpolitical aspect of which already shows up, here and there, in a number of major accidents or minor incidents, mostly remaining as yet unexplained.

How can we get a purchase on this transitional situation, this 'phase transition', as the physicists would say?

Here is a rather ancient piece of philosophical analysis from Nicholas of Cusa:

> The accident ceases to exist when the substance is removed, and its ceasing to exist in that instance is due to the fact that to inhere is of the nature of an accident and that its subsistence is the subsistence of the substance. Yet it cannot be said that an accident is nothing. . . . An accident gives something to a substance . . .; in fact, an accident gives so much to a substance that, although the accident had its being from the substance, *the substance cannot exist without any accident.*[2]

Today, as we have seen, the problem of the accident has shifted from the space of matter to the time of light.

The accident is, first, an *accident of transfer* of the limit-speed of electromagnetic waves, a speed that now allows us not only to hear and see at a distance, as we were already able to do with the telephone, radio, or television, but actually to act at a distance. Hence the necessity of the third type of interval (neutral sign) to try and grasp the *place of the no place* of a teleaction that is no longer the same as the here and now of immediate action.

So interactivity's *accident of transfer* opens not only on to the *technology transfer* between delayed time communication and real-time commutation, but more particularly on to a political transfer that undermines precisely those notions that lie at the heart of our age: the notion of *service* and the notion of *public*.

What, indeed, is left of the notion of *service* when you are automatically controlled? Similarly, what is left of the notion of *public* when the (real-time) public image prevails over public space?

Already the notion of public transport is gradually giving way to the idea of a *transit corridor*, the continuous prevailing over the discontinuous. What can one say about the wired household of electronic domesticity, with houses that have computers wired into them, controlling the house systems, or of the smart building, indeed the intelligent and interactive city such as Kawasaki? The crisis in the notion of physical dimensions thus hits politics and the administration of public services head on in attacking what was once geopolitics.

If the classic interval is giving way to the interface, politics in turn is shifting within exclusively *present time*. The question is then no longer one of the **global** versus the **local**, or of the **transnational** versus the **national**. It is, first and foremost, a question of the sudden temporal switch in which not only inside and outside disappear, the expanse of the political territory, but also the before and after of its duration, of its history; all that remains is a **real instant** over which, in the end, no one has any control. For proof of this, one need look no further than the inextricable mess geostrategy is in thanks to the impossibility of clearly distinguishing now between offensive and defensive – instantaneous, multipolar strategy now being deployed in 'preemptive' strikes, as they say in the military.

And so the age-old *tyranny of distance* between beings geographically distributed in different places is gradually yielding to the *tyranny of real time* which is not the exclusive concern of travel agents, as optimists claim, but a special concern of the employment agency, since the greater the speed of exchanges, the more unemployment spreads and becomes mass unemployment.

Redundancy of man's muscular strength in favour of the 'machine tool' from the nineteenth century on. Now redundancy, permanent unemployment, of his memory and his consciousness, with the recent boom in computers, in 'transfer machines', and the automation of postindustrial production combining with the automation of perception, and finally with computer-aided design, enabled by the software market, ahead of the coming of the artificial intelligence market.

To gain real time over delayed time is thus to commit to a quick way of physically eliminating the object and the subject and exclusively promoting the journey. But a journey without a trajectory and hence fundamentally uncontrollable.

The real-time interface then once and for all replaces the interval that once constituted and organized the history and geography of human societies, winding up in a true culture of the paradox in which everything arrives not only without needing physically to move from one place to another but, more particularly, without having to leave.

Surely we cannot fail to foresee the future conditioning of the human environment behind this critical transition.

If last century's transport revolution already brought about a mutation in urban territory throughout the continent, the current revolution in (interactive) transmission is in turn provoking a commutation in the urban environment whereby the image prevails over the thing it is an image of what was once a city becoming little by little a paradoxical agglomeration, relationships of immediate proximity giving way to remote interrelationships.

The paradoxes of acceleration are indeed numerous and disconcerting, in particular, the foremost among them: getting closer to the 'distant' takes you away proportionally from the 'near' (and dear) – the friend, the relative, the neighbour – thus making strangers, if not actual enemies, of all who are close at hand, whether they be family, workmates, or neighbourhood acquaintances. This inversion of social practices, already evident in the development of communication equipment (ports, stations, airports), is further reinforced, radicalized, by the new telecommunications equipment (teleports).

Once more we are seeing a reversal in trends: where the motorization of transport and information once caused a *general mobilization* of populations, swept up into the exodus of work and then of leisure, instantaneous transmission tools cause the reverse: *a growing inertia*; television and especially remote control action no longer requiring people to be mobile, but merely to be mobile on the spot.

Home shopping, working from home, online apartments and buildings: 'cocooning', as they say. The urbanization of real space is thus being overtaken by this urbanization of real time which is, at the end of the day, the urbanization of the actual body of the city dweller, this *citizen-terminal* soon to be decked out to the eyeballs with interactive prostheses based on the pathological model of the 'spastic', wired to control his/her domestic environment without having physically to stir: the catastrophic figure of an individual who has lost the capacity for immediate intervention along with natural motricity and who abandons himself, for want of anything better, to the capabilities of captors, sensors, and other remote control scanners that turn him into a being controlled by the machine with which, they say, he talks.[3]

Service or servitude, that is the question. The old public services are in danger of being replaced by a domestic enslavement whose crowning glory would surely

be home automation. Achieving a domiciliary inertia, the widespread use of techniques of *environmental control* will end in behavioural isolation, in intensifying
the insularity that has always threatened the town, the difference between the
(separate) 'block' and the (segregated) 'ghetto' remaining precarious.

Curiously, papers given at an international symposium on disability recently
held in Dunkirk in many ways echoed the critical situation evoked here, as though
recent technological and economic imperatives to produce *continuity and
networks*, wherever *discontinuities* still exist, failed to distinguish between the
various kinds of urban mobility. Hence the above-mentioned idea of scrapping the
notion of public transport and opting instead for the broader one of transit corridors.

This was Francois Mitterrand's noble conclusion to the Dunkirk conference:
'Cities must adapt to their citizens and not the other way round. Let's open up the
city to the physically challenged. I ask that an overall policy on the disabled be a
firm axis of Europe as a social institution.'

Though none of us would dispute the inalienable right of the disabled to live
the same way as everyone else, and therefore with everyone else, it is none the less
revealing to note the convergences that now exist between the reduced mobility of
the well-equipped disabled person and the growing inertia of the overequipped
able-bodied person, as though the transmission revolution always yielded an identical result, no matter what the bodily condition of the patient, this **terminal citizen**
of a teletopical City that is going up faster and faster.

At the end of the century, there will not be much left of the expanse of a planet
that is not only polluted but also shrunk, reduced to nothing, by the teletechnologies of generalized interactivity.

1   G. Cohen Tannoudji and M. Spiro, *La Matiere-espace-temps*.
2   Nicholas of Cusa (Nicolas Cusanos), *Of Learned Ignorance*, trans. Father Germain Heron, New
     Haven, CT 1954, pp. 78–9. See also Giuseppe Bufo, *Nicolas de Cues*, Paris 1964.
3   Paul Virilio, *L'Inertia polaire*, Paris 1990.

1999

Ben van Berkel and Caroline Bos

Techniques:
Network Spin, and
Diagrams

Ben van Berkel was born in Utrecht in 1957 and studied architecture at the Rietveld Academy, Amsterdam and the Architectural Association, London, where he completed a Graduate Diploma in 1987. He worked for a time with Zaha Hadid and Santiago Calatrava before returning to Amsterdam in 1988 to set up an architectural practice with Caroline Bos. Van Berkel has also taught at several architectural schools around the world including Columbia and Harvard Universities and the Architectural Association (1996–99). He is currently Professor of Conceptual Design at the Städelschule in Frankfurt. Caroline Bos was born in Rotterdam in 1959 and studied art history at Birkbeck College, University of London. She has been a tutor at the Berlage Institute in Amsterdam and has also taught at Princeton University.

Their best known built projects to date are the Erasmus Bridge in Rotterdam (1990–96), and the Möbius House (1993–95) which uses the idea of a continuous coiled surface to accommodate a fluid sequence of interconnected living spaces integrated with the landscape. In 1998 van Berkel and Bos established UN Studio (United Net), intended as a network of specialists in architecture, urban design, and infrastructure. Recently realized projects include: the refurbishment and interior design for the Galleria Hall West, Seoul, South Korea (2003–04); La Defense office building in Almere (1999–2004); and an exhibition pavilion for *Living Tomorrow* in Amsterdam (2000–03).

As with the work of Rem Koolhaas and OMA, UN Studio have produced a hybrid output of buildings, exhibitions, and publications – most notably the three-volume collection of projects and writings entitled *Move* (1999) from which the following extract is taken. The first part of the essay describes the process by which technologies combine to form ever larger networks of interdependent elements. Here this idea is scaled up to include the celestial orbits of communication satellites, suggesting a new "mediated cosmology" that architecture must somehow situate itself within. In the second part of the essay, the use of the diagram is discussed as a means of mapping a series of forces which impact upon a design problem – setting up a process by which these forces can exert their effects over a period of time. The resulting trace of the interactions between (what Maneul Castells would call) "places" and "flows," is used to trigger the "breeding" of new kinds of structures and spaces, beyond the restrictions of conventional typologies (see De Landa, 2002; Castells, 2004).

> Techniques: network spin

Techniques, which are distinct from production methods or style, are the most neglected element of cultural production. Theories are not built around techniques and they are only disseminated in the most direct manner from one practitioner to another in institutions, workplaces, or through handbooks. Techniques are dry – even when there is a narrative attached their significance rests not in that story. Techniques are mostly a thing of the past. The techniques of today, on the other hand, are of interest only to nerds, obsessing over the interior of a computer. Yet techniques form the bridge between abstract thought and concrete production. This is a two-way bridge: techniques also form thought. Technology stimulates mental fabrication by means of the specific potential that it possesses.

Each new technology changes the world. Ontological and technological permutations are interwoven. In the twentieth century, an avalanche of new techniques has coursed through the sciences, industry, the arts, and communication, revealing the deep integration in the wider world of all forms of social/cultural production.

The turbulent expansion of the inventory of techniques is interactively related to social, economic, and scientific change.

From the practitioner's viewpoint, the advancement of new techniques is an essential part of conceptualising, rather than responding to change; the concrete, visual effects generated by the development of new techniques stimulate the imagination. The specific properties of the techniques themselves are instrumental in shaping the concept. You can already see new effects and new models of organisation in a new technology.

Computer and mediation techniques represent the latest development in the twentieth-century catalogue of new techniques. They enable the storage, combination, manipulation, and display of information. They make time visible and calculable. They breed new words and new procedures – numerous small technologies proceed from the invention of new techniques. Above all, computer and mediation techniques promote and reveal a world of multiplied communications, in which everything and everyone is connected through technologies that are based on flexible, mobile, operational systems. In the new, mediated world all things spin in an invisible network. All the information that we receive and transmit comes to us through technologies that have reorganised the world. The image of a suddenly small earth encircled by a dense band of revolving satellites is an emblem of our time. Together, earth and satellites form a completely novel organisational typology, a network with a virtual structure, containing solely power points in space. The immaterial nodes in this network are always changing; their instability is as great as their expansiveness. The endless multiplication of communications leads new media to a narcissistic reflection on themselves, resulting in a new type of global success and scandal. Mediation to the present day is what the sublime was to the Romantic era: triumphant, exalted, poignant, massive, and uncontrollable. The question is: which are the appropriate techniques for architecture to use to instrumentalise this new mediated cosmology for its own ends?

› Diagrams

Diagrammatic technique provides a foothold in the fast streams of mediated information. The meaninglessness that repetition and mediation create is overcome by diagrams which generate new, instrumental meanings and steer architecture away from typological fixation. What is a diagram? In general, diagrams are best known and understood as visual tools used for the compression of information. A specialist diagram, such as a statistics table or a schematic image, can contain as much information in a few lines as would fill pages in writing. In architecture, diagrams have in the last few years been introduced as part of a technique that promotes a proliferating, generating, and instrumentalising approach to design. The essence of the diagrammatic technique is that it introduces into a work qualities that are unspoken, disconnected from an ideal or an ideology, random, intuitive, subjective, not bound to a linear logic – qualities that can be physical, structural, spatial, or technical.

There are three stages to the diagram: selection, application, and operation, enabling the imagination to extend to subjects outside it and draw them inside, changing itself in the process.

Diagrams are packed with information on many levels. A diagram is an assemblage of solidified situations, techniques, tactics, and functionings. The arrangement of the eighteenth-century Panopticon prison plan is the expression of a number of cultural and political circumstances cumulating in a distinctive manifestation of surveillance. It conveys the spatial organisation of a specific form of State power and discipline. It incorporates several levels of significance and cannot be reduced to a singular reading; like all diagrams, the Panopticon is a manifold. Characteristically, when a diagram breeds new meanings, they are still directly related to its substance – its tangible manifestation. Critical readings of previous interpretations are not diagrammatic. Put in the simplest possible terms, an image is a diagram when it is stronger than its interpretations.

The diagram is not a blueprint. It is not the working drawing of an actual construction, recognisable in all its details and with a proper scale. No situation will let itself be directly translated into a fitting and completely correspondent conceptualisation. There will always be a gap between the two. By the same token concepts can never be directly applied to architecture. There has to be a mediator. The mediating ingredient of the diagram derives not from the strategies that inform the diagram, but from its actual format, its material configuration. The diagram is not a metaphor or paradigm, but an 'abstract machine' that is both content and expression. This distinguishes diagrams from indexes, icons, and symbols. The meanings of diagrams are not fixed. The diagrammatic or abstract machine is not representational. It does not represent an existing object or situation, but it is instrumental in the production of new ones. The forward-looking tendency of diagrammatic practice is an indispensable ingredient for understanding its functioning.

Why use diagrams? Diagrammatic practice delays the relentless intrusion of signs, thereby allowing architecture to articulate an alternative to a representational design technique. A representational technique implies that we converge on reality from a conceptual position and in that way fix the relationship between idea and form, between content and structure. When form and content are superimposed in this way, a type emerges. This is the problem with an architecture that is based on a representational concept: it cannot escape existing typologies. In not proceeding from signs, an instrumentalising technique such as the diagram delays typological fixation. Concepts external to architecture are introduced rather than superimposed. Instances of specific interpretation, utilisation, perception, construction, and so on unfold and bring forth applications on various levels of abstraction.

How is the diagram chosen and applied? The function of the diagram is to delay typology and advance design by bringing in external concepts in a specific shape: as figure, not as image or sign. But how do we select, insert, and interpret diagrams? The selection and application of a diagram involves the insertion of an element that contains within its dense information something that our thoughts can latch onto, something that is suggestive, to distract us from spiralling into cliché. Although the diagram is not selected on the basis of specific representational information, it is not a random image. The finding of the diagram is instigated by specific questions relating to the project at hand: its location, programme, and construction. For us, it becomes interesting to use a diagram from the moment that it starts to relate specifically to organisational effects. Among our collection of diagrams are flow charts, music notations, schematic drawings of industrial buildings, electrical switch diagrams . . . all maps of worlds yet to be constructed, if only as a detail. To suggest a possible, virtual organisation, we have used ideograms, line diagrams, image diagrams, and finally operational diagrams, found in technical manuals, reproductions of paintings, or random images that we collect. These diagrams are essentially infrastructural; they can always be read as maps of movements, irrespective of their origins. They are used as proliferators in a process of unfolding.

How do diagrams become operational? The abstract machine of the diagram needs triggering. It has to be set in motion so that the transformative process can begin, but where does this motion originate? How is the machine triggered? What exactly is the principle that effectuates change and transformation? Furthermore, how can we isolate this principle and give it the dimensions that make it possible to grasp and use it at will? The insertion of the diagram into the work ultimately points to the role of time and action in the process of design. Interweaving time and action makes transformation possible, as in novels where long narrative lines coil around black holes within the story. If there were no black holes for the story's protagonist to fall into, the landscape of the narrative would be a smooth and timeless plane, in which the hero, whose character and adventures are formed by this landscape, cannot evolve. The story is an intrinsic combination of character, place, event, and duration. The landscape of the story, the black holes, and the character become one. Together they trigger the abstract machine. In architecture, it goes something like this: the project is set on its course. Before the work diverts into typology a diagram, rich in meaning, full of potential movement and loaded with structure, which connects to some important aspect of the project, is found. The specific properties of this diagram throw a new light onto the work. As a result, the work becomes un-fixed; new directions and new meanings are triggered. The diagram operates like a black hole, which radically changes the course of the project, transforming and liberating architecture.

1999

Ken Yeang

A Theory of
Ecological Design

Born in Penang, Malaysia in 1948, Ken Yeang trained as an architect at the Architectural Association in London from 1966 to 1971. After a period of study in the Department of Landscape Architecture at the University of Pennsylvania he completed a PhD at Cambridge University in 1975 with a dissertation entitled *A Theoretical Framework for the Incorporation of Ecological Considerations in the Design of the Environment*. In 1976 he founded the architectural practice T. R. Hamzah and Yeang International, specialising in ecologically responsive high-density urban projects and pioneering a new genre of tall buildings in tropical climates, known as "bioclimatic skyscrapers." Examples include: the UMNO Tower, Penang (1998); the MBF Tower, Penang (1993); and Menara Mesiniaga, Kuala Lumpur (1992). The firm currently employs over eighty staff in Kuala Lumpur in addition to associated offices in the UK, China, Singapore, Australia, and Germany. They have received numerous international prizes including the Prinz Claus Award (Netherlands), the Aga Khan Award for Architecture (Geneva), and the Royal Australian Institute of Architects International Award. Yeang has also taught at a number of schools of architecture internationally including: University of Hawaii, University of New South Wales, Curtin University, and the University of Malaya.

Paradoxically for a self-proclaimed "green" architect, Yeang accepts the skyscraper as an inevitable ingredient of high-density urban settlement patterns. He has therefore spent much of his career trying to refute the conventional wisdom that tall buildings are inherently destructive to the environment. His publications on this controversial topic include: *The Skyscraper Bioclimatically Considered: a Design Primer* (1996); and *Reinventing the Skyscraper: A Vertical Theory of Urban Design* (2002). He is currently working on a co-authored book entitled *The Encyclopedia of the Skyscraper*.

The Green Skyscraper (1999), the book from which the current extract is taken, is also intended as an "ecological design primer" for both architectural students and practitioners. In this chapter he describes the ecological building in relation to its surrounding environment – a kind of living organism responding dynamically to its context and forming part of a single interconnected system. Through a series of precise and measured arguments he sets out to define "the holistic and anticipatory properties of ecological design."

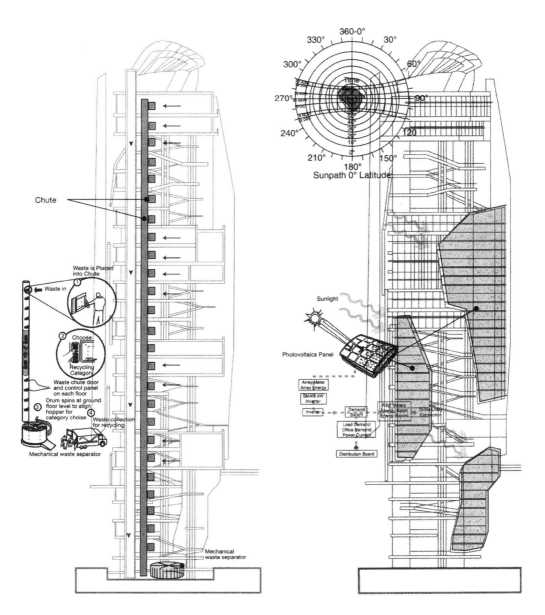

Fig 5
Ecological building systems: solid waste and ambient energy. Ken Yeang, 1999

For ecological design to be durable, we must have a theory – a general theoretical basis that will enable our design work to be environmentally holistic and anticipatory. What is generally found to be inadequate in current theoretical constructs is their incompleteness or failure to include an environmentally holistic property (e.g. "connectedness") crucial to the ecological approach.

This chapter will establish the fundamental criteria for ecological design and show how these are all interrelated. Therefore, any approach to design that does not take into account these aspects or the interactions between them cannot be considered holistic, and hence must either be not ecological or at best an incomplete approach to ecological design.

To begin with, we should first acknowledge that ecological design is complex, and certainly considerably more complex than is currently recognised by many ecological designers today. More specifically, it involves the incorporation of a complex set of "interdependent interactions" or connections with the environment (both global and local) which furthermore must be regarded dynamically (i.e. over time). This explains the need for establishing the holistic and anticipatory properties of ecological design.

To achieve environmentally sustainable objectives, ecological architecture must minimise (and at the same time be responsive to) the negative impacts that it has on the earth's ecosystems and resources. As mentioned earlier, we should be aware that ecological design is not a retreating battle, but on the contrary a designed system that can contribute productively to the environment (e.g. through production of energy using photovoltaics) as well as restore and repair damaged ecosystems.

› A general systems framework for design

For the purpose of developing a theory for ecological design, we can regard our building as a system (i.e. a designed system or a built system) that exists in an environment (including both the man-made and natural environments). The general systems concept is fundamental to the ecosystem concept in ecology. Briefly stated, in the analysis of the relationship of any system to its environment, there is essentially no limit to the number of variables that we can include in the analysis or in the description of the design problem. In fact, this applies to all design endeavours, for no matter how fortunate our choice may be of inputs and outputs to describe a system, they cannot be expected to constitute a complete description. The crucial task in design – and similarly in any theory – is therefore to pick the right variables to be included, which are those we find essential to our resolution of the design process.

What we need is a simple, general framework that structures the entire set of ecological interactions between a designed system and the earth's ecosystems and resources. This framework must be able to identify for the designer those impacts that are undesirable and which need to be minimised or altered in the design process. The theoretical basis for ecological design must provide the designer with a set of structuring and organising principles to carry out this goal. It can take the form of an open structure with which the selected and relevant design constraints (e.g., ecological considerations) can be holistically and simultaneously organised and identified. Furthermore, the open structure must facilitate the selection, consideration, and incorporation of design objectives in our subsequent design synthesis.

The open structure can be simply a conceptual or theoretical framework and should enable the designer to decide which ecological considerations to incorporate into his design synthesis, while at the same time ensuring a basis for a comprehensive check of other interdependent factors influencing design. Crucially, it must also demonstrate the interrelationships of all factors, which is an essential property of the connectedness of all ecological systems in the biosphere. By using the open structure as a design map, the designer can also include any other related and pertinent disciplines that are similarly concerned with the problems of environmental protection and conservation in design solutions (e.g. waste disposal, resource conservation, pollution control, applied ecology, etc.). Thus, the essential properties of the theory of ecological design are that it be inclusive, comprehensive, and open.

› Green design as environmental impact statement

As mentioned earlier, ecological design is prognostic and anticipatory, and as a result the design process becomes essentially the preparation of a "statement" of anticipated environmental impacts and benefits. From the earlier examination of ecology and ecological concepts, we have determined that the extent of the environmental consequences of any built system can be seen as the net result of its dependencies (i.e. demands and contributions) on the earth's ecosystems and processes and on the earth's energy and material. These dependencies are both global (for example, the use of non-renewable resources) as well as local (the impact on local ecology). If the designer is aware of the ecological consequences (both detrimental and beneficial) of his design, then this knowledge represents in effect a summation of the extent of the design's impacts on the environment, which thereby are accepted and anticipated by the designer.

Defining the design task in this way should not imply the exploitation of the biosphere (Bookchin, 1973). On the contrary, this approach emphasises the extent of human dependency (and of built structures) on the earth's resources and ecosystems. Such a viewpoint helps direct our attention to those aspects of the designed system that have ecological implications and indicates the critical areas where the undesirable impacts might be eliminated, reduced, or remedied. The ecological approach embodies the realisation that any designed system is dependent directly or indirectly upon the biosphere for specific elements and processes, which can be identified as including the following:

- Renewable and non-renewable resources including minerals, fossil fuels, air, water, and food.
- Biological, physical, and chemical processes, for example decomposition, photosynthesis, and mineral cycling.

- End point or processing of waste and discharges resulting from human activities, including life processes as well as the functioning of manmade systems (example: landfill waste disposal).
- Physical space in which to live, work, and build.

These various functions and aspects of the environment and our human use of them are interrelated and overlapping; they meet each other at "transfer points", where the designed system interacts with the surrounding ecosystem. Transfer points are vitally important to green design precisely because bad design at the points where exchange occurs frequently results in damage to the ecosystem.

It should be remembered, however, that it is impossible to design a system in which none of these linkages results in an impact on the ecosystem. The mere physical existence of the building, as we have seen, causes some spatial displacement (it takes up space) of the ecosystem, and our use of land represents a loss to the biosphere volumetrically. Absolute ecological compatibility is physically impossible because of this most basic impact on the environment. However, we can produce buildings that have greater or lesser destructive effects on the environment, and they can even have some results that are beneficial. It is the job of green design to minimise negative impacts and maximise beneficial interactions between built systems and natural ecosystems.

> The theoretical structure of ecological design

The theoretical framework of green design must be developed in line with a number of concepts, which will be briefly summarised here.

A building exists both in terms of its physical being (form, siting, and structure) and its functional aspects, i.e. the systems and operations that sustain it during its useful life. Both aspects involve the built structure in relationships with the natural environment which take place over time. The building acts like a living organism; in place of food, uses energy and materials, and also produces outputs into its environment. Our theoretical structure should model these exchanges.

Three components are essential for an ecological model of the designed system. Our framework must include a description of the built system itself, a description of its environment including the ambient ecosystem and natural resources, and a mapping of the interactions between these two components (i.e. between the building and its environment).

The first step is to systematically take account of the internal processes of the designed system. The second step is to measure, based on a thorough knowledge of the building's physical and functional requirements, its interactions with the earth's ecosystems in the form of the energy and resources removed from the environment by the construction and ongoing operation of the structure, and in addition amounts of matter and energy that are sent back into the natural environment as a result of the functioning of the building's internal systems (the "metabolism" that makes it function as a built environment).

A supplementary question is the relationship of the built structure as an element in the spatial configuration of the environment. Its existence as a built environment within the natural one implies further interactions and effects on biosphere. Analysis of any such impacts will also have to be factored into the theoretical framework.

An open general systems framework can be used to visualise "sets of interactions" taking place between the designed system and its environment.

The concept of an open system in contact with its environment as formulated in general systems theory is useful here. Based on the above analysis of the fundamental interactions of the built and natural environments, the interactions can be grouped into four sets.

Set 1: External interdependencies, consisting of the designed system's relations to the external environment.
Set 2: Internal interdependencies, being the designed system's internal relations.
Set 3: External-to-internal exchanges of energy and matter – i.e., system inputs.
Set 4: Internal-to-external exchanges of energy and matter – i.e., system outputs.

The four sets also usefully describe the "transfer points" between the built and natural environments which were discussed above. Green design must take account of all four sets as well as the interactions between them. In this way, our framework allows us to determine how architecture impinges on terrestrial ecosystems and natural resources whenever we address any design task.

Elsewhere (Yeang, 1995) I have developed a "partitioned matrix" (LP) which unifies these sets of interactions in a single symbolic form. The figure demonstrates this conceptualisation of the relationship of the designed system and its environment (suffix 1 stands for the system, suffix 2 the environment). If the letter L stands for the interdependencies within the framework, then four types of interaction can be identified (Tolman and Brunswick, 1935; Emery and Trist, 1965; Walmsley, 1972). In the partitioned matrix, they are identified as L_{11}, L_{12}, L_{21}, and L_{22}:

$$(LP) = \begin{array}{c|c} L_{11} & L_{12} \\ \hline L_{21} & L_{22} \end{array}$$

Remembering that "1" represents the built system and "2" the environment in which it is situated, we can map the four sets of interactions listed above onto the partitioned matrix. L_{11} represents processes that occur within the system (internal interdependencies), L_{22} represents activities in the environment (external interdependencies), and L_{12} and L_{21} refer to system/environment and environment/system exchanges, respectively. Thus, internal and external relations and transactional interdependencies are all accounted for.

The partitioned matrix is itself a complete theoretical framework embodying all ecological design considerations. The designer can use this tool to examine interactions between the system to be built and its environment holistically and inclusively, taking account of all the environmental interdependencies described by the above four sets.

› The law of ecological design

If a fundamental "law" for ecological design can be asserted, then this partitioned matrix constitutes the Law of Ecological Design. In ecological design, this "law" then requires the designer to look at his designed system in terms of its component parts, i.e. inputs, outputs, and internal and external relations, and then to see how these interact with each other (both statically and dynamically over time, these being the four components of the partitioned matrix).

In effect, the designer can then further ascertain which of their ecological impacts need to be given priority and which need to be taken into account or adjusted in the process of design. In this way, any designed system can be conceptually broken down and analysed based on these four sets of interactions as follows:

L_{22}

These interactions describe the designed system's external interdependencies or "external relations". By this is meant the totality of the ecological processes of the ambient ecosystem, which as we have seen interacts with other ecosystems; hence L_{22} takes in not only local but global environments and terrestrial resources in their totality. It therefore also includes the processes by which earth's resources are created (for example, the formation of fossil fuels and non-renewable resources), which may be affected by, and themselves affect, the built structure's functioning. These external resources will be modified, depleted or added to by the creation and functioning of the built system.

L_{11}

The internal interdependencies are the internal environmental relations of the built system. This means the sum of all the activities that go on inside the building, including all of its operations and functions. The functioning of the built structure's internal metabolism will have larger effects, extending to the ecosystem where it is sited; these effects, by the principle of connectivity, will in turn affect other ecosystems and the biosphere's totality of resources. The L_{11} effects take place over and describe the whole life cycle of the building.

L_{21}

This quadrant of the matrix describes the total inputs into the built system, including all of the exchanges of energy and matter that go into its construction. System inputs of a designed system include all of the resources that make up its component parts and the matter and energy upon which its operations and processes depend. Securing these resources that make the building "run" (the extraction of infrastructural materials and energy from the earth) often causes damage to the biosphere and its ecosystems.

L_{12}

The total outputs from the built environment into the natural one are the most obvious concern of the ecological designer, but they are only one quarter of the total interactions discussed here. These outputs, however, include not only discharges of waste and exhaust from the building's construction and operation, but also the physical matter of the structure itself, which must be disposed of at the end of the building's planned lifespan. Obviously, if these outputs cannot be assimilated by the natural environment they result in ecological harm.

Any design approach that claims to be ecological and does not take into account these four components and their interactions over time cannot be considered a complete and holistic ecological design, as interconnectedness is a crucial characteristic of ecosystems. Failure to take this factor into consideration is non-ecological.

2000

Bernard Cache

Digital Semper

Bernard Cache (born 1958) is a French architect, philosopher, and digital pioneer. He studied at the Ecole Polytechnique Fédérale de Lausanne, then with Gilles Deleuze at the Institut de Philosophie, and at the Ecole Supérieure des Sciences Economiques et Commerciales. He has taught at the University of Toronto, the University of Paris, the Institut Français de la Presse and the Berlage Institute in Amsterdam. He founded the Digital Production Workshop at the Escuela Superior d'Arquitectura in Barcelona in 2000, and was a visiting scholar at the Center of Canadian Architecture in 2005. In 2002 Cache founded the Architecture Office Objectile, a digital architecture lab creating tools and technologies needed for non-standard design and manufacture.

In Cache's first book, *Earth Moves: The Furnishing of Territories* (1995), written at the end of his studies in Lausanne (*Terre Meuble*, 1983), he gives a non-representational account of the architectural image following Deleuze and Henri Bergson. Since then Cache has delved deeply into both contemporary techniques and the long history of architectural treatises, revealing surprising affinities and patiently reminding the architecture profession of its own deep fascinations and techniques.

Gottfried Semper's master work, *Der Stil*, was highly influential among architects of the late nineteenth and early twentieth centuries, though because of the enthusiasms of his followers it was often misunderstood as a simple kind of materialism. Cache summarizes key elements of Semper's treatise, focusing on the principle of Stoffwechsel or "material transformation," which Cache traces to the dispute in biology between Cuvier and St. Hilaire.[1] On the one hand, the issue is the translation of forms from one material to another, especially the translation of Greek temples from wood to stone, which has been debated since the first architectural treatise written by the Roman architect Vitruvius. On the other hand, this involves the longstanding search for forming principles in architecture, from modules to systems of proportion, which draws the form–function debate directly from biology in the early nineteenth century. In that light, Cache connects Semper's reading of St. Hilaire to the contemporary architectural reading of works like Stuart Kauffman's *Origins of Order: Self Organization and Selection in Evolution*. A reading that suggests not just the search for the form of building systems, but its metabolism as well.

1   Isidore Geoffroy Saint-Hilaire, *Histoire générale et particulière des anomalies de l'organisation chez l'homme et les animaux, ou, Traité de tératologie* (Paris: J.-B. Baillière, 1832–36).

"Digital Semper." To put these two words together seems like a contradiction in terms. Starting with an analysis of the first word, however, I will try and dismantle the apparent contradiction.

L'atelier Objectile, which I created with Patrick Beauce, experiments with technologies in architecture by focusing primarily on software development in order to digitally design and manufacture building components. Beginning with a period dedicated to building research, furniture design, and sculpture, we worked for more than ten years with the French company TOPCAD in designing complex surfaces in order to debug our developing software. Three years ago we created Objectile and started to focus our work on flat and supposedly simpler components like panels or doors in order to tackle the problems generated by the industrial production of varying elements.

Industrial production forces us to confront many basic problems like zero-error procedures and stress-free MDF panels. A key element in digital manufacturing is to avoid bending a panel when machining one of its faces. Our experience now enables us to think of a fully digital architecture like our museum project and the pavilion we recently built on the occasion of the Archilab conference in Orleans. The four elements of this pavilion are the result of previous experiments with screens, panels, and tabletops. In that process, we noticed that our approach had a clear affinity to Gottfried Semper's theory as he articulated it in *Der Stil* (1863) not only because we come to architecture through the technical arts, or because we came to invent new materials in order to create new designs, but because our interest in decorative wooden panels is consistent with Semper's Bekleidung Prinzip (cladding principle). Even our investigations into the generation of software to map key elements of modern topology, like knots and interlacing, consist of a contemporary transposition of Semper's Urmotive or primitive pattern.

What does it mean today to refer to Gottfried Semper? Why, in 1999, should we look back to the nineteenth century just as everybody claims the twenty-first will be digital? And why focus on Semper, whose architecture seems to reveal nothing but the Renaissance historicism rejected by the Moderns?[1] Are we not in a very different period? We live in an age not of iron but silicon. Why would we need to reconnect the end of our iron, concrete, and glass century to the history of wood, stone, clay, and textiles? Do we not run the risk of a new technological determinism, by which the information age, the so-called "third wave," would create a second break with the past, definitively negating any historical experience, leaving us with no alternative other than a choice between the dinosaurs and the space shuttle? Or should we not instead be reminded that information technologies themselves are deeply rooted in the past? The computer is not an Unidentified Flying Object that landed one day in a California garage.

Let us recall a few examples of current computing issues that lead us back to the nineteenth century, if not earlier. We could begin with the Fast Fourier Transform Integrated Circuits, which you can find in any digital television set. Joseph Fourier, who discovered the mathematical method for coding the picture of the future, had worked alongside Champolion, his companion on Napoleon's expedition to Egypt, and who had found a way to decode hieroglyphs (1822) with the help of Quatremere de Quincy's cousin. We could also cite Sadi Carnot, inventor of the

Second Law of Thermodynamics (1824) that Richard Feynman, the great physicist hired by IBM, paired with Maxwell's Demon (1867) in his 1996 lectures about "Reversible computation and the thermodynamics of computing"[2] – lectures whose topic is the long-term future of the computer. We could also discuss the all too familiar, yet often misinterpreted, chaos theory, the mathematics of which were worked out by Henry Poincare. Recently, Michel Serres has brilliantly demonstrated how these mathematics share a common structure with Claude Debussy's musical composition, as well as with Charles Peguy's book on history, *Clio, ou dialogue de l'histoire et de l'ame paienne*.[3] And, last but not least, we could mention the integration of Desargues's geometry into modern CAD software.

Even if computer science cannot contemplate its future without returning to old debates, we should certainly expect architecture to benefit from reacquainting itself with its past in order to take advantage of information technology. We believe that cyberspace need not lead to cultural amnesia. We believe that innovation can be linked to history, without the return of a prehistory or the advent of a science fiction.

Our interest in Semper stems from his concise articulation of technology and history in architecture. But today we will leave aside the anthropological aspect of Semper's conception of history in order to focus instead on the structure of his theory, which I will summarize in the following four propositions:

1   Architecture, as with the other fine arts, finds its fundamental motivation in the technical arts.
2   The four major technical arts are: textiles, ceramics, tectonics, and stereotomy.
3   Among these four technical arts, textiles lend many aspects to the other three techniques.
4   The knot is the fundamental mode of textiles, and therefore of architecture, inasmuch as this monumental art is subordinated to the cladding principle (Das Bekleidung Prinzip).

This, of course, is too straightforward a summary, for when we get into the matter and read Semper's text carefully, we quickly realize that these summations are complicated further. Immediately following his introduction, Semper enunciates the four categories of materials according to physical criteria. Materials can be pliable, like fabrics; soft, like clay; elastic, sticklike elements, like wood; or dense, like stone. Semper then immediately switches to the second enunciation of four categories, no longer distinguishing materials but activities, or what we will call procedures. These are the famous technical arts: textiles; ceramics; tectonics (i.e., carpentry); and stereotomy (i.e., masonry).

At first sight, the second list seems to be redundant, given the first one. But then Semper makes a series of puzzling remarks. He tells us that "Not only have the four categories to be understood in a wider sense, but one should be aware of the numerous and reciprocal relations that link them together." To be sure, each of the four techniques applies to a privileged material from which originated the primeval motives. Nevertheless, procedures were also developed for each of the other materials. Thus, ceramics should not be restricted to earthenware but should also include objects made out of all kinds of materials, like metal, glass, and stone. Equally, brick-work, tiles, and mosaics, although made out of clay, should not be directly related to ceramics but rather to stereotomy and textiles, considering the fact that they are used both to compose masonry works and as cladding materials for the walls themselves. Semper follows with various other examples, creating a rather disconcerting impression. Textiles can no longer be considered to be made out of fabrics, and clay does not suffice to explain ceramics. All four technical arts remain abstract categories elaborated throughout the pioneering *Der Stil*.

However, things become clearer if we switch from a linear reading to a tabular one. The two lists of materials and procedures, far from paralleling one another, should rather be articulated as a table (Table 1), since we can find occurrences of each type of material for every type of procedure. Semper himself briefly explains his methodology in the second chapter of the introduction. Each technical art must be analyzed from two perspectives. One should first look at what he calls "the general-formal aspect" (Allgemein Formelles), which accounts for the intention or the purpose (Zwecklichkeit) of the works. Only afterward should one consider "the technical-historical aspects" (Technisch-Historisches) and analyze how these intentions or purposes have been realized throughout history, according to various local factors. Semper illustrates these two approaches by taking the example of the strip which holds an object in the shape of a ring. Semper points out that in the general-formal analysis,

> One should only consider certain characteristics, so to speak, abstract, which relates to the strip as something that links, while the question of its various forms, according to the expression of the concept within linen, silk, wool, and in wood, baked clay, stone or metal, should be relegated to the historical-technical analysis.[4]

We have then, on one side, the general procedure of linking, and on the other, the various materials through which this procedure is applied. But then Semper warns us that one cannot expect too much rigor in the distinction of the two approaches, since the function of a product also requires the use of appropriate materials. Hence the abstract procedure cannot be thought of in isolation from the historical material.

The table prevents an oversimplified reading of Semper's system; a reductiveness that I was lucky enough to avoid in two, opposing ways. While Alois Riegl complained about the materialist reading of the so-called Semperians, according to

Table 1 Historical and traditional materials

Abstract Procedures	Textile	Ceramics	Tectonics or Carpentry	Stereotomy or Masonry
Fabric				
Clay				
Wood				
Stone				

Table 2 Historical and traditional materials

Abstract Procedures	Textile	Ceramics	Tectonics	Stereotomy
Fabric	Carpets, rugs, flags, curtains	Animal skin flask, ex: Egyptian situla		Patchwork?
Clay	Mosaic, tiles, brickwork cladding	Vase-shape earthenware, ex: Greek hydria		Brickwork masonry
Wood	Decorative wooden panels	Barrels	Furniture, carpentry	Marquetry
Stone	Marble and other stone cladding	Cupola	Trabeated system	Massive stonework, Goldsmith's

which art forms would be strictly determined by materials, Otto Wagner, in his *Modern Architecture* (1896), criticized Semper's symbolist approach (although he could not help placing wreath-bearing angels on the top of his 1903 modern Post Office in Vienna). Riegl focused on the material, while Wagner isolated the abstract procedure from the material, each considering only one single part of Semper's system. More generally, this tabular structure explains why the style of *Der Stil* makes reading Semper so complex and consequently so prone to oversimplification. Semper approaches and returns from his major themes like the weaver's shuttle passes over and under weft threads (Table 2). His thinking on the surface is hard to account for in the linearity of his writing, which would explain why two of Semper's key arguments, that of the knot and that of the mask, find their ultimate development only in footnotes.

In a way, we could argue that the structure of this table has been worked out by Semper himself, inasmuch as we could consider that writing is only one of many modes of thinking – a type of intellectual activity among which we could also posit architecture (Ut scriptura architectura (Table 3)). Indeed, in the proposal given to the approval committee, Semper explained how the two museums built in Vienna, the Art History Museum and the Natural History Museum, illustrate the dualities of his practice and his theory, between his writings and his buildings. Harry Francis Mallgrave has brilliantly commented on the structure of the facades of these monumental pieces of architecture.[5] In the Art History Museum, each of the three stories was assigned a specific iconography related to one of the three factors affecting the development of primitive patterns according to Semper's definition of style. The ground story is dedicated to the materials of the various technical arts (what Semper called the external factors). Moving upward, the main floor is dedicated to the social and religious conditions of art, and the statues on the roof reify the individuals who opened significant new paths in art (the two types of internal factors). Hence, we have a vertical organization of the facade progressing from the material

Table 3 The built book: *The Art History Museum* (1869–91)

Roof: Creative individuals

Main story: Social and religious conditions of art Ground story: Technical arts

Artistic tendencies: Classical, Romantic

The written building: *Der Stil* (1860–63)

Materials of contemporary architecture

Materials of modern architecture

Materials of traditional architecture

Abstract procedures: Textile, Ceramics, Tectonics, Stereotomy

conditions of art toward its spiritual achievements; in other words, a progression from the material to the immaterial. This vertical progression is articulated with a horizontal opposition between classical and romantic tendencies in each of the arts, Doric being opposed to Ionic, Raphael to Michelangelo, Mozart to Beethoven; oppositions that anticipate later writings like *Renaissance und Barok* by Heinrich Wolfflin, or *Abstraktion und Einfuhlung* by Wilhelm Worringer, not to mention *The Birth of Tragedy*, in which Nietzsche opposed Apollo to Dionysus. So, each facade is organized as a table, and this tabular structure is applied to the four historical epochs allocated to each side of the building: Antiquity, the Middle Ages, the Renaissance, and Semper's present.

We could argue that the museum in Vienna constitutes the third volume, built rather than written, of the unfinished *Der Stil*. At first sight, the museum presents itself as a three-dimensional composition of tables whose height, length, and depth each make up one type of analysis. Actually, the two volumes of *Der Stil* could themselves be understood as a two-dimensional sub-table of the facade organization. As I already mentioned, in one direction we could find the technical arts, whereas orthogonally, we could find the progression from the material to the immaterial. These two directions are of very different natures. The leitmotiv of Semper is that there are a limited number of abstract procedures,[6] which is why he shows himself to be very parsimonious in counting them (Table 4). Thus, metal is introduced as a material in itself and, rightly or wrongly, Semper did not associate a specific procedure with it. Metal only provides another medium for the development of each of the four abstract procedures, especially that of textiles. Semper goes back to embossed Greek statuary to advocate metal as cladding or, at least, as a hollow structure that provides an alternative to the thin, cast-iron columns of Joseph Paxton's Crystal Palace. In this way, the fact that Semper fails to mention any procedure specific to this material should be taken as an indication that he disregarded it. Behind his claim for monumental columns lies a very modern conception of metal construction as hollow structure.

In the other direction, that of materials, the number of cases seems, on the contrary, to be limitless. Not only does Semper dedicate a full chapter to metal, in addition to the four privileged materials, but there are also many references to various other materials such as glass. Therefore, the openness of Semper's theory

Table 4 Historical and traditional materials (including metal)

Abstract Procedures	Textile	Ceramics	Tectonics	Stereotomy
Fabric	Carpets, rugs, flags, curtains	Animal skin flask, ex: Egyptian situla		Patchwork?
Clay	Mosaic, tiles, brickwork cladding	Vase-shape earthenware, ex: Greek hydria		Brickwork masonry
Wood	Decorative wooden panels	Barrels	Furniture, carpentry	Marquetry
Stone	Marble and other stone cladding	Cupola '	Trabeated system	Massive stonework, Goldsmith's
Metal	Hollow metal cladded statuary; Olympian Jupiter reconstituted by Quatremere de Quincy; metal roofing; articulated metal structures; curtain wall	Metal vases or shells	Cast iron columns	Forge, ironworks

is due to the possibility of introducing new materials (Table 5). It would be very interesting to see how reinforced concrete would fit in Semper's scheme and how much his theory would be able to account for the modern triumverate of metal, concrete, and glass. Even more pertinent would be an evaluation of Semper's theory with regard to those materials that he would have called more spiritual and that we would simply designate as more immaterial, in the sense that they deal with lower energies. In this category would fall both biology and information technologies.

As for biology, I will only briefly mention the fact that Semper was in Paris in 1830, at the key moment of the debate opposing Baron Georges Cuvier to Geoffrey Saint-Hilaire. The core of the problem was Cuvier's refusal of Saint-Hilaire's establishing a continuity between his four animal categories.

If one looks closely at Cuvier's four categories – the mollusks, the radiates, the vertebrates, and the articulated – they would appear to share a common geometric structure with Semper's four corresponding abstract procedures. It would be too involved to get into this matter here, but I would like to emphasize the fact that the abstract procedures should not be thought of as Platonic ideals, independent from the materials to which they are applied. On the contrary, it would be in the nature of these procedures to look relentlessly for more "immaterials" in order to

Table 5 Materials of Modern and Contemporary Architecture

Abstract Procedures	Textile	Ceramics	Tectonics	Stereotomy
Metal	Hollow metal cladded statuary; Olympian Jupiter reconstituted by Quatremere de Quincy; metal roofing; articulated metal structures; curtain wall	Metal vases or shells	Cast iron columns	Forge, ironworks
Concrete	Prefabricated concrete screens; light warps; curtain wall	Ruled surfaces, like: hyperbolic paraboloid	Slabs on stilts	
Glass Biology	Thermoformed glass; curtain wall Mollusks	Blown glass Radiates, D'AT: Surfaces de Plateau	System glued glass (pictet) Vertebrates, D'AT: squeletons and bridge structures	Glass bricks Articulated D'AT: bees' cells
Information	Modulation interlacing (Eurythmy)	Revolving solid, polar coordinates	Translation, Cartesian coordinates	Boolean operation, tiling algorithms

find a new occasion for their progressive abstraction.[7] Thus, information technologies would not simply be accidentally accounted for by Semper's theory; it would be in their very nature to fit into his system as the best vehicle to push the abstraction of the four technical procedures further. Far from being limited to fabrics, textiles could be the procedure of going alternatively over and under, what in terms of information technology is called modulation. In turn, distinct from pottery, ceramics could deal with revolving solids and operations in radial coordinates as opposed to tectonics, which could deal with nonrotational transformations adequately described in Cartesian coordinates. And finally, stereotomy could be the art of tiling and paving as it results from Boolean operations. Taken as a whole, we would have described the interface of a Semperian computer-aided-design software.

Of course, this remains a hypothesis so long as this software has yet to materialize. And it is not necessary to offer an uncritical acceptance of the closed number of four procedures. But we can find enough interesting arguments in

Semper's text itself. The definition of ceramics as a revolving operation rests much less on the technical gesture of the lathe than on the classification table that Semper borrowed from Jules Ziegler. As Mallgrave reminds us, Ziegler was a painter who worked for several years on the murals of La Madeleine church and established a classification of ceramics on the basis of the rotation and deformation of two simple geometric figures: the square and the circle. Interestingly enough, Ziegler conceived his Etudes Ceramiques as twenty-four Cartesian meditations.

More relevant to our own practice is the concept of modulation, which is not at all foreign to *Der Stil*. One could even say that it is the key concept of the *Prolegomena*. It is what, under the name of Eurythmy, Semper conceived of as the Gestaltungsprinzip. Eurythmy was first introduced as the principle of all regular closed figures, like snowflakes or crystals, but also architectural frames and cornices. Semper articulated a more general definition: "Eurythmy consists in the sequencing of spatial intervals displaying analogue configuration." This very general definition benefited from further specification. Sequences may be mere repetition, as in the dentils on Greek temples, or an alternation, as when a minor element is inserted between the repeating major elements like the triglyphs and the metopes, again on the frieze of Greek temples. Here, the principle of alternation becomes the rhythmic repetition of unequal parts. Beyond these two sequences, the eye would accept that the simple repetition and alternation be periodically interrupted, as in the Renaissance balustrade. There Semper points out additional levels of complication used when one wants to get rid of rigid architectural sequences to achieve the delightful confusion of lace and interlacing. At that point the reader is directed to the chapters dedicated to textiles.

This suggests that a close reading of Semper allows us at least to test the hypothesis of an identification of textiles with modulation when the former deals with electronic materials instead of fabrics. This association of textiles to modulation occurs through the concept of eurythmy, which is nothing other than the description of modulation techniques (with their various parameters of amplitude, frequency, and phase), techniques which provide the basis of the algorithms that we use in our practice, for example, to design our Semper Pavilion. The Semperian eurythmy provides us with a mathematical understanding of the concept of concinnitas which, as Caroline van Eck reminds us, was the keystone of Alberti's *De re aedificatoria*. Renaissance aesthetics should certainly not be reduced to a theory of proportions; but the fact that the concept of concinnitas encapsulates the sub-concepts of oppositio and varietas does not preclude a mathematical interpretation of Alberti's *De re aedificatoria* if we understand the encapsulated specifications as oppositions or variations of amplitude, frequency, and phase. This is precisely what Semper is hinting at in the fourth section of his book, dedicated to stereotomy. Commenting on the architectural proportions of the Doric order, he underscores the fact that the interval of the intercolumniation decreases from the middle toward the corners of the temples as a continuous variation. Let us not forget that the word modulation comes from the Latin modulum, which originally designated the diameter of the column to be used as a reference unit in the relationships of proportion. Therefore, by reading Semper's *Der Stil* on an abstract plane, rather than literally, we can draw many lessons from architectural history in view of a contemporary practice. In other words, Semper allows us to confront Euclid and Vitruvius. Could we not then look at the Renaissance style of Semper's architecture outside of mere historicism?

Against all claims of Semper himself, it seems that the German architect kept the very heart of the treatises of his Latin predecessors. What is so surprising in Vitruvius is his concept of transposition. Regardless of whether the motifs in stone, such as triglyphs, have their origin in wood, as Vitruvius argued, or in fabrics, as Semper would propose, the general principle is that the forms and proportions of the architectural orders are technically determined. Nevertheless, this determination does not come from the actual material but via procedures associated with another material, which then have to be transposed (Table 6). There is, then, a material determination in architecture, but it only appears through a process of transposition, a process which manifests itself in the stone pediment ending the series of wooden trusses that support the roof of Greek temples. The pediment transposes the wooden structure of the trusses in stone. The word transposition is the translation of what Semper termed Stoffwechsel in German – "material transformation" in English – which brings us back to biology, since this was the word used by Semper's friend Moleschott in describing the metabolism of plants and animals.

So, rather than contradicting Vitruvius's theory, Semper raised it to a higher level. The origin of architecture is no longer unique, since it comes from four technical arts, and, we might add, is no longer Greek. We could even say that there are no more origins at all, but instead a composition of several lineages of transposition by which the four abstract procedures constitute themselves by switching from one material to the other. Ut pictura architectura. Vitruvius invents the transposition principle, but its application to tectonics in stone as a transposition from wood is only one step within Semper's general table. Architecture emerges in the move from one technology to another. Hence, textiles would today be the abstract procedure emerging within the transposition process that leads us from primitive fabrics to contemporary modulation techniques, while continuously emulating mosaic cladding, wooden panels, and embossed metal. Technical art is a contracting memory as opposed to an engramme.

Table 6 Vitruvian and Semperian transposition

Abstract Procedures	Textile	Ceramics	Tectonics or Carpentry	Stereotomy or Masonry
Fabric	(Semperian)			
Clay	( )			
Wood	( )		(Vitruvian)	
Stone	( )		(transposition)	
Metal	( )			
Concrete	( )			
Glass	( )			
Information	(transposition)			

1   By moderns, I designate the architects of the twentieth century who took a strong position against history and ornamentation, which in practice did not prevent them from caring about surfaces and cladding, but actually limited their ornamental design to material textures, washed out coatings (a-plat), and rectilinear geometric patterns.

2   Richard P. Feynman, *Lectures on Computation* (Reading, MA: Addison-Wesley, 1996).

3   Michel Serres, *Eloge de la philosophie en langue francaise* (Paris: Fayard, 1995).

4   Gottfried Semper, *Der Stil 1860–63*, partially translated in *Gottfried Semper: The Four Elements of Architecture and other Writings*, trans. Harry Francis Mallgrave and Wolfgang Hermann (Cambridge: Cambridge University Press, 1988).

5   In Harry Francis Mallgrave, *Gottfried Semper, Architect of the Nineteenth Century* (New Haven, CT: Yale University Press, 1996). Generally speaking, all of our analyses rely on the considerable and remarkable work done by Mallgrave and Wolfgang Hermann to unearth the history of nineteenth-century architectural practice and theory.

6   It is true that in the museum, the number of technical arts is six, and that none of them is explicitly textiles, although metal is presented as "hollow metal and embossing," while sculpture in marble could be assimilated into the fine texture of Corinthian as opposed to the rough forms of Doric. Also note that casting could be understood as the imprint.

7   Up to the point of negating matter itself. See Semper in Mallgrave and Hermann: only by fully mastering the technique can an artist forget the matter.

Manuel De Landa (born 1952 in Mexico City) is a writer, philosopher, and artist who has lived and worked in New York City since 1975. He began his career as an independent filmmaker, showing his work in cine-clubs and art galleries. In 1980 he acquired an industrial-grade computer and became a programmer, writing his own software and working as a computer artist. He is currently an Adjunct Professor at the Graduate School of Architecture, Planning and Preservation, Columbia University (New York), a Professor for Contemporary Philosophy and Science at the European Graduate School in Saas-Fee, Switzerland, and he also teaches at the University of Pennsylvania. He is perhaps best known as the author of the books *War in the Age of Intelligent Machines* (1991), *A Thousand Years of Nonlinear History* (1997), and *Intensive Science and Virtual Philosophy* (2002).

De Landa's writing builds on the theories of the French philosopher Gilles Deleuze combined with the insights of contemporary science and chaos theory. He addresses topics such as nonlinear dynamics, theories of self-organization, artificial intelligence and artificial life, economics, architecture, and the history of science. His research into the principles of "morphogenesis" – the production of semi-stable structures out of the material flows found in both the natural and social worlds – has been read closely by theorists from many academic and professional fields, especially among the architects who have been his students and colleagues. De Landa's particular interpretations have made Deleuze the philosopher of system effects.

In the current essay De Landa discusses the architectural possibilities of the computer programs known as "genetic algorithms" – evolutionary simulations that might replace traditional design methods and result in the "breeding" of new forms. He uses the genetic algorithm to question determinist notions of linear causality by proposing alternative non-linear, bottom-up, and process-driven design techniques. In this proposal architects are invited to explore and play with the building-as-an-emergent-system, though its implications in this case are limited to the structural aspects of buildings.

2002

Manuel De Landa

Deleuze and the Use of the Genetic Algorithm in Architecture

The computer simulation of evolutionary processes is already a well-established technique for the study of biological dynamics. One can unleash within a digital environment a population of virtual plants or animals; and keep track of the way in which these creatures change as they mate and pass their virtual genetic materials to their offspring. The hard work lies in defining the relation between the virtual genes and the virtual bodily traits that they generate. The remaining tasks – keeping track of who mated with whom, assigning fitness values to each new form, determining how a gene spreads through a population over many generations – are performed automatically by computer programs known as 'genetic algorithms'. The study of the formal and functional properties of this type of software has now become a field in itself, quite separate from the applications in biological research which these simulations may have. In this essay I will not deal with the computer science aspects of genetic algorithms, or with their use in biology, but will focus instead on the applications which these techniques may have as aids in artistic design.

In a sense, evolutionary simulations replace design, since artists can use this software to breed new forms rather than to merely design them. However, as I argue below, there remains a part of the process in which deliberate design is still a crucial component. Since the software itself is relatively well known, and easily available, users may get the impression that breeding new forms has become a matter of routine. But it must also be noted that the space of possible designs in which the algorithm searches must be sufficiently rich for the evolutionary results to be truly exceptional. As an aid to design, these techniques would be rather useless if the designer could easily predict which forms would be bred.

Genetic algorithms will only serve as useful visualisation tools if virtual evolution can be used to explore a space in which it is impossible for the designer to consider all potential configurations in advance, and only if what results shocks, or at least surprises. In the task of designing fertile search spaces, certain philosophical ideas, traced to the work of Gilles Deleuze, play a crucial role. I would argue that the productive use of genetic algorithms necessitates the deployment of three philosophical forms of thought: populational, intensive, and topological. Deleuze did not invent these but he brought them together for the first time, and made this the basis for a new concept of the genesis of form.

In order to utilise genetic algorithms, a particular field of art must first be used to solve the problem of representation in the final product in terms of the process that generated it. Then one must figure out how to represent this process itself as a well-defined sequence of operations. It is this sequence or, rather, the computer code that specifies it, that becomes the 'genetic material' of the object in question. This problem can be simplified through the use of computer-aided design, given that a CAD model of an architectural structure is already defined by a series of operations. For example, a round column is produced with the following directions:

Draw a line defining the profile of the column.
Rotate this line to yield a surface of revolution.
Perform a few 'Boolean subtractions' to carve out detail in the body of the column.

Some software packages store this sequence and may even make available the actual computer code corresponding to it. In this case, the code itself becomes the 'virtual DNA' of the column. (A similar procedure is followed to create each of the other structural and ornamental elements of a building.)

In order to understand the next step in the process, one must apprehend the basic tenets of 'population thinking'. This method of reasoning was employed in the 1930s by the biologists who synthesised the theories of Darwin and Mendel, thereby creating the modern version of evolutionary theory. This method of thinking can be encapsulated in the brief phrase, 'never think in terms of Adam and Eve, but always in terms of larger reproductive communities'. That is to say, although at any time an evolved form is realised in individual organisms, the population, not the individual, is the matrix for the production of form. Any given animal or plant architecture evolves slowly as genes propagate in a population – at different rates and at different times – so that a new form is slowly synthesised within the larger reproductive community.[1] The lesson for computer design is simply that once the relationship between the virtual genes and the virtual bodily traits of a CAD building has been worked out, as articulated above, an entire population – not just a 'couple' – of such buildings must be unleashed within the computer. The architect should add points at which spontaneous mutations may occur to the CAD sequence of operations. For example, in the case of a column, one should take: the relative proportions of the initial line; the centre of rotation; and the shape with which the Boolean subtraction is performed; and allow these mutant instructions to propagate and interact collectively over many generations.

To population thinking Deleuze adds another cognitive style, 'intensive thinking', which, in its present form, is derived from thermodynamics but has roots as far back as late medieval philosophy. The modern definition of an intensive quantity becomes clear when contrasted with its opposite, extensive quantity, which includes familiar magnitudes such as length, area, and volume. These are defined as magnitudes that can be spatially subdivided, that is, a volume of water divided in half comprises two half volumes. The term 'intensive' on the other hand, refers to quantities like temperature, pressure, or speed, which cannot be subdivided as such; that is, two halves of a volume of water at 90 degrees of temperature do not become two half volumes at 45 degrees of temperature, but rather two halves at the original 90 degrees. Although for Deleuze this lack of divisibility is important, he also stresses another feature of intensive quantities: a difference of intensity, which spontaneously tends to cancel itself out and, in the process, drives fluxes of matter and energy. In other words, differences of intensity are productive differences since they drive processes in which the diversity of actual forms is produced.[2] For example, the process of embryogenesis, which produces a human body out of a fertilised egg, is a process driven by differences of intensity (differences of chemical concentration, of density, of surface tension).

What does this mean for the architect? It means that unless one brings to a CAD model the intensive elements of structural engineering, basically distributions of stress, a virtual building will not evolve as a building. In other words, if the column I described above is not linked to the rest of the building, as a load-bearing element, by the third or fourth generation this column may be placed in such a way that it can no longer perform its function of carrying loads in compression. The only way to ensure that structural elements do not lose their function, and hence that the overall building does not lose viability as a stable structure, is to attempt to represent the distribution of stresses. One must show which types of concentrations, during the process that translates virtual genes into bodies, will endanger the structure's integrity. For example, in the case of real organisms if a developing embryo becomes structurally unviable it won't even reach reproductive age where it would be subject to the process of natural selection. It gets selected out prior to that. A similar process would have to be simulated in the computer to make sure that the products of virtual evolution are viable in terms of structural engineering prior to being selected by the designer in terms of their 'aesthetic fitness'.

Now, let us assume that these requirements have indeed been met, perhaps by an architect-hacker who takes existing software (a CAD package and a structural engineering package) in order to write a code that brings the two together. If the individual now sets out to use virtual evolution as a design tool, he or she may be disappointed by the fact that the only role left for a human is to be the judge of aesthetic fitness. The role of design has now been transformed into (some would say downgraded to) the equivalent of a racehorse breeder. There remains, clearly, an aesthetic component, but hardly the kind of creativity that one identifies with the development of a personal artistic style. Although today slogans about the 'death of the author' and attitudes against the 'romantic view of the genius' are in vogue, I expect this to be a fad and that questions of personal style will return to the spotlight. Will these future authors be content in the role of virtual form breeders? Not that the process, thus far, is routine in any sense. After all, the original CAD model must be endowed with mutation points at just the right places. This involves design decisions and much creativity will still be needed to link ornamental and structural elements in just the right way. Nevertheless, this remains far from a design process by which one develops a unique style.

There is, however, another part of the process where stylistic questions are still crucial, although in a different sense than in ordinary design. Explaining this involves bringing in the third element in Deleuze's philosophy of the genesis of form: topological thinking. One way to introduce this style of thinking is to contrast the results artists have so far obtained with the genetic algorithm to those achieved by biological evolution. When one looks at current artistic results the most striking fact is that, once a few interesting forms have been generated, the evolutionary process seems to run out of possibilities. New forms do continue to emerge but they seem too close to the original ones, as if the space of possible designs, which the process explores, had been exhausted.[3] This stands in sharp contrast to the incredible combinatorial productivity of natural forms like the thousands of original architectural 'designs' exhibited by vertebrate or insect bodies. Although biologists do not have a full explanation for this, one possible way of approaching the question is through the notion of a 'body plan'.

As vertebrates, the architecture of our bodies (which combines bones bearing loads in compression and muscles bearing them in tension) makes us part of the phylum Chordata. The term 'phylum' refers to a branch in the evolutionary tree (the first bifurcation after animal and plant 'kingdoms'), but it also carries the idea of a shared body plan. By this I mean an 'abstract vertebrate' which, if folded and curled in particular sequences during embryogenesis, yields, for example, an elephant that, when twisted and stretched in another sequence, yields a giraffe, and in yet other sequences of intensive operations yields snakes, eagles, sharks, and humans. To put this differently, there are 'abstract vertebrate' design elements, such as the tetrapod limb, which may be realised in structures as different as the single digit limb of a horse, the wing of a bird or the hand with an opposing thumb of a human. Given that the proportion of each of these limbs, as well as the number and shape of digits, is variable, their common body plan cannot include any of these details. In other words, the form of the final product (an actual horse, bird, or human) does have specific lengths, areas, and volumes. But the body plan cannot possibly be defined in these terms and must be abstract enough to be compatible with many different combinations of these extensive quantities. Deleuze uses the term 'abstract diagram', or 'virtual multiplicity', to refer to entities akin to the vertebrate body plan, but his concept also includes the 'body plans' of nonorganic entities like clouds or mountains.[4]

What kind of theoretical resources do we need in order to analyse these abstract diagrams? In mathematics, those spaces in which terms like 'length' or 'area' constitute fundamental notions are called 'metric spaces'. The familiar Euclidean geometry is one example of this class, whereas non-Euclidean geometries, using curved instead of flat spaces, are also metric. On the other hand, there are geometries where these notions are not basic, since these geometries possess operations that do not preserve lengths or areas unchanged. Architects are familiar with at least one of these, projective geometry, as in the use of perspective projections. In this case, the operation 'to project' may extend or shrink lengths and areas so these cannot be basic notions. In turn, those properties which do remain fixed under projections may not be preserved under yet other forms of geometry, such as differential geometry or topology. The operations allowed in the latter, such as stretching without tearing, and folding without gluing, preserve only a set of abstract invariant properties. These topological invariants – such as the dimensionality of a space, or its connectivity – are precisely the elements we need in order to begin thinking about body plans or, more generally, abstract diagrams. It is clear that the kind of spatial structure defining a body plan cannot be metric, since embryological operations can produce a large variety of finished bodies, each with a different metric structure. Therefore body plans must be topological.

To return to the genetic algorithm: if evolved architectural structures are to enjoy the same degree of combinatorial productivity as biological ones, they must also begin with an adequate diagram, an 'abstract building', corresponding to the 'abstract vertebrate'. And it is at this point that design goes beyond mere breeding, with different artists designing different topological diagrams bearing their signature. The design process, however, will be quite different from the traditional one, which operates within metric spaces. It is indeed too early to say precisely what kind of design methodology will be necessary when one cannot use fixed lengths or even fixed proportions as aesthetic elements, but must rely instead on pure connectivities (and other topological invariants). But it is clear that without this the space of possibilities in which virtual evolution blindly searches will be too impoverished to be of any use. Thus, architects wishing to use the new tool of genetic algorithms must not only become hackers (so that they can create the code needed to bring extensive and intensive aspects together) but also be able 'to hack' biology, thermodynamics, mathematics, and other areas of science to tap into the necessary resources. As fascinating as the idea of breeding buildings inside a computer may be, it is clear that mere digital technology without populational, intensive, and topological thinking will never be enough.

1   First . . . the forms do not preexist the population, they are more like statistical results. The
         more a population assumes divergent forms, the more its multiplicity divides into
         multiplicities of a different nature . . . the more efficiently it distributes itself in the milieu,
         or divides up the milieu . . . Second, simultaneously and under the same conditions . . .
         degrees are no longer measured in terms of increasing perfection . . . but in terms of
         differential relations and coefficients such as selection pressure, catalytic action, speed of
         propagation, rate of growth, evolution, mutation . . . Darwinism's two fundamental
         contributions move in the direction of a science of multiplicities: the substitution of
         populations for types, and the substitution of rates or differential relations for degrees.
                                                        Gilles Deleuze and Felix Guattari, *A Thousand Plateaus*,
                                                        University of Minnesota Press (Minneapolis), 1987, p. 48

2   Difference is not diversity. Diversity is given, but difference is that by which the given is
         given . . . Difference is not phenomenon but the nuomenon closest to the phenomenon . . .
         Every phenomenon refers to an inequality by which it is conditioned . . . Everything which
         happens and everything which appears is correlated with orders of differences: differ-
         ences of level, temperature, pressure, tension, potential, difference of intensity.
                                                        Gilles Deleuze, *Difference and Repetition*,
                                                        Columbia University Press (New York), 1994, p. 222

3   See, for example, Stephen Todd and William Latham, *Evolutionary Art and Computers*, Acade-
         mic Press (New York), 1992.

4   'An abstract machine in itself is not physical or corporeal, any more than it is semiotic; it is
         diagrammatic. (It knows nothing of the distinctions between the artificial and the natural
         either). It operates by matter, not by substance; by function, not by form . . . The abstract
         machine is pure Matter-Function – a diagram independent of the forms and substances,
         expressions and contents it will distribute.' Deleuze and Guattari, op. cit., p. 141

David Leatherbarrow (born 1953) is an American architect, scholar, and writer, currently Chair of the Graduate Group in Architecture at the University of Pennsylvania's School of Design. He trained as an architect at the University of Kentucky and then moved to England to take a PhD in Art History at the University of Essex. He taught architectural theory and design at the Polytechnic of Central London (now University of Westminster) as well as at Cambridge and other British and American universities. In 1997–98 he was the recipient of a Visiting Scholar Fellowship from the Canadian Center of Architecture in Montreal. His recent books include *The Roots of Architectural Invention: Site Enclosure, Materials* (1993); *Uncommon Ground: Architecture, Technology and Topography* (2000); and *Topographical Stories: Studies in Landscape and Architecture* (2004).

Mohsen Mostafavi (born 1953) is an Iranian architect, writer, and educator, currently Dean and Professor of Architecture at the College of Architecture, Art and Planning at Cornell University. He received a Diploma in Architecture from the Architectural Association in London, and then undertook postgraduate research at the University of Essex and at Cambridge University. He directed the Master of Architecture program at the Graduate School of Design, Harvard University and has also taught at the University of Pennsylvania, Cambridge University, and the Frankfurt Academy of Fine Arts (Städelschule). From 1995 to 2004, he was Chairman of the Architectural Association School of Architecture in London. His recent publications include: *Delayed Space* (with Homa Fardjadi, 1994); *Approximations* (2002); and *Landscape Urbanism: A Manual for the Machinic Landscape* (2004).

In addition to their individual writings, they have also collaborated on a number of co-authored books, including *On Weathering: The Life of Buildings in Time* (1993) which received the American Institute of Architects' prize for writing on architectural theory. The following essay is taken from *Surface Architecture* (2002), a book that continues their earlier preoccupation with the spatial and temporal dimensions of materiality and tectonic form. This extract develops Martin Heidegger's discussion of technology as both a productive and a destructive force, exploring the question of whether "systems" of building are a particular threat to the lives lived within them. Like so many philosophers of technology, they clearly understand and describe the elements and operation of the systems they distrust and use it to formulate tactics for working within modern systems of construction. To counter the apparent estrangement between mass-produced building elements and unique sites and locations, the authors consider the ways in which technological objects – whether systems, devices or individual components – can become "situated" or assimilated within a specific cultural context through interaction with the local environment and particular patterns of user appropriation.

In a time when almost all of the elements used in the building process are pre-made in a factory or workshop, architectural construction has become a process of assembly. No longer does site labor involve the cutting, joining, and finishing of "raw materials"; instead it entails the installation of components that have been preformed and prefabricated somewhere other than the building site. Construction these days tends to be largely a dry not a wet process, the elements of which are not only precise and exact but meant for specific assembly procedures.

These techniques intend the construction of a *system*, an integrated unity that is characterized by (1) the functional interdependence of parts, (2) internal intentionality, and (3) independence from territorial obligations. The functioning of a glazing system, for example, depends on these conditions: first, on the interdependence of its mechanisms of operation – fasteners, sealants, sheets of glazing, etc; second, compatible performance standards for each of the parts; and, third, the relative autonomy of the ensemble, which allows it to be used in different locations. Although relatively recent as an achievement of building construction, the idea of such a composition has ample precedent in both architectural theory and its philosophical and scientific sources.

One of the most influential early formulations of the idea that the elements of a system are dependent on one another was set forth by Immanuel Kant:

> I consider a system to be the unity of manifold knowledge under one idea. This is the idea formed by the reason of the form of a whole, in so far as such a concept determines *a priori* both the size and the position of the parts in respect to each other. The scientific concept of the reason, therefore, contains the end [or functional purpose] and the form of the whole. . . . Thus, the whole is articulated, not heaped together; it can grow from within, like the body of an animal, whose growth does not add a limb, but makes every limb stronger and fitter for its purposes without changing the proportion.[1]

The difference between this conception of ensemble and an aggregate is that the interdependence of the parts is governed by an idea. Wholeness in this sense resembles Leon Battista Alberti's notion of *concinnitas*, a unity from which no part could be taken without weakening or destroying the whole. But for Kant the regulative principle, the functional or purposive nature of the system, was key: "purposive unity (*Zweckmässigkeit*) was a regulative principle in nature ... [as if or assuming that] the idea of unity had been her [nature's] foundation."[2] Later writers on art and architecture argued similarly about works of artifice. The functioning, performance, or operations of a building were seen to depend on the coordination and internal cooperation of its component parts.

In functionalist arguments, the idea of a system's purpose received great stress, so much so that these parts were said to "determine" nonpurposive concerns, such as style or figuration. This did not always mean that form was to follow function, but that overall shape, like purposiveness, was to be integral and consistent. The aesthetic qualities of an "organic" composition necessitated the perfect unity of parts.

Yet the organic approach was not without its critics in the late nineteenth and twentieth centuries. Aesthetic unity and the closure it implies have been criticized by writers and designers in favor of fragmentation and the *opera aperta* ["open work"]. The philosophical foundation for key aspects of this debate was set out by Martin Heidegger. His role is somewhat ironic, for many now see him as a conservative in matters of art and politics. Nevertheless, for Heidegger this desire for "gathering together," this dedication to unity, is precisely what typifies the "enframing" that has come to be characteristic of our technological age – an age that not only he but Herbert Marcuse, Jurgen Habermas, and Arnold Gehlen have seen as a threat to human existence.[3]

Are "systems" of building construction a "threat" to the lives that are meant to be accommodated in buildings?

When explaining his rather unusual sense of the German word *Gestell* (enframing), Heidegger first elaborated the implications of its prefix, comparing the gathering of elements in such a framework to the collection of mountains in a mountain range (*Gebirg*) and of feelings in one's disposition (*Gemüt*). The prefix *ge-* is important because it signifies "gathering together." The collection and integration of elements on a bookshelf or in a skeleton are both signified by the word *Gestell*.[4] But it is not the unity of parts in such an ensemble that makes Heidegger's sense of enframing unusual and difficult, nor is this what makes the *Gestell* threatening. For Heidegger the term *Stell* means a setting upon or standing forth. Every *Stellen*, he observed, is a standing forth, a placement, positioning, or imposition. This sense of the word is apparent in the German word *Dar-stellen*, meaning presentation. The peculiarity of technological enframing is that it is a positioning or standing that is also a "challenging forth," a production, that draws or tears "out of concealment" the resources of the earth, conferring upon them the status of a "standing reserve." This transforms the earth into material, a commodity, which makes the constructed work less the outcome of care or cultivation than of exploitation.

Even if we accept this sense of technology, the difficulty of Heidegger's account is not overcome, however, for he also maintains that the enframing that defines modern technology is related to the "bringing forward" that occurs in art and poetry, because both poetic and technological disclosure are productive and revealing. Yet, unlike poiesis, enframing is a "challenging which puts to nature the unreasonable demand that it supply energy that can be extracted and stored as such," exposing and ordering it to "stand by," ready to be used and eventually used up. This is the "danger" of technology.

For many readers, these observations have prompted a reactionary response: to avoid the danger of exhausting the environment, the forward march of modern technology must be halted. This sentiment is particularly evident in contemporary ecological theory and environmental ethics, where Heidegger's arguments on technology have surprising currency and are used to buttress the alternately alarmist and pious arguments for conservation and "letting be" (Heidegger's *Gelassenheit*). Whether or not one shares this ethics and politics, it clearly poses some difficulty for architecture, because a new building cannot result from "letting things be." Heidegger himself stressed the *productive* character of art, particularly architectural art. As long as architecture is understood to augment reality by establishing what it naturally lacks, architecture must be understood as essentially akin to technology. Even though most architects realize that the two cannot be separated, many take a position for or against technology. Even when the altern-ative is avoided, we are often presented with an odd mixture of confidence and doubt: blind enthusiasm for the newest methods and techniques coupled with pro-found inability to agree on the limits, even aims, of their use. Viewed more broadly, the continued faith in the advances of modern technology is paralleled by a corre-sponding doubt about its use. It is not that most doubt technology, or reject its results (for few critics are willing to relinquish the newest means), but many doubt our ability to deploy these results and methods responsibly, which is why it is impossible to ignore the current debate on environmental ethics.

The other reason that architecture and technology are difficult to separate is that both involve *foresight*; both involve the intelligence implied in the non-graphic sense of the term "making plans." Plans of this sort are made by architects, politi-cians, and tourists – by all of us. Does our tendency to rely on plans, forecasts, and predictions pose a problem?

In his criticism of the common assumption that technology is essentially a matter of contrivances or instruments, Heidegger introduced the related concepts of cause, "occasioning," and "bringing forth." Technology is not essentially a matter of devices or machines, but the knowledge that exists in advance of some-thing "coming forth," serving in part as its agency. This knowledge also exists in architecture, for part of the intellectual labor of project-making is the understand-ing that proposes and governs building construction. The building itself is anticipa-tory; its parts are prepared for some occurrence in natural or human affairs. Because anticipatory thinking cannot be divorced from human understanding, the alternative between technology and anthropology proposed in much reactionary criticism cannot be sustained.

› Technique

The centrality of technological thought to both human understanding and architec-tural imagination has been the subject of much speculation. One story of the origin of the arts, which is also the origin of human existence, suggests that they came into being as a result of the same series of mythical events.

All arts and techniques were given to the human race by Prometheus. Of the versions of this story that survive from antiquity, the account set forth by Plato is for our purposes the most useful:

> Once upon a time, there existed gods but no mortal creatures. When the appointed time came
> for these also to be born, the gods formed them within the earth out of a mixture of earth and fire
> and the substances which are compounded from earth and fire. And when they were ready to
> bring them to the light, they charged Prometheus and Epimetheus with the task of equipping
> them and allotting suitable powers to each kind. Now Epimetheus begged Prometheus to allow
> him to do the distribution himself – and "when I have done it, you can review it." . . . In his
> allotment he gave to some creatures strength without speed, and equipped the weaker kinds
> with speed . . . [Thus, giving to each its powers,] he made his whole distribution on a principle of
> compensation, being careful by these devices that no species should be destroyed. . . .

> Now Epimetheus was not a particularly clever person, and before he realized it he had used
> up all the available powers on the brute beasts, and being left with the human race on his hands
> unprovided for, did not know what to do with them. While he was puzzling about this,
> Prometheus came to inspect the work, and found the other animals well off for everything, but
> man naked, unshod, unbedded, and unarmed, and already the appointed day had come, when
> man too was to emerge from within the earth into the daylight. Prometheus therefore, being at a
> loss to provide any means of salvation for man, stole from Hephaestus and Athena the gift of
> skill in the arts, together with fire – for without fire it was impossible for anyone to possess or
> use this skill – and bestowed it on man.[5]

The story begins with the topic of proportion: Epimetheus was to give to each species its due power, speed to some, strength to others, and so on. In the allotment, however, he forgot the human race. This forgetting is the first introduction of the problem of knowledge into the story. At this stage, three points merit emphasis: (1) humans are similar to animals because they are formed "out of the earth," (2) humans are different from animals because they were left out of the initial proportioning of "natural" powers, and (3) before they were given fire and the arts, humans, as such, did not exist. Next came Prometheus, and following his arrival a rapid series of decisive events. After the fault of neglect came the theft, which resulted in a gift, the gift of that power which distinguishes the human race: art, technique, or "sly thought." Here the account of origins collapses on itself; for the gift was presupposed in the giving: the theft itself was an instance of "clever" thought, of foresight, or of technique.

The tale concludes with a series of consequences of the fault and gift. First, the use of fire and of the arts led the human race to develop religious practices, language, clothing, and houses – each of which was an artifact that in some ways covered the nakedness that defined humans before they were humans. The next consequence was that shortly after the human race settled itself in groups they learned they could not defend themselves against attack, for they had not learned the art of war, nor of politics. They tried to protect themselves by forming cities, but failed again because they lacked political skill. This led to another gift, perhaps the greatest of all. Observing their failures and fearing their destruction, Zeus sent Hermes to the human race in order to teach them a sense of justice, which would allow them to bring order to their cities. Concluding his story, Protagoras stressed that the sense of justice was bestowed upon all individuals equally, no matter what art they practiced, whether farming or financing.

The story describes two faults and two gifts: Epimetheus and Prometheus were responsible for the first two, Prometheus and Zeus for the second. Why this division, or how are we to understand the promise and weakness of technical know-how? Epimetheus is a name that signifies "knowing after the fact," hindsight. This understanding was demonstrated by the not-so-clever realization of the need for humans to be given their "powers." Prometheus, by contrast, is a name that signifies foresight (*pro-mathein*). He is the hero of prescience, of anticipation, of what is to come. In this narrative, his foresight was apparent in his grasp of what the underprovisioned humans would require to satisfy their needs, fire and the arts. Although beneficial, his foresight was incomplete, for even with his gift human existence was still wanting in its lack of political skill. Thus the gift of justice (from Zeus through Hermes), by which cities were established. This story suggests that cities and by implication architecture will not result from technical modalities of understanding alone.

In what way is technical knowledge incomplete, incompletely human and incompletely prescient? On first thought, it would seem that the first gift was a supplement: neglected by Epimetheus, people were naked. Le Corbusier, too, imagined "original" nakedness in his account of the interdependent beginnings of the human race, arts, and language. In his *Talks with Students*, he observed that "folklore," or prescientific knowledge, "shows us 'man naked,' dressing himself, surrounding himself with tools and objects, with rooms and a house, reasonably satisfying his minimum requirements and coming to terms with the surplus to permit him the enjoyment of his great material and spiritual well-being."[6] Does this mean that at their origin – when naked – humans were already fully human, complete as such? This idea would suggest that the instruments of dressing were both supplementary and accessory, which is to say nonessential, like ornaments draped or inscribed on a building's primary enclosure or "white walls." The Prometheus story implies, however, that the thesis of original nakedness is false; before the gift, before art, humans had not yet appeared "out of the earth." In the time before art, the human race was not only neglected but unknown. Fire, and with it the arts, allowed mankind to come into existence, to emerge from within the earth and stand up or out in the daylight. On this point the Promethean myth parallels Plato's parable of the cave. Heidegger observed that technological understanding uncovers, is a matter of disclosing or revealing. Because the human race is not entirely part of the animal or natural world, manufactured instruments are the means by which the human race comes into being.

Thus, what may seem to be a supplement is actually constitutive: the arts and instruments of language, dressing, and architecture are not added to naked bodies but constitutive of them. Before language, clothing, and architecture, the human race was not yet human, only potentially so. And like the bodies they constitute, these artifacts *stand out* in order to define what is human. This relationship is expressed clearly in the word *prosthesis*, which signifies something that is placed in front of, or outside of, something else. In this instance, however, what is outside the body is also what makes it what it is. There is thus no "original" nakedness in human existence. The life of a human being involves being outside oneself. Before technique, humans were "without qualities," with neither prospect nor possibility.

Insofar as prosthetics enable art, they also inaugurate human temporality. Instruments anticipate, prepare, or propose something to be done, enacted, or performed. Prometheus was the hero of foresight, of expectation. The same is true of the instruments he gave the human race: tools and the know-how their use assumes enable all manner of performances, for each tool allows one to expect some result or consequence. And no occurrence of this kind is natural. Thus, artworks, instruments, and prosthetics compensate for nakedness, whereas premature existence is defined by a lack of qualities. Epimetheus's effort was compensatory.

Prometheus lacked access to the citadel of Zeus because he lacked understanding of the whole or of justice, which would contribute to the good of the whole. Thus, the origin of art is also the prompting of labor, for the many arts are all examples of knowledge that involves actual projects or work. And work can be successful or not. Prometheus was aware of this, because his foresight was a promise of failure: "I knew – and yet not all." Technique inaugurates human history because the foresight of its proposals is always somewhat lacking; each eventually fails. Every failure, great or small, gives rise to a new task, and thus to the history of similar performances. Every tool of language, clothing, and building is both a memory and a project. Technical operations demonstrate foresight and planning, but this foresight recognizes past traces of neglect. Failure prompts projects, and every new production proposes a recuperation.

Bernard Stiegler has described the "posterity" of technical objects as that of sustaining their own "naturalization" or appropriation into what has arisen non-technologically.[7] One agency of this "naturalization" is the "territory" neglected by the "autonomy" or internal definition of the technical system – an agency of obvious potential and inevitability in the making of an architectural "system."

In the last pages of Heidegger's reflections on enframing, he describes technological modalities of thought as a matter of "danger." These modalities block the appearance of truth, the subject matter of poetic disclosure. But this observation does not prompt him to reject technology. Enframing, he said, is a "prelude" to its opposite, to "the event of appropriation."

› Appropriation

What would characterize such an "event of appropriation" in a technological age? In fact, its characteristics have already been indirectly proposed: it would be non-propositional or improbable, and it would be territorial or territorially specific. Also characteristic of such an event would be the discovery of similarities, which had been deferred by technical knowing. This would be the discovery of unforeseen relationships among individuals, among ensembles or systems, with one another and with what had not been planned.

Architectural surfaces consist not only of instances of "internal intentionality" or autonomy – for example, Mies's primitivism, Albert Kahn's Fordism, or Jean Prouvé's "closed systems" – but also of situations in which similarly systematic premade objects and processes are appropriated into nontechnical conditions. In many cases there are instances of movement back and forth between technical and nontechnical conditions. In these cases, elements or systems enter into *play* with conditions of human praxis and location: parts that were planned to "work" independently are reworked in dialogue with the conditions of the project. The premade parts that were selected to perform certain roles are reworked in order to play different roles: what had been deferred in design is acknowledged in construction and use, as if what had been premade were of necessity remade. So, too, with the materials of construction. If one puts in abeyance assumptions about "the nature of materials" and about their natural meanings (which are certainly as much associational as natural), one can see how construction can be a process whereby improbable uses are discovered to yield new, and newly significant, meanings. Misuse can lead to reuse and new use, as is apparent not only in the work of Lewerentz and de la Sota but also in that of the Smithsons and Jean Nouvel. Here, as before, design projection is only a partial determination; equally effective or participant in the historical process are improbabilities of place and performance. And these "agencies" of appropriation unfold through time, a time of partial successes and partial failures.

Each device or technical apparatus, then, is a chronicle of its own modification, a proposal that is discovered to be "not so clever," which is followed by a recuperative and appropriating reproposal that is itself, eventually, discovered once again to be incomplete. Such a discovery is a consequence of both partial foresight and changed circumstances, no matter whether one judges these circumstances to be a punishment or a gift, threatening or emancipating. Regardless of such a judgment, this "history" cannot be escaped. Nor is it insignificant, for it is precisely this history that gives to the surface its identity as the site of a performance, of both a people and a place. This history also gives to the construction its signifying substance, as a prominently visible evidence of care (in construction and reconstruction), which in architecture can be defined as the tragic labor of reconciling foresight with neglect.

1   Immanuel Kant, *Critique of Pure Reason* (Riga, 1781), B 860–1; cited and translated in Caroline van Eck, *Organicism in Nineteenth-Century Architecture* (Amsterdam, 1994), 122.

2   Ibid., 123.

3   The basic texts include Martin Heidegger, "The Question Concerning Technology," in *The Question Concerning Technology and Other Essays* (New York, 1977), 3–35; Herbert Marcuse, "From Negative to Positive Thinking: Technological Rationality and the Logic of Domination," in *One Dimensional Man* (Boston, 1964), 144–69; Jurgen Habermas, *Toward a Rational Society* (Boston, 1971), chapters 4–6; and Arnold Gehlen, *Man in the Age of Technology* (New York, 1980).

4   Heidegger, "The Question Concerning Technology," 19–21; the same comparison is set forth in Martin Heidegger, "The Principle of Identity," in *Identity and Difference* (New York, 1969), and is interconnected with his arguments about both "sameness" and "belonging together."

5   Plato, *Protagoras* (Princeton, 1938), 320d–1d.

6   Le Corbusier, *Le Corbusier Talks with Students* (New York, 1961), 60.

7   Bernard Stiegler, *Technics and Time* (Stanford, 1998), 76–81.

William McDonough (born 1951 in Tokyo) is an American architect specializing in sustainable design. He was winner of the Presidential Award for Sustainable Development (1996), the National Design Award (2004), and the Presidential Green Chemistry Challenge Award (2003), and in 1999 *Time* magazine recognized him as a "Hero for the Planet." He is well known among designers for *The Hannover Principles: Design for Sustainability*, the official design guidelines for the 2000 World's Fair, which the City of Hannover commissioned him to write in 1991, and which have become a benchmark for environmental design.

Dr. Michael Braungart is an industrial chemist and dedicated environmentalist. In 1992, he collaborated with McDonough on the drafting of the Hannover Principles. In 1995 they formed McDonough Braungart Design Chemistry, now known as MBDC, a practice dedicated to the "Next Industrial Revolution." Together they co-authored *Cradle to Cradle*, from which this excerpt is drawn, and have extended the arguments of that book to develop a "comprehensive design protocol."

Among environmental designers, the work of McDonough and Braungart is significant because of their focus on industry and its processes, which has led them to reformulate the design project as the redesign of industrial production itself. Architecture is then one-product-among-others of the system of manufacturing.

Imagine that you have been given the assignment of designing the Industrial Revolution – retrospectively. With respect to its negative consequences, the assignment would have to read something like this:

Design a system of production that
- puts billions of pounds of toxic material into the air, water, and soil every year
- produces some materials so dangerous they will require constant vigilance by future generations
- results in gigantic amounts of waste
- puts valuable materials in holes all over the planet, where they can never be retrieved
- requires thousands of complex regulations – not to keep people and natural systems safe, but rather to keep them from being poisoned too quickly
- measures productivity by how few people are working
- creates prosperity by digging up or cutting down natural resources and then burying or burning them
- erodes the diversity of species and cultural practices.

Of course, the industrialists, engineers, inventors, and other minds behind the Industrial Revolution never intended such consequences. In fact, the Industrial Revolution as a whole was not really designed. It took shape gradually, as industrialists, engineers, and designers tried to solve problems and to take immediate advantage of what they considered to be opportunities in an unprecedented period of massive and rapid change.

It began with textiles in England, where agriculture had been the main occupation for centuries. Peasants farmed, the manor and town guilds provided food and goods, and industry consisted of craftspeople working individually as a side venture to farming. Within a few decades, this cottage industry, dependent on the craft of individual laborers for the production of small quantities of woolen cloth, was transformed into a mechanized factory system that churned out fabric – much of it now cotton instead of wool – by the mile.

This change was spurred by a quick succession of new technologies. In the mid-1700s cottage workers spun thread on spinning wheels in their homes, working the pedals with their hands and feet to make one thread at a time. The spinning jenny, patented in 1770, increased the number of threads from one to eight, then sixteen, then more. Later models would spin as many as eighty threads simultaneously. Other mechanized equipment, such as the water frame and the spinning mule, increased production levels at such a pace, it must have seemed something like Moore's Law (named for Gordon Moore, a founder of Intel), in which the processing speed of computer chips roughly doubles every eighteen months.

In preindustrial times, exported fabrics would travel by canal or sailing ships, which were slow and unreliable in poor weather, weighted with high duties and strict laws, and vulnerable to piracy. In fact, it was a wonder the cargo got to its destination at all. The railroad and the steamship allowed products to be moved more quickly and farther. By 1840 factories that had once made a thousand articles a week had the means and motivation to produce a thousand articles a day. Fabric workers grew too busy to farm and moved into towns to be closer to factories, where they and their families might work twelve or more hours a day. Urban areas spread, goods proliferated, and city populations increased. More, more, more jobs, people, products, factories, businesses, markets – seemed to be the rule of the day.

Like all paradigm shifts, this one encountered resistance. Cottage workers afraid of losing work and Luddites (followers of Ned Ludd) – experienced cloth makers angry about the new machines and the unapprenticed workers who operated them – smashed labor-saving equipment and made life difficult for inventors, some of whom died outcast and penniless before they could profit from their new machines. Resistance touched not simply on technology but on spiritual and imaginative life. The Romantic poets articulated the growing difference between the rural, natural landscape and that of the city – often in despairing terms: "Citys . . . are nothing less than over grown prisons that shut out the world and all its beauties," wrote the poet John Clare. Artists and aesthetes like John Ruskin and William Morris feared for a civilization whose aesthetic sensibility and physical structures were being reshaped by materialistic designs.

There were other, more lasting problems. Victorian London was notorious for having been "the great and dirty city," as Charles Dickens called it, and its unhealthy environment and suffering underclasses became hallmarks of the burgeoning industrial city. London air was so grimy from airborne pollutants, especially emissions from burning coal, that people would change their cuffs and collars at the end of the day (behavior that would be repeated in Chattanooga during the 1960s, and even today in Beijing or Manila). In early factories and other industrial operations, such as mining, materials were considered expensive, but people were often considered cheap. Children as well as adults worked for long hours in deplorable conditions.

But the general spirit of early industrialists – and of many others at the time – was one of great optimism and faith in the progress of humankind. As industrialization boomed, other institutions emerged that assisted its rise: commercial banks, stock exchanges, and the commercial press all opened further employment opportunities for the new middle class and tightened the social network around economic growth. Cheaper products, public transportation, water distribution and sanitation, waste collection, laundries, safe housing, and other conveniences gave people, both rich and poor, what appeared to be a more equitable standard of living. No longer did the leisure classes alone have access to all the comforts.

The Industrial Revolution was not planned, but it was not without a motive. At bottom it was an economic revolution, driven by the desire for the acquisition of capital. Industrialists wanted to make products as efficiently as possible and to get the greatest volume of goods to the largest number of people. In most industries, this meant shifting from a system of manual labor to one of efficient mechanization.

Consider cars. In the early 1890s the automobile (of European origin) was made to meet a customer's specifications by craftspeople who were usually independent contractors. For example, a machine-tool company in Paris, which happened to be the leading manufacturer of cars at the time, produced only several hundred a year. They were luxury items, built slowly and carefully by hand. There was no standard system of measuring and gauging parts, and no way to cut hard steel, so parts were created by different contractors, hardened under heat (which often altered dimensions), and individually filed down to fit the hundreds of other parts in the car. No two were alike, nor could they be.

Henry Ford worked as an engineer, a machinist, and a builder of race cars (which he himself raced) before founding the Ford Motor Company in 1903. After producing a number of early vehicles, Ford realized that to make cars for the modem American worker – not just for the wealthy – he would need to manufacture vehicles cheaply and in great quantities. In 1908 his company began producing the legendary Model T, the "car for the great multitude" that Ford had dreamed of, "constructed of the best materials, by the best men to be hired, after the simplest designs that modern engineering can devise . . . so low in price that no man making a good salary will be unable to own one."

In the following years, several aspects of manufacturing meshed to achieve this goal, revolutionizing car production and rapidly increasing levels of efficiency. First, centralization: in 1909 Ford announced that the company would produce only Model T's and in 1910 moved to a much larger factory that would use electricity for its power and gather a number of production processes under one roof. The most famous of Ford's innovations is the moving assembly line. In early production, the engines, frames, and bodies of the cars were assembled separately, then brought together for final assembly by a group of workmen. Ford's innovation was to bring "the materials to the man," instead of "the man to the materials." He and his engineers developed a moving assembly line based on the ones used in the Chicago beef industry: it carried materials to workers and, at its most efficient, enabled each of them to repeat a single operation as the vehicle moved down the line, reducing overall labor time dramatically.

This and other advances made possible the mass production of the universal car, the Model T, from a centralized location, where many vehicles were assembled at once. Increasing efficiency pushed costs of the Model T down (from $850 in 1908 to $290 in 1925), and sales skyrocketed. By 1911, before the introduction of the assembly line, sales of the Model T had totaled 39,640. By 1927, total sales reached fifteen million.

The advantages of standardized, centralized production were manifold. Obviously, it could bring greater, quicker affluence to industrialists. On another front, manufacturing was viewed as what Winston Churchill referred to as "the arsenal of democracy," because the productive capacity was so huge, it could (as in the two world wars) produce an undeniably potent response to war conditions. Mass production had another democratizing aspect: as the Model T demonstrated, when prices of a previously unattainable item or service plummeted, more people had access to it. New work opportunities in factories improved standards of living, as did wage increases. Ford himself assisted in this shift. In 1914, when the prevailing salary for factory workers was $2.34 a day, he hiked it to $5, pointing out that cars cannot buy cars. (He also reduced the hours of the workday from nine to eight.) In one fell swoop, he actually created his own market, and raised the bar for the entire world of industry.

Viewed from a design perspective, the Model T epitomized the general goal of the first industrialists: to make a product that was desirable, affordable, and operable by anyone, just about anywhere; that lasted a certain amount of time (until it was time to buy a new one); and that could be produced cheaply and quickly. Along these lines, technical developments centered on increasing "power, accuracy, economy, system, continuity, speed," to use the Ford manufacturing checklist for mass production.

For obvious reasons, the design goals of early industrialists were quite specific, limited to the practical, profitable, efficient, and linear. Many industrialists, designers, and engineers did not see their designs as part of a larger system, outside of an economic one. But they did share some general assumptions about the world.

2002

William J. Mitchell

E-Bodies,
E-Buildings,
E-Cities

William J. Mitchell (born 1944) holds a Bachelor of Architecture from the University of Melbourne (1967), a Master of Environmental Design from Yale University (1969), and an MA from the University of Cambridge (1977). He is a Fellow of the Royal Australian Institute of Architects and a Fellow of the American Academy of Arts and Sciences. He is currently Professor of Architecture and Media Arts and Sciences at the School of Architecture and Planning at the Massachusetts Institute of Technology, where he also directs the Media Lab's *Smart Cities* research group. He moved to MIT in 1992 from the Graduate School of Design at Harvard University, where he was Professor of Architecture and director of the Masters in Design Studies program. From 1970 to 1986 he was on the faculty of the School of Architecture and Planning at the University of California, Los Angeles, where he headed the Architecture/Urban Design program. He has also taught at Yale, Carnegie-Mellon, and Cambridge Universities.

His most recent book, *Placing Words: Symbols, Space, and the City* was published by the MIT Press (2005). His earlier publications include: *ME++: The Cyborg Self and the Networked City* (2003); *E-Topia: Urban Life, Jim – But Not As We Know It* (1999); the edited volume *High Technology and Low-Income Communities* (with Donald A. Schon and Bish Sanyal) (1999); *City of Bits: Space, Place, and the Infobahn* (1995); *The Reconfigured Eye: Visual Truth in the Post-Photographic Era* (1992); and *The Logic of Architecture: Design, Computation, and Cognition* (1990).

In the essay reprinted here (originally presented at the RIBA *Future Studies* conference in 2001) Mitchell sets out a picture of the near-future implications of digital communications technologies on the patterns of everyday living and working. As these networks and systems become more deeply embedded within the fabric of buildings and urban spaces it is, paradoxically, their gradual "disappearance" that threatens to exert the biggest impact on the architect's traditional approach to design. As Mitchell notes at the conclusion of the essay, with regard to just one of a number of possible side-effects of these developments: "as personal mobility increases with the growing use of portable wireless devices, we will see a decline in the power of the program to organize architectural form."

It is now a commonplace observation (to the point of weary cliché) that the explosive combination of tiny, inexpensive electronic devices, increasingly ubiquitous digital networking, and the world's rapidly growing stock of digital information is dramatically changing our daily lives.[1] But what does this condition suggest, concretely, for architectural and urban design strategy in the twenty-first century?

› The cost of being there

In order to develop some useful answers to this question, I shall begin by adopting a rather brutally reductionist perspective. Specifically, I shall assume that there are three types of costs associated with assigning particular activities to specific urban locations: *fixed* costs, *interactive* costs, and *churn* costs.[2] And, in each case, there are corresponding benefits, which for mathematical simplicity can just be treated as negative costs. The new technological context affects all of these (but in different ways and to different extents), and changes the balance among them. The ultimate result is a new mix of space types in the city, together with new spatial patterns at all scales.

Fixed costs, such as rent, are intrinsic to a location.[3] The corresponding benefits, such as the pleasures of climatic and scenic attractions, are valuable advantages that cannot be changed by transportation or telecommunication connections. (You cannot pump the sybaritic attractions of a beach through a wire.) Very often, location decisions are made by trading off these fixed costs and benefits against other types of costs and benefits. For example, you might choose to live in the reasonably priced, leafy outer suburbs to gain quiet and greenery, but you might pay heavily for this in terms of the time and cost of commuting to a job in the central city. In general, consideration of fixed costs and benefits produces spatial patterns in which activities cluster at locations characterised by unusual local attractions or by invitingly low rents.

Interactive costs of assigning an activity to a location are those that result from interactions with other activities. For example, there may be substantial flows of goods between a factory and some warehouses. If the warehouses are nearby, then the resulting yearly transportation costs are low. However, if the warehouses are distant, the costs will be higher. Since everything cannot be adjacent to everything else, consideration of interactive costs and benefits usually produces location patterns in which highly interactive activities are located centrally, minimally interactive activities go to the periphery, and closely interlinked activities are as near to each other as possible.

Churn costs are those that result, over time, from moving activities around.[4] For example, if you move an office to another floor there will be associated transportation, renovation. and transaction costs. If you rely on a lot of heavy, fixed equipment, then churn costs will be high and you will have an incentive not to move – even if your location is not ideal from other viewpoints. Conversely, if you work with only a portable laptop and cell phone, your costs of picking up and moving will be low. In general, high churn costs produce stable spatial patterns, while low churn costs encourage a more nomadic condition.

Actual architectural, urban, and regional spatial patterns result largely from overlays, interactions and balances of patterns produced by fixed, interactive, and churn costs and benefits.[5] On an isotropic plane, interactive costs would dominate to produce patterns that responded very directly to traffic flow and accessibility considerations. But in complex and differentiated topography, the intrinsic advantages and disadvantages of particular places play a bigger role. And the high churn costs associated with permanent construction and sunk investments tend to lock in established urban spatial patterns; nomads think little of relocating and reconfiguring their tent encampments when they need greener pastures.

› The revenge of place

Now, what are the effects of electronic interconnection on these costs, benefits and associated patterns?

First, the effects on fixed costs and benefits are minimal; network connections do not change the climate or the scenery. It follows that, when other types of costs are reduced by electronic interconnection, fixed costs and benefits begin to dominate. In other words, if you can locate anywhere you will locate where it's particularly attractive in some way. I shall refer to this phenomenon as the *revenge of place.*

One extreme manifestation of the revenge of place is the much-hyped (and even occasionally instantiated) 'electronic cottage' located deep in the woods, high in the mountains, or on some idyllic island. Electronic connectivity provides necessary economic, social, and cultural linkages to the wider world, while self-contained power generation, water collection, and waste recycling systems keep everything working. It is the sort of thing that Robert Louis Stevenson had in mind, albeit relying upon the earlier interconnection technologies of the steamship and the international mail, when he moved his dwelling and the site of his work to Apia in the South Pacific.

A more common and practical manifestation is the affluent telecommuter village, such as is now being seen in the vicinity of Paris or in resort settings such as Aspen or Camden, Maine. Here the attraction is the picturesque, charming, and generally exclusive small community. The inhabitants are very high-end knowledge workers (stock traders, software wizards, script writers) who can now work electronically to a large extent, and who can afford very high-quality personal transportation (limousines, light aircraft) when they need it.

› Fragmentation and recombination

The most obvious effects of electronic interconnection are on interactive costs and benefits. The whole point of digital telecommunications systems is that they reduce spatial and temporal interdependencies among activities; they make it possible to do things at a distance, and to conduct transactions asynchronously. This does not mean (as some have suggested) that the 'friction of distance' simply disappears so that you can locate anything anywhere. Rather, it means that spatial and temporal linkages among activities are *selectively* loosened. Internet distribution drastically reduces the cost of getting recorded music from producer to consumer for example, and it eliminates trips to the record store, but you still have to drag your body from your residence to the dental surgery when you need a filling.

This selective loosening allows latent demands for proximity to manifest themselves; proximity requirements that could not previously be satisfied, since they were dominated by other requirements, now come to the fore. Furthermore, latent demands for quality of place also begin to take over; if you can telecommute, for example, you might relocate to a scenic but hitherto hopelessly inaccessible location. The resulting phenomenon, as activities regroup, is *fragmentation and recombination* of building types and urban patterns.[6]

Fragmentation and recombination processes sometimes result in decentralisation to reach larger markets, to get closer to customers, and so on. But they can also produce centralisation, motivated by efforts to achieve economies of scale, or to take advantage of knowledge spillover effects. And they can yield mobilisation of certain activities, as these activities float free of traditional locational ties and thereby become easier to relocate in response to dynamic conditions such as changes in labour markets. (It all depends, of course, on what latent proximity demands are lurking.) All these things, and more, can take place simultaneously, producing complicated and sometimes apparently contradictory spatial outcomes.

The death of the branch bank vividly illustrates the complexities of fragmentation and recombination. Not so long ago, branch banks were a prominent building type on any high street. Then came automated teller machines (ATMs), followed by electronic home banking. Face-to-face interactions with a teller during banking hours were replaced by electronically mediated remote interactions at any time of the day or night. The space for retail banking systems fragmented and decentralised; ATMs are now found on street corners, in airport terminals, in gambling casinos – in short, wherever people may need cash – while electronic home banking transactions can be conducted from anywhere there is internet connectivity. Simultaneously, electronic commerce technology allowed back-office functions to cluster for efficiency and to relocate to places where the labour market was attractive – often offshore. The old branch banks were shut down in their thousands, and a radically new spatial pattern, involving different building types (large-scale back-office facilities, call centres) emerged. Banking organisations were no longer represented by their dignified high street facades, but by screen logos on ATM and personal computer screens.

Electronically mediated retailing of books and similar articles has generated parallel effects, with the additional twist of restructuring transportation as well as spatial patterns. A traditional urban bookstore is a place to store books, to advertise them, to allow customers to browse among them and make selections, to conduct purchase transactions across a counter, and to pursue necessary back-office activities such as ordering and inventory tracking. All these activities need to be clustered tightly together within a well-defined spatial envelope, because of requirements for face-to-face interaction, efficient circulation of stock and customers, visual supervision, and physical security. And there are some inherent contradictions; it is desirable, for instance, to offer customers the largest stock of titles possible, but capacity to do this is limited by expensive, highly constrained urban real-estate. But an operation like Amazon.com changes the rules of the game. It virtualises and radically decentralises the browsing and purchasing functions, it shifts book storage from local storage points to huge, highly automated warehouse and distribution centres at national airline hubs (where huge numbers of titles can be kept in stock economically), and it mobilises back-office work by exploiting electronic commerce technology for maintaining relationships with customers and suppliers. Amazon back-office employees could be located just about anywhere, but it turns out that they are largely located in downtown Seattle – because that's where they want to be. Overall, under this pattern, the local bookstore disappears, residences and offices become decentralised sites for retail transactions, the national distribution centre emerges and is located to minimise interaction costs, and the back-office work gravitates to attractive urban locations (for higher-level employees) or to rural locations where the land is cheap and the labour market is depressed (for lower-level employees).

It is immediately obvious that new transportation patterns will result from all this. Where books were once delivered in medium-sized shipments to intermediate storage points provided by local retailers, then carried home by customers, they are now delivered in large shipments from publishers to national distribution centres, from where they travel in small packages, by air and van, directly to homes and offices. Some places get more traffic, others get less. Certainly, as a result of electronic retailing, I make fewer trips to go shopping but there are now more delivery vans on my street.

Of course, that is not the only way to change the rules of the book retailing game. An increasingly attractive alternative is to store books on online servers, and to download them on demand to sophisticated machines that do high-quality printing and binding. Instead of physically distributing after manufacture, you electronically deliver before manufacture. This fragments traditional, centralised factory space and recombines it with retail space.

› Erasing incompatibilities

The converse to spatial attraction produced by high interaction between activities is spatial repulsion resulting from incompatibility. One of the major moves of nineteenth- and twentieth-century urban planning was to separate residential suburbs from the noise, traffic, and pollution of urban industrial areas. In general, the central idea of land use zoning has been to cluster compatible activities together and to separate them from incompatible ones. (Of course, there is lots of room for contention about the definitions of 'compatible' and 'incompatible'.)

But the information work that is such a crucial part of today's economy, supported by networked electronic devices, is not like factory work. It does not generate noise and pollution, it does not necessarily require large concentrations of workers in one place, and it does not generate large amounts of delivery traffic. Therefore, the incompatibilities with residential land uses are greatly reduced or eliminated, and reintegration of the home and the workplace becomes an increasingly attractive possibility. Fine-grained, mixed-use neighbourhoods created from live/work dwellings can begin to re-emerge.

Similarly, the activities associated with online retailing are not like those associated with suburban shopping malls. When you order goods online, you do not need access to a big-box facility stocked with goods and serviced by big trucks, nor do you need a large parking lot. You do need space in your home to conduct the transactions, and you do need some convenient way to receive deliveries – either by being at home, by providing some sort of secure delivery locker (maybe a refrigerated one in the case of food), or by making arrangements with a neighbourhood delivery point such as a mom-and-pop corner store. This potentially reintegrates fine-grained retail activities with the neighbourhood, separates them from large-scale storage and distribution functions, and displaces those big-box functions to regional transportation nodes, unvisited by customers, on the urban periphery.

This electronic erasure of long-standing incompatibilities creates conditions for re-establishing traditional neighbourhood patterns, as advocated by the American New Urbanists, and as suggested by the Urban Villages proposals of Prince Charles.[7] We have seen these patterns emerging in certain high-tech hotspots, such as the SoHo area of Manhattan, and the South of Market (SoMa) district of San Francisco. This is not a matter of ignoring the genuine problems addressed by traditional zoning strategies, nor one of sentimental hankering after the virtues of pre-industrial small-town life, but a realistic response to emerging post-industrial conditions.

> Tunnel effects

In addition to producing fragmentation and recombination, and allowing new spatial patterns to emerge (or traditional patterns to re-emerge) by erasing incompatibilities, radical reductions in interactive costs can generate profoundly antispatial interdependencies among towns, cities, and regions.

Traditionally, there have been strong interdependencies among geographically clustered activities, but much weaker interdependencies among widely separated activities. A city might have strong economic, social, and cultural linkages to its agricultural hinterland and to nearby provincial towns, for example, but much more tenuous relationships to distant corners of the world. The development of transportation networks enabled stronger linkages and interdependencies at a distance – among trading cities, in particular – but the 'tyranny of distance' remained potent.[8] There was substantial congruence between place and community.

Now, in contexts where interactions among activities can effectively be supported by electronic interconnection, very strong interdependencies can develop at a distance. Thus the information technology clusters of Silicon Valley and Bangalore are very closely linked to one another and highly interdependent. Hollywood (a world centre for film production) and London's Soho (a remarkable concentration of postproduction facilities and talent) have become increasingly symbiotic as high-speed electronic interconnections have linked them ever more effectively. And, as everyone knows, the world's major financial centres are now strongly interconnected through sophisticated electronic linkages to form a global system. Such 'tunnel effects', which unevenly warp accessibility surfaces, are becoming increasingly common.[9]

As many have noted, this condition generates dramatic slippages, and discontinuities within the urban fabric. A high-rise office building in Jakarta may function as a node in the global financial networks, while the surrounding urban kampongs belong to a completely different economic, social, and cultural order. A campus workplace in Bangalore may be almost indistinguishable from one in Palo Alto, but the cows amble down the dusty road outside. A telephone call centre in Sydney may exist to serve customers in Hong Kong. Being in the right time zone can now be far more important than being in the right neighbourhood.

At worst, it is easy to imagine these distortions, slippages, and discontinuities becoming chasms, destroying any sense of cohesive local community, and producing an urban fabric of juxtaposed but socially and culturally disconnected fragments held in a matrix of common physical infrastructure.[10] At best, one might imagine combining the virtues of small-town cohesiveness (provided spatially) with the opportunities and excitement of cosmopolitan connections (provided electronically) to a wider world. Finding ways to get the balance right will be one of the great design and planning challenges of the coming years.

We will not achieve this goal by conceiving of electronic interaction as a direct (though perhaps inferior) substitute for face-to-face. Nor will we get there by treating social interaction as a zero-sum game in which time devoted to one mode is time subtracted from another. We will do better to consider some of the subtler ways in which digital and physical space may intersect. Consider, for example, the common problem of providing network access on a university campus. One approach is to network desktop computers in dormitory rooms; in this case, spatial organisation and network configuration clearly conspire to produce fragmentation and isolation. Some students will almost never come out of their rooms. An alternative approach is to combine wireless laptop computers with a system of inviting, informal public study spaces: sidewalk cafes, common rooms, nooks and crannies off public spaces, shady spots under trees, and so on; this combination of digital and physical arrangements activates social spaces, promotes accidental encounters, and allows students to create informal study groups as they wish, while retaining all the advantages of electronic connectivity.

› Electronic mobilisation

Whereas these various effects of the digital revolution on interactive costs derive from loosening of spatial and temporal linkages among activities, the effects on churn costs follow from miniaturisation and dematerialisation. For example, office work used to require filing cabinets filled with paper, a desktop and a typewriter; now the desktop has virtualised and shrunk to a laptop screen, files are accessible online, the typewriter has transmuted into word-processing software, and the telephone fits into a pocket. It was slow and expensive to move all your stuff from one office to another, but now it is effortless to pick up your laptop and your cell phone to relocate. And you can work just about anywhere: not only in an 'official' work-space, but also at home, in an aeroplane seat, at a customer location, on a park bench, or in a cafe. It isn't that we all turn into full-time telecommuters; face-to-face interaction still has its important uses. But work hours and locations become far more fluid and adaptable to changing circumstances.[11]

Simply put, wireless networking increases mobility, reduces churn costs, and provides flexibility to reorganise and regroup rapidly and efficiently in response to changing conditions. The effects are felt in a wide range of contexts, from design offices rearranging themselves to take on new projects to kids in the street with cell phones organising raves and protests on the fly.

› Attentive architecture

A second significant consequence of electronically enabled miniaturisation and dematerialisation is the increasing prevalence of electronic tags, sensors, and sophisticated control systems in buildings. HVAC and lighting systems have long had electronic sensors and controls, of course, and electronic security systems are commonplace, but the tags and sensors are now getting smaller, cheaper, more versatile, and more ubiquitous. Some of them are wireless. And they are being integrated into standard IP networks (that is, they become part of the internet) rather than operating as specialised proprietary systems.

Potentially, we can think of all the devices and appliances in a building as smart objects that can sense and respond to their changing environments, and can operate as servers in peer-to-peer networks (a sort of architectural Napster) within buildings.[12] Even a single light bulb might incorporate sensors, intelligence, network connectivity, and TCP/IP capability; you could send email to it and get a reply. The ultimate consequences of this will be profound, and they are probably not yet fully imaginable, but the first-order outcome will surely be to enhance the versatility of spaces. A given space, through electronic intelligence and functionality, will not only be more responsive and efficient, it will also be programmable for wider ranges of activities.

We should be careful to distinguish flexibility and multifunctionality achieved through electronic reprogramming of services from the 1960s and 1970s strategy of providing modular, reconfigurable spaces and partition and furniture systems. This older strategy tended to produce characterless architecture, and it often foundered on the inconvenience and high labour costs of actually moving things around to accommodate new requirements. But electronic reconfiguration can be swift and effortless. In a classroom or conference space, for example, speakers might define pre-sets for lighting and audio-visual equipment, and simply invoke their personal configurations as they take their turns at the podium. In a hotel room or office cubicle, you might download your complete personal work environment as you entered.

› Rethinking programming, design, and construction

These reductions in interactive and churn costs, together with reductions in special-isation and enhancements of the versatility of spaces, challenge the characteristic modernist practice of beginning an architectural project by developing a detailed space program.[13] Such programs typically enumerate the specialised spaces that will be required in a building, tabulate their floor areas and technical requirements, and specify their proximity requirements. But increasingly, under the conditions I have described, the need is less for specialised spaces providing fixed-in-place resources, and more for electronically serviced, diverse, interesting, and humane habitats that can support a nomadic style of habitation. The boundaries among different building types are blurring, spaces are becoming more multifunctional, and satisfaction with complex adjacency and proximity requirements is becoming less critical.

All these conditions also come close to home for architects. They apply to the activities of design and construction, just as they do to other professional and produc-tion fields. They regroup and restructure design and construction tasks, redistribute them spatially, and ultimately change the material processes and formal languages of architecture. CAD/CAM digital models replace paper documentation, electronic telecommunication supports geographically distributed design and construction teams, and electronically mediated mass-customisation techniques supplant strat-egies of component standardisation and industrial mass production as manifested, in the extreme, in industrialised component building.

Frank Gehry's Bilbao Guggenheim was the first great architectural triumph to emerge from these new conditions.[14] Digital modelling was at the heart of the design, fabrication, and onsite assembly processes, the design and construction team was spread across the globe from Santa Monica to the Veneto, and the complex, non-repeating forms were made feasible through clever exploitation of advanced CAD/CAM production capabilities.

Unfortunately, some of the post-Gehry blob projects that we have seen can be dismissed as fairly mindless NURBS-mongering. But this should not obscure the fact that an important new direction is vigorously emerging, particularly among stu-dents and the more adventurous younger practitioners.

› Summary: new conditions and strategies

I do not mean to suggest, of course, that these new material conditions deter-mine architectural and urban form in any simple way. But they are powerful current realities, independently of whatever techno-enthusiasts or techno-sceptics may wish. They create new ground for generation of socio-spatial systems in particular contexts. And they open up new opportunities for respond-ing to particular cultural and political goals.

Here then, as a brief guide for the perplexed, is my checklist of the concrete architectural and urban consequences most worthy of critical consideration, design investigation, and debate. First, as spatial and temporal linkages among activities selectively loosen, we will see fragmentation and recombination of familiar building types and urban patterns. Second, with the electronic erasure of some traditional incompatibilities, it will make increasing sense to recombine the home and the workplace, and to favour fine-grained, mixed-use neighbourhood patterns rather than coarse-grained, single-use zoning. Third, as unique local advantages (such as a beachfront location of historic significance) gain in relative importance compared to the diminishing benefits of mere accessibility, we will encounter the revenge of place. Fourth, as tunnel effects radically warp time and space, and as local and remote interactions continually compete for attention, we will have to find effective, electronic/spatial strategies for getting the balances and complementarities right. Fifth, as places become more versatile through electronic augmentation, as adjacency and proximity requirements become less critical, and as personal mobility increases with the growing use of portable wireless devices, we will see a decline in the power of the program to organise architectural form. And finally, as CAD/CAM design and construction replace paper-based processes, and as design and construction processes globalise, we will see ways of making places that privilege variety, complexity, and local responsiveness rather than the standardisation, repetition, and tight spatial disciplines characteristic of the industrial era.

1   A wide ranging, insightful introduction to these changes is provided by Manuel Castells in *The Rise of the Network Society.* Oxford: Basil Blackwell, 1996.

2   This approach is directly based upon a classic formulation of location-allocation problems. See Tjalling C. Koopmans and Martin Beckmann, 'Assignment problems and the location of economic activities'. *Econometrica* 1957, 25(l): 53–76.

3   This standard terminology may be slightly confusing. Fixed costs, such as rents, clearly may vary over time. They are fixed in the sense that they are independent of interaction effects with activities at other locations.

4   Facility managers in large organisations are acutely aware that churn costs can be very significant over time. And real estate agents and removal companies largely make their livings from dealing with churn.

5   Given a set of activities to be assigned, a set of available locations and relevant cost data, the task of assigning activities to locations in the least costly way can be formulated as a quadratic assignment problem. Such problems are difficult to solve for large numbers of activities and locations, but computer software exists for generating good solutions in reasonable time. By running this software, it is possible to explore the spatial effects of varying the relative magnitudes of fixed, interactive, and churn costs. See Robin S. Liggett, 'Optimal spatial arrangement as a quadratic assignment problem'. In John S. Gero (Ed.) *Design Optimization.* New York: Academic Press, 1985, pp. 1–40.

6   The phenomenon of fragmentation and recombination is explored in more detail in William J. Mitchell, *City of Bits: Space, Place, and the Infobahn*, Cambridge, MA: MIT Press, 1995, and William J. Mitchell, *E-topia: Urban Life, Jim – But Not As We Know It*, Cambridge, MA: MIT Press, 1999. See also Thomas Horan, *Digital Places: Building Our City of Bits*, Washington, DC: Urban Land Institute, 2000, and Joel Kotkin, *The New Geography: How the Digital Revolution is Reshaping the American Landscape*, New York: Random House, 2000.

7   See Andreas Duany, Elizabeth Plater-Zyberg, and Jeff Speck, *Suburban Nation: The Rise of Sprawl and the Decline of the American Dream*, New York: North Point Press, 2000; Peter Calthorpe and William Fulton, *The Regional City: Planning for the End of Sprawl*, Washington, DC: The Island Press, 2001; and Urban Villages Forum, *Urban Villages: A Concept for Creating Mixed-Use Urban Developments on a Sustainable Scale* (2nd edn), London: The Urban Villages Group, 1992.

8   Geoffrey N. Blainey, *The Tyranny of Distance.* Melbourne: Sun Books, 1966.

9   Stephen Graham and Simon Marvin, *Splintering Urbanism: Networked Infrastructures, Technological Mobilities and the Urban Condition.* London: Routledge, 2001.

10 A version of this dystopian scenario is developed in Martin Pawley, *Terminal Architecture.* London: Reaktion Books, 1997.

11 Jack M. Nilles, *Managing Telework: Strategies for Managing the Virtual Workforce.* New York: John Wiley, 1998.

12 For an introduction to the relevant technology see Neil Gershenfeld, *When Things Start to Think.* New York: Henry Holt, 1999.

13 John Summerson, 'The case for a theory of modern architecture'. *RIBA Journal* 1957, 64: 307–10.

14 William J. Mitchell, 'Roll over Euclid: How Frank Gehry designs and builds'. In *Frank Gehry, Architect*, New York: Guggenheim Museum, 2001, pp. 353–63.

The following excerpt was taken from *Smart Architecture*, a publication prepared by SLA – Ed van Hinte, Marc Neelen, Jacques Vink, and Piet Vollaard – a group based in Rotterdam. As they explain on their web site, SLA is an acronym for *Slimme Architectuur*, the Dutch translation for Smart Architecture. But "SLA in Dutch also means lettuce, a main product of our agricultural industry known for its 'greenness' and 'freshness.' SLA aims to be as green and fresh as a lettuce-leaf."

The seven system-based layers described by SLA further expand on those originally described by Brand (1994), who was expanding on earlier observations by Frank Duffy (1997). See Figure 6. The earlier observations about relative rates-of-change have turned into simple advice: "be careful when mixing systems together." That advice would apply to this whole collection.

2003

SLA

Changing Speeds

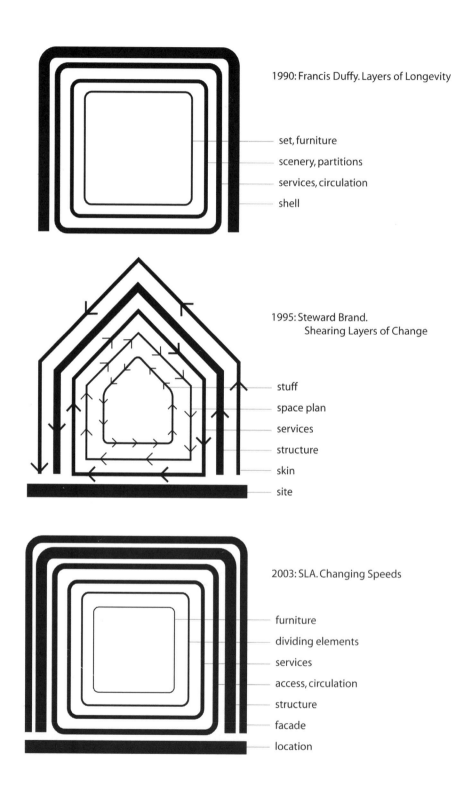

1990: Francis Duffy. Layers of Longevity

set, furniture
scenery, partitions
services, circulation
shell

1995: Steward Brand.
        Shearing Layers of Change

stuff
space plan
services
structure
skin
site

2003: SLA. Changing Speeds

furniture
dividing elements
services
access, circulation
structure
facade
location

Buildings aren't just buildings. They can be divided up into seven system-based layers. Each of these has its own lifespan, all the way from centuries down to a couple of years.

1   **Location.** Generally speaking the geographic location has a very long lifespan. Amsterdam and New York, to name just two examples, have maintained the same grid of streets and roads for many years.
2   **Structure.** It is quite costly to change the foundation and the main carrying structure of buildings. Therefore their quality determines the architectural endurance of a building. The structure usually lasts between thirty and 300 years.
3   **Access.** Stairs, escape routes, escalators, and lifts have a long life, but not as long as lift shafts that are part of the main structure. Changing these can be a far-reaching process. Emergency and secondary stairs on the other hand may be replaced more quickly because of changing regulations.
4   **Facade.** If the facade has not been designed to last, it usually has to be replaced or renovated after some twenty years. This is mostly a technical matter but fashion can be a consideration.
5   **Services.** Systems for climate control, wiring, sprinklers, water, and sewers are outdated after seven to fifteen years.
6   **Dividing elements.** In a commercial context it is common practice to renew doors, inside walls, elevated floors, and lowered ceilings as often as every three years.
7   **Furniture** is replaced fairly quickly.

For a flexible building, by and large, the dynamics of these layers have to be taken into consideration. If, for example, the facade is part of the main structure, the resulting building may be too rigid, because to change the facade the whole building has to be taken apart. The same holds true for a service that is too 'deeply rooted' in the building. Integration of different parts, the destiny of technological development, may hamper flexibility, which is a different kind of development. Like scale (should energy be provided to a city by a power plant or should every building or even every home have its own generator?) flexibility is a complex issue to decide on. Be careful when mixing systems together.

Fig 6
Changing speeds, after Duffy, Brand, and SLA

2004

Manuel Castells

Space of Flows,
Space of Places:
Materials for a
Theory of Urbanism
in the Information
Age

Manuel Castells was born in Spain in 1942 and brought up mainly in Valencia and Barcelona. He studied law and economics at the University of Barcelona from 1958 to 1962. Following his student activism against the Franco dictatorship he moved to Paris, where he subsequently received doctorates in both sociology and the human sciences from the University of Paris-Sorbonne. Between 1967 and 1979 he taught sociology at the University of Paris and was then appointed Professor of City and Regional Planning and of Sociology at the University of California, Berkeley. Castells currently holds the Wallis Annenberg Chair in Communication Technology and Society at the Annenberg School for Communication, University of Southern California, Los Angeles. He is also Research Professor at the Open University of Catalonia in Barcelona. He has served as an advisor to a number of international governments, as well as a consultant to major international organizations such as US AID, the European Commission, the World Bank, and UNESCO.

Castells is a prolific writer and has published over twenty books and 100 articles in the areas of urban sociology, new technologies, and political economy. His major work is the three-volume study, *The Information Age: Economy, Society and Culture* (1996–98), hailed by the sociologist Anthony Giddens as "the most compelling attempt yet to map the contours of the global information age."

The essay included here was originally published in a collection entitled *The Cybercities Reader,* and offers a summary of Castells' current thinking on the relationship between technology, the city, and society. His main argument is that a new form of capitalism has emerged at the turn of the millennium: global in its character and more flexible than previous forms. This system is also being challenged by a multitude of smaller scale social movements trying to maintain a sense of cultural identity and a measure of local political control. This tension between local and global provides the central dynamic of the three volumes of *The Information Age*, summarized in Castells' claim that: "our societies are increasingly structured around the bipolar opposition of the Net and the Self."[1] The Net in this formulation stands for the new organizational structures based on the pervasive use of networked communication media. The Self symbolizes the practical activities through which people try to reaffirm their identities under the conditions of global change and instability that accompany this reorganization of social and economic networks. On the broadest level, Castells regards social development as inseparable from the changes in the technological infrastructure through which most of society's activities are carried out, "since technology is society and society cannot be understood or represented without its technological tools."[2]

His bipolar formulation can also be seen as the latest in a series of attempts throughout the twentieth century to understand the competing forces driving technological progress: the persistent dichotomy between technological determinism and social constructivism.

1   Manuel Castells (1996), *The Rise of the Network Society, Volume I: The Informa-tion Age* (Oxford: Blackwell, 1996) p. 3.
2   Ibid. p. 5.

We have entered a new age, the Information Age. Spatial transformation is a fundamental dimension of the overall process of structural change. We need a theory of
spatial forms and processes, adapted to the new social, technological, and spatial
context where we live. I will attempt here to propose some elements of this theory, a
theory of urbanism in the information age. I will not develop the analysis of the
meaning of the information age, taking the liberty to refer the reader to my trilogy on
the matter (Castells, 1996–2000).

I will not build theory from other theories, but from the observation of social
and spatial trends in the world at large. Thus, I will start with a summary characterization of the main spatial trends at the onset of the twenty-first century. Then I will
propose a tentative theoretical interpretation of observed spatial trends. Subsequently I will highlight the main issues arising in cities in the information age, with
particular emphasis on the crisis of the city as a socio-spatial system of cultural
communication. I will conclude by drawing some of the implications of my analysis
for planning, architecture, and urban design.

› The transformation of urban space in the early twenty-first century

Spatial transformation must be understood in the broader context of social transformation: space does not reflect society, it expresses it, it is a fundamental dimension of society, inseparable from the overall process of social organization and
social change. Thus, the new urban world arises from within the process of formation of a new society, the network society, characteristic of the Information Age. The
key developments in spatial patterns and urban processes associated with these
macro-structural changes, can be summarized under the following headings (Scott,
2001):

- Because commercial agriculture has been, by and large, automated, and a
  global economy has integrated productive networks throughout the planet, the
  majority of the world's population is already living in urban areas, and this will
  be increasingly the case: we are heading towards a largely urbanized world,
  which will comprise between two-thirds and three-quarters of the total population by the middle of the century (Freire and Stren, 2001).
- This process of urbanization is concentrated disproportionately in metropolitan
  areas of a new kind: urban constellations scattered throughout huge territorial
  expanses, functionally integrated and socially differentiated, around a multi-
  centered structure. I call these new spatial forms metropolitan regions
  (Garreau, 1991; Hall, 2001; Nel.Lo, 2001; Dunham-Jones, 2000).
- Advanced telecommunications, Internet, and fast, computerized transportation
  systems allow for simultaneous spatial concentration and decentralization,
  ushering in a new geography of networks and urban nodes throughout the
  world, throughout countries, between and within metropolitan areas (Wheeler
  et al., 2000).
- Social relationships are characterized simultaneously by individuation and

communalism, both processes using, at the same time, spatial patterning and online communication. Virtual communities and physical communities develop in close interaction, and both processes of aggregation are challenged by increasing individualization of work, social relationships, and residential habits (Russell, 2000; Wellman, 1999; Putnam, 2000).

- The crisis of the patriarchal family, with different manifestations depending on cultures and levels of economic development, gradually shifts sociability from family units to networks of individualized units (most often, women and their children, but also individualized co-habiting partnerships), with considerable consequences in the uses and forms of housing, neighborhoods, public space, and transportation systems.

- The emergence of the network enterprise as a new form of economic activity, with its highly decentralized, yet coordinated, form of work and management, tends to blur the functional distinction between spaces of work and spaces of residence. The work–living arrangements characteristic of the early periods of industrial craft work are back, often taking over the old industrial spaces, and transforming them into informational production spaces. This is not just New York's Silicon Alley or San Francisco's Multimedia Gulch, but a phenomenon that also characterizes London, Tokyo, Beijing, Taipei, Paris, or Barcelona, among many other cities. Transformation of productive uses becomes more important than residential succession to explain the new dynamics of urban space (Mitchell, 1999; Horan, 2000).

- Urban areas around the world are increasingly multi-ethnic, and multicultural. An old theme of the Chicago School, now amplified in terms of its extremely diverse racial composition (Waldinger, 2001).

- The global criminal economy is solidly rooted in the urban fabric, providing jobs, income, and social organization to a criminal culture, which deeply affects the lives of low-income communities, and of the city at large. It follows rising violence and/or widespread paranoia of urban violence, with the corollary of defensive residential patterns.

- Breakdowns of communication patterns between individuals and between cultures, and the emergence of defensive spaces, leads to the formation of sharply segregated areas: gated communities for the rich, territorial turfs for the poor (Blakely and Snyder, 1997; Massey, 1996).

- In a reaction against trends of suburban sprawl and the individualization of residential patterns, urban centers and public space become critical expressions of local life, benchmarking the vitality of any given city (Hall, 1998; Borja and Zaida, 2001). Yet, commercial pressures and artificial attempts at mimicking urban life often transform public spaces into theme parks where symbols rather than experience create a life-size, urban virtual reality, ultimately destined to mimic the real virtuality projected in the media. It follows increasing individualization, as urban places become consumption items to be individually appropriated (Fernandez-Galiano, 2000).

- Overall, the new urban world seems to be dominated by the double movement of inclusion into transterritorial networks, and exclusion by the spatial separation of places. The higher the value of people and places, the more they are connected into interactive networks. The lower their value, the lower their connection. In the limit, some places are switched off, and bypassed by the new geography of networks, as is the case of depressed rural areas and urban shanty towns around the world. Splintering urbanism operates on the basis of segregated networks of infrastructure, as empirically demonstrated by Graham and Marvin (2001).

- The constitution of mega-metropolitan regions, without a name, without a culture, and without institutions, weakens the mechanism of political accountability, of citizen participation, and of effective administration (Sassen, 2001). On the other hand, in the age of globalization, local governments emerge as flexible institutional actors, able to relate at the same time to local citizens and to global flows of power and money (Borja and Castells, 1997). Not because they are powerful, but because most levels of government, including the nation states, are equally weakened in their capacity of command and control if they operate in isolation. Thus, a new form of state emerges, the network state, integrating supra-national institutions made up of national governments, nation-states, regional governments, local governments, and even non-governmental organizations. Local governments become a node of the chain of institutional representation and management, able to input the overall process, yet with added value in terms of their capacity to represent citizens at a closer range. Indeed in most countries, opinion polls show the higher degree of trust people have in their local governments, relative to other levels of government. However, institutions of metropolitan governance are rare and when they exist they are highly centralized, with little citizen participation. There is an increasing gap between the actual unit of work and living, the metropolitan region, and the mechanisms of political representation and public administration. Local governments compensate for this lack by cooperating and competing. Yet, by defining their interests as specific subsets of the metropolitan region, they (often unwillingly) contribute to further fragmentation of the spatial framing of social life.

- Urban social movements have not disappeared, by any means. But they have mutated. In an extremely schematic representation they develop along two main lines. The first is the defense of the local community, affirming the right to live in a particular place, and to benefit from adequate housing and urban services in their place. The second is the environmental movement, acting on the quality of cities within the broader goal of achieving quality of life: not only a better life but a different life. Often, the broader goals of environmental mobilizations become translated into defensive reactions to protect one specific community, thus merging the two trends. Yet, it is only by reaching out to the cultural transformation of urban life as proposed by ecological thinkers and activists that urban social movements can transcend their limits of localism. Indeed, enclosing themselves in their communities, urban social movements may contribute to further spatial fragmentation, ultimately leading to the breakdown of society.

It is against the background of these major trends of urban social change that we can understand new spatial forms and processes, thus re-thinking architecture, urban design and planning in the twenty-first century.

› A theoretical approach to spatial transformation

To make the transition from the observation of urban trends to the new theorization of cities, we need to grasp, at a more analytical level, the key elements of socio-spatial change. I think the transformation of cities in the information age can be organized around three bipolar axes. The first relates to function, the second to meaning, the third to form.

Function
Functionally speaking the network society is organized around the opposition between the global and the local. Dominant processes in the economy, technology, media, institutionalized authority are organized in global networks. But day-to-day work, private life, cultural identity, political participation, are essentially local. Cities, as communication systems, are supposed to link up the local and the global, but this is exactly where the problems start since these are two conflicting logics that tear cities from the inside when they try to respond to both, simultaneously.

Meaning
In terms of meaning, our society is characterized by the opposing development of individuation and communalism. By individuation I understand the enclosure of meaning in the projects, interests, and representations of the individual, that is, a biologically embodied personality system (or, if you want, translating from French structuralism, a person). By communalism I refer to the enclosure of meaning in a shared identity, based on a system of values and beliefs to which all other sources of identity are subordinated. Society, of course, exists only in between, in the inter-face between individuals and identities mediated by institutions, at the source of the constitution of civil society which, as Gramsci argued, does not exist against the state but in articulation with the state, forming a shared public sphere, à la Habermas.

Trends I observe in the formative stage of the network society indicate the increasing tension and distance between personality and culture, between indi-viduals and communes. Because cities are large aggregates of individuals, forced to coexist, and communes are located in the metropolitan space, the split between personality and commonality brings extraordinary stress upon the social system of cities as communicative and institutionalizing devices. The problematique of social integration becomes again paramount, albeit under new circumstances and in terms radically different from those of early industrial cities. This is mainly because of the role played in urban transformation by a third, major, axis of opposing trends, this one concerning spatial forms.

Forms
There is a growing tension and articulation between the space of flows and the space of places.

The space of flows links up electronically separate locations in an interactive network that connects activities and people in distinct geographical contexts. The space of places organizes experience and activity around the confines of locality.

Cities are structured and destructured simultaneously by the competing logics of the space of flows and the space of places. Cities do not disappear in the virtual networks. But they are transformed by the interface between electronic communication and physical interaction, by the combination of networks and places. As William Mitchell (1999), from an urbanist perspective, and Barry Wellman (1999), from a sociological perspective, have argued, the informational city is built around this double system of communication. Our cities are made up, at the same time, of flows and places, and of their relationships. Two examples will help to make sense of this statement, one from the point of view of the urban structure, another in terms of the urban experience.

Turning to urban structure, the notion of global cities was popularized in the 1990s. Although most people assimilate the term to some dominant urban centers, such as London, New York, and Tokyo, the concept of global city does not refer to any particular city, but to the global articulation of segments of many cities into an electronically linked network of functional domination throughout the planet. The global city is a spatial form rather than a title of distinction for certain cities, although some cities have a greater share of these global networks than others. In a sense, most areas in all cities, including New York and London, are local, not global. And many cities are sites of areas, small and large, which are included in these global networks, at different levels. This conception of global city as a spatial form resulting from the process of globalization is closer to the pioneering analysis by Saskia Sassen (1991) than to its popularized version by city marketing agencies. Thus, from the structural point of view, the role of cities in the global economy depends on their connectivity in transportation and telecommunication networks, and on the ability of cities to mobilize effectively human resources in this process of global competition. As a consequence of this trend, nodal areas of the city, connecting to the global economy, will receive the highest priority in terms of investment and management, as they are the sources of value creation from which an urban node and its surrounding area will make their livelihood. Thus, the fate of metropolitan economies depends on their ability to subordinate urban functions and forms to the dynamics of certain places that ensure their competitive articulation in the global space of flows.

From the point of view of the urban experience, we are entering a built environment that is increasingly incorporating electronic communication devices everywhere. Our urban life fabric, as Mitchell (1999) has pointed out, becomes an *e-topia*, a new urban form in which we constantly interact, deliberately or automatically, with online information systems, increasingly in the wireless mode. Materially speaking, the space of flows is folded into the space of places. Yet, their logics are distinct: online experience and face-to-face experience remain specific, and the key question then is to assure their articulation in compatible terms.

These remarks may help in the re-configuration of the theory of urbanism in response to the challenges of the network society, and in accordance to the emergence of new spatial forms and processes.

› The urban themes of the information age

The issue of social integration comes again at the forefront of the theory of urban-ism, as was the case during the process of urbanization in the industrial era. Indeed, it is the very existence of cities as communication artefacts that is called into question, in spite of the fact that we live in a predominantly urban world. But what is at stake is a very different kind of integration. In the early twentieth century the quest was for assimilation of urban sub-cultures into the urban culture. In the early twenty-first century the challenge is the sharing of the city by irreversibly dis-tinct cultures and identities. There is no more dominant culture, because only global media have the power to send dominant messages, and the media have in fact adapted to their market, constructing a kaleidoscope of variable content depending on demand, thus reproducing cultural and personal diversity rather than overimposing a common set of values. The spread of horizontal communication via the Internet accelerates the process of fragmentation and individualization of sym-bolic interaction. Thus, the fragmented metropolis and the individualization of communication reinforce each other to produce an endless constellation of cultural subsets. The nostalgia of the public domain will not be able to countervail the structural trends towards diversity, specification, and individualization of life, work, space, and communication, both face to face, and electronic (Russell, 2000; Putnam, 2000). On the other hand, communalism adds collective fragmentation to individual segmentation. Thus, in the absence of a unifying culture, and therefore of a unifying code, the key question is not the sharing of a dominant culture but the communicability of multiple codes.

The notion of communication protocols is central here. Protocols may be phys-ical, social, and electronic, with additional protocols being necessary to relate these three different planes of our multidimensional experience.

Physically, the establishment of meaning in these nameless urban constella-tions relates to the emergence of new forms of symbolic nodality which will identify places, even through conflictive appropriation of their meaning by different groups and individuals (Dunham-Jones, 2000).

The second level of urban interaction refers to social communication patterns. Here, the diversity of expressions of local life, and their relationship to media culture, must be integrated into the theory of communication by doing rather than by saying. In other words, how messages are transmitted from one social group to another, from one meaning to another in the metropolitan region requires a redefin-ition of the notion of public sphere moving from institutions to the public place, away from Habermas and towards Kevin Lynch. Public places, as sites of sponta-neous social interaction, are the communicative devices of our society, while formal, political institutions have become a specialized domain that hardly affects the private lives of people, that is, what most people value most. Thus, it is not that politics, or local politics, does not matter. It is that its relevance is confined to the world of instrumentality, while expressiveness, and thus communication, refers to social practice, outside institutional boundaries. Therefore, in the practice of the city, its public spaces, including the social exchangers (or communication nodes) of its transportation networks become the communicative devices of city life (Borja

and Zaida, 2001; Mitchell, 1999). How people are, or are not, able to express themselves, and communicate with each other, outside their homes and off their electronic circuits, that is, in public places, is an essential area of study for urbanism. I
call it the sociability of public places in the individualized metropolis.

The third level of communication refers to the prevalence of electronic communication as a new form of sociability. Studies by Wellman, by Jones, and by a
growing legion of social researchers have shown the density and intensity of electronic networks of communication, providing evidence to sustain the notion that
virtual communities are often communities, albeit of a different kind than face to
face communities (Wellman and Haythornthwaite, 2002; Jones, 1998). Here again,
the critical matter is the understanding of the communication codes between
various electronic networks, built around specific interests or values, and between
these networks and physical interaction. There is no established theory yet on
these communication processes, as the Internet as a widespread social practice is
still in its infancy. But we do know that online sociability is specified, not downgraded, and that physical location does contribute, often in unsuspected ways, to
the configuration of electronic communication networks. Virtual communities as
networks of individuals are transforming the patterns of sociability in the new metropolitan life, without escaping into the world of electronic fantasy (Castells, 2001).

Fourth, the analysis of code sharing in the new urban world requires also the
study of the interface between physical layouts, social organization, and electronic
networks. It is this interface that Mitchell considers to be at the heart of the new
urban form, what he calls e-topia. In a similar vein, but from a different perspective,
Graham and Marvin's (2001) analysis of urban infrastructure as splintered networks, reconfigured by the new electronic pipes of urban civilization, opens up the
perspective of understanding cities not only as communication systems, but as
machines of deliberate segmentation. In other words, we must understand at the
same time the process of communication and that of in-communication.

The contradictory and/or complementary relationships between new metropolitan centrality, the practice of public space, and new communication patterns
emerging from virtual communities, could lay the foundations for a new theory of
urbanism – the theory of cyborg cities or hybrid cities made up by the intertwining
of flows and places (see Part 3).

Let us go farther in this exploration of the new themes for urban theory. We
know that telecommuting – meaning people working full time online from their
home – is another myth of futurology (Gillespie and Richardson, 2000; see Andrew
Gillespie and Ranald Richardson, p. 212). Many people, including you and me, work
online from home part of the time, but we continue to go to work in places, as well
as moving around (the city or the world) while we keep working, with mobile connectivity to our network of professional partners, suppliers, and clients. The latter is
the truly new spatial dimension of work. This is a new work experience, and indeed
a new life experience. Moving physically while keeping the networking connection
to everything we do is a new realm of the human adventure, on which we know
little (Kopomaa, 2000; see Zac Carey, p. 133; Timo Kopomaa, p. 267). The analysis
of networked spatial mobility is another frontier for the new theory of urbanism. To
explore it in terms that would not be solely descriptive we need new concepts. The

connection between networks and places has to be understood in a variable geometry of these connections. The places of the space of flows, that is, the corridors and halls that connect places around the world, will have to be understood as exchangers and social refuges, as homes on the run, as much as offices on the run. The personal and cultural identification with these places, their functionality, their symbolism, are essential matters that do not concern only the cosmopolitan elite. Worldwide mass tourism, international migration, transient work, are experiences that relate to the new huddled masses of the world. How we relate to airports, to train and bus stations, to freeways, to customs buildings, are part of the new urban experience of hundreds of millions. We can build on an ethnographic tradition that addressed these issues in the mature industrial society. But here again, the speed, complexity, and planetary reach of the transportation system have changed the scale and meaning of the issues. Furthermore, the key reminder is that we move physically while staying put in our electronic connection. We carry flows and move across places.

Urban life in the twenty-first century is also being transformed by the crisis of patriarchalism. This is not a consequence of technological change, but I have argued in my book *The Power of Identity* (Castells, 1997) that it is an essential feature of the information age. To be sure, patriarchalism is not historically dead. Yet, it is contested enough, and overcome enough so that everyday life for a large segment of city dwellers has already been redefined vis-à-vis the traditional pattern of an industrial society based on a relatively stable patriarchal nuclear family. Under conditions of gender equality, and under the stress suffered by traditional arrangements of household formation, the forms and rhythms of urban life are dramatically altered. Patterns of residence, transportation, shopping, education, and recreation evolve to adjust to the multi-directionality of individual needs that have to share household needs. This transformation is mediated by variable configurations of state policies. For instance, how child care is handled by government, by firms, by the market, or by individual networking largely conditions the time and space of daily lives, particularly for children.

We have documented how women are discriminated against in the patriarchal city. We can empirically argue that women's work makes possible the functioning of cities – an obvious fact rarely acknowledged in the urban studies literature (Borja and Castells, 1997; Susser, 1996). Yet, we need to move forward, from denunciation to the analysis of specific urban contradictions resulting from the growing dissonance between the de-gendering of society and historical crystallization of patriarchalism in the patterns of home and urban structure. How do these contradictions manifest themselves as people develop strategies to overcome the constraints of a gendered built environment? How do women, in particular, re-invent urban life, and contribute to re-design the city of women, in contrast to the millennial heritage of the city of men (Castells and Servon, 1996)? These are the questions to be researched, rather than stated, by a truly postpatriarchal urban theory.

Grassroots movements continue to shape cities, as well as societies at large. They come in all kinds of formats and ideologies, and one should keep an open mind on this matter, not deciding in advance which ones are progressive, and which ones are regressive, but taking all of them as symptoms of society in the making. We should also keep in mind the most fundamental rule in the study of social movements. They are what they say they are. They are their own consciousness. We can study their origins, establish their rules of engagement, explore the reasons for their victories and defeats, link their outcomes to overall social transformation, but not to interpret them, not to explain to them what they really mean by what they say. Because, after all, social movements are nothing else than their own symbols and stated goals, which ultimately means their words.

Based on the observation of social movements in the early stage of the network society, two kinds of issues appear to require privileged attention from urban social scientists. The first one is what I called some time ago the grassrooting of the space of flows, that is the use of Internet for networking in social mobilization and social challenges (Castells, 2000). This is not simply a technological issue, because it concerns the organization, reach, and process of formation of social movements. Most often these online social movements connect to locally based movements, and they converge, physically, in a given place at a given time. A good example was the mobilization against the World Trade Organization meeting in Seattle in December 1999, and against subsequent meetings of globalizing institutions, which, arguably, set a new trend of grass-roots opposition to uncontrolled globalization, and redefined the terms of the debate on the goals and procedures of the new economy. The other major issue in the area of social movements is the exploration of the environmental movement, and of an ecological view of social organization, as urban areas become the connecting point between the global issues posed by environmentalism and the local experience through which people at large assess their quality of life. To redefine cities as eco-systems, and to explore the connection between local eco-systems and the global eco-system lays the ground for the overcoming of localism by grass-roots movements.

On the other hand, the connection cannot be operated only in terms of ecological knowledge. Implicit in the environmental movement, and clearly articulated in the deep ecology theory, as reformulated by Fritjof Capra (1996), is the notion of cultural transformation. A new civilization, and not simply a new technological paradigm, requires a new culture. This culture in the making is being fought over by various sets of interests and cultural projects. Environmentalism is the code word for this cultural battle, and ecological issues in the urban areas constitute the critical battleground for such struggle.

Besides tackling new issues, we still have to reckon in the twenty-first century with the lingering questions of urban poverty, racial and social discrimination, and social exclusion. In fact, recent studies show an increase of urban marginality and inequality in the network society (HDR, 2001). Furthermore, old issues in a new context, become in fact new. Thus, Ida Susser (1996) has shown the networking logic underlying the spread of AIDS among the New York poor along networks of destitution, stigma, and discrimination. Erie Klinenberg (2000), in his social anatomy of the devastating effects of the 1995 heat wave in Chicago, shows why dying alone in the city, the fate of hundreds of seniors in a few days, was rooted in the new forms of social isolation emerging from people's exclusion from networks of work, family, information, and sociability. The dialectics between inclusion and exclusion in the network society redefines the field of study of urban poverty, and forces us to consider alternative forms of inclusion (e.g. social solidarity, or else, the criminal economy), as well as new mechanisms of exclusion and technological apartheid in the era of Internet.

The final frontier for a new theory of urbanism, indeed for social sciences in general, is the study of new relationships between time and space in the information age. In my analysis of the new relationships of time and space I proposed the hypothesis that in the network society, space structures time, in contrast to the time-dominated constitution of the industrial society, in which urbanization, and industrialization were considered to be part of the march of universal progress, erasing place-rooted traditions and cultures. In our society, the network society, where you live determines your time frame of reference. If you are an inhabitant of the space of flows, or if you live in a locality that is in the dominant networks, timeless time (epitomized by the frantic race to beat the clock) will be your time as in Wall Street or Silicon Valley. If you are in a Pearl River Delta factory town, chronological time will be imposed upon you as in the best days of Taylorism in Detroit. And if you live in a village in Mamiraua, in Amazonia, biological time, usually a much shorter lifespan, will still rule your life. Against this spatial determination of time, environmental movements assert the notion of slow-motion time, the time of the long now, in the words of Stewart Brand, by broadening the spatial dimension to its planetary scale in the whole complexity of its interactions, thus including our great-grand children in our temporal frame of reference (Brand, 1999).

Now, what is the meaning of this multi-dimensional transformation for planning, architecture, and urban design?

› Planning, architecture, and urban design in the reconstruction of the city

The great urban paradox of the twenty-first century is that we could be living in a predominantly urban world without cities – that is without spatially based systems of cultural communication and sharing of meaning, even conflictive sharing. Signs of the social, symbolic, and functional disintegration of the urban fabric multiply around the world. So do the warnings from analysts and observers from a variety of perspectives (Kuntsler, 1993; Ascher, 1995; Davis, 1992; Sorkin, 1997; Russell, 2000).

But societies are produced, and spaces are built, by conscious human action. There is no structural determinism. So, together with the emphasis on the economic competitiveness of cities, on metropolitan mobility, on privatization of space, on surveillance and security, there is also a growing valuation of urbanity, street life, civic culture, and meaningful spatial forms in the metropolitan areas around the world. The process of reconstruction of the city is under way. And the emphasis of the most advanced urban projects in the world is on communication, in its multidimensional sense: restoring functional communication by metropolitan planning; providing spatial meaning by a new symbolic nodality created by innovative architectural projects; and reinstating the city in its urban form by the practice of urban design focused on the preservation, restoration, and construction of public space as the epitome of urban life.

However, the defining factor in the preservation of cities as cultural forms in the new spatial context will be the capacity of integration between planning, architecture, and urban design. This integration can only proceed through urban policy influenced by urban politics. Ultimately, the management of metropolitan regions is a political process, made of interests, values, conflicts, debates, and options that shape the interaction between space and society. Cities are made by citizens, and governed on their behalf. Only when democracy is lost can technology and the economy determine the way we live. Only when the market overwhelms culture and when bureaucracies ignore citizens can spatial conurbations supersede cities as living systems of multidimensional communication.

Planning

The key endeavor of planning in the metropolitan regions of the information age is to ensure their connectivity, both intra-metropolitan and intermetropolitan. Planning has to deal with the ability of the region to operate within the space of flows. The prosperity of the region and of its dwellers will greatly depend on their ability to compete and cooperate in the global networks of generation/appropriation of knowledge, wealth, and power. At the same time planning must ensure the connectivity of these metropolitan nodes to the space of places contained in the metropolitan region. In other words, in a world of spatial networks, the proper connection between these different networks is essential to link up the global and the local without opposing the two planes of operation.

This means that planning should be able to act on a metropolitan scale, ensuring effective transportation, accepting multinodality, fighting spatial segregation by acting against exclusionary zoning, providing affordable housing, and desegregated schooling. Ethnic and social diversity is a feature of the metropolitan region, and ought to be protected. Planning should seek the integration of open space and natural areas in the metropolitan space, going beyond the traditional scheme of the greenbelt. The new metropolitan region embraces a vast territorial expanse, where large areas of agricultural land and natural land should be preserved as a key component of a balanced metropolitan territory. The new metropolitan space is characterized by its multifunctionality, and this is a richness that supersedes the functional specialization and segregation of modernist urbanism. New planning practice induces a simultaneous process of decentering and recentering of population and activities, leading to the creation of multiple subcenters in the region.

The social and functional diversity of the metropolitan region requires a multimodal approach to transportation, by mixing the private automobile/highway system with public metropolitan transportation (railways, subways, buses, taxis), and with local transportation (bicycles, pedestrian paths, specialized shuttle services). Furthermore, in a post-patriarchal world, childcare becomes a critical urban service, and therefore must be integrated in the schemes of metropolitan planning. In the same way that some cities require additional housing and transportation investment per each new job created in certain areas, childcare provision should be included in these planning standards.

Overall, most metropolitan planning nowadays is geared towards the adaptation of the space of places of the metropolitan region to the space of flows that conditions the economic competitiveness of the region. The challenge would be to use planning, instead, to structure the space of places as a living space, and to ensure the connection and complementarity between the economy of the metropolitan region and the quality of life of its dwellers.

Architecture

Restoring symbolic meaning is a most fundamental task in a metropolitan world in crisis of communication. This is the role that architecture has traditionally assumed. It is more important than ever. Architecture, of all kinds, must be called to the rescue in order to recreate symbolic meaning in the metropolitan region, marking places in the space of flows. In recent years, we have observed a substantial revival of architectural meaningfulness that in some cases has had a direct impact in revitalizing cities and regions, not only culturally but economically as well. To be sure, architecture *per se* cannot change the function, or even the meaning, of a whole metropolitan area. Symbolic meaning has to be inserted in the whole fabric of the city, and this is, as I will argue below, the key role of urban design. But we still need meaningful forms, resulting from architectural intervention, to stir a cultural debate that makes space a living form. Recent trends in architecture signal its transformation from an intervention on the space of places to an intervention on the space of flows, the dominant space of the information age by acting on spaces dedicated to museums, convention centers, and transportation nodes. These are spaces of cultural archives, and of functional communication that become transformed into forms of cultural expression and meaningful exchange by the act of architecture.

The most spectacular example is Frank Gehry's Guggenheim Museum in Bilbao, that symbolized the will of life of a city immersed in a serious economic crisis and a dramatic political conflict. Calatrava's bridges (Seville, Bilbao), telecommunication towers (Barcelona), airports (Bilbao) or Convention Centers (Valencia) mark the space of flows with sculpted engineering. Bofill's Barcelona airport, Moneo's AVE railway station in Madrid and Kursaal Convention Center in San Sebastian, Meier's Modern Art Museum in Barcelona, or Koolhaas's Lille Grand Palais, are all examples of these new cathedrals of the information age, where the pilgrims gather to search for the meaning of their wandering. Critics point at the disconnection between many of these symbolic buildings and the city at large. The lack of integration of this architecture of the space of flows into the public space would be tantamount to juxtaposing symbolic punctuation and spatial meaninglessness. This is why it is essential to link up architecture with urban design, and with planning. Yet, architectural creation has its own language, its own project that cannot be reduced to function or to form. Spatial meaning is still culturally created. But their final meaning will depend on its interaction with the practice of the city organized around public space.

Urban design

The major challenge for urbanism in the information age is to restore the culture of cities. This requires a socio-spatial treatment of urban forms, a process that we know as urban design. But it must be an urban design capable of connecting local life, individuals, communes, and instrumental global flows through the sharing of public places. Public space is the key connector of experience, opposed to private shopping centers as the spaces of sociability.

Borja and Zaida (2001), in a remarkable book supported with case studies of several countries, have shown the essential role of public space in the city. Indeed it is public space that makes cities as creators of culture, organizers of sociability, systems of communication, and seeds of democracy, by the practice of citizenship. This is in opposition to the urban crisis characterized by the dissolution, fragmentation, and privatization of cities. Borja and Zaida document, on a comparative basis, the projects of reconstruction of cities and of the culture of cities around the (re)construction of public space: the synthesis between places and flows is realized in the public space, the place of social cohesion and social exchanges (Borja and Zaida, 2001, 35).

This is in fact a long tradition in urban design, associated with the thinking and practice of Kevin Lynch, and best represented nowadays by Allan Jacobs. Jacobs' work on streets, and, with Elizabeth McDonald, on boulevards as urban forms able to integrate transportation mobility and social meaning in the city, shows that there is an alternative to the edge city, beyond the defensive battles of suburbanism with a human face (Jacobs, 1993). The success of the Barcelona model of urban design is based on the ability to plan public squares, even mini-squares in the old city, that bring together social life, meaningful architectural forms (not always of the best taste, but it does not matter), and the provision of open space for people's use. That is, not just open space, but marked open space, and street life induced by activities, such as the tolerance of informal trade, street musicians, etc.

The reconquest of public space operates throughout the entire metropolitan region, highlighting particularly the working-class peripheries, those that need the most attention to socio-spatial reconstruction. Sometimes the public space is a square, sometimes a park, sometimes a boulevard, sometimes a few square meters around a fountain or in front of a library or a museum. Or an outdoor café colonizing the sidewalk. In all instances what matters is the spontaneity of uses, the density of the interaction, the freedom of expression, the multi-functionality of space, and the multi-culturalism of the street life. This is not the nostalgic reproduction of the medieval town. In fact, examples of public space (old, new, and renewed) dot the whole planet, as Borja has illustrated in his book. It is the dissolution of public space under the combined pressures of privatization of the city and the rise of the space of flows that is a historical oddity. Thus, it is not the past versus the future, but two forms of present that fight each other in the battleground of the emerging metropolitan regions. And the fight, and its outcome, is of course, political, in the etymological sense: it is the struggle of the polis to create the city as a meaningful place.

› The government of cities in the Information Age

The dynamic articulation between metropolitan planning, architecture, and urban design is the domain of urban policy. Urban policy starts with a strategic vision of the desirable evolution of the metropolitan space in its double relationship to the global space of flows and to the local space of places. This vision, to be a guiding tool, must result from the dynamic compromise between the contradictory expression of values and interests from the plurality of urban actors. Effective urban policy is always a synthesis between the interests of these actors and their specific projects. But this synthesis must be given technical coherence and formal expression, so that the city evolves in its form without submitting the local society to the imperatives of economic constraints or technological determinism.

The constant adjustment between various structural factors and conflictive social processes is implemented by the government of cities. This is why good planning or innovative architecture cannot do much to save the culture of cities unless there are effective city governments, based on citizen participation and the practice of local democracy. Too much to ask for? Well, in fact, the planet is dotted with examples of good city government that make cities livable by harnessing market forces and taming interest groups on behalf of the public good. Portland, Toronto, Barcelona, Birmingham, Bologna, Tampere, Curitiba, among many other cities, are instances of the efforts of innovative urban policy to manage the current metropolitan transformation (Borja and Castells, 1997; Verwijnen and Lehtovuori, 1999; Scott, 2001). However, innovative urban policy does not result from great urbanists (although they are indeed needed), but from courageous urban politics able to mobilize citizens around the meaning of their environment.

› Conclusion

The new culture of cities is not the culture of the end of history. Restoring communication may open the way to restoring meaningful conflict. Currently, social injustice and personal isolation combine to induce alienated violence. So, the new culture of urban integration is not the culture of assimilation into the values of a single dominant culture, but the culture of communication between an irreversibly diverse local society connected/disconnected to global flows of wealth, power, and information.

Architecture and urban design are sources of spatio-cultural meaning in an urban world in dramatic need of communication protocols and artefacts of sharing. It is commendable that architects and urban designers find inspiration in social theory, and feel as concerned citizens of their society. But first of all, they must do their job as providers of meaning by the cultural shaping of spatial forms. Their traditional function in society is more critical than ever in the information age, an age marked by the growing gap between splintering networks of instrumentality and segregated places of singular meaning. Architecture and design may bridge technology and culture by creating shared symbolic meaning and reconstructing public space in the new metropolitan context. But they will only be able to do so with the help of innovative urban policy supported by democratic urban politics.

Ascher, F. (1995), *La Metapolis. Ou L'Avenir de la Ville*, Paris: Odile Jacob.

Blakely, E. and Snyder, M. (1997), *Fortress America: Gated Communities in the United States*, Washington, DC: The Brookings Institution.

Borja, J. and Castells, M. (1997), *Local and Global: The Management of Cities in the Information Age*, London: Earthscan.

Borja, J. with Zaida, M. (2001), *L'Espai Public. Ciutat I Ciutadania*, Barcelona: Diputacio de Barcelona.

Brand, S. (1999), *The Clock of the Long Now*, New York: Basic Books.

Capra, F. (1996), *The Web of Life*, New York: Doubleday.

Castells, M. (1989), *The Informational City*, Oxford: Blackwell.

Castells, M. (1996), *The Rise of the Network Society, Volume I: The Information Age*, Oxford: Blackwell (revised edition, *2000*).

Castells, M. (1997), *The Power of Identity, Volume II: The Information Age*, Oxford: Blackwell.

Castells, M. (1998), *The End of Millennium, Volume III: The Information Age*, Oxford: Blackwell.

Castells, M. (2000), 'Grassrooting the space of flows'. In J. Wheeler, Y. Aoyama, and B. Warf (eds), *Cities in the Telecommunications Age: The Fracturing of Geographies*, London: Routledge, *18–30.*

Castells, M. (2001), *The Internet Galaxy*, Oxford: Oxford University Press.

Castells, M. and Servon, L. (1996), 'The feminist city: a plural blueprint', Berkeley, CA: University of California, Department of City Planning, unpublished.

Davis, M. (1992), *City of Quartz*, New York: Vintage Books.

Dunham-Jones, E. (2000), 'Seventy-five per cent', *Harvard Design Magazine*, Fall, *5–12.*

Fernandez-Galiano, L. (2000), 'Spectacle and its discontents', *Harvard Design Magazine*, Fall, *35–8.*

Freire, M. and Stren, R. (eds) (2001), *The Challenge of Urban Government: Policies and Practices*, Washington, DC: The World Bank Institute.

Garreau, J. (1991), *Edge City: Life on the New Frontier*, New York: Doubleday.

Gillespie, A. and Richardson, R. (2000), 'Teleworking and the city: myths of workplace transcendence and travel reduction'. In J. Wheeler, Y. Aoyama, and B. Warf (eds), *Cities in the Telecommunications Age: The Fracturing of Geographies*, London: Routledge, *228–48.*

Graham, S. and Marvin, S. (2001), *Splintering Urbanism: Networked Infrastructures, Technological Mobilities, and the Urban Condition*, London: Routledge.

Hall, P. (1998), *Cities in Civilization*, New York: Pantheon.

Hall, P. (2001),'Global city-regions in the 21st century'. In A. Scott (ed.), *Global City-Regions. Trends, Theory, Policy*, New York: Oxford University Press, *59–77.*

Horan, T. (2000), *Digital Place: Building Our City of Bits*, Washington, DC: The Urban Land Institute.

Human Development Report, United Nations Development Programme (2001), *Technology and Human Development*, New York: Oxford University Press.

Jacobs, A. (1993), *Great Streets*, Cambridge, MA: MIT.

Jones, S. (ed.) (1998), *Cybersociety 2.0*, London: Sage.

Klinenberg, E. (2000),'The Social Anatomy of a Natural Disaster: The Chicago Heat Wave of 1995', Berkeley, CA: University Of California, Dept of Sociology, PhD Dissertation (unpublished).

Kopomaa, T. (2000), *The City in Your Pocket: Birth of the Mobile Information Society*, Helsinki: Gaudemus.

Kotkin, J. (2000), *The New Geography: How the Digital Revolution is Reshaping the American Landscape*, New York: Random House.

Kuntsler, G. (1993), *The Geography of Nowhere*, New York: Simon & Schuster.

Massey, D. (1996), 'The age of extremes: inequality and spatial segregation in the 20th century', Presidential Address, Population Association Of America.

Mitchell, W. (1999), *E-Topia*, Cambridge, MA: MIT Press.

Nel.Lo, O. (2001), *Ciutat De Ciutats*, Barcelona: Editorial Empuries.

Putnam, R. (2000), *Bowling Alone: The Collapse and Revival of American Community*, New York: Simon & Schuster.

Russell, J. (2000), 'Privatized lives', *Harvard Design Magazine*, Fall, 20–9.

Sassen, S. (1991), *The Global City: London, Tokyo, New York*, Princeton, NJ: Princeton University Press.

Sassen, S. (2001), 'Global cities and global city-regions: a comparison'. In A. Scott (ed.), *Global City-Regions: Trends, Theory, Policy*, New York: Oxford University Press, 78–95.

Scott, A. (ed.) (2001), *Global City-Regions: Trends, Theory, Policy*, New York: Oxford University Press.

Sorkin, Michael (1997), *Variations on a Theme Park: The New American City and the End of Public Space*, New York: Hill and Wang.

Susser, I. (1996), 'The construction of poverty and homelessness in US Cities', *Annual Reviews of Anthropology*, 25, 411–35.

Verwijnen, J. and Lehtovuori, P. (eds) (1999), *Creative Cities*, Helsinki: University of Art and Design.

Waldinger, R. (ed.) (2001), *Strangers at the Gate: New Immigrants in Urban America*, Berkeley, CA: University of California Press.

Wellman, B. (ed.) (1999), *Networks in the Global Village*, Boulder, CO: Westview Press.

Wellman, B. and Haythornthwaite, C. (eds) (2002), *The Internet in Everyday Life*, Oxford: Blackwell.

Wheeler, I, Aoyama, Y. and Warf, B. (2000), *Cities in the Telecommunications Age: The Fracturing of Geographies*, London: Routledge.

BIBLIOGRAPHY

Alexander, Christopher, *Notes on the Synthesis of Form*, Cambridge, MA: Harvard University Press, 1964.

Banham, Reyner, "A Home is Not a House," *Art in America* 2, April, 1965, 70–9.

Banham, Reyner, *Theory and Design in the First Machine Age*, 2nd edn, Cambridge, MA: MIT Press, 1980.

Berkel, Ben van and Caroline Bos, *Move: Techniques*, Amsterdam: UN Studio & Goose Press, 1999.

Boyce, James R, "What is the Systems Approach," *Progressive Architecture*, November, 1969, 118–21.

Brand, Stewart, *How Buildings Learn: What Happens After They're Built*, New York: Viking, 1994.

Cache, Bernard, "Digital Semper," *Anymore*, edited by Cynthia Davidson, Cambridge, MA: MIT Press, 2000, pp. 190–9.

Castells, Manuel, "Space of Flows, Space of Places: Materials for a Theory of Urbanism in the Information Age," *The Cybercities Reader*, London: Routledge, 2004, pp. 82–93.

Collins, Peter, "The Biological Analogy," *Architectural Review* 126, 1959, 303–6.

Colquhoun, Alan, *Essays in Architectural Criticism: Modern Architecture and Historical Change*, Cambridge, MA: MIT Press, 1981.

Cook, Peter, *Experimental Architecture*, New York: Universe Books, 1970.

Cowan, Ruth Schwartz, "The 'Industrial Revolution' in the Home: Household Technology and Social Change in the Twentieth Century", *Technology and Culture* 17, 1976, 1–23.

De Landa, Manuel, "Deleuze and the Use of the Genetic Algorithm in Architecture," *Contemporary Techniques in Architecture, Architectural Design* 72, January, 2002, 9–12.

Duffy, Francis, *The New Office*, London: Conran Octopus, 1997.

Fernández-Galiano, Luis, *Fire and Memory: On Architecture and Energy*, Cambridge, MA: MIT Press, 2000.

Fuller, Richard Buckminster, *Operating Manual for Spaceship Earth*, New York: Pocket Books/Simon and Schuster, 1970.

Fuller, Richard Buckminster, *4D Time Lock*, Albuquerque, NM: Lama Foundation/Biotechnic Press, 1972.

Geddes, Patrick, Sir, *Cities in evolution; an introduction to the town planning movement and to the study of civics.* New York: H. Fertig, 1968.

Giedion, Siegfried, *Space, Time and Architecture: The Growth of a New Tradition*, 3d ed. Cambridge, MA: Harvard University Press, 1959.

Giedion, Siegfried, *Mechanization Takes Command: A Contribution to Anonymous History*, New York: W. W. Norton, 1948.

Giedion, Siegfried, *Building in France, Building in Iron, Building in Ferro-Concrete*, Santa Monica, CA: Getty Center for the History of Art and Humanities, 1995.

Guattari, Félix, *Chaosmosis: An ethico-aesthetic paradigm*, Bloomington, IN: Indiana University Press, 1995.

Häring, Hugo, "The House as an Organic Structure," in *Programs and Manifestoes on 20th-Century Architecture*, edited by Ulrich Conrads, Cambridge, MA: The MIT Press, 1970, pp. 126–7.

Honzík, Karel, "Biotechnics: Functional Design in the Vegetable World," *Architectural Review* 81, January, 1937, 21–2.

Katavolos, William, "Organics" in "Chemical Architecture," in *Programs and Manifestoes on 20th-Century Architecture*, edited by Ulrich Conrads, Cambridge, MA: MIT Press, 1970, pp. 163–4.

Kiesler, Frederick, "On Correalism and Biotechnique: A Definition and Test of a New Approach to Building Design," *Architectural Record*, September, 1939, 60–75.

Kohr, Leopold, *The Inner City: From Mud to Marble*, Talybont, Dyfed, Wales: Y Lolfa Cyf., 1989.

Koolhaas, Rem, "Speculations on Structures and Services," *S, M, L, XL*, New York: Monacelli Press, 1995, pp. 232–9.

Kurokawa, Kisho, *Metabolism in Architecture*, Boulder, CO: Westview Press, 1977.

Latour, Bruno (Jim Johnson), "Mixing Humans and Nonhumans Together: The Sociology of a Door-Closer," *Social Problems* 35, June, 1988, 298–310.

Le Corbusier [Jeanneret, Charles-Edouard], "Architecture, the Expression of the Materials and Methods of our Times," *Architectural Record* LXVI, August, 1929, 123–8.

Le Corbusier (Jeanneret, Charles-Edouard), *Towards a New Architecture*, New York: Praeger Publishers, 1960.

Leatherbarrow, David and Mohsen Mostafavi, *Surface Architecture*, Cambridge, MA: MIT Press, 2002.

Lönberg-Holm, Knud, "Architecture in the Industrial Age," *Arts and Architecture* 84, April, 1967, 22.

McCleary, Peter, "Some Characteristics of a New Concept of Technology," *Journal of Architectural Education* 42, Fall, 1988, 4–9.

McDonough, William and Michael Braungart, *Cradle to Cradle: Remaking the Way we Make Things*, New York: North Point Press, 2002.

McLuhan, Marshall, *Understanding Media: the Extensions of Man*, 2nd edn, New York: New American Library, 1964.

Mies van der Rohe, Ludwig, "Technology and Architecture," in *Programs and Manifestoes on 20th-Century Architecture*, edited by Ulrich Conrads, Cambridge, MA: MIT Press, 1970, p. 154.

Mitchell, William, "E-Bodies, E-Buildings, E-Cities," in *Designing for a Digital World*, edited by Neil Leach, London: Wiley-Academic, 2002, pp. 50–6.

Mumford, Lewis, *Technics and Civilization*, New York: Harcourt Brace & World, 1934.

Neutra, Richard, *Survival Through Design*, New York: Oxford University Press, 1954.

Pawley, Martin, "Technology Transfer," *Architectural Review*, September, 1987, 31–9.

Rykwert, Joseph, "Organic and Mechanical," *Res: Anthropology and Aesthetics* 22, Autumn, 1992, 11–18.

Sant' Elia, Antonio and Filippo Tommaso Marinetti, "Manifesto of Futurist Architecture," In *Programs and Manifestoes on 20th-Century Architecture*, edited by Ulrich Conrads, Cambridge, MA: MIT Press, 1970, pp. 34–8.

SLA (Ed van Hinte, Marc Neelen, Jacques Vink, and Piet Vollaard), "Changing Speeds," *Smart Architecture*, Rotterdam: 010 Publishers, 2003, p. 24.

Soleri, Paolo, *The Bridge Between Matter and Spirit is Matter Becoming Spirit: The Arcology of Paolo Soleri*, Garden City, NY: Anchor Press/Doubleday, 1973.

Steadman, Philip, *The Evolution of Designs: Biological Analogy in Architecture and the Applied Arts*, New York: Cambridge University Press, 1979.

Superstudio: Adolfo Natalini, Cristiano Toraldo di Francia, Alessandro Magris, Roberto Magris, and Piero Frassinelli, *Italy: The New Domestic Landscape*, New York: MOMA, 1972.

Team 10 Primer, edited by Alison Smithson. Cambridge, MA: MIT Press, 1968.

Ternoey, Steven *et al.*, "The Patterns of Innovation and Change," in *The Design of Energy Responsive Commercial Buildings*, Solar Energy Research Institute, New York: Wiley-Interscience, 1985. pp. 5–8.

Virilio, Paul, "The Third Interval," In *Open Sky*, London; New York: Verso, 1997, pp. 9–21.

Wachsmann, Konrad, "Seven Theses," In *Programs and Manifestoes on 20th-Century Architecture*, edited by Ulrich Conrads, Cambridge, MA: MIT Press, 1970, p. 156.

Wright, Frank Lloyd, *Frank Lloyd Wright Collected Writings*, edited by Bruce Brooks Pfeiffer, New York: Rizzoli; Frank Lloyd Wright Foundation, 1992.

Yeang, Ken, *The Green Skyscraper: the Basis for Designing Sustainable Intensive Buildings*, Munich; New York: Prestel, 1999.

INDEX

Aalto, Alvar 348
Adopters 291
Adventure of Ideas (A. N. Whitehead) 59
advertisers as link between social and
    technological change 223–4
ageing 78n6
Ageing Couple (P. Klee) 102
aircraft industry, influence on construction
    306–7n12
Akrich, Madeleine 313, 318
Alberti, Leon Battista 283, 347, 404
Album (V. de Honnecourt) 271
analogy, biological: applied to art theory 132;
    and architectural evolution 133; as
    associated with Frank Lloyd Wright 279;
    classification of organic functions 133–4;
    confusion in 131; correlation between organs
    134–5; evolution 130–1, 137; form and
    function 132; and the functional analogy
    135; importance of environment 134, 282–3;
    morphology 131, 260–2; origin of 130;
    process of artistic creation 132–3; term
    biology invented 131; *see also* artefacts,
    changes in
Analytical Archaeology (D. Clarke) 246
anthropomorphism 314–16
Archigram Group 301
architects: diminishing role of 297; and
    engineers 56, 355
Architectural Digest 351
architecture: after invention of printing 4–5; as
    a changing concept 40–1; chemical 149–50;
    and clients of 53; and crisis 43; defined 43;
    and energy 77; and engineering 34; futurist
    18–21; health as only criterion of 73–4; and
    industry 52–4; innovations in industrial
    buildings 82; knowledge of not scientific
    255–6; lack of innovation 44–5;
    masterpieces of 65; material determination
    in 405; modern, symbolic expression in
    266–9; and new materials and methods 19,
    43; office buildings 6–7; omnipresence of
    biological and mechanical analogies 271;
    organic 149–50, 340, 342, 346–8; origin of
    405; as the principle art 3; professional
    organizations 53; of rehabilitation 283–4; as
    representative of traditional art 3; search for
    newness in 356–7; as selfconscious
    discipline 152–8, 159n5; and speed of
    technological development 295; and

technology 114, 145–7; of technology
    transfer, visibility of 303–4; three reminders
    35; urge to invent not apparent 81; use of
    genetic algorithms in design 408–12
arcology 209–11
Ariès, Phillippe 215
Aristotle 338
Arnheim, Rudolf 274
art and the Machine 9, 14, 16
Art History Museum, Vienna 400–1, **401**, 406n6
Art Nouveau movement 347
artefacts, changes in: abrupt and radical,
    reasons for 250–1; cathedral construction
    249; chronological evidence needed 246;
    constraints of evolutionary process 262;
    cookery example 251; and the designer 249;
    due to competition in fashion 247–8; due to
    technical improvements 248–50;
    evolutionary or devolutionary change 246–7;
    geographical diffusion 245; reasons for
    directional change 247–50; sequential
    development 246; situational logic 249–50,
    256; through copying 245, 247; women's
    clothing 248
artificial, sciences of 257
artist and the Machine 2–3
arts, origin of 416–19
Asplund, Gunnar 348
assembly line: biscuit-making 89; in the
    eighteenth century 85–8; Ford cars 104–6;
    growth of 84–5; human aspect of 107–9;
    machine tool factory 90–1; meaning of
    110n1; railway influences 84;
    slaughterhouses 90, 92–4; total automation
    of 106–7; *see also* motion, study of;
    scientific management
automata as replicas of living beings 273,
    284n13, 285n10, 287n35, 341
automatization of human workers 96, 107–9
automobiles *see* cars

Babbage, Charles 103
Bacon, Francis 338
Balmond, Cecil 356
Banham, Peter Reyner 193, 298
banking, fragmentation and recombination of
    429
bathrooms 217, 226n12
Bedaux, Charles 103
Being and Time (M. Heidegger) 329–30

Also available from Routledge...

Housing and Dwelling
Perspectives on Modern Domestic Architecture
Edited by Barbara Miller Lane

Housing and Dwelling collects the best in recent scholarly and philosophical writings that bear upon the history of domestic architecture in the nineteenth and twentieth centuries. Lane combines exemplary readings that focus on and examine the issues involved in the study of domestic architecture. The extracts are taken from an innovative and informed combination of philosophy, history, social science, art, literature and architectural writings. The readings address, among other issues, the relation between the public and the private sphere, the gendering of space, notions of domesticity, the relation between domesticity and social class, the role of builders and prefabrication, and the relationship between architects and the inhabitants of dwellings.

Uniquely, the readings in *Housing and Dwelling* underline the point of view of the user of a dwelling and assess the impact of varying uses on the evolution of domestic architecture.

Housing and Dwelling is a valuable asset for students, scholars, and designers alike. The book explores the extraordinary variety of methods, interpretations and source materials now available in this important field. For students, it opens windows on the many aspects of domestic architecture. For scholars, it introduces new, interdisciplinary points of view and suggests directions for further research. It acquaints practising architects in the field of housing design with history and methods and offers directions for future design possibilities.

ISBN: 978-0-415-34655-9 (Hb)
      978-0-415-34656-6 (Pb)

For ordering and further information please visit:
http://www.routledge.com/builtenvironment

Narrating Architecture
A retrospective anthology
Edited by James Madge and Andrew Peckham

The Journal of Architecture is jointly published by The Royal Institute of British Architects (RIBA) and Routledge. An international journal committed to advancing architectural discourse in its widest sense, its aim is to seek diverse views of the past, present and future practice of architecture, and to attract a wide variety of perspectives from the architectural and related professions, as well as from academics.

This anthology brings together in one volume a selection of papers that stand out after ten years of publication. The editors give readers access to international contributions in a carefully structured book, bringing coherence to a wide range of topics. The book is divided into seven parts: Architects and the practice of design; Architecture and the discourses of science; Issues of materiality; Narratives of domesticity; Problems of building; The sociology of architectural practice and Identity and the appropriation of place. It offers those teaching or running seminars in this subject area a readily available collection of recent research in several key areas.

ISBN: 978-0-415-37435-4 (Hb)
        978-0-415-38564-0 (Pb)

For ordering and further information please visit:
http://www.routledge.com/builtenvironment

Rethinking Architecture
A reader in cultural theory
Edited by Neil Leach

Brought together for the first time – the seminal writing
on architecture by key philosophers and cultural
theorists of the twentieth century.

Issues around the built environment are increasingly
central to the study of the social sciences and
humanities. The essays offer a refreshing take on the question of
architecture and provocatively rethink many of the accepted tenets of
architecture theory from a broader cultural perspective.

The book represents a careful selection of the very best theoretical writings
on the ideas which have shaped our cities and our experiences of
architecture. As such, *Rethinking Architecture* provides invaluable core
source material for students on a range of courses.

ISBN: 978-0-415-12825-4 (Hb)
       978-0-415-12826-1 (Pb)

For ordering and further information please visit:
http://www.routledge.com/builtenvironment